TIGERS OF THE WORLD

NOYES SERIES
IN
ANIMAL BEHAVIOR, ECOLOGY, CONSERVATION AND MANAGEMENT

Benjamin B. Beck, Ph.D., Consulting Editor

A series of professional and reference books in ethology devoted to the better understanding of animal behavior, ecology, conservation, and management.

WOLVES OF THE WORLD: Perspectives of Behavior, Ecology, and Conservation.
Edited by *Fred H. Harrington* and *Paul C. Paquet*

IGUANAS OF THE WORLD: Their Behavior, Ecology, and Conservation.
Edited by *Gordon M. Burghardt* and *A. Stanley Rand*

HORSE BEHAVIOR: The Behavioral Traits and Adaptations of Domestic and Wild Horses, Including Ponies.
By *George H. Waring*

GAZELLES AND THEIR RELATIVES: A Study in Territorial Behavior.
By *Fritz R. Walther, Elizabeth Cary Mungall*, and *Gerald A. Grau*

THE MANAGEMENT AND BIOLOGY OF AN EXTINCT SPECIES: *PERE DAVID'S DEER*.
Edited by *Benjamin B. Beck* and *Christen Wemmer*

APES OF THE WORLD: Their Social Behavior, Communication, Mentality and Ecology.
By *Russell H. Tuttle*

TIGERS OF THE WORLD: The Biology, Biopolitics, Management, and Conservation of an Endangered Species.
Edited by *Ronald L. Tilson* and *Ulysses S. Seal*

TIGERS
OF THE WORLD

The Biology, Biopolitics, Management, and Conservation of an Endangered Species

Edited by

Ronald L. Tilson

Minnesota Zoological Gardens
Apple Valley, Minnesota

Ulysses S. Seal

AAZPA/SSP Tiger Coordinator
Veterans Administration Research Service
Minneapolis, Minnesota

 NOYES PUBLICATIONS
Park Ridge, New Jersey, U.S.A.

Library of Congress Cataloging-in-Publication Data

Tigers of the world.

    Papers presented at the symposium held in Minneapolis
on Apr. 13-17, 1986; sponsored by the Minnesota Zoologi-
cal Garden and the IUCN/SSC Captive Breeding and Cat
Specialist Groups.
    Includes bibliographies and indexes.
    1. Tigers--Congresses. 2. Wildlife management--
Congresses. 3. Wildlife conservation--Congresses.
I. Tilson, Ronald Lewis. II. Seal, Ulysses S.
III. Minnesota Zoological Garden. IV. IUCN/SSC Captive
Breeding Group. V. IUCN/SSC Cat Specialist Group.
QL737.C23T475 1987   639.9'7974428   87-12204
ISBN 0-8155-1133-7

Foreword

This symposium on world conservation strategies for tigers has, for the first time, brought together all those involved with tigers, from the pure scientist and the laboratory worker, to the most applied field worker, reserve manager, zoo director and keeper.

As chairman of the IUCN Species Survival Commission I have found the many papers fascinating, and I look forward to seeing the Global Tiger Survival Plan as a most valuable output both from my and IUCN's point of view.

I am sure that if the Global Tiger Survival Plan is not generated here and by you then it will never become a reality and a model for many of the other specialist groups facing identical and equally difficult problems.

All significant tiger populations in the wild are represented here by the reserve managers except, sadly, those in the Soviet Union. The total support of the "wild population" is, I am sure, the key to the tiger's long-term survival. I am equally sure, however, that all these skills and techniques exposed here, in the zoo and laboratory world, hold the key as to how we can improve and ensure the diversity of this supreme creature and all its subspecies both in the "wild" and in captivity.

The papers from our colleagues managing free-ranging populations reassure me that we are more than holding our own with regard to short-term conservation solutions, but we must not be seduced by our present successes—we must be sure that we have the long-term solution, and I believe that with the partnership we have at this meeting both the "wild" and the zoo world can give us such long-term solutions and success.

As the "wild" reserve areas become more and more restricted, managerial techniques must be improved to ensure the necessary diversity for the long-term.

This will cost money, for further land purchase and reserve management on the one hand and, on the other, for captive breeding within the "zoo" situation. The money does not always come from the same sources, and so is not necessarily competing.

In both "wild" and captive situations the tiger is always a drawer—a flag bearer for its habitat in the wild whilst in the city—in the zoo context—it represents many creatures and their environments in educational programs.

We have also to remember the political world in which we live. The politics between zoos, and the "real" world must all be subsumed in the Global Tiger Survival Plan if we are to succeed in our ultimate aim, the tiger's long-term survival!

This is a first step. The next must be a new Cat Red Data Book, and then further cat specialist plans.

This meeting is creating a template from which many specialists will benefit, and I congratulate our host, the Minnesota Zoological Garden, on the foresight and activities in ensuring the success of this most exciting meeting.

Gren Lucas
Chairman, IUCN/SSC

Preface

This symposium grew out of our concern that tigers, with a global distribution extending from India across China to the Soviet Far East, and south through Peninsular Malaysia to Indonesia, are steadily declining in many parts of their range. Three of the recognized eight subspecies are now extinct; a fourth is near extinction. The remaining tiger populations, with the possible exception of the Sundarbans population in Bangladesh, are both too fragmented and too small for long-term survival.

Guidelines emerging from conservation biology theory suggest that for a species like *Panthera tigris* a minimal population size of about 2,000 animals is necessary for the long-term survival of each subspecies. This implies that if management units of tigers are to center around the currently recognized five subspecies, then a world population of 10,000 wild tigers is needed. This is not likely to occur in the near future. Alternatively, if some of the management units are to survive with retention of even 90% of their remaining genetic diversity for 200 years, then population sizes of 350 to 500 animals must be maintained. Only one wild population of one tiger subspecies possibly fits this criterion. The others are fragmented into small disjunct populations with little opportunity for genetic exchange or recolonization if the population is lost. Populations subdivided into many small populations—say of 10 to 100 animals—are fragile and face a high probability of extinction. This means we must be prepared to provide periodic recolonization of habitat vacated by local extinctions as well as exchange of genetic material between locally decimated populations. This will require development of strategies for interactive management of the fragmented wild populations and use of captive populations for backup and support.

The objective of this symposium was to contribute to the development of a Global Tiger Survival Plan that would facilitate a multinational agreement for the sustained conservation of the world's remaining tigers. The contents of this book were designed to help encourage this process. We believed that the necessary first step toward addressing strategic long-term goals of tiger conservation was to establish the current status of each subspecies, the subpopulations that comprise them, and the extent and integrity of their habitat. More precisely, we wanted to identify, as closely as possible, population numbers and their spatial relationship to each other relative to political boundaries. The revised distribution map of the world's remaining free-ranging tigers and the available data on their numbers suggest that the tiger is not as wide ranging as commonly perceived, and does not exist in sufficient numbers to be considered a self-sustaining species (see map).

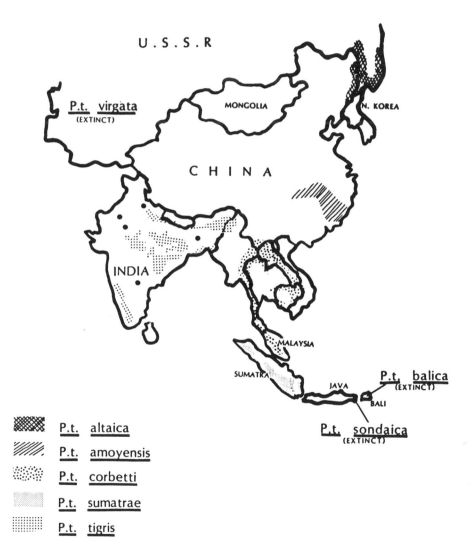

A second necessary step toward developing and implementing comprehensive reproductive and management programs for the world's captive population was to request managers of these populations to share their knowledge. The declining availability of replacement animals from the wild has forced zoos to manage their widely dispersed animals as single biological populations if they are to be self-sustaining. Since many traditional exhibit species have become endangered in the wild, the possible contribution of the captive populations to survival of these species has become important. Thus, the demographic and genetic concepts derived from conservation biology that serve to guide the management of small, fragmented captive populations are directly applicable to the management of remnant populations of the same species in the wild.

The third necessary step was to assemble the managers of both the wild and captive tiger populations to facilitate the international and interdisciplinary exchange of ideas and data. The chapters included in this book reflect this diversity of nationality, language and culture. To retain their perspectives we edited the chapters for format only and left intact the nuances of the individual and idea.

We thank the many people who labored to make the symposium and this book a reality. The symposium, held in Minneapolis on April 13-17, 1986, was sponsored by the Minnesota Zoological Garden and the IUCN/SSC Captive Breeding and Cat Specialist Groups. We are particularly indebted to Peter Jackson, Chairman of the Cat Specialist Group, Siegfried Seifert, Keeper of the International Tiger Studbook, Hemendra Panwar, past Director of India's Project Tiger, and Chris Wemmer, from the Smithsonian-Nepal Tiger Ecology Project, for helping with the initial planning of the symposium. They and Betsy Dresser, Mohammed Khan bin Momin Khan and Steve O'Brien were responsible for chairing their respective sessions. Gren Lucas, the IUCN/Species Survival Commission Chairman, kindly consented to deliver the keynote address. Steven Kohl helped translate for our Chinese colleagues, Fritz Jantschke and James Dolan helped translate for our German colleagues, while Tom Foose and Nate Flesness helped plan travel arrangements for the overseas delegates. We gratefully acknowledge the unselfish dedication of thirteen Minnesota Zoo Volunteers, coordinated by Kate Anderson, who kept the symposium agenda on track and on schedule.

Funds to support international travel and cost for the delegates were received from the Smithsonian's National Museum Act, the Windstar Foundation's Web of Life Fund, the United Nations' Environment Programme, the World Wildlife Fund, the U.S. Fish and Wildlife Service, the Minnesota Zoological Garden and from private donations. We thank our friends at Meyers Printing, Minneapolis, for producing the poster and brochure commemorating the symposium, Rebecca Becker for the artwork, Adele Smith for the graphics, Tom Cajacob for the photographs, Carole Stead for assisting with the symposium arrangements, and G. Allen Binczik for formatting and editing the many drafts of the book. We would also like to thank Rebecca Noyes for assisting in final manuscript preparation. We are especially grateful to our families and colleagues for their patience in understanding our absence from other duties.

Apple Valley, Minnesota   Ronald L. Tilson
April 1987   Ulysses S. Seal

Contributors

This list contains the names of the speakers who presented papers at the symposium, as well as those persons who did not attend the Symposium but submitted papers to be included in the published proceedings. Additional contributors to each chapter are listed under their respective chapter titles.

Kenneth R. Ashby
Zoology Department
University of Durham
Durham, England

Jonathon D. Ballou
National Zoological Park
Washington, D.C.

G. Allen Binczik
Minnesota Zoological Garden
Apple Valley, Minnesota

R.L. Brahmachary
Indian Statistical Institute
Calcutta, India

Mitchell Bush
National Zoological Park
Washington, D.C.

Karen G. Cronquist-Jones
International Species Inventory
  System
Apple Valley, Minnesota

Ellen S. Dierenfeld
New York Zoological Society
Bronx, New York

Betsy L. Dresser
Cincinnati Zoological Garden
Cincinnati, Ohio

David A. Ferguson
U.S. Fish & Wildlife Service
Department of the Interior
Washington, D.C.

Nathan R. Flesness
International Species Inventory
  System
Apple Valley, Minnesota

Thomas J. Foose
American Association of
  Zoological Parks & Aquariums
Apple Valley, Minnesota

Anna M. Goebel
Biology Department
University of Texas
Arlington, Texas

Edmund Graham*
College of Veterinary Medicine
University of Minnesota
St. Paul, Minnesota

Michael K. Hackenberger
Canadian Zoological Systems Ltd.
Cambridge, Ontario
Canada

Helmut Hemmer
Institut fur Zoologie
Johannes Gutenberg-Universitat
  Mainz
Mainz-Ebersheim, West Germany

Hugh C. Hensleigh
Obstetrics and Gynecology
  Department
University of Minnesota
Minneapolis, Minnesota

Sandra J. Herrington
Museum of Natural History
University of Kansas
Overland Park, Kansas

Peter Jackson
IUCN/Cat Specialist Group
Bougy-Villars, Switzerland

Jia Xianggang
Division of Nature Conservation
National Environmental
  Protection Agency
Beijing, Peoples Republic of China

K. Ullas Karanth
Centre for Wildlife Studies
Mysore, India

Mohammad Ali Reza Khan
Al Ain Zoo
Abu Dhabi, United Arab Emirates

Mohammed Kahn bin Momin Khan
Wildlife and National Parks
Kuala Lumpur, Malaysia

Steven G. Kohl
Fish & Wildlife Service
Department of the Interior
Washington, D.C.

Katherine Latinen
Detroit Zoological Park
Box 39
Royal Oak, Michigan

Lu Houji
Biology Department
East China Normal University
Shanghai, Peoples Republic of China

Gren Lucas
Species Survival Commission
IUCN/World Conservation Center
Gland, Switzerland

Lynn A. Maguire
School of Forestry and Environ-
  mental Studies
Duke University
Durham, North Carolina

Edward J. Maruska
Cincinnati Zoological Garden
Cincinnati, Ohio

Charles W. McDougal
Tiger Tops Jungle Lodge
Kathmandu, Nepal

Hemanta R. Mishra
National Parks & Wildlife
    Conservation
Kathmandu, Nepal

Peter Muller
Zoologischer Garten Leipzig
Leipzig, East Germany

Stephen J. O'Brien
Section of Genetics
National Cancer Institute
Frederick, Maryland

Hemendra S. Panwar
Wildlife Institute of India
Uttar Pradesh, India

S.V. Popov
Moscow Zoo
Moscow, U.S.S.R.

Gerald S. Post
College of Veterinary Medicine
University of Minnesota
St. Paul, Minnesota

P.N. Romanov
Moscow Zoo
Moscow, U.S.S.R.

A.K. Roychoudhury
Biometry and Population
    Genetics Unit
Bose Institute
Calcutta, India

Charles Santiapillai
Forest Protection and Nature
    Conservation
Bogor, Indonesia

Pranabes Sanyal
Sundarbans Tiger Reserve
West Bengal, India

Marsha Schmitt*
College of Veterinary Medicine
University of Minnesota
St. Paul, Minnesota

Ulysses S. Seal
V.A. Medical Center
Minneapolis, Minnesota

Bernd Seidel
Tierpark Berlin
Berlin, East Germany

John Seidensticker
Mammal Department
National Zoological Park
Washington, D.C.

Siegfried Seifert
Zoologischer Garten Leipzig
Leipzig, East Germany

Lee G. Simmons
Henry Doorly Zoo
Omaha, Nebraska

E.N. Smirnov
Moscow Zoo
Moscow, U.S.S.R.

James L. David Smith
Fisheries & Wildlife Department
University of Minnesota
St. Paul, Minnesota

Vladimir V. Spitsin
Moscow Zoo
Moscow, U.S.S.R.

Tan Bangjie
Beijing Zoological Garden
Beijing, Peoples Republic of China

J. Andrew Teare
Archer, Florida

Ronald L. Tilson
Minnesota Zoological Garden
Apple Valley, Minnesota

Christen Wemmer
Conservation & Research Center
National Zoological Park
Front Royal, Virginia

David E. Wildt
National Zoological Park
Smithsonian Institute
Washington, D.C.

Xiang Peilon
Chongqing Zoological Gardens
Chongqing, Peoples Republic of China

*Speaker at Symposium. Paper not available at time of publication.

Contents

PART III
STATUS IN CAPTIVITY

PART V
CAPTIVE MANAGEMENT

PART VI
WHITE TIGER POLITICS

PART VII
CONSERVATION STRATEGIES

1

Bearing Witness: Observations on the Extinction of *Panthera tigris balica* and *Panthera tigris sondaica*

John Seidensticker

BEARING WITNESS

If you go to Bali Barat or Meru-Betiri national parks on the Indonesian islands of Bali and Java and talk with the people who live near them and who use the forest, you almost certainly will hear stories of recent encounters with tigers. When I worked there in the late 1970's, people had many stories to tell of encountering a tiger drinking near a temple, or of a tiger that regularly came and rested along a forest stream at a place called Pondok Macan (Tiger Place), and there were many other tiger stories. Animals as metaphysically important as tigers live on in our minds after they are gone. No one wants to be the bearer of the bad news; I too hoped. Why else would I travel half way around the world to be there asking these people about tigers? The lingering image of tigers and the hope they remain a part of the landscape make it difficult to date extinction events and to discover the proximal causes of the extinction.

Some of our oldest tiger fossils have been found on Java (Hemmer 1971) and tigers and man have mixed on these Sunda Shelf islands for about that long. Java is one of the most densely-populated areas on earth (Biro Pusat Statistik 1977) and yet the tiger was only recently lost and another great beast, the Javan rhino (<u>Rhinoceros</u> <u>sondaicus</u>), still lives in Ujung Kulon, a rain forest jewel. A.R. Wallace (1962) described Java in 1861 as:

> "Taking it as a whole, and surveying it from every point of view, Java is probably the very finest and most interesting tropical island in the world . . . scattered through the country, especially in the eastern part of it, are found buried in lofty forests, temples, tombs and statues of great beauty and grandeur; and the remains of extensive cities, where the tigers, the rhinoceros, and the wild bull now roam undisturbed" (pp. 75-76).

When I worked in Java in the mid-1970's, I was intrigued and impressed that tigers were still extant, a condition that seemed contrary to the dogma of extinction-prone species on which I had been weaned. A good system of nature reserves located in tiger

Fig. 1.   A photograph of a Javan Tiger in the Ujung Kulon
National Park taken in 1938 by A. Hoogerwerf and published as
Plate 60 in Udjung Kulon, The Land of the Last Javan Rhinoceros.
Reproduced with permission of E.J. Brill, Leiden.

habitat on Java (Fig. 1) and Bali was established in the mid-
1930's and early 1940's (Hoogerwerf 1970), but in the end this
did not suffice for the Javan tiger.

In addressing the primary task of this symposium, "A Global
Tiger Survival Plan", it will be useful to keep in mind what we
know about the extinction of two of the three Sunda Island tiger
subspecies.

DATING THE EXTINCTIONS

Java

My Indonesian colleagues and I (Seidensticker and Suyono
1980) found tracks of at least three tigers living in the
Meru-Betiri National Park (then reserve) in 1976. We found no
evidence to suggest that "effective" reproduction was occurring.
Tigers were not confining their movements to the reserve, nor
where they using all the reserve area. We found no evidence that
suggested any tiger had been killed recently by man, but most
people living in the area were misinformed about the plight of
the Javan tiger and its survival needs. A track count indicated

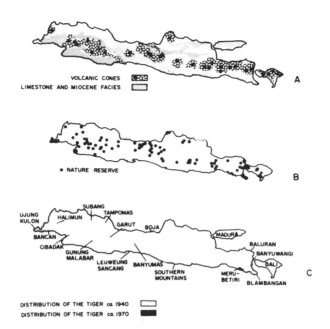

Fig. 2.   Islands of Java and Bali, Indonesia: A) Landforms of
Java.   B) Systems of nature reserves on Java established mostly
during the 1920's and 1930's.   C) Decline of _Panthera_ _tigris_
sondaica: it was distributed over most of Java below 900 m before
1830; by 1940, it was found only is a few remote forest areas and
reserves; by 1970, it lived only in the area in and around Gunung
Betiri.   Reproduced with the permission of the author and the
National Wildlife Federation, Washington, DC.

there may have been three tigers still alive in 1979.   After
1979, there has been no confirmable evidence of tigers (Blouch
1982).

     I have attempted to trace the tiger's decline on Java (Fig.
2).   Two hundred years ago the tiger ranged over most of the
island, and as late as the 1850's, tigers were considered a
nuisance in some populated areas (Harper 1945).   By 1940, tigers
were found only in the most remote mountain and forested areas
(Treep 1973).   By 1970, the only known tigers were in the Gunung
Betiri complex on the eastern, south coast (Fig. 2).   This is an
isolated region of the Southern Mountains that has been protected
in the past from extensive habitat alterations by precipitous and
dissected topography.   While the last tigers managed to survive
there, it was not a habitat where tigers ever occurred at a very
high density.

Bali

     If we use crude density estimates for tigers in good habit of

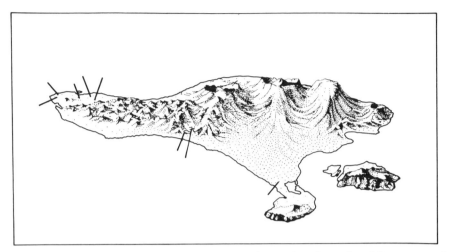

Fig. 3. Island of Bali, Indonesia, showing landforms and approximate locations of sites where museum specimens of <u>Panthera tigris balica</u> were collected between 1909 and 1939.

one adult/40 sq. km (Sunquist 1981) as the basis for an estimate, there were never more than about 125 adult Bali tigers at best at any one time on this 5500 sq. km island.

As I outline below, the Bali tiger population could have been regularly supplemented with tigers from Java swimming the Bali Straits (Seidensticker 1986).

We can not be as precise in reconstructing what happened on Bali as we have been on Java. Most of the known specimens of the Bali tiger (Mazak et al. 1978) entered the world's museums in the 1920's and 1930's (Fig. 3), and all but one came from western Bali. There are reports that tigers existed in Bali until the 1950's, but we have no specimens from that period. None of Clifford Geertz' (pers. comm.) informants told him about the presence of tigers when he worked in towns along the southwest coast in the late 1950's, and Lee Talbot did not find evidence of extant tigers when he surveyed Bali in 1960 (pers. comm.). The Bali Barat Game Reserve (20,000 ha) was established in 1941, and in the 1960's and 1970's, much of the adjacent forest land was planted in forest plantations (Seidensticker 1978). I suspect that most, if not all, Bali tigers were eliminated from the Island by the end of World War II. It is possible that the stragglers reported in the 1950's were immigrants swimming the 2.5 km wide Bali Straits from the Baluran Reserve on Java. Tigers in the Sundarbans swim much larger tidal rivers (Hendrichs 1975). The Bali tiger was extinct about half a century after it was first described to science by Schwarz (1912).

ENVIRONMENTAL STRESSES AND TIGER EXTINCTION

Java

Wet-rice or sawah agriculture is the basis of the high human populations on Java, and in some areas, Java supports 2,000 people/sq. km (Geertz 1963a). Sawah cultivation developed in Java in the rich alluvial basins surrounded by volcanoes and un-irrigable limestone hills (Fig. 2). The alluvial coastal plains that comprise about half of the island were malaria-infested and posed technical irrigation problems; they were largely ignored for cultivation until the mid-1800's. From the mid-1800's through the beginning of the World War I, the Culture System efficiently and systematically brought nearly all remaining cultivatable lands in Java under production (Geertz 1963a).

The period between the Wars saw little increase in cultivat-able lands. Just before World War II, it was estimated that 23% of the island remained under forest cover (Seidensticker 1986). During World War II, there was widespread deforestation without replanting (B. Galstaum pers. comm.). From 1950 through 1970, many of the remaining forest tracts were converted to plantations of teak (Tectona grandis), especially in east Java. These teak forests are generally depauperate of wildlife. By 1975, 85 million people lived in Java with less than eight percent of the land under forest cover (Seidensticker 1986).

Tigers and other wildlife declined as forested areas, alluvial plains, and river basins were converted for use in agriculture (Hoogerwerf 1970). As habitat contracted rapidly, set-guns and poison were used to remove unwanted tigers, and Hoogerwerf, (1970) also reported that ". . . many tigers fell victim to eating poisoned wild boar" (p. 244).

During the 1920's and 1930's, a system of reserves was established in Java. By the mid-1960's tigers survived only in Ujung Kulon, Leuweng Sancang, and Baluran (Fig. 2). Tigers did not survive in these reserves after mid-1960's when major civil unrest rocked the island and the reserves were sometimes used as sanctuaries by armed groups. Also during this period, disease reduced the rusa (Cervus timorensis) population in reserves and in many forest areas on Java (B. Galstaum pers. comm.).

By 1970, tigers survived only in a rugged area on the Southeast coast known as Meru-Betiri. This was established as a 50,000 ha game reserve only in 1972. Large scale plantation agriculture came to this and much of the surrounding area after World War II. Most of the areas below 1,000 m surrounding the Meru-Betiri reserve were planted in teak, coffee and rubber trees. Tiger density and home range size are strongly and positively correlated with biomass of large cervid prey (Sunquist 1981, Seidensticker 1986). Essential tiger habitat in the Meru-Betiri Reserve where we would expect good numbers of cervids, including the lower alluvial river flood plains, had been converted to plantations. By 1976, an estimated 7,000 people lived there. A few banteng (Bos javanicus) used the forest/plantation edge, but there were no rusa surviving in the reserve (Seidensticker and Suyono 1980). The tiger did not survive under these conditions.

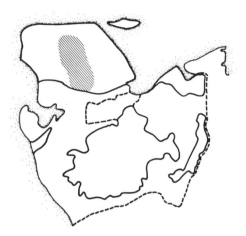

Fig. 4. Located on the western tip of Bali, the Bali Barat
National Park includes the areas where the last Bali tigers were
collected in the late 1930's. The 20,000 ha area of the park
established as a game reserve in 1941 is shown by the dashed
line; the solid lines enclose areas of the reserve not influenced
by forest plantations, and the hatched areas where reasonably
intact forest still remained at lower elevations in 1978. This
damaged ecosystem remains an essential conservation area and is
the last remaining habitat for the endemic and endangered Bali
starling (<u>Leucopasar</u> <u>rothschildi</u>).

    When the system of small reserves was established on Java in
the 1930's, they were nestled in expanses of forest that still
covered about 25% of the island. These reserves became
increasingly fragmented and isolated through the next decades as
forest was removed and plantations expanded. The areas where the
last Javan tigers did hang on were in the largest, most diverse,
and remote blocks of contiguous habitat on the island (Fig. 2):
Gunung Betiri and associated river systems, Ujung Kulon-Halimun,
and the Southern Mountains. With fragmentation, the largest
contiguous forest blocks are 500 sq. km or less, and tigers did
not survive in blocks of tropical high rain forest of this size.
We would not expect these areas to support over 5-10 adult
tigers, at best, at any one time (Sunquist 1981). Today there is
not room on Java for tigers.

Bali

    The Balinese culture and political system developed around
intensive wet-rice agriculture on palm-fringed terraces up to 700
m, especially on the southern slopes of the volcanoes (Geertz
1963b). Major land-use changes were later in coming to Bali.
When plantation and small-labor agriculture for exportdid become
established in the late 1880's, development focused on the
northern slopes of the volcanoes and the narrow alluvial strip
around the island. The relatively barren and unproductive
southern peninsula and eastern end of the island were largely
ignored. The Dutch did not establish colonial control until

about 1910. The major gorges and spurs and the instability
wrought by frequent earthquakes made establishing a road net for
the Island difficult, and it has only recently been completed.
The collection sites for the last tigers were largely at the end
of the road system as it existed in western Bali in the 1930's.
Dutch tourist literature describes tiger hunting in western Bali
in the 1930's (Dressen 1937) and apparently that is where the
Dutch hunters, A. and B. Ledeboer, killed most of their tigers
(Sody 1933).

     The Bali Barat National Park, located on the western tip of
Bali, includes the area where the last Bali tigers were collected
in the late 1930's. This is an important conservation area. The
area of the Park that includes the former Bali Barat Reserve has
been fragmented (Fig. 4), and there is little natural forest
remaining. On Bali, as in Java, there is not room today for
tigers.

COUP DE GRACE

     The coup de grace to tigers in these small areas was
stochastic processes or the human condition, depending on your
point of view: 1) wide spread poisoning during the period while
habitat was being rapidly reduced, 2) uncontrolled fragmentation
of the forest during the social disruption of World War II and
events following, 3) loss of critical ungulate prey populations
to disease, and 4) civil unrest of the 1960's resulting in tigers
killed by armed groups seeking sanctuary in these reserves.
There is not much a wildlife manager can do about this class of
problem. However, we can learn the lessons from the extinction
of these Sunda Island tigers: 1) it is dangerous to rely on
small, isolated reserves as a means to assure the long-term
survival of wild tigers; 2) large tracts of contiguous habitat
are essential to assure the long-term survival of wild tigers.
What will become of other wild tiger populations if these lessons
do not become principles in long-term tiger conservation efforts?

ACKNOWLEDGEMENTS

     I worked in Indonesia at the invitation of the government
through the World Wildlife Fund-Indonesia Program. I thank Ir.
Suyono, J.A. McNeely, B. Galstaun, and C. Santiapillai for their
assistance with this paper. Our report, The Javan Tiger and the
Meru-Betiri Reserve: A Plan for Management (Seidensticker and
Suyono 1980) summarized the results of our Javan surveys. The
work from Bali and the conclusions presented here were reported
in Large Carnivores and the Consequences of Habitat
Insularization: Ecology and Conservation of Tigers in Indonesia
and Bangladesh (Seidensticker 1986). The views expressed are my
own.

REFERENCES

Biro Pusat Statistik. 1977. Statistical Pocketbook of
    Indonesia, Jakarta: Biro Pusat Statistik.
Blouch, R.A. 1982. Proposed Meru-Betiri National Park
    Management Plan 1983-1988. Bogor, Indonesia: World Wildlife
    Fund.
Dressen, W. 1937. Hundred Tag auf Bali. Hamburg: Verlag
    Broschek and Co.
Geertz, C. 1963a. Agriculture Involution, The Process of
    Ecological Change in Indonesia. Berkeley: Univ. Calif. Pr.
Geertz, C. 1963b. Peddlers and Princes, Social Development and
    Economic Changes in Two Indonesian Towns. Chicago: Univ.
    Chicago Pr.
Harper, F. 1945. Extinct and Vanishing Mammals of the Old
    World. New York: Amer. Comm. Intern. Wldlf. Prot., New York
    Zool. Soc.
Hemmer, H. 1971. Zur Fossilgeschichte des Tigers (Panthera
    tigris, L.) in Java. Koninkl. Nederl. Akad, Wetensch. Proc.
    (Ser. B) 74:37-49.
Hendrichs, H. 1975. The status of the tiger Panthera tigris
    (Linne, 1785) in the Sundarbans Mangrove Forest (Bay of
    Bengal). Saugetierd. Mitt. 23:161-99.
Hoogerwerf, A. 1970. Udjung Kulon, The Land of the Last Javan
    Rhinoceros. Leiden: E.J. Brill.
Mazak, V., C.P. Groves and P.J.H. van Bree. 1978. On a skin and
    skull of the Bali tiger and a list of prepared specimens of
    Panthera tigris balica. Zeit. Saugetierd. 43:108-113.
Schwarz, E. 1912. Notes on Malay tigers, with description of a
    new form from Bali. Ann. Mag. Nat. Hist. Ser. 8 10:324-26.
Seidensticker, J. and Ir. Suyono. 1980. The Javan Tiger and The
    Meru-Betiri Reserve, A Plan for Management. Gland,
    Switzerland: IUCN.
Seidensticker, J. 1978. Bali Barat Reserve 1978. Bogor,
    Indonesia: World Wildlife Fund.
Seidensticker, J. 1986. Proceedings of the International Cat
    Symposium, ed. S.D. Miller and D.D. Everett. Washington DC:
    Nat. Wldlf. Fed.
Sody, H.J.V. 1933. The Balinese tiger Panthera tigris balica
    (Schwarz). J. Bombay Nat. Hist. Soc. 36:233-34.
Sunquist, M.E. 1981. The social organization of tigers
    (Panthera tigris) in Royal Chitwan National Park, Nepal.
    Smithson. Contrib. Zool. 336:1-98.
Treep, L. 1973. On the Tiger in Indonesia (with Special
    Reference to its Status and Conservation). Wageningen,
    Netherlands: Agric. Univ.
Wallace, A.R. 1962. The Malay Archipelago, New York: Dover
    Pub., Inc.

PART I

SYSTEMATICS AND TAXONOMY

2

Setting the Molecular Clock in Felidae: The Great Cats, *Panthera*

Stephen J. O'Brien, Glen E. Collier, Raoul E. Benveniste, William G. Nash,
Andrea K. Newman, Janice M. Simonson, Mary A. Eichelberger,
Ulysses S. Seal, Donald Janssen, Mitchell Bush and David E. Wildt

A special fascination of zoologists and naturalists with the cultural, aesthetic and scientific aspects of the family Felidae has produced a rich literature on the taxonomy of felid species. A variety of classification schemes, which have been based on morphological, ethological, and physiological considerations of the 37 generally recognized extant species, range from a "lumping" of all felids into two genera (Felis and Acinonyx) to a "splitting" of as many as twenty genera (de Beaumont 1967, Ewer 1973, Hemmer 1978, Leyhausen 1979, Simpson 1982, Herrington 1983, Nowak and Paradisio 1983). For example, Walker's Fourth Edition of Mammals of the World (Nowak and Paradisio 1983) lists no less than four distinct classification schemes which illustrates a number of unresolved phylogenetic inconsistencies concerning the evolutionary relationship of the family.

Several years ago we decided to apply a different approach to the study of felid taxonomy. Essentially we have employed a series of "molecular metrics" to estimate the evolutionary distance between different cat species. These molecular techniques have formed the basis of data collection in the field of molecular evolution, an exciting way to study phylogenetic relationships by looking at molecular differences in living species. Our biological material was derived from blood samples and skin biopsies converted to tissue culture lines. The tissues were collected from 34 of the 37 species with the gracious cooperation of many zoological parks and wildlife preserves throughout the world. By measuring the extent of protein and DNA differences between cat species, we were able to draw phylogenetic inferences from our results.

Before we summarize our findings, however, we would like to explain a basic tenet of molecular evolution, the theory of the "molecular clock". This concept, originally articulated by E. Zuckerkandl and L. Pauling (1962), is exquisitely simple, but dramatically powerful. As biological species evolve their genes accumulate nucleotide substitutions in the DNA backbone in a steady "clocklike" manner. These genetic substitutions are sometimes biologically consequential, but more often are selectively neutral. When the mutation occurs in a structural gene, the primary amino acid sequence becomes changed. As species evolve, these substitutions continue to accumulate and with increased time, the extent of sequence divergence becomes greater. The mutational differences are mostly evolutionary "noise", but have the advantage of being proportional to the time elapsed since the existence of a common ancestor of two extant species. Further, for any individual molecular metric, the time of divergence can be calibrated by measuring the same metric between species whose time of divergence has been established biogeographically (for example, the time of separation of Old and New World can be dated to the geologic separations of those continents). Thus, by measuring one or more molecular metrics between two taxa, a relative estimate of their evolutionary distance and their time of divergence can be obtained. The use of the molecular clock has been elegantly applied to primate radiations and other evolutionary groups by V. Sarich, A. Wilson and their colleagues (Sarich and Wilson 1967, Wilson et al. 1977) and now is an established method of evolutionary analysis. The reader is referred to Wilson et al. (1977) and Thorpe (1982) for a technical discussion of the molecular clock hypothesis and to Gribbin and Cherfas (1982) for an excellent popular description of the contributions of molecular evolution to the deciphering of the evolutionary history of mankind.

In our approach to the molecular phylogeny of felids, we employed four molecular and genetic methodologies (Table 1). Each of these methods has been shown to have certain advantages (and disadvantages) in other mammalian orders, specifically in primates and carnivores. We recently have used these same techniques to resolve the phylogenetic relationship of the giant panda and the lesser panda with respect to ursids and procyonids (O'Brien et al. 1985a). As had been the case in earlier primate studies, the panda results were concordant and permitted a confident reconstruction of evolutionary divergence in that group. The cat family seems at the onset to be somewhat more complex for two reasons. First, there are many more species involved; and second, the timing of divergence was more recent so the sensitivity of the metrics are approaching the limits of resolving power.

ALBUMIN IMMUNOLOGICAL DISTANCE

The first method we employed was that of albumin immuno-logical distance (AID). Our results have been published in detail elsewhere (Collier and O'Brien 1985) so here we shall simply summarize the results and conclusions. The procedure measures immunological distance between different species based on amino acid substitutions in homologous proteins. Substitutions were detected by the displacement of titration curves in the microcomplement fixation assay. Briefly, several rabbits were immunized with serum albumin from species A. The sera were pooled and titered against albumin from species A (homologous reaction). In the evolutionary distance determination, albumin from species B was incubated with antisera against species A (heterologous reaction). Then the degree of cross reactivity was quantified by determining the difference between the two species. When several antisera are prepared, a matrix of immunological distances can be used for estimating phylogenetic relationships.

Albumin was purified from the serum of ten cat species which were thought to be at extreme evolutionary distances by previous

Table 1.   Methods used for the four evolutionary metrics.

Evolutionary Metrics	Methodology
Immunological Distance	Titration curves using microcomplement fixation
Genetic Distance	Electrophoretic comparison of 50 isozyme systems
DNA hybridization	Shift in melting profile of unique sequence DNA
G-banded karyology	Comparative alignment of high resolution G-banded chromosomes

taxonomic classifications. A matrix of reciprocal AID values was generated (Table 2) and this matrix was tested for efficacy in tree construction using three different procedures. The first test measures the similarity of reciprocal measurements. Serological reagents which accurately reflect evolutionary divergence should show similar values for species A anti-B and species B anti-A sera. Our data passed this test well since the standard deviation was low and the "mean percent deviation of the mean" was 9.9%, a rather low value compared to previously reported AID matrices (Daugherty et al. 1982, Uzzell 1982).

A second test of such a matrix is the low incidence of violation of the "triangle inequality" (Farris 1981). This means that for distances between any three taxa, one distance cannot exceed the sum of the remaining two. The data set in Table 2 included 165 triads and there were only nine violations of the triangle inequality (95% conformance). A final component of an adequate matrix of evolutionary distances is the relative rate test (Sarich and Wilson 1967, Wilson et al. 1977). This means that the distance of each group of closely related cats from a distant relative or "outgroup" (a hyena in our data set) should be equal. When the distances are very similar, this is a signal that the molecular clock is "ticking" at about the same rate in all the cat lineages, an important component of a robust data set. In the feline AID matrix, this test worked well with three progressively more closely related "out groups", hyena, ocelot and domestic cat (see below). Successful conformance of our AID matrix to each of the three tests demonstrated the constancy of the albumin molecular clock within the Felidae radiation and supported the application of these data to phylogenetic tree construction.

The transformation of the AID matrix into an evolutionary topology was achieved by the application of several computer algorithms which have been developed for phylogenetic applications of evolutionary distance data (Table 3). The theoretical basis for each of these is extensive and have been published in detail, so we shall not review them here. It is sufficient to mention that each of the algorithms constructs trees based on minimizing leg distances, clustering species with

Table 2. Reciprocal immunological distances among ten species of Felidae and spotted hyena*.

Antisera	Lion	Tiger	Caracal	Cheetah	Serval	Clouded Leopard	Puma	Asian Golden Cat	Domestic Cat	Ocelot	Hyena
						Antigen					
Lion	---	0.6	1.7	2.3	4.3	6.0	5.3	7.5	14.4	17.0	44.4
Tiger	1.7	---	2.1	2.7	5.6	5.3	7.7	6.0	12.5	16.3	50.0
Caracal	1.8	0.6	---	2.3	3.3	5.5	6.0	6.0	12.1	15.3	46.4
Cheetah	4.0	2.8	1.2	---	1.7	5.9	4.5	9.0	11.0	16.1	45.1
Serval	6.3	6.8	3.0	3.3	---	9.8	6.1	7.2	13.4	14.8	43.4
Clouded Leopard	5.8	4.7	3.5	4.4	5.4	---	11.5	8.8	10.6	16.1	46.4
Puma	5.3	5.9	6.9	5.2	6.4	10.4	---	11.3	16.5	19.0	42.3
Asian Golden Cat	6.0	3.7	5.0	4.0	4.8	8.6	9.3	---	16.1	16.9	41.3
Domestic Cat	13.2	12.3	12.9	12.4	13.0	12.3	15.6	16.1	---	17.0	52.0
Ocelot	14.7	17.7	15.2	14.5	15.2	16.0	16.8	15.9	16.7	---	44.7
Hyena	40.8	45.4	43.3	43.5	43.5	46.8	43.9	50.2	47.7	48.3	---

* Data from Collier and O'Brien (1985).

Table 3.   Phylogenetic algorithms.

Algorithms	Reference
(1) UPGM	Sneath and Sokal 1973
(2) Fitch-Margoliash	Fitch and Margoliash
(3) Distance-Wagner	Farris 1972
(4) MATTOP	Dayhoff 1976
(5) Neighborliness	Fitch 1982

small distances, and/or minimizing the standard deviations and sum of squares of the derived phylogenetic trees.

Placing a time scale with some degree of confidence on the derived molecular topologies is a difficult problem. Although the carnivore fossil record is substantial and has been studied intensively, even the commonly accepted dates for divergence of these families can vary by as much as 25%-50%. For example, the best geological dates for the time of procyonid-ursid divergence vary from 30 to 50 M.Y. ago. A strategy we have used for setting our molecular clock was to take advantage of the demonstration that the primate and carnivore clocks appear to run at the same rate (Sarich 1969, Kohne et al. 1972, Benveniste 1985, O'Brien et al. 1985a). Because the primate radiations have been studied rather extensively, we reasoned that the species pairs that had the same distance for a particular metric within the two orders shared a common ancestor at approximately the same time in their evolutionary history. In Fig. 1 we employ a time scale which sets one AID unit equal to 0.6 M.Y. as has been used previously in primate and carnivore studies (Wilson et al. 1977, Gribbin and Cherfas 1982, Thorpe 1982, O'Brien et al. 1985a).

The phylogenetic relationship between 34 species of cats based on Table 2 using the UPGM algorithm is presented in Fig 1. The 10 index species (presented in bold print) form the basis of the tree. A series of one-way measurements using the 10 antisera and those 24 species for which albumins were available (but antisera were not) permitted the addition of "species' limbs" to the previously derived phylogenetic tree.

The major conclusion derived from the molecular topology was the resolution of felid evolution into three major lineages. The earliest branch of the felid radiation occurred approximately 12 M.Y. before the present (b.p.) and led to the small South American cats (ocelot, LPA; margay, LWI; Geoffroy's, LGE, etc.). The second branching occurred eight to ten M.Y. b.p. and included the close relatives of the domestic cat (wildcats, FSI, FLI: jungle cat, FCH; sand cat, FMA; and blackfooted cat, FNI) plus the Pallas cat, OMA. The pantherine lineage began to radiate four to six M.Y. ago and included several early branches (cheetah, AJU; serval, LSE; clouded leopard, NNE; golden cats, PTE; and puma, PCO) and a very recent (two M.Y. b.p.) split between the lynxes and the modern great cats (Panthera).

One surprise in the molecular scheme was the occurrence of the cheetah in the middle of the pantherine lineage. Morphological considerations had often placed the cheetah as an early

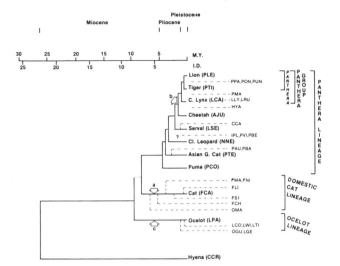

Fig. 1.   UPGM phylogenetic tree of 34 species of cat family using
reciprocal immunological distances from ten antisera prepared
against indicated species (Collier and O'Brien 1985).   The
additional species (three-letter code) were added to the tree
based upon "one-way" immunological distances.   Arrows indicate:
a(position of introduction of endogenous retroviral families,
FeLV and Rd-114, in the cat family.   Species appearing after this
point are all positive for endogenous FeLV and RD-114 retroviral
sequences (see text); b) Panthera group all have an identical
karyotype distinct from all others; c) New World small cats
(ocelot lineage) all have a 2N = 36 karyotype.   All other cat
species have 2N = 38 chromosomes.   Abbreviations: CCA, caracal;
FCH, jungle cat; FLI, African wild cat; FMA, sand cat; FNI,
blackfooted cat; FSI, European wild cat; HYA, jagaurundi; IPL,
flat-headed cat; LWI, margay; LTI, tiger cat or little spotted
cat; LGE, Geoffroy's cat; LCO, pampas cat; LLY, lynx; LRU,
bobcat; OGU, kodkod; OMA, Pallas's cat; PON, jaguar; PPA,
leopard; PUN, snow leopard; PMA, marbled cat; PBA, bay cat; PBE,
leopard cat; PVI, fishing cat; PAU, African golden cat.

divergence because of its many derived and specialized characters
associated with its high speed (Nowak and Paradisio 1983).
Additional genetic data, specifically that from karyology and the
distribution of endogenous retroviral families, lent support to
the major features of the topology presented in Fig. 1.

KARYOLOGY

Thirty-two of the thirty-seven extant species of cats have
been karyotyped by Doris Wurster-Hill in the last decade
(Wurster-Hill and Gary 1975, Wurster-Hill and Centerwall 1982).
Our laboratory has confirmed many of her karyologic observations
using retroviral transformed skin fibroblasts to obtain high
resolution karyotypes (Fig. 2).   In general, the karyotype of the
Felidae is highly conserved.   The domestic cat has 19 chromosome
pairs, 15 of which are invariant in all cat species.   The limited

rearrangements which have occurred, however, seem to be clustered according to the presented phylogeny (Wurster-Hill and Gary 1975, Wurster-Hill and Centerwall 1982). All the small South American cats (six species of the ocelot lineage) have a diploid number of 36 while all other cats have 38. The ocelot lineage cats share a derived fusion chromosome C3 which is unique to that group. Similarly, all the cats in the domestic cat lineage except the earliest line, Pallas's cat, have a unique karyotype which is distinct from other felids. In addition, the five species of Panthera have an identical karyotype which is also found in the lynxes and in marbled cats (PMA). Most of the karyologic variation in the Felidae occurs along the early phase of the pantherine lineage. The homogeneity of the karyotype of three groups (ocelot lineage, domestic cat lineage, and Panthera) lends additional support to the veracity of the lineages presented in Fig. 1.

ENDOGENOUS RETROVIRAL SEQUENCES

Endogenous retroviruses are sequences of genomic DNA found in a number of mammalian species which are homologous to RNA containing viruses isolated from neoplastic tumors. Every domestic cat (healthy or not) has at least two families of these viruses called RD-114 and FeLV encoded in their chromosomal DNA (Benveniste 1985). Both of these retroviral DNA sequences were introduced into the cat family from other mammalian families subsequent to the divergence of Felidae but prior to certain of the felid speciation events (arrow a in Fig. 1) (Kohne et al. 1972). RD-114 and FeLV sequences are present only in the DNA of six species of Felidae (domestic cat, FCA; European and African wild cat, FSY and FLI; sand cat, FMA; jungle cat, FCH; and blackfooted cat, FNI). When one examines the position of these species in Fig. 1, this result not only supports our molecular inference but also permits the placement of the time of introduction of viral sequences into the Felidae at the internode between Pallas's cat (OMA) and other species on the domestic cat lineage (approximately six to eight M.Y. ago).

DNA HYBRIDIZATION

The next molecular method we used was hybridization of genomic DNAs from heterologous cat species. Non-repetitive DNA (high c_ot DNA) was extracted from several species of cats and hybridized with radiolabeled DNA from an index species. Two types of measurements can be derived from these experiments: the extent (%) of reassociation and the thermal stability of DNA hybrids ($\hat{}$Tm). The extent of hybridization of the radioactive DNA and the "cold" heterologous species DNA (relative to the homologous hybridization value) is inversely proportional to the time elapsed since the two species shared a common ancestor (Kohne et al. 1972, Benveniste 1985). The hybridization of imperfectly matched DNA strands also results in the formation of hybrids that melt at a lower temperature than the homologous DNA hybrids; the extent of this difference ($\hat{}$TmR of $\hat{}$T50H) is linearly related to the percent hybridization obtained and to the time since two species diverged (Benveniste 1985).

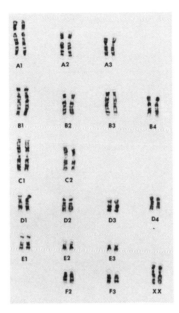

Fig. 2.   G-banded karyotype of P. leo.   Skin fibroblast cultures were established from biopsies of each species, transformed in vitro with feline sarcoma virus (Snyder-Theilen strain) and G-trypsin banded as described previously (Nash and O'Brien 1982). Every member of Panthera plus marbled cat and the three lynx species (Siberian lynx, Canadian lynx and bobcat) have an identical karyotype (Wurster-Hill and Gary 1975, Wurster-Hill and Centerwall 1982).

Table 4 lists the thermal stability difference obtained after hybridizing the radioactively labeled DNAs from two index species (domestic cat and lion) to DNA from a number of cat species.   The results show that five species of small cats from the Mediterranean region (Ewer's genus Felis) (1973) are closely related to the domestic cat (^TmR range = 0.3-1.4° C), while other felids tested are more distantly related (^TmR range = 2.2-3.2° C).   When lion DNA was used as a probe, the close relationship of the other great cats tested (tiger, leopard, and snow leopard) to the lion was evident.   The largest thermal stability difference obtained between cat species tested (3.2°C) is consistent with a radiation of the living members of this family within the last 12 to 14 M.Y. and supports the AID data.   This small ^TmR difference makes determination of relationships within this family difficult by these cellular DNA hybridization techniques.   Nevertheless, the data do support the Aid data and fossil evidence presented elsewhere in this volume which suggest that the various Panthera species radiated within the Pleistocene.

Table 4.   Thermal stability (^TmR) of DNA hybrids between
domestic cat and lion unique sequence DNA and the DNA of other
cat species (derived from Benveniste 1985).

Species	Radioactive DNA Probe	
	Domestic Cat	Lion
Domestic cat	0.0	2.6
European wildcat	0.3	2.9
Sand cat	1.0	2.9
Jungle cat	1.1	NT
African wildcat	1.1	NT
Black-footed cat	1.4	2.3
Caracal	2.2	2.7
Leopard cat	2.2	2.5
Margay	2.5	2.6
Lion	2.8	0.0
Tiger	3.2	1.2
Leopard	2.8	1.1
Snow leopard	3.2	1.2

NT - not tested.

ISOZYME GENETIC DISTANCE

The third molecular metric we used was isozyme genetic
distance, a statistical calculation developed by M. Nei for
estimating the degree of allelic substitutions at a group of loci
between populations (or species) based upon the electrophoretic
mobility of soluble proteins (Nei 1972, 1978).   The distance
estimate, D, is defined as the average number of gene differences
per locus between individuals from two test populations.   Within
the limits of certain assumptions relating to electrophoretic
resolution and the relative rates of nucleotide substitution, the
genetic distance estimates increase proportionately with the
amount of time the compared populations have been reproductively
isolated.   The method is designed to handle outbred populations
which contain polymorphic allozyme loci and has been used
extensively in evolutionary studies of several taxonomic groups
including several studies of the Hominidae (Bruce and Ayala 1979,
Avise and Aquadro 1981, O'Brien et al. 1985a).   In general, the
method does appear to conform to the evolutionary clock
hypothesis so long as enough loci are sampled and the studied
taxa are within the linear range of the estimates (Nei 1972,
1978, 1979, Bruce and Ayala 1979, Avise and Aquadro 1981).

In order to resolve a large number of loci for each species,
isozyme extracts of erythrocytes, lymphocytes and cultured
fibroblasts were prepared and analyzed for 50 enzyme systems
(Fig. 3).   Thus, for most measurements, duplicate gels were
compared permitting confirmation of electrophoretic differences.
We report here the results of analysis with ten species,
including the five great cats plus five cat species also shown to
be in the Pantherine lineage (Fig. 1).   In addition, we compared

three subspecies of tigers _Panthera_ _tigris_ _tigris_ (Bengal tiger)
and _P._ _tigris_ _altaica_ (Siberian tiger) and _P._ _tigris_ _sumatrae_
(Sumatran tiger). Animal tissues were collected from a variety
of zoos and wildlife preserves over the last five years. The
frequency of allelic variation at polymorphic isozyme loci has
been estimated in seven of the ten species (O'Brien 1980, et al.
1985b, Newman et al. 1985). The allelic designations of the ten
species at 50 isozyme loci is presented in Table 5.

Genetic distance was computed and the derived distance
matrix (Table 6) was employed to produce phylogenetic topologies
using the same algorithms discussed above (Table 3).

A tree derived from the Fitch-Margoliash algorithm with
the assumption of contemporaneous species tips is presented in
Fig. 4.

To set the time scale in Felidae, the primate and carnivore
clocks were assumed to have evolved at the same rate. The timing
of two divergence nodes in primate evolution were selected as
standards; namely, the human-chimp-gorilla split 4.5 M.Y. b.p.,
and the orangutan-human divergence at 12 M.Y. b.p. We then
compared the D values computed here with the D values obtained
for primate and carnivore species previously reported from our

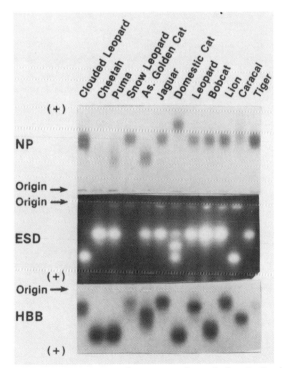

Fig. 3. Electropherogram of a starch gel loaded with extracts of
indicated species and developed histochemically purine nucleoside
phosphorylase (NP), for esterase-D (ESD), and hemoglobin (HBB).
Polymorphic allozyme frequencies were estimated in seven species
(O'Brien 1980, et al. 1985b, Newman et al. 1985).

Table 5.   Isozyme composition of selected feline species*.

Gene Symbol	Tiger	Caracal	Lion	Leopard	Jaguar	Asian Golden Cat	Snow Leopard	Puma	Cheetah	Clouded Leopard
ACP1	A	A	A	A	A	A	-	A	A	B
ACP2	A	A	A	A	A	A	A	A	A	A
ADA	AB	B	AB	AB	A	-	A	A	C	AB
AK1	A	C	A	A	A	B	A	C	B	D
ALB	A	A	A	A	A	A	A	A	A	B
APRT	A	BD	A	BA	A	B	C	C	C	A
CAT	A	BB	A	A	A	A	A	A	A	C
CPKB	A	A	A	A	A	A	A	A	A	A
DIAB1	A	AB	A	A	A	A	A	-	C	A
DIAB2	A	-	A	A	A	B	-	-	CB	A
DIAC	A	E	A	A	A	B	A	C	C	A
ESD	A	B	A	A	A	B	A	A	AC	B
ESU2	-	A	A	A	A	A	A	A	A	A
ESA1	A	E	BA	AB	BC	A	BC	C	AD	AE
ESA2	B	A	B	A	A	A	A	AB	A	AB
CA2	A	BD	A	A	A	B	A	B	B	A
FUCA	A	AB	A	A	A	A	C	D	A	A
GALB	A	A	A	A	A	A	A	A	A	A
G6PD	A	A	A	A	A	A	A	A	B	A
GLO1	A	A	A	A	A	A	A	A	A	A
GOT2	A	A	AB	A	A	AC	AB	B	A	A
GPI	A	A	AB	A	A	A	A	A	A	A
GPT	AC	B	A	A	A	B	A	B	B	B
GSR	A	BA	B	A	A	B	A	C	B	A
GUSB	AB	B	B	A	A	B	A	C	CD	D
HBB	A	B	A	AE	A	B	A	C	C	A

Table 5 Continued.

Gene Symbol	Tiger	Caracal	Lion	Leopard	Jaguar	Asian Golden Cat	Snow Leopard	Puma	Cheetah	Clouded Leopard
HK1	A	A	A	A	A	A	A	A	A	A
HPRT	A	B	A	A	A	A	A	-	-	-
IDH1	A	AB	A	A	A	A	A	A	A	A
IDH2	A	A	A	A	A	A	B	A	A	C
LDHA	A	A	A	A	A	A	A	A	A	A
LDHB	A	A	A	A	A	A	A	A	A	A
MDH1	A	A	A	A	A	A	A	A	A	A
MDH2	A	A	A	A	A	A	A	A	A	A
ME1	A	A	A	A	A	A	A	A	B	A
MP1	B	A	AB	A	A	B	A	B	A	A
NP	A	A	A	A	A	A	A	A	B	A
PEPA	A	A	A	A	A	A	A	A	A	A
PEPB	A	AB	A	A	A	A	A	AB	AB	A
PEPC	A	AB	A	A	A	A	A	A	B	B
PEPD	A	AB	A	FA	A	A	A	A	B	E
PGD	A	CG	A	A	A	B	A	D	AD	A
PGM1	AB	A	C	A	A	A	A	B	A	A
PGM3	A	A	A	A	A	A	A	B	B	C
PK	A	B	C	A	C	B	A	AB	-	A
PP	A	A	A	A	A	A	-	B	B	B
SOD1	A	B	A	A	A	B	A	F	A	C
TF	AG	B	A	A	A	B	E	A	B	A
TPI	A	A	A	A	A	A	A	A	A	A

* Abbreviations for isozymes follow human enzyme convention (McAlpine et al. 1985). Letters are arbitrary allele designations based on electrophoretic mobility (eg., see Fig. 3).

Table 6. Genetic distance between cat species based on isozyme analysis*.

	Sumatran Tiger (3)	Siberian Tiger (14)	Bengal Tiger (6)	Lion (20)	Leopard (18)	Jaguar (3)	Snow Leopard (3)	Clouded Leopard (20)	Cheetah (55)	Puma (3)	Asian Golden Cat (2)	Caracal (16)
Sumatran Tiger	----	.003	.010	.117	.073	.095	.180	.370	.588	.588	.419	.549
Siberian Tiger	.00	----	.007	.126	.064	.085	.169	.366	.576	.546	.415	.552
Bengal Tiger	.00	.00	----	.121	.068	.088	.173	.353	.564	.534	.385	.516
Lion	.093	.112	.100	----	.152	.096	.243	.349	.604	.581	.135	.456
Leopard	.036	.037	.034	.125	----	.055	.123	.358	.476	.487	.309	.432
Jaguar	.073	.072	.070	.084	.030	----	.124	.325	.497	.475	.380	.474
Snow Leopard	.155	.155	.152	.229	.095	.111	----	.431	.581	.461	.441	.557
Clouded Leopard	.346	.353	.334	.336	.332	.314	.417	----	.585	.661	.504	.443
Cheetah	.554	.552	.534	.580	.440	.475	.557	.562	----	.324	.238	.425
Puma	.535	.534	.515	.568	.461	.464	.448	.649	.301	----	.319	.416
Asian Golden Cat	.395	.406	.370	.426	.287	.373	.431	.417	.219	.312	----	.252
Caracal	.507	.520	.478	.424	.387	.444	.525	.413	.386	.386	.226	----

* Above diagonal, Nei (1972) genetic distance; below diagonal Nei unbiased genetic distance corrected for small sample size (Nei 1978). Matrix was computed using BIOSYS program kindly provided by R. Swofford, University of Illinois. In parenthesis, number of individuals typed in estimating allelic polymorphisms frequencies. Allele frequencies were taken from O'Brien (1980), et al. (1985b) and Newman et al. (1985).

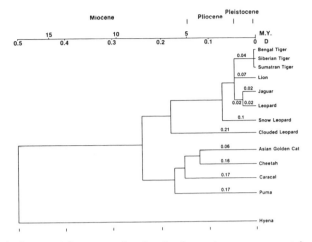

Fig. 4. Phylogenetic tree derived from isozyme genetic distance
data in Table 6 using the Fitch-Margoliash algorithm. The actual
tree was drawn to scale using the KITSCH subroutine of the PHYLIP
program, developed and generously provided by J. Felsenstein
(Univ. Washington). This program computes a rooted topology
based on the assumption of an evolutionary molecular clock
rendering all terminal species as contemporaneous. The numbers
are the leg lengths of the unrooted tree generated by the Fitch-
Margoliash algorithm in the absence of the above assumptions.

laboratory using the same methology (O'Brien et al. 1985a).
The chimp-gorilla-human distances average 0.26. This value
actually corresponds to 9.0 M.Y.; i.e., 4.5 M.Y. back to the
common divergence node for man and for chimpanzee. The average D
value of the orangutan from African apes is 0.38 so D=0.19
corresponded to 12 M.Y. These two points do not define a line
with an intercept of zero for reasons that are unclear; however,
they may reflect the fact that although the molecules evolve at a
rate proportionate to elapsed time, the proportionality is not
precisely linear with time. In Fig. 4 we adopt the human-
gorilla-chimp node value (i.e., 4.5 M.Y. is equivalent to a D
value of 0.13), recognizing that the older dates may be
understated and less accurate.

     Examination of the topology generated in Fig. 4 reveals
several important suggestions about the Felidae radiation. In
agreement with the AID and the DNA results , as well as the
fossil record, the five Panthera species have diverged rather
recently during the Pleistocene. There is some evidence that the
spotted cats, leopard and jaguar, may have diverged more recently
from a common ancestor, although both are rather close to the
lion and tiger. The earliest _Panthera_ leg leads to the snow
leopard. In the early pliocene, or possibly in the miocene, the
clouded leopard diverged from the line leading to the modern
Panthera. The _Panthera_-clouded leopard genetic distance is about
the same as the orangutan-human distance (O'Brien et al. 1985a).
The second lineage supports the placement of the cheetah, golden
cat, puma and caracal as having evolved from the Pantherine
lineage much earlier (six to twelve M.Y. ago). The apparent

Table 7.   Comparison of Nei-genetic distance between selected species and subspecies*.

	D	No. Loci	Reference
Interspecies comparisons			
Homo sapiens-Pan troglodytes	0.21	44	O'Brien et al. 1985a
Gorilla gorilla-P. troglodytes	0.21	44	O'Brien et al. 1985a
H. Sapiens-G. gorilla-H. sapiens	0.25	44	O'Brien et al. 1985a
Ursus arctos-Ursus malayanus	0.21	50	O'Brien et al. 1985a
U. arctos-Tremarctos ornatus	0.45	50	O'Brien et al. 1985a
U. malayanus-T. ornatus	0.39	50	O'Brien et al. 1985a
Intraspecies comparisons			
H. sapiens (black)-H. sapiens (Asian)	0.029	62	Nei and Roychoudhury 1982
H. sapiens (black)-H. sapiens (Caucasian)	0.027	62	Nei and Roychoudhury 1982
H. sapiens (Asian)-H. sapiens (Caucasian)	0.010	62	Nei and Roychoudhury 1982
Mus musculus domesticus-M. m. molossinus	0.400	51	O'Brien et al. 1982
M. m. domesticus 2 California populations	0.019	46	Rice and O'Brien 1980
Swiss mice (3 outbred factory colonies)	0.020-0.025	46	Rice and O'Brien 1980
Swiss mice (4 inbred strains)	0.045-0.067	46	O'Brien et al. 1982
Swiss outbred strains-4. Swiss derived inbred strains	0.039-0.081	46	O'Brien et al. 1982
Leontopithecus rosalia (3 subspecies)	0.007-0.030	47	Forman et al. In Press
Acinonyx jubatus jubatus-A. j. raineyi	0.004	49	O'Brien et al. In Prep.
Panthera tigris (3 subspecies)	0.003-0.010	50	This Study

* All values except the human racial groups were estimated in our laboratory using essentially the same group of 50 isozyme systems.

association of cheetah, golden cat, caracal and puma species should be considered as tentative because of the variance in these larger D values. (For example, for reasons relating to loci which were technical failures and not scored in the golden cats, their distance from all other species appeared low; see Table 6). Thus, although each of the four species seem to be equidistant from the modern _Panthera_ radiation, their relative relationship could not be clearly resolved with the D values derived in Table 6. The out-group hyena was two to three times the distance from any of the felids as they were from each other (data not shown), which is consistent with a valid distance matrix for this group. The three subspecies of tiger were very similar in molecular terms with a mean genetic distance of 0.007 (Table 6). This distance is 30 times less than the distance between chimpanzee and man, four times less than the distance between human racial groups and three times less than the distance between two isolated demes of _Mus musculus domesticus_ separated by 100 miles in California (see Table 7).

The tiger subspecies values are 10 to 35 times less than the distance between different species of great cats and are comparable to sub-species distances of cheetahs (_A. jubatus_) or lion tamarins (_Leontopithecus rosalia_) (Foreman et al. in press). We interpret these results as consistent with the isolation of these geographic subspecies of tigers within the past 10,000 years, most probably within historic times. No subspecies-specific allozyme markers were detected which supports this conclusion, and there is no evidence to suggest that sufficient time has elapsed for specific genetic isolating mechanisms to have developed.

The molecular and cytological results presented here have provided limited insight into the evolutionary history of the Felidae and specifically the _Panthera_. The recent radiation of the great cats (< two M.Y. ago) is evident from molecular and paleontological results. The relative relationship between the specfics within the ocelot lineage, within the domestic cat lineage and in the Pantherine lineage cannot be definitively reconstructed with the present molecular data. We expect that the results of anticipated molecular studies combined with morphological and zoogeographical considerations from other laboratories should ultimately converge on a consensus phylogeny of the Felidae.

ACKNOWLEDGEMENTS

We are grateful to the directors and veterinary staff of the following zoological parks for generous provision of tissues from animals which made these studies possible: Henry Doorly Zoo (Omaha, NE); National Zoological Park (Washington,D.C.); Blijdorp Zoo (Rotterdam, The Netherlands); Detroit Zoological Park (Royal Oak, MI); San Diego Wild Animal Park (Escondido, CA); Minnesota Zoological Gardens (Apple Valley, MN); Walter D Stone Memorial Zoo (Stoneham, MA); San Antonio Zoological Gardens (San Antonio, TX); Lincoln Park Zoological Gardens (Chicago, IL); Exotic Feline Breeding Colony (Rosamond,CA); Carnivore Evolution Research Institute (Pittsboro, NC). Tissue culture specimens of exotic animals were collected in full compliance with specific Federal Fish and Wildlife permits (CITES; Endangered and Threatened

Species; Captive Bred) issued to the National Cancer Institute,
National Institutes of Health, principal officer Stephen J.
O'Brien, by the U.S. Fish and Wildlife Service of the Department
of the Interior.

REFERENCES

Avise, J.C. and C.F. Aquadro. 1981. A comparative summary of
    genetic distance in vertebrates. Evol. Biol. 14:114-26.
Benveniste, R.E. 1985. The contributions of retroviruses to the
    study of mammalian evolution, in Molecular Evolutionary
    Genetics (Monographs in Evolutionary Biology Series), ed.
    R.J. MacIntyre. New York: Plenum Press.
Bruce, E.J. and F.J. Ayala. 1979. Phylogenetic relationships
    between man and the apes: Electrophoretic evidence.
    Evolution 33:1040-56.
Collier, G.E. and S.J. O'Brien. 1985. A molecular phylogeny of
    the Felidae: Immunological distance. Evolution 39:473-87.
Daugherty, C.H., L.R. Maxson and B.D. Bell. 1982. Phylogenetic
    relationships within the New Zealand frog genus Leiopelma:
    Immunological evidence. New Zealand J. Zool. 9:239-42.
Dayhoff, M.O. 1976. Atlas of Protein Sequence and Structure.
    Washington, DC: National Biomedical Research Foundation.
    Vol. 5, Suppl. 2.
de Beaumont, G. 1967. Remarques sur la classification des
    Felidae. Ecolog. Geol. Helv. 57:837-45.
Ewer, R.F. 1973. The Carnivores. New York: Cornell Univ. Pr.
Farris, J.S. 1972. Estimating phylogenetic trees from distance
    matrices. Am. Nat. 106:645-88.
Farris, J.S. 1981. In Advances in Cladistics, ed. V.A. Funk and
    D.R. Brooks. New York: The New York Botanical Garden.
Fitch, W.M. and E. Margoliash. 1967. Construction of
    phylogenetic trees. Science 155:279-84.
Fitch, W.M. 1982. A non-sequential method for constructing
    trees and hierarchal classifications. J. Mol. Evol. 18:30-
    37.
Forman, L., D.G. Kleiman, R.M. Bush, J.M. Dietz, J.D. Ballou, L.
    Phillips, A.F. Coimbra-Filho and S.J. O'Brien. In press.
    Genetic variation within and among lion tamarins. Am. J.
    Phys. Anthrop.
Forman, L., D. Kleiman, M. Bush, J. Ballou, L.Phillips, J.L.
    Dietz, A. Coimbra-Filho and O'Brien. In press. Genetic
    variation within and among lion tamarins. Am. J. Phys.
    Anthropol.
Gribbin, J. and J. Cherfas, eds. 1982. The Monkey Puzzle:
    Reshaping the Evolutionary Tree. New York: Pantheon Books.
Hemmer, H. 1978. The evolutionary systematics of living
    Felidae: Present status and current problems. Carnivore
    1:71-79.
Herrington, S.J. 1983. Systematics of the Felidae: A
    quantitative analysis. M.A. Thesis, University of Oklahoma.
Kohne, D.E., J.A. Chisow and B.H. Hoyer. 1972. Evolution of
    primate DNA Sequences. J. Hum. Evol. 1:627-44.
Leyhausen, P., ed. 1974. Cat Behavior. New York: Garland.
McAlpine, P.J., T.B. Shows, R.L. Miller and A.J.Pakstis. 1985.
    The 1985 catalog of mapped genes and report of the
    nomenclature committee. Eighth International Workshop on
    Human Gene Mapping. Cytogenet. Cell. Genet. 40:8-66.

Nash, W.G. and S.J. O'Brien. 1982. Conserved subregions of homologous G-banded chromosomes between orders in mammalian evolution: Carnivores and primates. Proc. Natl. Acad. Sci. U.S.A. 79:6631-35.

Nei, M. 1972. Genetic distance between populations. Am. Nat. 106:283-92.

Nei, M. 1978. Estimation of average heterozygosity and genetic distance from a small number of individuals. Genetics 89:583-90.

Nei, M., ed. 1979. Molecular Population Genetics and Evolution, Amsterdam: North Holland Publ. Co.

Nei, M. and A. Roychoudhury. 1982. In Evolutionary Biology, ed. M.K. Hecht, B. Wallace, and G.T. Prance. New York: Plenum Press.

Newman, A., M. Bush, D.E. Wildt, Dirk van Dam, M. Frankehuis, L. Simmons, L. Phillips and S.J. O'Brien. 1985. Biochemical genetic variation in eight endangered feline species. J. Mammal. 66:256-67.

Nowak, R.M. and J.L. Paradisio, eds. 1983. Walker's Mammals of the World, 4th ed. Baltimore: Johns Hopkins Univ. Pr.

O'Brien, S.J. 1980. The extent and character of biochemical genetic variation in the domestic cat (Felis catus). J. Hered. 71:1-8.

O'Brien, S.J., J.L. Moore, M.A. Martin and J.E. Womack. 1982. Evidence for the horizontal acquisition of murine AKR virogenes by recent horizontal infection of the germ line. J. Exp. Med. 155:1120-32.

O'Brien, S.J., W.G. Nash, D.E. Wildt, M.E. Bush and R.E. Benveniste. 1985a. A molecular solution to the riddle of the giant panda's phylogeny. Nature 317:140-44.

O'Brien, S.J., M.E. Roelke, L. Marker, A. Newman, C.W. Winkler, D. Meltzer, L. Colly, J. Everman, M. Bush and D.E. Wildt. 1985b. Genetic basis for species vulnerability in the cheetah. Science 227:1428-34.

Rice, M.C. and S.J. O'Brien. 1980. Genetic variance of laboratory outbred Swiss mice. Nature 283:157-67.

Sarich, V.M. and A.C. Wilson. 1967. Immunological time scale for hominid evolution. Science 158:1200-03.

Sarich, V.M. 1969. Pinniped origins and the rate of evolution of carnivore albumins. Syst. Zool. 18:286-95.

Simpson, G.G. 1982. Am. Mus. Nat. Hist. 22:451-60.

Sneath, P.H.A. and R.R. Sokal, eds. 1973. Numerical Taxonomy: The Principles and Practice of Numerical Classification. San Francisco: W.H. Freeman and Co.

Thorpe, J.P. 1982. The molecular clock hypothesis: Biochemical evolution, genetic differentiation and systematics. Ann. Rev. Ecol. Sys. 13:139-68.

Uzzell, T. 1982. Immunological relationship of western Palearctic water frogs (Salientia: Ranidae). Amphibia Reptilia 3:135-43.

Wilson, A.C., S.S. Carlson and T.J. White. 1977. Biochemical evolution. Ann. Rev. Biochem. 46:573-639.

Wurster-Hill, D.H. and C.W. Gary. 1975. The interrelationship of chromosome banding patterns in procyonids, viverrids, and felids. Cytogenet. Cell Genet. 15:306-31.

Wurster-Hill, D.H. and W.R. Centerwall. 1982. The interrelationships of chromosome banding patterns in canids, mustelids, hyena and felids. Cytogenet. Cell Genet. 34:178-92.

Zuckerkandl, E. and L. Pauling. 1962. In Horizons in Biochemistry, ed. M. Kasha and B. Pullman. New York: Academic Press.

3

The Phylogeny of the Tiger
(Panthera tigris)

Helmut Hemmer

The phylogeny of a species involves its roots in ancestral populations as well as its later diversification into different populations. Therefore, to study a species' history it is necessary to look at its relationships with other species and at its status in time and space.

The tiger (*Panthera tigris*) belongs to a group of cat species called pantherines (Hemmer 1966). The lion, first thought to be the nearest tiger relative (e.g. Haltenorth 1936/37), is now placed closer to the leopard and jaguar (Leyhausen 1950, Hemmer 1966). Characteristics of structure and function can be understood as being derived in common, connecting P. leo, P. onca and P. pardus, excluding P. tigris (Leyhausen 1950, Hemmer 1966, 1978a, 1981, Peters 1978). It is not possible to split the former three species into successive dichotomies by the use of Hennig's phyletical principles, thus pointing to one last radiation in the pantherine speciation processes giving rise to the lion, jaguar and leopard. Therefore, the branching off of the tiger line from the common stem must be interpreted as an earlier event (Hemmer 1981). There are differing views of what other species were derived at the same time with the tiger from this earlier stem species. Leyhausen (1973) classified the tiger, then named Neofelis tigris, with the clouded leopard, N. nebulosa. Hemmer (1978a) presented two different possibilities; the tiger line, not in common with the clouded leopard's ancestry, split off alone from the lion/jaguar/leopard stem species, or branched off within a radiation from which the snow leopard (Uncia uncia) originated. Later arguments favor the branching of the tiger line alone before the last pantherine radiation. The stem species of this dichotomy should be understood as derived from an earlier pantherine radiation, from which the snow leopard and clouded leopard also may have originated (Hemmer 1981). The last species of living pantherines, the marbled cat (Pardofelis marmorata), first aligned with this group of cats (Hemmer 1978a), has been shown to share derived features with the clouded leopard (Groves 1982), suggesting that a common stem species of clouded leopard and marbled cat arose from the basic pantherine radiation. The isozyme genetic distances noted by O'Brien (this volume) confirm that the clouded leopard line branched off first from the pantherine stem, followed by the snow leopard branch, as first

suggested on morphological and ethological grounds (Hemmer 1966).
The isozyme distance data of O'Brien (this volume) are not in
accordance with his DNA hybridization data as far as the ultimate
ramifications of the pantherine phylogeny are concerned. The DNA
data agree with the above results, suggesting two subsequent
branching events; the first splitting of the tiger line before
finally giving rise to the lion/jaguar/leopard group (Hemmer
1966, 1978a, 1981).

The fossil documentation does not allow determination of
when the tiger dichotomy occurred. The oldest reliable remains
of the lion/jaguar/leopard group date back less than two million
years (upper Blancan in North America, upper Villafranchian in
Europe, East and South African correlates; Hemmer 1976, Kurten
1973). As no undisputed _Panthera_ fossils are represented in the
otherwise well known carnivore communities of Europe, Africa and
North America before this time (East African fossils mentioned as
pantherines from early australopithecine sites need comparative
studies), and as the ecological niche of the later _Panthera_ has
been kept by the genus _Dinofelis_, this time level of
approximately two million years may be the time of the last
Panthera radiation and dispersal, suggesting their origin should
have been in Asia. The dating of this event using a biological
clock procedure agrees with the procedures using isozyme genetic
distances (O'Brien this volume). Thus, the dichotomy of the
tiger line and the lion/jaguar/leopard stem pieces should have
occurred earlier, presumably parallel to early Blancan times,
somewhere in Asia. A biogeographic analysis points to East Asia
(Hemmer 1981).

The most complete and most important fossil of early tiger
history has been dated in the lower Pleistocene or upper Pliocene
(Zdansky 1924, Pei 1934, Hemmer 1967). Its complete skull has
been described as _Felis palaeosinensis_ (Zdansky 1924). A later
analysis suggested it was a primitive tiger, having a majority of
morphognostic osteological features typical for _P. tigris_ but
also shared with several jaguar features, considered
symplesiomorphic, the jaguar most resembling the common stem
species in the lion/jaguar/leopard group (Hemmer 1967, 1981).
When this 1967 analysis of the Chinese paleotiger was performed,
the significance of brain size for the evaluation of the
evolutionary level for mammals other than man was not yet clearly
recognized. Working with an accurate cast provided by Zdansky
from the Upsala museum for the osteological analysis, some way
had to be found to obtain a crude estimate of brain size in the
palaeosinensis skull. Direct morphognostic comparison with the
smallest Balinese and Sumatran tiger skulls (Hemmer 1969),
equaling in absolute size the Chinese paleotiger fossil, shows a
flatter profile, a less widened braincase in the temporal region
and a weak but complete sagittal crest. The temporal muscle
lines more or less diverge in the former group. These features
indicate a less vaulted and thus smaller braincase in the
paleotiger. Balinese and Sumatran tiger braincase volumes were
measured at about 260 cu. cm, the smallest Sumatran tiger brain
volume was 228 cu. cm. On the other hand, morphognostic
comparison with large jaguar skulls points to a larger brain in
the _palaeosinensis_ tiger skull. Braincase volumes of such large
jaguars have been measured as 168 - 204 cu. cm. This indicates a
volume less than 250 but more than 200 cu. cm in the Chinese
paleotiger. The cross section of the foramen magnum of the
latter (greatest height x greatest breadth) is approximately 20%
smaller than in the _balica_ holotype skull, the smallest cross

section measured in modern tiger skulls so far compared. There
is no significant correlation of foramen magnum cross section and
brain volume intraspecifically in tigers, but such a correlation
does appear interspecifically in the lion/jaguar/leopard group.
Parallel to tigers, a brain volume in the same range as estimated
by the morphognostic comparison results for the palaeosinensis
skull.

Having a brain size between jaguars and modern Indonesian
tigers, the Chinese paleotiger occupies an advanced position
along the tiger branch above the dichotomy, for at this point no
higher cephalization should be expected than the lowest found in
the offspring species, using the principles of progressive
cephalization in mammalian evolution other than domestication.
This conclusion allows use of the paleotiger as an object of
outgroup comparison in the modern populations.

The oldest studied tiger material besides the Chinese skull
comes from the Jetis beds of Java dated approximately 1.3 to 2.1
million years old (tigers: Dubois 1908, Brongersma 1935, Hemmer
1971; stratigraphic correlation and dating: Ninkovich and
Burckle 1978, De Vos et al. 1982, Hooijer and Kurten 1984). The
mandible and teeth remains are relatively close in absolute size
as well as in morphometric and morphognostic features compared to
the Chinese paleotiger, although they differ in some details
(Hemmer 1971). The dispersion of tigers over East and Southeast
Asia must have been completed two million years ago. An
attribution of large felid femora from the earlier Javanese Kali
Glagah complex (Von Koenigswald 1933) to a paleotiger is
disputable (Hemmer 1971).

Unfortunately, there has been no chance to study original
Chinese materials of upper Pliocene or lower Pleistocene levels.
The Pleistocene evolution of tiger populations in Java does not
correspond to a straight line from early Pleistocene to modern
tigers, but rather to a mosaic pattern of some gene pool changes
and possibly exchanges in the succession of marine regression and
progression phases in Indonesia during the northern Ice Age
(Hemmer 1971). This is in accordance with the low genetic
distances found among modern populations (O'Brien this volume).

This long period of tiger differentiation in the
Pleistocene, probably longer than the time available for
intraspecific differentiation in the history of lions, jaguars
and leopards, resulted in a magnitude of morphological, at least
osteological distance between contemporary geographic tiger
populations, that are reminiscent of species differences in other
cats. To distinguish Javanese and Sumatran, or old male Indian
and middle Asian tiger skulls is easier than distinguishing
jaguar and leopard skulls of comparable size or even lion and
tiger skulls in some cases.

Descriptive work in intraspecific tiger taxonomy in the last
two decades is voluminous (Mazak 1967, 1968, 1976, 1983, Mazak
and Volf 1967, Mazak et al. 1978). Unfortunately, Mazak, working
mainly on static taxonomy, did not comprehend dynamic
evolutionary changes in tiger populations that occurred during
intraspecific phylogeny. Two central points may be contrasted:
1) evolutionary specializations for enhanced effectiveness in
functional environmental adaptations and; 2) general evolutionary
progression for enhanced behavioral plasticity by progressive
cephalization.

The first point involves changes in size, in fur structure and in the masticatory apparatus in tigers. There is a size decline for many mammalian species from temperate to tropic regions (Bergman's rule). This decline is obvious in tigers from USSR populations to Indonesian populations. Support for this decline in size in tigers is given by Pleistocene fossils of Java, showing large tigers instead of the small tropic ones in a middle or upper Pleistocene fauna that also contained the northern crane (<u>Grus grus</u>), indicative of a cooler climate. The size trend is paralleled by a trend of fur thickness that provides tigers with their main heat isolation mechanism. Consequently a small Indonesian tiger with short hair (despite special mane tendencies) is better adapted to the tropics than the large long-haired Siberian tiger, and vice versa. The southern tigers are of lighter build than the northern ones, but greater in linear measurements; 295 to 349 mm skull length in male Sumatran and Javanese tigers compared to between 280 and 318 mm in Siberian tigresses (Mazak 1983).

Taking the small and short first lower molar and upper last premolar carnassial complex in the Chinese paleotiger and the earliest Javanese tigers as primitive, there is a trend toward specialization parallel to that observed in other large cat species. The relative enlargement of this highly specialized carnassial complex provides enhanced cutting efficiency. There may be two independent centers of such a parallel evolution; the longest carnassials (relative size as a ratio of premolar to skull length) are found in modern tigers in Indonesia, and the population in the Manchuria/Amur region also show especially long carnassials (Hemmer 1971).

In modern South China tiger populations, located between these two areas and largely unchanged from early times (Hemmer 1978c, Herrington this volume), the original short carnassials may have been preserved to a greater extent (Hilzheimer 1905). These middle Asian tigers have kept the original, shorter molar. This evolution cannot have begun later than the beginning of the Trinil faunal complex in Java, dated 0.7 to 1.3 million years ago (Ninkovitch and Burckle 1978), as Trinil tigers present this characteristic (Hemmer 1971). This evidence suggests a remote onset of tiger population differentiation more than one million years ago.

The cephalization is a useful tool for measuring the level of progressive cephalization within a given mammalian taxon (Hemmer 1978b for carnivores). Basic parameters are body and brain weight. Tiger body weights were compiled by Mazak (1983) and brain weight estimates were made by measuring skull braincase volumes and converting to brain weights (using a regression of both parameters across different mammalian orders). Capacities of one Balinese, five Javanese and five Sumatran tiger skulls (Senckenberg Museum Frankfurt/M.) were examined (by Petra Jung). Estimates of the cephalization constant for each of these island populations was approximately 19. Using a body weight between 65 and 75 kg for the Chinese paleotiger, its cephalization constant would be between 15 and 18. A few available braincase volumes of Siberian tigers suggest a lower cephalization constant than in Indonesian tigers (presumably around 18). Some modern tigers of unknown populational origin have a cephalization value at approximately 17, while the cephalization value of jaguars is near 13-14 (Hemmer 1978b). Great morphognostic differences in braincase external surfaces, as indicated by extreme differences in sagittal crest shape and size at comparable skull size and individual age in different tiger populations, points to more than one cephalization level in modern tigers.

Multiple differences in specializations, as well as in evolutionary progression in the main geographic populations of tigers, suggest that specimens from different ecosystems are not exchangeable. Unfortunately, these differences offer little information on special coadapted gene complexes, nor do they offer any information on population boundaries, where local adaptations are obvious from taxonomic work but cannot be taxonomically differentiated on a regional or local subpopulations level. The search for clear population boundaries will be no problem in island populations, but will be especially difficult where there is a continuous distribution. The detection of coadapted gene complexes cannot presently rely on studies of natural populations, as ecoethological and ecophysiological studies are lacking in tigers. Other detection schemes for such evolutionarily important gene complexes (for a compilation on these issues see Templeton et al. in press) involve an assessment of how genetic or phenotypic variation is distributed within versus between populations. Unfortunately, such studies have only been done in tigers for geographic populations identified at the subspecific level, and are lacking at the regional level. Even at the subspecific level, it is not clear where the boundaries of tiger populations exist, especially in South and East Asian tiger populations. Thus, although subspecies such as P. t. tigris, corbetti, and amoyensis clearly code for central ranges of phenotypes that are quite different - and such differences indeed should partially result from coadapted gene complexes - all classic taxonomic work based on a static concept is of no use in understanding what happens in transition zones.

The multiple distribution pattern of tigers in middle Asia along great river systems and mountain woodlands separated by vast desert areas suggests there are multiple coadapted gene complexes in regional populations from east of Turkey to the Lob Nor region of China's Sinkiang province. Differences in fur color and pattern may be a result of genetic drift between widely scattered regional populations that underwent several genetic bottlenecks during primary dispersal and subsequent climatic changes during the Pleistocene Ice Age (Hemmer 1978c). The usual taxonomic lumping of all middle Asian tigers as the virgata subspecies may mask a great differentiation in coadapted gene complexes between these regional populations, just as the taxonomic splitting of South and Southeast Asian tigers into tigris and corbetti subspecies may mask a more continuous biological entity. The classic attachment of all Indian tigers to the type subspecies tigris, regardless of their differing environments, may have hindered studies capable of revealing differences on the coadapted gene level within this large subcontinent.

In summary, basic tiger phylogeny began differentiating after the primary dispersal of the species approximately two million years ago from East Asia in two separate directions. One was over north central Asian woodlands and river systems to the west and then to Southwest Asia. The other was from China east of the central Asian mountains to Southeast Asia and the Indonesian islands, finally reaching India from the east, nearly closing the circle in the total distributional pattern (Hemmer

1981, Mazak 1983). Regional genetic differentiation selected by different environmental conditions, but not in parallel with the stochastic processes of isozyme molecular evolution (O'Brien 1986) proceeded from this early time on. To protect genetic adaptive integrity, any artificial exchange of tigers between separated populations should be limited to the distance that an individual tiger of the population can cover, taking into account ecological barriers that would hinder free access to the other populations.

Siberian and Caspian tigers are known to travel up to 1,000 km (Heptner and Sludskii 1980). This figure should mark the exchange limit over ecologically unbroken country. It provides a useful frame for a global tiger conservation plan. This implies that existing regional populations of the species must be protected. Tigers have phylogenetically developed populational differences, but man has developed the concept of subspecific taxonomy (Hemmer 1978c). Thus, conservation strategies must not rely primarily on such man-made concepts, but on nature's existing populations. Two examples, the success of Project Tiger in India and the success of tiger conservation after World War II in the Soviet Union Far East region, suggest that wild populations can be protected even in very small and more or less widely scattered relief populations.

If some still surviving Caspian tiger micropopulation in eastern Turkey or southern Soviet Union, and a Lob Nor tiger micropopulation in Chinese Sinkiang province were discovered, they should receive the uppermost care as discrete taxonomic units. For captive breeding entities (geographic populations, Hemmer 1978c) an approximately 1,000 km radius could be established. This would establish Manchurian and Amur tigers (altaica subspecies), South China tigers (amoyensis subspecies), and Southeast Asian tiger populations as still existing (corbetti subspecies). The Sumatra population would be defined by their island isolation (sumatrae subspecies), and finally, most of the surviving Indian tigers would be regional populations (tigris subspecies).

REFERENCES

Brongersma, L.D. 1935. Notes on some recent and fossil cats, chiefly from the Malay Archipelago. Zool. Mededel. 8:1-89.
Dubois, E. 1908. Das geologische Alter der Kendeng- oder Trinil-fauna. Tijdschr. Koninkl. Nederlandsch Aard. Gen., 2 Ser., 4.
Groves, C. 1982. Cranial and dental characteristics in the systematics of Old World Felidae. Carnivore 2:28-39.
Haltenorth, T. 1936-37. Die verwandtschaftliche stellung der groskatzen zueinander. Zeit. Saugetierkd. 1:32-105 and 2:97-240.
Hemmer, H. 1966. Untersuchungen zur stammesgeschichte der Pantherkatzen (Pantherinae). Veroffentl. Zool. Staatssamml. Munchen 1:1-121.
Hemmer, H. 1967. Wohin gehort "Felis" palaeosinensis Zdansky, 1924, in systematischer hinisicht? N. Jb. Geol. Palaont. Abh. 2:83-96.
Hemmer, H. 1969. Zur stellung des tigers (Panthera tigris) der insel Bali. Zeit. Saugetierkd. 4:216-23.

Hemmer, H.   1971.   Fossil mammals of Java. II: Zur fossil-
    geschichte des tigers (Panthera tigris (L.)) in Java.
    Koininkl. Nederl. Adad. Wetensch., Proc., Ser. B, 4:35-52.
Hemmer, H.   1976.   Fossil history of living Felidae. World's
    Cats. 3(2):1-14.
Hemmer, H.   1978a.   The evolutionary systematics of living
    Felidae: present status and current problems. Carnivore
    1(1):71-79.
Hemmer, H.   1978b.   Socialization by intelligence. Social
    behaviour in carnivores as a function of relative brain size
    and environment. Carnivore 1(1):102-05.
Hemmer, H.   1978c.   Zur intraspezifischen, geographischen
    Variabilitat des Tigers (Panthera tigris L.) nebst
    anmerkungen zu taxonomischen fragen. [Congress Report 1st
    Int. Symp. Management and Breeding of the Tiger, pp. 60-64.]
    Int. Tiger Studbook. Leipzig Zool. Gart.: Leipzig.
Hemmer, H.   1981.   Die evolution der pantherkatzen. Modell zur
    uberprufung der brauchbarkeit der hennig, schen prinzipien
    der phylogenetischen, systematik fur wirbeltier-
    palaontologische studien. Palaont. Zeit. 5:109-16.
Heptner, V.G. and A.A. Sludskij.   1980.   Die saugetiere der
    sowjetunion. Vol. 3: Raubtiere (Feloidea). VEB Gustav
    Fischer: Jena.
Hilzheimer, H.   1905.   Uber einige tigerschadel aus der
    strabburger zoologischen sammlung. Zool. Anz. 8.
Hooijer, D.A. and B. Kurten.   1984.   Trinil and Kedungbrubus: the
    Pithecanthropus-bearing fossil faunas of Java and their
    relative age. Ann. Zool. Fenn. 1:135-41.
Koenigswald, G.H.R. von.   1933.   Beitrag zur kenntnis der
    fossilen wirbeltiere Javas. Wetenschapp. Mededel. Bandoeng
    3.
Kurten, B.   1973.   Pleistocene jaguars in North America. Comment.
    Biol. 2:1-23.
Leyhausen, P.   1950.   Beobachtungen an Lowen-Tiger-Bastarden mit
    einigen bemerkungen zur systematik der grobkatzen. Zeit.
    Tierpsychol. 7:46-83.
Leyhausen, P.   1973.   Verhaltensstudien an Katzen. 3d ed.
    Berlin/Hamburg: Parey.
Mazak, V.   1967.   Notes on Siberian long-haired tiger, Panthera
    tigris altaica (Temminck, 1844), with a remark on Temminck's
    mammal volume of the "Fauna Japonica." Mammalia 1:537-573.
Mazak, V.   1968.   Nouvelle sous-espece de tigre provenant de
    l'Asie du sud-est. Mammalia 2:104-12.
Mazak, V.   1976.   On the Bali tiger, Panthera tigris balica
    (Schwarz, 1912). Vest. Cesk. S. Zool. 10:179-95.
Mazak, V.   1983.   Der Tiger. 3d ed. Wittenberg Lutherstadt:
    Ziemsen.
Mazak, V., C.P. Groves and P.J.H. van Bree.   1978.   On a skin and
    skull of the Bali Tiger, and a list of preserved specimens
    of Panthera tigris balica (Schwarz, 1912). Zeit.
    Saugetierkd. 3:108-13.
Mazak, V. and J. Volf.   1967.   Einige bemerkungen uber den
    Sibirischen tiger, Panthera tigris altaica Temminck, 1845,
    und seine zucht in dem Zoologischen Garten Prag. Acta So.
    Zool. Bohemoslov. 1:28-40
Ninkovich, D. and L.H. Burckle.   1978.   Absolute age of the base
    of the hominid-bearing beds in Eastern Java. Nature
    75:306-08.
Pei, W.C.   1934.   On the carnivora from locality 1 of
    Choukoutien. Palaeont. Sinica (C) 1 (5).

Templeton, A., H. Hemmer, G. Mace, U.S. Seal, W.M. Shields and
    D.S. Woodruff. In press. Coadapted gene complexes and
    population boundries.
Vos, J. de, S. Sartono, S. Hardia-Sasmita and P.Y. Sondaar.
    1982. The fauna from Trinil, type locality of Homo erectus:
    a reinterpretation. Geol.en Mijnbouw 11:207-11.
Zdansky, O. 1924. Jungtertiare Carnivoren Chinas. Palaeont.
    Sinica (C) 1.

4

Use of Electrophoretic Data in the Reevaluation of Tiger Systematics

Anna M. Goebel and Donald H. Whitmore

The systematic value of electrophoretic data is well documented (Avise 1974, 1976, Avise and Aquadro 1982, Ayala 1975, Nei et al. 1975). Electrophoretic analysis is particularly useful in systematic determination for tigers because genic variation (measured with electrophoretic techniques) may or may not give results similar to systematic evaluations using morphological variation alone (Cherry et al. 1978, King and Wilson 1978). Genetic differentiation as determined from electrophoretic data is frequently measured by genetic distance (D) or genetic identity (I) (Nei 1972, 1978). The identity is a probability measure and estimates the proportion of genes that are identical in any two populations. The distance measure estimates the average number of alleles that have been altered, for each locus. Both values take into account not only allelic differences at each locus but also the different frequencies of identical alleles between different populations. By comparing the results from a large number of studies, identity and distance values that represent local populations, subspecies and species are apparent. Other measures of similarity (S), such as Rogers' (Rogers 1972), correlate closely with the identity values described by Nei.

Generally, electrophoretic data can distinguish between species, and subspecies, and potentially down to the individual level, by giving each organism a unique "finger print" that is the basis of paternity testing in humans. The potential discriminatory power of electrophoresis is related to the amount of variation (number of polymorphisms and number of alleles) found. It is becoming increasingly apparent that large mammals exhibit lower levels of variation than mammals in general, and much less than other classes and phyla (Baccus et al. 1983, Nevo 1978, Pemberton and Smith 1985, Wooten and Smith 1985). For this reason, electrophoretic studies of large mammals are increasing in complexity. Higher numbers of gene loci (soluble proteins) are being examined and additional techniques such as sample heat treatments, two-dimensional electrophoresis and isoelectric focusing are being employed to elucidate variation.

The following study was designed to determine the amount and kinds of genic variation in the North American captive population of Amur tigers. It is hoped that the data will facilitate current management strategies within the Amur tiger management plan (See Foose this volume) and will help in the current reevaluation of tiger systematics.

MATERIALS AND METHODS

During the months of July and August 1984, letters requesting blood samples were sent to all zoos listed in the American Association of Zoological Parks and Aquariums (AAZPA) Species Survival Plan for the Siberian (Amur) Tiger (Seal 1983) and to institutions listed in the International Species Inventory System (ISIS 1985) as housing any subspecies of tiger. Institutions were requested to draw blood into heparinized tubes, to separate plasma and cells and to wash the erythrocyte (RBC) pellet twice with mammalian saline (0.9% NaCl). Both plasma and RBC's were to be frozen immediately and shipped on dry ice by an overnight mailing service. Upon arrival at the University, any deviations from the instructions, and the previous storage conditions of the samples, were noted. Before electrophoresis, the RBC pellet was thawed, resuspended with twice its volume of deionized water, and centrifuged (20 min, 12,000 x g) to remove cellular debris. The cell lysate supernatant and the plasma samples were divided into 100 ul aliquots and refrozen at -60 to -70°C until use.

Forty-two proteins were initially examined. Twenty-six were successfully resolved. All proteins were named according to Shows and McAlpine (1978). Electrophoretic buffer systems and staining procedures were taken directly or modified from literature sources (Allen et al. 1981, Brewer and Sing 1970, Harris and Hopkinson 1976, Mueller et al. 1962, Nichols and Ruddle 1973, Poulik 1957, Selander et al. 1971, Shaw and Prasad 1970, Siciliano and Shaw 1976). Between five and twelve different buffer systems for starch and polyacrylamide gels were examined for each protein in preliminary studies to ensure proper interpretation of banding patterns. Both tiger and human RBC and plasma samples were examined. Final determinations for buffer systems were based, first, on resolving power and, second, on the minimum number of buffer systems required to adequately resolve all the enzymes of choice. Details on the scored loci, blood fraction, matrix composition, buffer system, running conditions for each locus scored, additional buffer systems examined, and exact staining procedures used, are described elsewhere (Goebel 1986).

Horizontal gel electrophoresis was performed using 12% starch: 6% Electrostarch (lot # 392 Electrostarch Co., Madison, Wisconsin) and 6% Connaught starch (lots #376-1 and #396-1 Connaught Laboratories Limited, Willodale, Ontario, Canada). Gels were cast in 18 cm wide by 15.5 cm long plastic molds and samples were applied using filter paper wicks.

Vertical polyacrylamide slab gel electrophoresis was performed using the apparatus described by Ogita and Markert (1979). Separating gels (1.05M Tris-HCl buffer, pH 8.8, 0.002M ammonium persulfate, 0.05% TEMED) were 6.5% (0.17% bis), 10% (0.27% bis) or 15% (0.4% bis) acrylamide solutions. Stacking gels (0.5M Tris-HCl buffer, pH 6.8, 0.004M ammonium persulfate, 0.1% TEMED) were 5.5% (0.15% bis) acrylamide solutions. Samples

were layered under the electrode buffer within formed wells using
a Hamilton syringe (Hamilton Co., Reno, Nevada). For analysis of
AMY, GDH, G6PD and GOT, samples were diluted with a glycerin-
bromophenol blue solution (1.5:1). Twenty microliters were
applied to gels. For albumin electrophoresis, 5 l of a plasma-
glycerin-bromophenol blue solution (1:20) were applied. Samples
for transferrin determination were treated with rivanol as in
Chen and Sutton (1967) to remove most of the proteins except
transferrin from the sample (Horejsi and Smetana 1956). Initial
identification of transferrin bands was determined with iron
binding stains (Allen et al. 1981, Newman 1983).

Isoelectric focusing was performed at 7°C on an flatbed
Isobox (Hoefer Scientific Instruments, San Francisco, CA). Gels
contained 1% agarose, 10% sorbitol and 0.8% ampholyte. They
measured 14 x 15 cm. For Hb determination, the anode solution
was 0.05M acetic acid, pH 2.6, and cathode solution was Serva
precoat, pH 7.0, (Serva Fine Biochemicals Inc, Garden City Park,
N. Y.). For Es, anode solution was the same but the cathode
solution was 1.0M NaOH, pH 13.0. Electrode buffers were applied
on filter paper wicks on either end of the gel. Samples were
applied into wells of an applicator mask placed on the gel
surface and allowed to soak into the gel five minutes before
applying power (M158 Automatic Power Supply, MRA Corporation,
Clearwater, FL). The applicator was removed after 45 minutes,
surface fluid was blotted with filter paper, and the power was
restarted. Gels were run for 10 minutes at 200 V, then at 5 W,
for two hours.

Polymorphic loci were identified based on variation in
electrophoretic mobility. Banding patterns were interpreted
using literature descriptions of patterns in other cats and/or in
humans (Allen et al. 1981, Auer and Bell 1981, Brown 1982, Brown
and Brisbin 1983, Harris and Hopkinson 1976, Kohn and Tamarin
1973, Newman et al. 1985, O'Brien 1980, Ritte et al. 1981,
Thuline et al. 1972, Thuline et al. 1967, Van de Weghe et al.
1981). The subunit composition of each enzyme was assumed to be
the same as for humans. Age of both the blood sample and the
animal (given in years and calculated to the nearest month), the
condition of the sample upon arrival at the University and any
known variation from standard protocol were carefully recorded to
identify non-genetic factors that might cause pattern variations.
Lastly, pedigree analysis was performed to confirm that
segregation followed Mendelian patterns of inheritance. Due to
the small number of animals examined and the high proportion of
family groupings, all loci with two or more alleles are termed
polymorphic in this study.

Two statistics are used to measure genetic variation and are
discussed by Hedrick et al. (in press). These are P, the
proportion of polymorphic loci, and heterozygosity H, calculated
for the whole population.

A systematic evaluation of the data is presented using Nei's
genetic distance (D) and genetic identity (I) along with Rogers'
similarity (S). All were calculated using a computer program
corrected for small sample sizes (Green 1979, 1984).

RESULTS AND DISCUSSION

Approximately 110 letters requesting blood were written. A positive response was received from 60% of all institutions solicited and 22% sent blood within a 14 month period. Of 101 samples received from 24 institutions, 46 were sent from two zoos, Minnesota and Henry Doorly. All Amur tigers tested in this study are identified by studbook number (Seifert 1976-1984), and all non-Amur tigers are identified by ISIS (1985) number of the source institution and/or by name and source institution. Studbooks were used as the only absolute identification of subspecies. All animals that have not been assigned a studbook number were considered to be of suspect or known mixed subspecies origin. Two thirds of the tigers examined in this study are listed in studbooks.

Due to the variation in sample treatment before arrival at the University, problems concerning interpretation arose. The variation in sample age and treatment was considered for each deviation in normal banding pattern or staining ability. For example, a single sample (Amur tiger 860) exhibited light staining of a number of enzymes, no staining for GDH and ES, and a unique cathodal pattern for Tf. Since all related animals exhibited normal isozyme patterns these irregularities were considered to be due to the age of the blood sample (4 years kept at -5°C before arrival at the University).

The amount of protein and/or enzyme activity varied considerably between and within samples. All samples in which there was no staining were repeated twice, with a more concentrated staining solution or agar overlay. All enzymes tested in both RBC and plasma were less active in plasma. Peptidases (PEPA, PEPB, and PEPD) showed the greatest variation in strength, however, sample age did not appear to be a major factor in the level of peptidase activity.

A number of variant isozyme patterns could be directly attributed to sample age. Harris and Hopkinson (1976) have noted numerous isozyme changes in human tissues associated with in vitro storage and which are especially prominent in lysed red cells or tissues contaminated by red cells. These changes have been attributed to the interactions of oxidized glutathione, which accumulates in tissue samples on storage, and the sulphydryl groups of cysteine residues in the enzymes tested. Typically, discrete stepwise changes occur with time, increasing the anodal mobility of the enzyme without affecting the enzyme activity. The rate of these changes varies among enzymes and among individual samples (Harris and Hopkinson 1976). The effect of oxidized glutathione can be reversed by the addition of various thiol reagents to the sample such as B-mercaptoethanol.

The effects of oxidized glutathione were noted in several samples which were received in a thawed and warm condition. Anodal isozymes were noted for cell lysate enzymes ACP, GOT, GPT, PEPB, and PEPD. Plasma samples from these animals exhibited normal patterns for PEPB and PEPD. In addition, all related animals exhibited normal patterns for all red cell enzymes. All enzymes were restored to normal isozyme patterns by the addition of B-mercaptoethanol to the sample (1:150 l of sample) except GOT which was not reexamined. The phenotypes and explanations of banding patterns for monomorphic loci are given elsewhere (Goebel

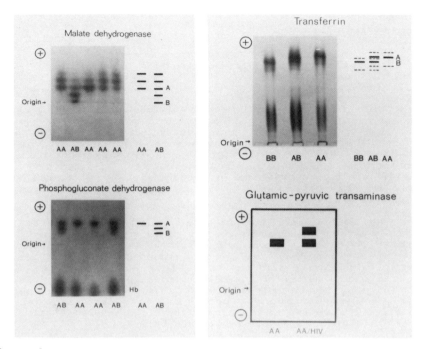

Fig. 1. Phenotypes of Polymorphic Loci and GPT. MDH(soluble)--MDH has a dimeric subunit structure and exhibits anodal secondary bands as in humans (Leakey et al. 1972). A five banded heterozygote (AB) indicates subunit interaction between alleles A and B. PGD--This enzyme is a dimer. A three banded heterozygote (AB) indicates subunit interaction. Tf - each allele (A and B) exhibit 3 bands possibly due to the number of bound Fe ions (0, 1, or 2). Heterozygotes are six-banded indicating no subunit interaction. GPT--see text.

1986). The phenotypes and explanations of banding patterns for polymorphic loci and GPT are given in Fig. 1. Allele frequencies for all polymorphic loci are given in Table 1. Additional confirmation of polymorphism was made by comparison of the results to a previous study by Newman, (1983). The loci that were found to be polymorphic in this study and in that of Newman are listed in Table 2. The results from eight tigers that were examined in both studies are given in Table 3.

MDH - A single heterozygote (non-Amur tiger ISIS 20) was scored. Owners of this animal believed it to be a direct offspring of wild-caught tigers (studbook numbers 17 and 18). However, it was not given a studbook number by Seifert, indicating inadequate proof of parentage. The MDH genotype of this animal suggests that the suspected lineage may be incorrect as no other animals related to 17 and 18 (17 animals) were found to have this allele. However, the possibility of a rare allele can not be excluded.

PGD - Only one heterozygote was observed. A similar three-banded pattern was reported in a single newborn tiger

Table 1. Allele frequencies of polymorphic loci of tigers.

Locus						
			Allele	Frequency		
	Amur Tigers			Non-Amur tigers		
	N	a	b	N	a	b
MDH	53	1.00	0.00	34	0.99	0.01
PGD	53	1.00	0.00	34	0.99	0.01
TF	59	0.49	0.51	38	0.80	0.20

examined by Thuline et al. (1967) for which no subspecies identification was given. The presence of other heterozygotes indicates that this enzyme needs to be examined in all subspecies of tiger.

TF - Both heterozygotes (AB) and homozygotes for the less frequent (B) allele were seen in the population. Tiger transferrin was also found to be polymorphic by Newman (1983) and she described a two-banded heterozygote pattern. However, contradictions in results are evident when patterns of non-Amur tigers ISIS 2931 and ISIS 3199 generated in the two studies are compared. Neither interpretation (compared in Table 3) could be refuted by pedigree analysis. Errors in animal identification or errors in scoring transferrin gels are possible. In both studies, transferrin followed normal patterns of Mendelian inheritance in all cases where it could be followed through pedigrees (Goebel 1986, Newman 1983).

GPT - When tigers were initially surveyed, two isozyme patterns were noted; a single-banded form and a double-banded form (see Fig. 1). The former pattern was interpreted as a homozygous common allele (AA) and the latter as a heterozygous (AB) form without subunit interaction. This interpretation was not contradicted by pedigree analysis. Comparison of the polymorphic loci in this study with those of Newman (1983) revealed similar results, however, non-Amur tiger 3199 (Table 3) was scored differently. Sample degradation was suspected and samples with two-banded patterns were reexamined after the addition of -mercaptoethanol. All were restored to the single AA pattern. Thus, GPT was not scored as a true polymorphic locus in this study.

Both studies found that only non-Amur tigers (except one sample, 527 which exhibited anodal isozymes in a large number of enzymes) exhibited the two-banded or degradation form of GPT. Because of its potential as a diagnostic locus to differentiate between Amur and non-Amur tigers, the genetic and epigenetic basis of this enzyme was examined further: a heat treatment of 22°C for 20 hours was applied to all samples. Thermal stability of tissue samples has been studied in Drosophila (Bernstein et al. 1973, Trippa et al. 1976), in mice (Bonhomme and Selander 1974) and in humans (Mohrenweiser and Neel 1981). While heat treatments in these studies were typically 20 minutes at high temperatures (60°C) and probably involved denaturation of the enzyme, heat treatments of tiger blood were intended to simulate errant sample treatments, or long-term storage. Heat treatment resulted in the production of two-banded patterns in additional

Table 2.   Reported polymorphic loci of tigers.

Enzyme	This Study	Newman 1985
ADA	N.D.	Polymorphic[a]
MDH	1 heterozygote	Monomorphic
GUS	N.D.	Polymorphic[a]
GPT	Heat variant	Polymorphic
PGD	1 heterozygote	Monomorphic
PP	Monomorphic	Polymorphic
Tf	Polymorphic	Polymorphic

[a]Both ADA and GUS were scored in WBC.

non-Amur tiger samples, while Amur tiger samples showed no effect. As a result, animals were then scored as homozygous for the frequent allele A or as a heat-induced variant (HIV) of the AA form.

A genetic basis of this GPT variation was not evident from the pedigrees analyzed. A high proportion of the offspring (12 of 14) of animal ISIS 2931 exhibit the HIV form. It is unlikely that all are heterozygotes. Perhaps the HIV form is a dominant allele. Similarly, Mendelian segregation for codominant alleles was not observed for all variation due to heat treatments found in other studies (Singh et al. 1976, Tsakas and Diamantopoulou-Panopoulou 1980). Additional numbers of tigers are needed to have confidence in ratios that indicate dominant/ recessive alleles. GPT does hold promise as a diagnostic enzyme and its analysis deserves further consideration.

PP - Newman reported the presence of a null allele for inorganic pyrophosphatase. I found it to be monomorphic with no indication of a null allele (Table 2). Results of individual tigers common to both studies were at variance (Table 3). When techniques were carefully compared (Martinson pers. comm.) it was discovered that similar buffer systems were used in both studies, but that B-mercaptoethanol was routinely added to samples to be stained for PP in this study (as in humans; Harris and Hopkinson 1976) and not by Newman. Sample ISIS 3198 was reexamined without the addition of B-mercaptoethanol and was found to exhibit normal activity. Several samples were then given heat treatments (22°C or 37°C for 20 hr). Samples treated at 22°C lost activity, which could be restored with the addition of B-mercaptoethanol. Samples treated at 37°C also lost activity, and it could not be restored. It is hypothesized that the absence of PP activity diagnosed as the result of a null allele by Newman is the result of sample deterioration. Like GPT, Newman found the less common null allele only in the non-Amur population. With careful heat treatments, it too might be a diagnostic enzyme.

GDH - Glucose dehydrogenase was interpreted as a dimer and exhibited two isozyme patterns in this study. Adults exhibited a single-banded pattern and juveniles exhibited a two-banded pattern (Fig. 2). All parents of the juvenile tigers examined exhibited the single banded pattern. Although a single locus for serum GDH was scored in this study, three loci have been scored

Table 3. Comparison of variant loci from eight tigers examined in this study (G&W) and by Newman (1983)[a].

Animal	PP		Tf		GPT	
Number	Newman	G&W	Newman	G&W	Newman	G&W
Non-Amur Tigers						
ISIS 2931	+	+	AB	AA	AB	AA/HIV
ISIS 3198	Null	+	AA	AA	AA	AA
ISIS 3199	+	+	BB	AB	AA	AA/HIV
Amur Tigers						
1303	+	+	AB	AB	AA	AA
1172	+	+	AA (or AB)	AA	AA	AA
1173	+	+	AB	AB	AA	AA
1179	+	+	AB	AB	AA	AA
1180	+	+	AA (or AB)	AA	AA	AA

[a] Data were supplied by Janice Martinson (pers. comm. 1985).

in kidney and RBCs in domestic cats (O'Brien 1980). A second allele at a single locus may have been more active in juveniles than in adults, or perhaps a second locus was involved. Examination of additional tissues might discriminate between the two possibilities.

Before the data can be used, several questions must be addressed. How well are each of the subspecies represented in this study, how well does the captive population represent the wild population, and, how will the results be affected in a systematic evaluation? The numbers of tigers available in the wild and available in captivity are tabulated in Foose, this volume.

Only one subspecies of tiger, the Amur, was adequately represented in this study. Approximately 20% of the total captive North American population was sampled. The current Amur population has descended from 17 wild caught founders. From analyzing pedigrees, the percentage of the genome that could have been inherited from each of the founders was calculated for each animal examined. This percentage was then averaged over the whole population tested. The founder representation of the animals in this study was compared to the founder representation of the total North American population. While it is clear that the founders are not represented equally within either group, the sampled animals do represent the actual captive North American population. How many founders are represented? Six pairs of founders were caught at the same time at the same age and may represent sibling pairs. In considering the worst possible case, all 6 pairs could have been identical twins. In addition, founder 19's input is negligible. This still leaves good representation from 10 individuals.

How will this affect results? An increasing number of zoo populations, like the Amur subspecies examined here, are

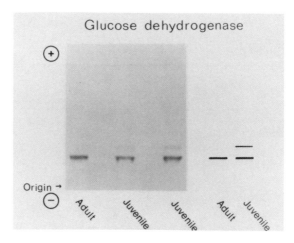

Fig. 2. Developmental Isozymes. Juveniles are nine months and younger, adults are older than 14 months.

represented by large populations derived from a few (perhaps 10-20) founders. Barring strong selection (See Flesness this volume) heterozygosity levels are probably representative of original populations (Gorman and Renzi 1979). Alleles with low frequencies will be lost with small founder numbers and will continue to be lost in subsequent generations if populations are kept small. Alleles with frequencies of 25% or greater should pass through a bottleneck of 10 animals. However, allele frequencies may be altered in subsequent generations. Any alleles that are fixed (show no variation within any given subspecies) will not be affected by selection factors or bottlenecks. If unique fixed alleles can be found in different subspecies, they can be used in positive subspecific identification.

When the data for the Amur population are examined, no fixed alleles unique to the Amur subspecies were found. In addition, no rare alleles were found. One common variant allele (for transferrin) passed through the bottleneck of 17 founders into the American population. It is unknown whether the allele frequencies found in this population represent frequencies in the wild population. It is highly probable that of the proteins examined, transferrin is the only one with a high frequency of both alleles (A and B) since animals from the Russian captive population and Chinese population also exhibited only this polymorphism.

The absence of rare alleles, or alleles of low frequency, from the Amur subspecies is not unique to this study but was also found by Newman (1983). Mazak (1981) suggests that the Russian Amur tiger may have gone through a population reduction (bottleneck) of 20 - 30 animals in 1930-1940. Similarly, estimates from northeast China in 1973 indicated that only "several tens of tigers" remained in China and Korea (Matjushkin 1977). Such population reductions would certainly reduce the number of low frequency alleles found in the Amur tiger. In addition, the habitat of the Amur subspecies can support fewer animals (Mazak 1981) than the more tropical Indian and Indochinese subspecies. The maintenance of small populations through evolutionary time may result in fewer unique alleles accumulating through mutation, the ultimate source of new variation. The lack of rare alleles may affect the future of Amur tigers through evolutionary time. As environments change the Amur tiger will have less variation for selection to act upon. We would not expect to see the extinction of the Amur tiger from lack of variation in our generation, but it is a real possibility in the future. The only source of new variation will be from the migration of tigers of other subspecies into the Amur population. The affects of this breeding strategy are still uncertain.

Other subspecies are more difficult to analyze. The Indian subspecies is represented in this study by a single studbook individual (ISIS 2931) and possibly 36 Indian/Amur hybrids. The Indian subspecies is represented in North America by five individuals. All five are siblings and stem from only two founding Indian individuals. These five have been highly selected for the white coat coloration and are highly inbred (inbreeding coefficients of 0.406)(Murtaugh 1980). How does this affect results? Due to the high level of inbreeding, even alleles with high frequencies may be lost. It is unknown whether the selection forces for the white coat coloration also selected for or against any specific alleles measured here. In addition, the allele frequencies are highly suspect. For example, 24 of the 38 non-Amur tigers examined were from a single family group. The GPT-AA/HIV polymorphism can be seen in only one of 14 individuals outside of this family group. The frequency of this allele may be much lower in the actual non-Amur captive population. The polymorphisms we found in the non-Amur population, however, most probably are of Indian subspecies derivation since Indian tigers predominate among untraceable tigers (Seifert pers. comm.).

The Sumatran tiger is not represented in this study except by a known Amur/Indian/Sumatran hybrid. Even unique fixed alleles could have been lost with three generations of hybridization. The Sumatran subspecies is represented in North America by 14 studbook Sumatran tigers which are traceable to two wild caught founders. If these 14 tigers could be examined, unique fixed alleles would be found, however, most alleles with low frequencies could be lost from two founders. Sampling the European captive population (N=58 from 12 founders) would give a much more accurate picture if wild tigers from Sumatra cannot be examined.

The Indochinese subspecies is represented in this study by a single tiger that is a known Amur/Indochinese hybrid. Again, even unique fixed alleles could have been lost and no substantial judgement can be made from this animal alone. Five Indochinese tigers in North America were identified by Schroeder (1983), but the purity of these lines has since become suspect. Indochinese tigers, preferably from the wild, must be sampled.

No known Chinese tigers (amoyensis) have been identified in North America and again, these animals must be sampled out of pure lineage from Chinese zoos or from the wild.

What systematic value then do the data have? No additional variation to the North American population of Amur tiger was noted in four animals representing the Russian captive population of Amur tigers (1306, 2430, 2431, 2432), or in three imported animals representing the Chinese population of Amur tigers (2456, 2468, 2469). The Amurs from the American, Chinese and Russian populations can certainly be considered one subspecies. Whether the Amur subspecies should be combined with other subspecies in a single breeding unit cannot be determined without further testing of other subspecies. There appear to be different alleles for several loci in the non-Amur populations. While their frequencies are not high in the current North American population, their presence indicates the need for further study. To measure heterozygosity levels, 8 to 12 tigers of each of the five extant subspecies must be examined at a minimum of 50 loci. Forty to fifty animals of each subspecies must be examined before 95% of alleles at low (0.05) frequency will be detected.

Knowing the limitation of the current data set, Nei's genetic distances (D) and genetic identity (I) values were calculated between the Amur and non-Amur catagories. In order to obtain the greatest distance value possible, the GPT heat-induced variants are included in the calculations as GPT heterozygotes. D then has a value of 0.0098 and I a value of 0.9902. The distance values are less and the identity values are greater than those expected for local populations (D = 0.031, I = 0.970) (Ayala 1975). When Rogers' coefficient of genetic similarity values for tigers (S = 0.9824) is compared to values for a number of rodent species (Avise 1974), the value for tigers falls within the ranges for conspecific populations. The distance values calculated from the electrophoretic data generated by this study do not indicate differences large enough to be considered subspecies based on precedents set by previous studies of Drosophila and small mammals. Electrophoretic variation is very low in large mammals, when compared to insects or small mammals. Perhaps the identity and distance values for subspecies will be much different for large mammals. There is evidence that chromosomal variation (Bush et al. 1977) may be much more important in the speciation of large mammals than the variation in soluble proteins. The need for examining additional animals of known subspecies is once again apparent.

Perhaps the most significant use of biochemical poly-morphisms will be as a measure of levels of genetic variation to monitor populations within established breeding units. Most breeding units will be made up of disjunct populations: the Amur tiger is already made up of six disjunct wild populations (Seal and Foose 1983) and at least three captive populations (North American, Russian and Chinese). Quantifiable variation between populations will help determine when, and how many migrants to exchange between populations within any breeding unit. Migrants will help offset drift that is inevitable in small populations.

The data obtained in this study will be most powerful when combined with a number of additional genetic analysis techniques. These include chromosomal analysis, DNA analysis and immuno-logical techniques. If blood and additional tissues can be obtained from other subspecies they should be examined by a variety of techniques which evaluate genetic polymorphisms. For electrophoresis, it is desirable to obtain red cells, plasma and white cells. In addition, it is desirable that the same team collect blood samples from all tigers in order to minimize variation due to handling technique, so that variations such as the GPT (HIV) and PP isozymes can be accurately determined. The new technique of examining mitochondrial DNA variation using restriction endonucleases could be employed using DNA prepared from the leukocytes or platelets of these blood samples. This technique is highly promising and may provide another dimension of information to tiger biochemical genetics.

Electrophoresis appears to be a useful tool for the genetic analysis of tigers. Six (perhaps seven) loci show electrophoretically detectable variations. For systematic analysis, future plans need to include the examination of animals from all subspecies in question. The current breeding plans for the Amur tiger need to be reexamined in light of the kinds and amounts of electrophoretic variation found. Other subspecies need to be similarly examined so that good captive breeding strategies can be initiated. The problems encountered with the tiger are not unique, but may have implications for many zoo populations. In addition, several things are apparent in the electrophoretic analysis of zoo populations of large mammals from this study. First, much can be learned from repeated sampling of the same individuals especially when different laboratories are employed. And second, high numbers of loci and increasingly sophisticated techniques are needed to elucidate variation, especially when studies are limited to blood samples alone.

ACKNOWLEDGEMENTS

We would like to thank Dr. U.S. Seal for his many helpful suggestions during the course of this study and for providing numerous blood samples. Nancy Manning provided valuable assistance by organizing blood shipments. We would like to thank Janice Martinson for information on previous electrophoretic techniques applied to tigers and for the data on individual animals. We gratefully acknowledge the help of many veterinarians and curators for organizing and collecting blood samples. Specifically we would like to thank the veterinary staffs at the Henry Doorly and the Detroit Zoos for repeated help in analyzing pedigrees and collection of blood samples. Funding for this project was provided by Phi Sigma, Sigma Xi, The Wildlife Preservation Trust and the Nixon Griffis Fund for Zoological Research. In addition, we would like to thank Dr. Jonathan Campbell, Dr. James Robinson and Dr. Steve O'Brien for their continued help throughout the course of this research.

REFERENCES

Allen, J., W. Putt and R.A. Fisher. 1981. An investigation of the products of 23 gene loci in the domestic cat, _Felis catus_. _Anim. Blood Groups Biochem. Gen._ 12:95-105.

Auer, L. and K. Bell. 1981. Phosphohexose isomerase polymorphism in the domestic cat. _Anim. Blood Groups Biochem. Gen._ 12:89-94.

Avise, J.C. 1974. Systematic value of electrophoretic data. _System. Zool._ 23:465-81.

Avise, J.C. 1976. Genetic differentiation during speciation. In

Molecular Evolution, ed. F.J. Ayala. Sunderland, MA: Sinauer.

Avise, J.C. and C.F. Aquadro. 1982. A comparative summary of genetic distances in the vertebrates, patterns and correlations. _Evol. Biol._ 15.

Ayala, F.J. 1975. Genetic differentiation during the speciation process. _Evol. Biol._ 8:9-78.

Baccus, R., N. Ryman, M.H. Smith, C. Reuterwall and D. Cameron. 1983. Genetic variability and differentiation of large grazing mammals. _J. Mam._ 64:109-20.

Bernstein, S., L.H. Throckmorton and J.L. Hubby. 1973. Still more genetic variability in natural populations. _Proc. Natl. Acad. Sci. U.S.A._ 70:3928-31.

Bonhomme, F. and R.K. Selander. 1974. Estimating total genic diversity in the house mouse. _Biochem. Gen._ 16:287-97.

Brewer, G.W. and C.F. Sing. 1970. _An Introduction to Isozyme Techniques_. New York: Academic Pr.

Brown, C.J. 1982. Genotypic and morphometric comparisons of pariah house cat populations from the southeastern United States. Masters Thesis. Univ. Georgia, Athens.

Brown, C.J. and I.L. Brisbin, Jr. 1983. Genetic analysis of pariah cat populations from the southeastern United States. _J. Hered._ 74:344-48.

Bush, G.L., S.M. Case, A.C. Wilson and J.L. Patton. 1977. Rapid speciation and chromosomal evolution in mammals. _Proc. Natl. Acad. Sci U.S.A._ 74:3942-46.

Chen S. and H.E. Sutton. 1967. Bovine transferrins: sialic acid and the complex phenotype. _Genetics_ 56:425-30.

Cherry, L.M., S.M. Case and A.C. Wilson. 1978. Frog perspective on the morphological difference between humans and chimpanzees. _Science_ 200:209-11.

Foose, T.J. and U.S. Seal. 1982. A species survival plan for the Siberian (Amur) tiger (_Panthera tigris altaica_) in North America. American Association of Zoological Parks and Aquariums, Wheeling, WV. Unpublished report.

Goebel, A. 1986. Biochemical genetic variation of the North American captive population of Amur tiger (_Panthera tigris altaica_). Masters Thesis, University of Texas, Arlington, TX.

Gorman, G.C. and J. Renzi, Jr. 1979. Genetic distance and heterozygosity estimates in electrophoretic studies: Effects of sample size. _Copeia_ 2:242-49.

Green, D.M. 1979. A BASIC computer program for calculating indices of genetic distance and similarity. _Hered._ 70:429-30.

Green, D.M. 1984. Calculation of indices of genetic distance and similarity: Unbiased estimation corrected for sample size in BASIC computer program. _Hered._ 75:415-16.

Harris, H. and D.A. Hopkinson. 1976. _Handbook of Enzyme Electrophoresis in Human Genetics_. Amsterdam: North Holland Publ. Co.

Hedrick, P., P. Brussard, F. Allendorf, J. Beardmore and S. Orzack. 1986. Protein variation, fitness and captive propagation. _Zoo Biol._ 5(2):91-100.

Horejsi, J. and R. Smetana. 1956. The isolation of gamma globulin from blood-serum by rivanol. _Acta Med. Scand._ 62:65-70.

International Species Inventory System. 1985. _Species Distribution Report_. Minnesota Zoological Gardens, Minneapolis, MN. June.

King, M.C. and A.C. Wilson. 1975. Evolution at two levels in humans and chimpanzees. _Science_ 188:107-16.

Kohn, P.H. and R.H. Tamarin. 1973. Isozyme activities in the domestic cat (Felis catus). Anim. Blood Groups Biochem. Gen. 4:59-62.

Leakey, T.E.B., R.A. Coward, A. Warlow and A.E. Mourant. 1972. The distribution in human populations of electrophoretic variants of cytoplasmic malate dehydrogenase. Human Hered. 22:542-51.

Matjushkin, E.N., V.I. Zhivotchenko and E.N. Smirnov. 1977. The Amur Tiger in the USSR. Monograph.

Mazak, V. 1981. Panthera tigris. Mam. Sp. No. 152. The American Society of Mammalogists.

Mohrenweiser, H.W. and J.V. Neel. 1981. Frequency of thermostability variants: Estimation of total "rare" variant frequency in human populations. Proc. Natl. Acad. Sci. U.S.A. 78:5729-33.

Mueller, J.O., O. Smithies and M.R. Irwin. 1962. Transferrin variation in Columbidae. Genetics 47:1385-92.

Murtaugh, J. 1980. A genetic analysis of the North American population of white tigers with recommendations for future management. National Zoological Park. Unpublished report.

Nei, M. 1972. Genetic distance between populations. Am. Nat. 106:283-92.

Nei, M. 1978. Estimation of average heterozygosity and genetic distance from a small number of individuals. Genetics 89:583-90.

Nei, M., T. Maruyama and T. Chakraborty. 1975. The bottleneck effect and genetic variability in populations. Evolution 29:1-10.

Nevo, E. 1978. Genetic variation in natural populations; patterns and theory. Theor. Pop. Biol. 13:121-77.

Newman, A. 1983. Estimating the extent of biochemical genetic variation in eight species of the felidae. Masters Thesis, Hood College, Frederick, MD.

Newman, A., M. Bush, D.E. Wildt, D. VanDam, M. Th. Frankenhuis, L. Simmons, L. Phillips and S.J. O'Brien. 1985. Biochemical genetic variation in eight endangered or threatened felid species. J. Mam. 66:256-67.

Nichols, E.A. and R.H. Ruddle. 1973. A review of enzyme polymorphism, linkage, and electrophoretic conditions of mouse and somatic cell hybrids in starch gels. J. Histochem. Cytochem. 21:459-62.

O'Brien, S.J. 1980. The extent and character of biochemical genetic variation in the domestic cat. J. Hered. 71:2-8.

Ogita, Z. and C.L. Markert. 1979. A miniaturized system for electrophoresis on polyacrylamide gels. Anal. Biochem. 99:233-41.

Pemberton J.M. and R.H. Smith. 1985. Lack of biochemical polymorphism in British fallow deer. Heredity 55:199-207.

Poulik, M.D. 1957. Starch gel electrophoresis in a discontinuous system of buffers. Nature 180:1477-79.

Ritte, V., E. Neufeld and K. Saliternik-Vardy. 1981. Electrophoretic variation in blood proteins in the domestic cat. Carnivore Gen. Newsl. 4:98-107.

Rogers, J.S., IV. 1972. Measures of genetic similarity and genetic distance. Studies in Genetics 7. University of Texas Publication 7213:45-153.

Schroeder, C. 1983. Subspecific origins of unregistered tigers alive in North America. American Association of Zoological Parks and Aquariums, Wheeling, WV. Unpublished report.

Seal, U.S. 1983. Species survival plan Siberian tiger (Panthera tigris altaica). American Association of Zoological Parks and Aquariums, Wheeling, WV.

Seal, U.S. and T. Foose. 1983. Development of a masterplan for captive propagation of Siberian tigers in North American zoos. Zoo Biol. 2:241-44.

Seifert, S. 1976-1984. Internationales Tigerzuchtbuch. Leipzig, East Germany: Zoologisher Garten Leipzig.

Selander, R.K., M.H. Smith, S.Y. Yang, W.E. Johnson and J.B. Gentry. 1971. Biochemical polymorphism and systematics in the genus Peromyscus. I. Variation in the old-field mouse (Peromyscus polionotus). Studies in Genetics 6. University of Texas Publication 7103:49-90.

Shaw, C.R. and R. Prasad. 1970. Starch gel electrophoresis of enzymes - a compilation ofrecipes. Biochem. Gen. 4:297-320.

Shows, T.B. and P.J. McAlpine. 1978. The catalog of human genes and chromosome assignments. A report on human genetic nomenclature and genes that have been mapped in man. Cytogen. Cell Gen. 22:132-45.

Siciliano, M.J. and C.R. Shaw. 1976. Separation and localization of enzymes in gels. In Chromatographic and Electrophoretic Techniques, 4th ed., ed. I. Smith. Vol 2, pp 184-209. London: W.M. Heinemann Medical Books.

Singh, R.S., R.C. Lewontin and A.A. Felton. 1976. Genetic heterogeneity within electrophoretic "alleles" of xanthine dehydrogenase in Drosophila pseudoobscura. Genetics 84:609-29.

Thuline, H.C., E. Giblett, J. Anderson and D.E. Norby. 1972. 6-Phosphogluconic dehydrogenase polymorphism in the bobcat (Felis rufa Schreber). Comp. Biochem. Physiol. 41B:277-79.

Thuline, H.C., A.C. Morrow, D.E. Norby and A.G. Motulsky. 1967. Autosomal phosphogluconic dehydrogenase polymorphism in the cat (Felis catus L.). Science 157:431.

Trippa, G., A. Loberse and A. Catamo. 1976. Thermostability studies for investigating non-electrophoretic polymorphic alleles in D. melanogaster. Nature 260:42-44.

Tsakas, S.C. and E. Diamantopoulou-Panopoulou. 1980. Does the "Hidden heat-sensitive polymorphism in electrophoresis of crude extracts involve loci other than the structural locus examined? Experiments with D. pseudobscura. Biochem. Gen. 18:1159-74.

Van de Weghe, A., Y. Bouquet, D. Mattheews and A. Van Zeveren. 1981. Polymorphism in blood substances of the cat. Comp. Biochem. Physiol. 69B:223-30.

Wooten, M.C. and M.M. Smith. 1985. Large mammals are genetically less variable? Evolution 39:210-12.

5

Subspecies and the Conservation of *Panthera tigris:* Preserving Genetic Heterogeneity

Sandra J. Herrington

INTRODUCTION

The concept of the subspecies is one means of formally recognizing the geographic variability of a species. A subspecies is considered to be a recognizably different geographic population or set of populations of a species (Futuyman 1979). Therefore, there are two basic aspects of the subspecies as a taxonomic category: morphological difference, a consequence of genetic differentiation, and geographical distribution. Species that are subdivided into subspecies are known as polytypic (Mayr 1969). The informed recognition of subspecies is useful, because it allows for recognition of genetic heterogeneity, while preserving the concept of overall genetic compatibility by uniting the populations within one recognized species.

Subspecies of mammals are para- or allopatric (Mayr 1969). However, not all geographic variation is recognized at this taxonomic level. If a species varies relatively uniformly over a latitudinal or longitudinal gradient, the variation may be considered clinal, and subspecies may not be recognized. However, this does not mean that intergradation of populations negates the recognition of subspecies under other conditions. Since the populations belong to the same species, parapatric distributions may result in zones of introgressive hybridization, and abrupt shifts, or "steps" in the pattern of variation. Therefore, two subspecies may have a zone of intergradation, yet remain taxonomically distinct at the subspecies level.

There are eight generally accepted subspecies of <u>Panthera</u> <u>tigris</u> (Fig. 1). Relatively isolated in the northwestern part of the geographic range of the tiger was <u>P. t. virgata</u> the now extinct Caspian tiger; the type locality for this population was northeastern Iran. The type locality of the Bengal tiger (<u>P. t. tigris</u>) was fixed by Thomas (1911) in Bengal, India. Toward southeast Asia, this subspecies gives way to Corbett's tiger (<u>P. t. corbetti</u>), the type locality of which is Vietnam. There are three insular populations recognized as subspecifically distinct. These are the Sumatran tiger (<u>P. t. sumatrae</u>), the extinct Javan tiger (<u>P. t. sondaicus</u>), and the extinct Bali tiger (<u>P. t. balica</u>). Northeast of the range of Corbett's tiger on the mainland is <u>P. t. amoyensis</u> the South China tiger (type locality Hupeh, China), and finally to the north is <u>P. t. altaica</u>, the Siberian tiger (type locality in northern Korea). Tigers generally follow Bergman's rule, since the northern <u>P. t. altaica</u> is the largest, while the southern insular populations are smallest. However, specimens of <u>P. t. tigris</u> average larger than the more northerly <u>P. t. virgata</u> or <u>P. t. amoyensis</u> contrary to Bergman's rule.

The probable center of origin of the species is southern to east-central China, within the present range of <u>P. t. amoyensis</u>. From there, the range may have spread to the north (<u>P. t. altaica</u>) and east (<u>P. t. corbetti</u>, <u>P. t. tigris</u>) and to the south (insular populations). The species apparently reached the area inhabited by Caspian tigers via the Middle East, avoiding the mountains of Central Asia (Heptner and Sludskii 1972).

The primary distinguishing characters given for subspecies of tigers tend to center on overall size, and variations in pelage color and markings. For example, the <u>P. t. corbetti</u> description specified ground color and marking pattern, and smaller skull dimensions (Mazak 1981) as distinguishing characters. However, the pelage is quite variable within each population, so such external characters may not be reliable indicators of distinctiveness. Furthermore, although bone epiphyses fuse with sexual maturity, individual variation in size also exists. Tigers continue to grow for a large part of their lives (e.g., continued development of skull crests; Heptner and Sludskii 1972), so using size as an identifier also poses problems.

The concept of subspecies has important implications for any captive breeding program aimed at preservation of a species. Not only must the species as a whole be salvaged, but ideally, its genetic heterogeneity must be protected as well. Preserving the genetic variability may be important to the long term survival of a species, since it increases the likelihood that at least some individuals will survive if a factor such as catastrophic epidemic kills off a large percentage of the species. In addition, we cannot truly claim to have "preserved" a species unless we preserve its naturally occurring variability as well. However, breeders of captive populations need to be assured that subspecies are more than figments in the imaginations of taxonomists. In addition they need to be able to accurately categorize animals to subspecies if they are to make use of the concept in their breeding programs.

The objectives of this study were to: examine representatives of different commonly recognized subspecies of <u>P. tigris</u> to determine whether they are distinguishable based on internal morphology (i.e., skeletal characters) and; provide specific recommendations for captive breeding programs.

MATERIALS AND METHODS

I studied all wild caught and selected captive individuals at the American Museum of Natural History, and the National Museum of Natural History (Smithsonian Institution; see Specimens Examined). These included representatives of six of the eight

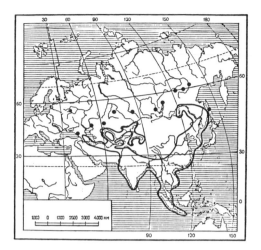

Fig. 1.   General historical distribution of Panthera tigris (enclosed area).   Some records of occasional individuals encountered well outside their normal range are represented by filled circles.   From Heptner and Sludskii (1972).

commonly recognized subspecies of P. tigris.   No Caspian or Bali tigers were available for study. I first performed a discriminant analysis of a large set of skeletal measurements on a smaller sample of tigers to arrive at a character set best able to distinguish among the groups (i.e., subspecies) of tigers.   The final character set consisted of 45 cranial measurements. Postcranial data were excluded, because most museum specimens lack skeletons.   A total of 45 tigers were measured, including eight zoo specimens bred in captivity or otherwise of unknown origin.   These individuals were included to assess the effect of past zoo breeding programs involving tigers of unknown origins at preserving the integrity of the subspecies.   In addition, some wild-caught individuals kept in zoos for long periods of time were included for comparison.   The data were also analyzed in a phenetic study, using cluster analysis and principal components analysis.   Finally, some fossil tigers were included in a smaller subset of characters as well.

RESULTS

Fossil History

    Fossils definitely assigned to P. tigris are from the lower to upper Pleistocene.   These are from central Asia (Brandt 1871, Tscherski 1892), eastern and northern China (Hooijer 1947, Loukashkin 1938, Teilhard de Chardin and Young 1936, Zdansky 1928), northern Siberia (Tscherski 1892), Sumatra (Brongersma 1937) and Java (Brongersma 1935) and Japan (Hemmer pers. comm.). Very late Pleistocene to early Holocene fossils are recorded from the Caucasus (Vereschagin 1959) and India (Lydekker 1886).   The fossils from northern Siberia ranged up to 70° latitude.   These

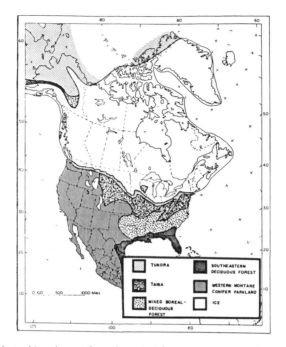

Fig. 2. Distribution of major habitat types during the height of the Wisconsin glaciation. Considerable local heterogeneity existed within each general habitat type.

specimens are of somewhat uncertain affinity, and have also been allied with cave lions (P. spelaea; Heptner and Sludskii 1972).

The possible presence of tigers this far north, at a time when the Beringian subcontinent linked northeastern Asia and northwestern North America, raises the question of why tigers did not cross the Bering land bridge, when so many other mammals did. Tigers appear to shun open country (Seidensticker 1976), and the Bering land bridge appears to have been primarily steppe tundra habitat (Fig. 2). However, there is evidence that a variety of habitats were to be found in Beringia, including wooded regions (Hopkins 1982).

There were large Panthera in North America during the Pleistocene (Fig. 3), and these were assigned to the Fossil lion species (P. atrox; now considered by some to be a subspecies of P. leo or consubspecific with cave lions; e.g., Kurten 1985). However, it is not always easy to distinguish between modern lions and tigers, based only on skeletal morphology, so the possibility remained that tigers were among the North American Ice Age Panthera in the past (e.g. Merriam and Stock 1932). These fossils were compared with lions and tigers to determine their affinities. However, the possibility that both lions and tigers were inadvertently included in P. atrox seems never to have been considered. In a recently completed study of these fossils, I developed a set of morphometric characters that

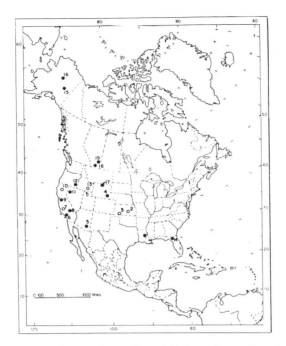

Fig. 3. Some North American localities where P. atrox has been found. Open circles, specimens of uncertain affinity. Filled circles, Rancholabrean sites. 1 Natchez, Mississippi; 1 Harvey Co., Kansas; 3 Meade Co., Kansas; 4 Little Box Elder Cave, Wyoming; 5 Ventana Cave, Arizona; 6 Rancho La Brea, California; 7 Carpenteria Asphalt, California; 8 McKittrick Tar Seeps, California; 9 Livermore Valley, California; 10 Potter Creek Cave, California; 11 Astor Pass, Nevada; 12 Malheur Co., Oregon; 13 American Falls, Idaho; 14 Sante Fe Ichetucknee, Florida; 15 Fairbanks, Alaska; 16 Lost Chicken Creek, Alaska; 17 Natural Trap Cave, Wyoming; 18 Medicine Hat, Alberta; 19 Bindloss, Alberta.

distinguished between lions and tigers with 100% accuracy in discriminant function analyses. In addition, there were several qualitative morphological characters that could distinguish P. leo and P. tigris with a high degree of accuracy (Fig. 4). I then compared modern lions and tigers with the fossil North American material (Fig. 5 and 6). All fossil material from the area of the contiguous United States, an area south of the continental ice sheet during the Wisconsin glaciation, represented lions. However, the material from Beringia included both lions and tigers. It appears that lions invaded eastern Beringia during the Illinoian glaciation, and arrived in the area of the contiguous United States during the Sangamonian interglacial. Tigers, and a different population of lions, penetrated as far as eastern Beringia during the Wisconsin glaciation in a separate dispersal event. These results suggest that other large fossil Panthera may have been misidentified, and additional evaluation may further increase the range of fossil tigers.

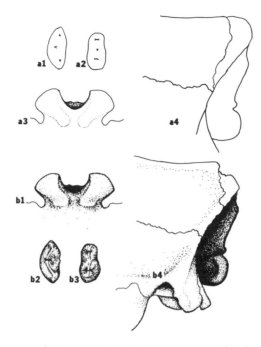

Fig. 4. Some cranial and dental characters distinguishing tigers (a) and lions (b). Shape of ml (al; crown view); shape of p4 (a2; crown view); shape of anterior border of occipital condyles (a3; on basicranium); appearance of parietal-occipital and parietal-temporal sutures (a4);shape of anterior border of occipital condyles (bl; on basicranium); shape of ml (b2; crown view); shape of p4 (b3; crown view); appearance of parietal-occipital and parietal-temporal sutures (b4).

Unlike the fossil lions, the fossil tigers of Beringia were not larger than their modern counterparts. Fossil tigers from southern China from the collections at the American Museum of Natural History (Hooijer 1947) appear to be larger than modern P. tigris but only slightly so.

Morphological Distinctiveness of Subspecies

Based on the discriminant analysis of 45 cranial measurements, the six subspecies studied were 100% distinguishable. However, in a highly conservative analysis, involving a jack-knifed classification of the specimens (in which the specimen under consideration is excluded from computation of the group mean), there was considerable overlap of subspecies tigris and corbetti and some overlap of subspecies corbetti and sumatrae.

Morphologically the most distinctive subspecies was amoyensis, the South China tiger. The first canonical axis separated amoyensis from the other five subspecies (Fig. 6).

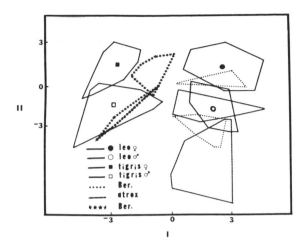

Fig. 5. Positions of lions, tigers and North American large fossil Panthera on canonical axes I and II of a discriminant function analysis of twelve dentary measurements. The first axis is positively related to relative width of ml and relative overall size and posterior width of p4, and negatively related to size. Pattern of distribution of the specimens, and group are shown.

Character loadings on this axis indicate that amoyensis has a shorter cranial region, orbits set further forward and closer together, larger internal naris, larger postorbital processes (or greater width across the processes), larger infraorbital canal, smaller paracone on P^4 relatively narrower m^1 and narrower condyloid process on the dentary. The remaining subspecies were separated along an axis reflecting overall size to an extent, but also reflecting other features. P. t. altaica had the smallest value on this axis, with tigris, corbetti, sumatrae, and finally sondaicus having larger values. A larger value indicates smaller overall size, and also relatively narrower lower canine, relatively larger P^4 more slender P^3 shorter premaxillary bone (possibly reflecting smaller external naris), smaller mastoid width (reflecting the laterally constricted occipital region in the insular populations; Mazak 1981), shallower skull, relatively longer cranial region, larger tympanic bulla and narrow postorbital constriction.

As discussed above, the most distinctive subspecies was amoyensis. The characters distinguishing this population reflect a primitive morphology, and amoyensis occupies a geographical region thought to represent the center of origin for P. tigris. This appears to represent a situation in which the center of origin is occupied by the most primitive member of a clade, a situation clearly demonstrated by other mammalian examples (e.g., the distribution of Priorailurus, the most primitive felid genus, includes Southeast Asia, a probable center of origin for the Felidae). Therefore, amoyensis may be regarded as a relict population of "stem" tigers.

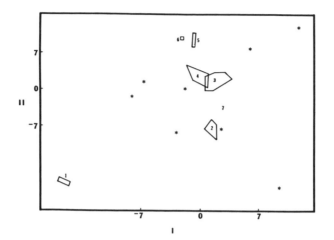

Fig. 6. Positions of tiger subspecies and zoo tigers of unknown
locality background on canonical axes I and II of a discriminant
function analysis of 45 cranial measurements. The first axis
represents characters reflecting a primitive morphology, and
seperates the South China tiger (1) from the other subspecies.
The second axis is negatively related to overall size, and
seperates the other subspecies. Panthera tigris amoyensis, 1; P.
t. altaica, 2; P. t. tigris, 3; P. t. corbetti, 4; P. t.
sumatrae, 5; P. t. sondaicus, 6; group mean for zoo tigers of
unknown locality background, 7; subspecies P. t. virgata and P.
t. balica are not represented.

     The situation with the zoo specimens of questionable origin
was complex. None showed any affinity with amoyensis. However,
few showed particular affinity with any of the other subspecies.
One specimen was close to altaica while another was close to
corbetti but the remainder were randomly distributed, with the
group mean roughly between the means for subspecies tigris and
altaica. The literature indicates that life in captivity results
in marked changes in skeletal morphology. However, specimens
caught in the wild, and held in captivity for many years
maintained positions very near their group means. Therefore, the
differences observed in the zoo-bred specimens appear to reflect
changes that are not merely environmental. The morphological
differences seen in these specimens are probably due to hybridization between subspecies, and they serve to illustrate the
dangers of breeding programs utilizing animals of unknown
origins.

     To summarize the morphological findings, the subspecies of
P. tigris included in this study were morphologically
distinctive, indicating that they are not mere geographical
constructs. However, the relationship between subspecies tigris,
corbetti, and the insular populations is probably largely a
clinal one. Subspecies amoyensis is very distinctive and may
represent a relict population of primitive tigers. The zoo
specimens studied appear to bear little morphological resemblance
to their wild counterparts, aside from belonging to the same
species. Therefore, we might refer to these as "generic" tigers.

RECOMMENDATIONS FOR BREEDING PROGRAMS

The results have strong implications for captive breeding programs. There are two basic classes of tigers that we need to consider: wild-caught individuals, or captive-bred individuals descended without question from individuals of known locality background, and; the "generic" tigers mentioned above -- those of unknown locality background. Overall size and external characters do not appear to be adequate as the sole features used to identify tiger subspecies, so the geographic locality data are absolutely critical. Even then, it may be difficult to deal with specimens from borderline localities. The majority of captive tigers will fall into the "generic" group, since locality data are often questionable.

For the first class of tigers with known locality background, I recommend a program designed to maintain locally adapted genotype. This is especially recommended for subspecies amoyensis due to its morphological distinctiveness, and probable status as a "stem" tiger. This type of breeding program would be difficult to carry out, but would be important in preserving the distinctiveness of the subspecies. In addition, the progeny of such a program would be better suited for release into the wild, since they would have the gene combination specifically adapted to the geographic range of their particular subspecies. I do not recommend breeding any tiger of known locality background with one of unknown background, since this would serve only to dilute the locally adapted genotype of the known species.

A different captive breeding strategy is most appropriate for the second class of tigers, those of uncertain locality background. It would be useless to attempt to assign these tigers to subspecies, since by and large, they bear no resemblance to the true geographic races of tiger. Assuming that skeletal morphology reflects the genotype, these tigers do display genetic variability. Indeed, they appear to be more variable than the tigers of known locality background. Therefore, using them in a breeding program will preserve genetic heterogeneity, so we need not be overly concerned with arriving at a "monoculture" of tigers, provided that measures are taken to prevent too much inbreeding. What these tigers lack are the locally adapted genotypes of the known locality tigers. Therefore, they are not as well suited to programs for future release into the wild, but are certainly suitable for preservation of the species as a whole. The primary concern with breeding in this group of tigers should be with inbreeding, and the problems of lethal or deleterious recessive genes that too much inbreeding invariably produces.

In summary, I recommend two breeding programs. The first would require utilizing an international studbook for each subspecies, with breeding only between tigers of the same subspecies. The aim would be to preserve locally adapted genotypes and the genetic integrity of the various subspecies. The possible goal of such a program might be the future release of the progeny, to repopulate areas in which tigers are now, or soon will be extinct. The second program involves breeding tigers of unknown locality background, and primarily requires safeguards against inbreeding, or breeding of genetically unfit individuals. This program would insure the preservation of the species, and would allow preservation of the genetic heterogeneity of tigers as well.

Finally, a word is necessary on two subjects. First, the urgent need to do everything possible to save the South China tiger from extinction cannot be overemphasized. The unique position of this subspecies as a stem population means that losing the South China tiger would mean far more than simply losing a piece of a cline. The population is the most distinctive of all living tiger subspecies. Furthermore, its loss would mean the loss of a unique part of the genetic heritage of P. tigris. Second, all specimens, regardless of origin, should be deposited in museums after their deaths. Like other areas of science, systematic studies require data, and data cannot be gathered without specimens. For example, those involved in biochemical studies should arrange for the specimens providing their samples to be deposited in museums as voucher specimens, so other researchers can refer to the source material when appropriate. A large sample size can serve only to improve any systematic study, providing an adequate data source for bettering our knowledge of the species, and strengthening arguments for intensive conservation efforts.

ACKNOWLEDGEMENTS

I sincerely thank Ron Tilson and Robert Hoffmann for their assistance in designing the study and preparation of the manuscript. In addition, I thank Guy Mussier and Robert Tedford (American Museum of Natural History), and Richard Thorington (National Museum of Natural History - Smithsonian Institution) for allowing me to examine specimens under their care. I also wish to thank Professor T. Bangjie (Beijing Zoo), and R. Tilson for the use of slides, including those of the South China tiger, in my presentation.

SPECIMENS EXAMINED

United States National Museum (Smithsonian Institution): 111982, 49726, 253289, 174981, 278470, 399096, 7549, 16144, 3804, 253286, 252287, 253290, 253285, 218321, 49470, 396137, 251789, 258210, 269320, 399556, 536895, 396272, 16145, 49773, 49728, 49799. American Museum of Natural History: 19680, 90016, 54458, 54459, 54460, 113744, 113743, 87348, 87349, 45520, 45519, 00061, 60771, 85405, 85404, 85396, 135846, 113748, 14030.

REFERENCES

Brandt, F. 1871. Neue Untersuchung uber die in dem altaischen Hohlen aufgefundenen Saugethierreste; Ein Beitrag zur quarternaren Fauna des Russuschen Reiches. Bull. Acad. Imp. Sci., St.-Petersburg Russia 15:147-202.
Brorgersma, L.D. 1935. Notes on some recent and fossil cats, chiefly from the Malay Archipelago. Zool. Medede. 18:1-89.

Brongersma, L.D. 1937. Notes on fossil and prehistoric remains of "Felidae" from Java and Sumatra. Comptes Rendus 12th Congr. Int. Zool., Lisbonne 1935, pp. 1855-65.

Heptner, V.G., and A.A. Sludskii. 1972. Mammals of the Soviet Union, Vol. 2, Part 2. Moscow: Vysshaya Shkola.

Hooijer, D.A. 1947. Pleistocene remains of Panthera tigris (Linnaeus) subspecies from Wanhsien, Szechwan, China, compared with Fossil and recent tigers from other localities. Am. Mus. Nov. 1346:1-17.

Hopkins, D.M. et al. 1982. Paleoecology of Beringia. New York: Academic Pr.

Kurten, B. 1985. The Pleistocene lion of Beringia. Ann. Zool. Fenn. 22:117-21.

Loukashkin, A.S. 1938. The Manchurian tiger. China J. Shanghai 28:127-33.

Lydekker, R. 1886. Preliminary note on the Mammalia of the Karnul Caves. Rec. Geol. Surv. India 19:120-22.

Mazak, V. 1981. Panthera tigris. Mam. Spec. No. 152.

Merriam, J.C. and C. Stock. 1932. The Felidae of Rancho la Brea. Carn. Inst. Wash. Publ. 422:1-232.

Seidensticker, J. 1976. On the ecological separation between tigers and leopards. Biotropica 8:225-34.

Teilhard de Chardin, P. and C.C. Young. 1936. On the Mammalian remains from the archaeological site of Anyang. Paleont. Sinica, ser. C 12:1-61.

Thomas, O. 1911. The mammals of the tenth edition of Linnaeus: an attempt to fix the types of the general and exact bases and localities of the specimens. Proc. Zool. Soc. London 1911:120-58.

Tscherski, J.D. 1892. Wissenschaftliche Resultate der von der Kaisertiche Akademic der Wissenschaften zur Erforschung des Jana-Landes und der neusibirischen Inseln in den Jahren 1885 und 1886 ausgesandten Expedition. Abt. IV. Beschreiburg der Ammlung post-tertiarer Saugetiere. Acad. Imp. Sci. St. Petersburg, Mem., ser. 7, v. 40, 511 pp.

Vereschagin, N.K. 1959. Mlokopitajusciie Kavkaza. Istorija formirovanija fauny. Moscow: Akad. nauk SSSR. 704 pp.

PART II
STATUS IN THE WILD

6

The Siberian Tiger (*Panthera tigris altaica*—Temminck 1844) in the USSR: Status in the Wild and in Captivity

V.V. Spitsin, P.N. Romanov, S.V. Popov, and E.N. Smirnov

STATUS OF WILD SIBERIAN TIGERS IN THE USSR

No less than one-third of the present range of P. t. altaica lies within the territory of the USSR (Fig. 1). Typical Siberian tiger habitat comprises broadleaved and broadleaved-coniferous forests, thickets covering hill slopes and rock and river valleys (Fig. 2-5). Secondary growth, which provides cover for prey animals such as wild boar (Sus scrofa), red deer (Cervus elaphus xanthopygus) and sikas (Cervus nippon hortulorum), is preferred. The home range of a female comprises about 200-400 sq. km, the male's larger home range comprises 800 to 1000 sq. km, overlapping that of two or three females. (Matjushkin 1973, 1977). The sex ratio in the wild is 1:2 (Yudakov and Nikolaev 1973).

Since the late 1930's, when the USSR tiger population numbered only 20 to 30 animals, urgent conservation measures were taken and the tigers' numbers began to increase. By the early 1980's the USSR population of P. t. altaica was estimated to be over 200 animals (Pikunov et al. 1982). An increase in the number of tigers and a simultaneous decrease in the amount of suitable habitat available for these animals results in the dispersal of immature tigers into territories cultivated by man. This in turn leads to increased tiger encounters by humans, resulting in an overestimation of the tigers' numbers. In recent years, tiger attacks on domestic livestock and dogs have become more frequent, and some specimens appeared to pose a threat to humans. Poaching has therefore increased, and licenced shooting of dangerous animals is more often allowed than in the past. In the end, this leads to stabilization, or even a decrease, in the tigers' numbers. Destruction of tiger habitat will proceed in the future. At present, the tiger population of Sikhote-Alin' is almost completely isolated from tiger populations living near the East-Manchurian Mountains. Each of these parts of the tiger range is in turn seperated by elements of anthropogenic landscape into more or less isolated micropopulations numbering from several to several dozens of animals.

The demographic processes at work in the populations of P.t. altaica have been most thoroughly studied in the Sikhote-Alinsky biosphere Reserve. A total of 44 cubs have been registered in

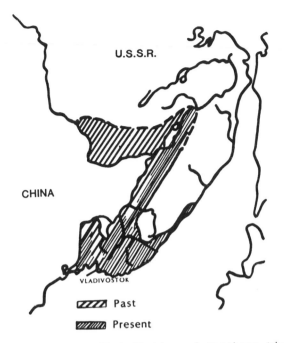

Fig. 1. Past and present distribution of <u>Panthera</u> <u>tigris</u> <u>altaica</u> in the Soviet Union.

the Reserve within the last 19 years. Considering that the mean number of females in the Reserve during these years was 4.6, the mean fertility index of <u>P. t. altaica</u> at the Sikhote-Alinsky Reserve would equal 0.5. Four litters consisted of three cubs each, six or seven litters of two cubs each, and 16 to 19 litters had one cub only. Out of 35 cubs born during the years 1966-82, 15 disappeared during the first year of life and five more died a year later. The loss of cubs due to natural causes was 57%. With an annual natural mortality of adult tigers estimated at 5%, the average annual increase of the population would be 8%. This enables the staff of the Reserve to expect a further increase in the tigers' numbers and expansion of its range.

CAPTURING TIGERS FOR ZOOS

Wild Siberian tigers are captured by teams of professional tiger-catchers. Having discovered tracks of a female with cubs, the catchers follow them until they reach the weakened animals. The female is driven away from the cubs, which are stopped by dogs, pinned to the ground and then secured with rope. Chemical immobilization is not used for capturing tigers in the USSR. Young tigers up to three years old, but mostly yearlings, are captured in this way. Generally, all the cubs of a litter are captured. Each catcher-team hunts within the same territory from

Fig. 2.   Siberian tiger habitat in the Soviet Far East.

year to year, so all tigers captured by one team are likely to be related.

At present, there are four major sites where Siberian tigers are captured in the USSR - two in the Primorsky and two in the Khabarovsky Kraj. The sites in the Khabarovsky Kraj are a little over 100 km apart and are situated within one watershed. Because Siberian tigers are known to disperse much further than 100 km, it is quite probable that all the animals captured in Khabarovsky Kraj belong to one local population with a common gene pool.

The two localities where the tigers are captured in Primorsky Kraj - the basin of the river Iman (B. Ussurka) and the basin of the river Rasdolnaya - are 300 km apart, and are separated by settlements, roads and railways. This enables us to suppose that no constant gene flow between the Primorsky subpopulations exists.

Thus, it is possible to single out three non-crossing populations of P. t. altaica within the territory of the Soviet Union which provide tigers for zoos; tigers from Khabarovsky Kraj; from Primorsky Kraj, basin of the river Iman; and from Primorsky Kraj, basin of the river Rasdolnaya. Unfortunately, it is at present not possible to learn the exact place of capture, and information on tigers captured from 1976-80 is lacking altogether. From 1967-86, excluding the forementioned years, 53 Siberian tigers were captured in the Soviet Far East. The sex ratios and localities of these animals are shown in Table 1.

The majority of Siberian tigers received by zoos originated from the populations of Khabarovsky Kraj and from that of the basin of the river Iman in Primorsky Kraj (Table 1). The sex ratio of the captured animals was approximately 1 male:2 females, which corresponds to that reported in the wild.

Table 1.   Tigers (P. t. altaica) captured for zoos in the Soviet Far East.

Sex	Khabarovsky Kraj	Iman Basin	Primorsky Kraj Rasdolnaya Basin	Unknown	Total
M	6	9	2	2	19
F	12	13	4	3	32
Unknown	2	–	–	–	2
Total	20	22	6	5	53

STATUS OF SIBERIAN TIGERS IN ZOOS OF THE USSR

In January 1985 there were 67 Siberian tigers in 25 USSR zoos. Soviet zoos have acquired immense experience in keeping and breeding this species. An analysis of veterinary work of the Moscow Zoo demonstrates that tigers in captivity suffered most from infectious stomatitis (29% of all the cases of Siberian tigers' diseases within the last 15 years) and from gastro-intestinal disturbance (24%), diseases easily cured. Also, there were single cases of catarrhal disease and dermatitis, which were also cured. Two tigers had malignant tumors and two more had epilepsy.

During the years 1960-85 post-mortem examinations of 23 Siberian tigers were performed at the Moscow Zoo. Fourteen of the tigers were zoo-born cubs under one month old. The major causes of cub mortality were after-effects of pathologic parturition: asphyxia, birth injuries of the skull and chest, and prematurity.

Fig. 3.   Siberian tiger habitat in the Soviet Far East.

Fig. 4.   Siberian tiger habitat in the Soviet Far East.

In five out of nine adult tigers, pathology of cardiac activity was discovered. Although this was not the principal cause of death in all cases, cardiac disorders, myocardial dystrophy in particular, is considered most dangerous for captive tigers. Two deaths occurred after epileptic fits (in both cases cardioplegia occurred), two tigers died from malignant tumors, one female died of pyo-necrotic osteitis, and one death resulted from immobilization.

Table 2.   Status of Siberian tigers (P. t. altaica) in zoos in the USSR.

Origin	No. Tigers	
	Males	Females
Wild-caught	13 (6)	14 (5)
Captive-born to wild-caught parents	10 (8)	13 (9)
Captive-born to one wild-caught, one captive-born parent	6 (4)	7 (7)
Captive-born to captive-born parents	3 (3)	1 (1)
Total	32 (21)	35 (22)

( ) Denotes number of tigers less than 12 years of age; i.e., of reproductive age according to the IUDZG Tiger Plan.

Fig. 5.   Male Siberian tiger on the coast of the Sea of Japan.

Maintenance of captive tigers may cause a predisposition to pathology of cardiac activity (this might also be one of the causes of pathologic parturition). The comparatively frequent occurrence of epilepsy may reflect the close relations between Siberian tigers received by zoos (also those from the wild).

In most USSR zoos Siberian tigers breed well. Breeding information is regularly reported to the International Tiger Studbook (Table 2).

The total number of tigers, 67 animals, approximates the number of spaces available for them in Soviet zoos; any further significant increase in the tigers' numbers in zoos in the USSR is improbable. Soviet zoos can cooperate in the IUDZG Tiger Plan by improving the structure of the captive population of P. t. altaica; forming new lines, increasing the degree of genetic diversity, exchanging sires with other participants and establishing breeding programs. Considering the great number of animals captured in the wild and those born from wild-caught parents, there is every possibility for this kind of work. Details of cooperation of Soviet Zoos in the IUDZG Plan must be discussed by representatives of Soviet zoos with the Species Coordinator for European zoos.

A restricted number of capture localities and a high probability of congeniality of tigers captured in one locality cause a considerable danger of inbreeding, even when pairs are formed of wild-caught partners. There would therefore be special

significance associated with the exchange of tigers originating
from definitely non-crossing populations, e.g. animals captured
within the territory of the USSR and that of China. Another
possible approach to solving the task of retention of maximum
genetic diversity within the captive population of P. t. altaica
is to establish the degree of congeniality between the real and
supposed founders of the population by cytogenetic and
biochemical methods. We regard all this as prospective aspects
of cooperation under the Global Tiger Recovery Plan.

REFERENCES

Matjushkin, E.N. 1973. Tiger and man: problems of neighborhood.
     Priroda (Russian). 2:82-88.
Matjushkin, E.N. 1977. Choice of way and use of territory by
     the Amur tiger (according to winter tracking). In Mammal
     Behavior, Moscow.
Pikunov D., et al. 1983. Report at the 3rd all-union meeting on
     rare species of mammals.
Yudakov A.G. and I.G. Nikolaev. 1973. Status of the population
     of the Amur tiger Panthera tigris altaica in Primorsky Kraj.
     Zool. zhurn. (Russian). 52(6):908-18.

7

Habitat Availability and Prospects for Tigers in China

Lu Houji

INTRODUCTION

In China, this is the year of the tiger. Everywhere in the media are mentions of the legendary tiger, but there are few references to the actual animal. There is a dearth of new data. The information obtained recently is sparse compared to that of a few years ago. Our studies show that in 1982 there were only 150-200 Chinese tigers (Panthera tigris amoyensis) surviving in subtropical areas (Lu and Sheng 1986). By contrast, only 150 Siberian tigers (P. t. altaica) were reported in the northeastern parts of China in the survey of 1974-76 (Ma 1979).

The lack of field work and data makes it extremely difficult to reliably estimate numbers of the Chinese and Siberian tigers. At present only a crude estimate based on interviews, hearsay or provincial fauna surveys can be inferred. For the present we estimate 50-80 Chinese tigers occur in subtropical regions and about 50 Siberian tigers in northeastern China. These estimates may be very optimistic.

Four factors are responsible for the decline of tigers; loss of habitat due to the clearing of land for agricultural purpose, timber cutting, overhunting and poaching. Of these, overhunting was formerly the primary cause for the decline.

There are two aspects to the people's regard of the tiger. From experience, people in rural areas dread the tiger as a killer--a man-eater, and a predator of livestock. From the early 1950's hunters were encouraged to kill tigers, and a bounty was paid for each one so dispatched. But in 1977 the Government belatedly awoke to the fact that the tiger population was decreasing alarmingly. In an effort to stop the decline, a law was passed forbidding the hunting of all tigers. Unfortunately, this law could not be strictly enforced and hunting continued. Now poaching is the principal cause for their decline. Poachers have stalked the animal so relentlessly that it has retreated deep into its natural habitat, and the black market continues to thrive.

Why is poaching a problem? Conversely, the tiger is a symbol of strength and courage in Chinese culture, and perhaps in no other country is the animal so closely linked with the life of the people. Tigers feature prominently in the epics of great heros, and in the arts and superstitions. Tiger parts are highly valued for their medicinal and rejuvenating properties. The Chinese have been using tiger organs for centuries; the greatest demand is for the skeleton, which is used in traditional Chinese medicine. Tiger bone wine is famous in China for rheumatism. Despite these cultural attributes, and in part because of the medicinal properties, the black market demand for tiger parts continues.

HABITATS ARE SCATTERED FROM NORTH TO SOUTH

The habitat of the tiger is diverse and extensive. China occupies a vast expanse. According to Chinese Government publications, 12% of China has a range of environments and variation in temperature from extremely temperate to extremely tropical. Vegetation types include needle-leaved and broad-leaved deciduous forest, broad-leaved evergreen forest, tropical semi-evergreen and rain forest in the extreme south (Hou 1983). Prey are abundant in these regions and include cervids and wild boar.

The Siberian tiger was once found throughout northeastern China, near the Korean and Soviet border, in mixed deciduous forest known as "forest sea." No doubt the Siberian tiger was once numerous in these forests; particularly in the Greater and Lesser Hinggan Mountains in the north; and the Zhanngucai Mountains and Changbai Mountains in the east. In the 19th century the Siberian tiger still had a wide distribution range, but by the 20th century it began to disappear from most habitats and has continued to decline, especially during the last 50-60 years.

In the early 1900's the Siberian tiger was already rare in the Greater Hinggan Mountains; only scattered reports were found in the Soviet literature about this tiger between 1925-70. No trace or hearsay about this tiger was noted in the Greater Hinggan Mountains in the survey of 1974-76 (Ma 1979).

Today Siberian tigers are found in the Lesser Hinggan Mountains and the Wanda Mountains in Heilongjian province. They also persist in the Changbai Mountains and Laoye Mountains that run through Heilongjiang province and Jilin province, including Lao mountain to the border of North Korea.

These areas in northeastern China consist of mixed broad-leaved deciduous and needle-leaved evergreen forest, with dense undergrowth on the hills and mountains. The tiger subsists on deer and wild boar among other prey.

Data concerning the location of tigers killed in the 1950's showed that the distribution range was from approximately 134° longitude in the east to 126° longitude in the west. The northern latitude was approximately 48°; the latitude in the south was 42°. In a 1980-81 survey there were no tigers reported in Yichan Prefecture within the Lesser Hinggan Mountains, once a tiger stronghold (Ma Yeching pers. comm).

During the Quing dynasty, exploitation of the forests was prohibited, but in 1870 the Government lifted the ban. The rapid opening up of primeval forest to cultivation gradually extended from low to higher altitudes. This caused a decline in large herbivorous animals. Human activities and lack of prey hastened the tiger's decline. After 1950 national developmental plans were made, but they were on economical rather than ecological basis, and again the forests were cut down.

A survey of 23,000 sq. km in the left bank of Tumen valley of Yanbian Chosian (Korean) Autonomous Prefecture was conducted in 1979-80. Tumen valley is an important part of the Changbai Mountain range with nearly one hundred peaks over 700 m high and many valleys, ravines, river basins and primeval forest. The study areas contained 14,185 sq. km of primeval forest in 1870; by 1956 it dropped to 4,124 sq. km. By 1980 only 573 sq. km of primeval forest remained. From 1951-57 there were 12 tigers killed in the study area; an average of 1.7 killed per year. Between 1958-80 there were 11 tigers reported killed; an average of 0.5 tigers annually. Tiger losses in northeastern China were attributed to uncontrolled hunting and loss of optimal forest for the species (Jin Lian pers. comm.).

According to historical annals, the Chinese tiger was once distributed over many provinces, especially in the east. Our survey in 1981-82 showed only 14 provinces with tigers. The subspecies' range in the past 30 years was from Zhejiang and Fujian province at approximately 119-120° longitude westward to Sichuan province at 100° longitude. The northern latitude was 35° N to Tsingling-Wangho (Yellow River) range, and as far as south as the tropic of Cancer in Guangdong, Guangxi and Yuann province at 24° latitude (Lu and Sheng 1986). In these areas the Chinese tiger has declined sharply from estimates of 4,000 in 1949 to 150-200 in the 1981-82 survey that we conducted (Lu and Sheng 1986). At present we estimate there are fewer than 50-80 Chinese tigers due to continued poaching.

Comprehensive data of tiger kills and their localities in the vast areas during the 1981-82 survey showed that the tiger range in the last decade was broken into three isolated areas. The eastern region includes the adjacent provinces of Jiangxi, Fujian and Zhejiang. The central region starts with the boundaries of Hunan province with adjoining boundaries of Guizhou, southeastern Sichuan, western Jiangxi, western Hubai, and the boundaries of northern Guangdong and Guangxi province. The third is north of Hubei province, southern Henan province and southern Shaanxi province. In these areas there is only hearsay about surviving tigers.

The central areas used to be considered important refugia for Chinese tigers, especially Hunan, Guizhou and Jiangxi province. At present the tigers are few and scattered.

After 1977 the Government banned the killing of tigers. Records were no longer kept in the furbearing department and no survey has been taken since 1982.

PROSPECTS FOR THE FUTURE ARE MIXED BUT HOPEFUL

From 1980-82 the Government established four tiger reserves; one in Qixianlize Nature Reserve within Huanan and Jixhen county, Heilongjiang province, with an area of 33,000 ha for the Siberian tiger. The other three are in the south; one for the Bengal tiger (P. t. tigris) named Nangun Nature Reserve in Changyuan county, Yunan province, with an area of 7,000 ha; the Bamianshan Nature Reserve in Guidong county, Hunan province, with an area of 20,000 ha; and in Jiangxi province, the Jiangganshan Nature Reserve, with an area of 15,873 ha, are both for the endemic Chinese tiger. Some other forest reserves also contain tigers.

The above-mentioned habitats, close to the hills and mountains along the boundaries of adjoining provinces, may be recolonized by tigers if the hunting pressure abates. There is a secondary growth of forest, and a few stands of semi-virgin forest are interspersed in patches along the border of adjacent provinces. Here, due to less human activity and predation, there is abundant prey including 17 species of cervids, comprising 85% of all the cervids species in China. Their population increased 23% since the late 1970's (Sheng and Lu 1985) due to fewer predators. The annual game counts report that there are about 100,000 tufted deer (Elaphodus cephalophus), 140-150,000 Indian muntjac (Muntiacus muntjak), 650,000 Chinese muntjac (Muntiacus Reevesi), and abundant wild boar (Sheng and Lu 1982, 1985).

However, there is a problem -- the remaining tigers are scattered. This may affect their breeding potential and make their comeback slow and difficult.

Now, in the year of the tiger, the Government plans to launch a vigorous campaign to save the tiger. We shall inaugurate a nationwide conservation effort to educate our people in the need for preservation of our diminishing tigers.

We shall establish new tiger reserves. We must increase and strengthen their protection by legal means. We must increase our vigilance against poachers and black marketeers. This will take time, but if we are successful, the future of the tiger will be assured.

REFERENCES

Hou Hsichyu. 1983. Vegetation of China with reference to its geographical distribution. Ann. Missouri Bot. Gard. 70:509-48.

Lu Houji and Sheng Helin. In Press 1986. Distribution and status of the Chinese tiger. In Proc. Int. Cat Symp. Caesar Kleberg Wildlf. Research Inst., Texas.

Ma Yiqing. 1979. Manchuria tiger in China. (In Chinese) Wildlife Conservation and Management (trail Publ.). Harbin. p. 22-26.

Sheng Helin and Lu Houji. 1982. Distribution, habits and resources status of the tufted deer (Elaphosus cepholophus). Acta Zool. Sinica. 28:307-11.

Sheng Helin and Lu Houji. 1985. Cervids resources in subtropical and tropical areas of China. J. East China Normal University. Natural Science Edition 1:96-104.

8

Tigers in Malaysia: Prospects for the Future

Mohammed Khan bin Momin Khan

<u>BACKGROUND</u>

The Malayan tiger has successfully co-existed with the aborigines in Peninsular Malaysia. Even today, when 51% of the forests in Malaya have been cleared for development, problems do not exist between them. Aborigines gather fruits and vegetables, hunt small mammals or fish in the rivers, and could be potential prey for tigers, but such mortalities are almost unknown. It is evident that man is not the normal prey of the tiger.

Spears, knives and blowpipes are some of the simple, common weapons used by aborigines. Their homes are flimsy shelters constructed on the ground, mainly from bamboo and leaves, which could not possibly withstand the attacks of a tiger. Occasional attacks are known to occur where a tiger has been maimed or wounded. Such casualties are small compared to deaths from disease, childbirth and infant mortality.

Once a tiger killed two aborigines within a week. The second man was reported to be the sixth aborigine to have been killed in the Banding area of Perak. The killings did not deter the aborigines, but made them more careful. After much persuasion, an aborigine guide was obtained for a visit to a salt lick in the vicinity. On the return trip to camp, the party found the fresh pug marks of a young tiger that had followed the group to the salt lick.

Banding is an aborigine outpost where the presence of the police field force upriver at Fort Tapong had upset the tiger population by the use of guns, in addition to the use of traps by the aborigines. Animals such as the gaur, deer, and pigs were killed for food. A man-eating tiger, hitherto unknown in the area, thus became a serious problem.

There was a time in the history of Peninsular Malaysia when man and tiger co-existed, each its own master, with the tiger occasionally killing a man and vice-versa. The situation has changed as the human population has increased. With modern machinery, forests are quickly cleared, resulting in habitats

75

being altered or completely destroyed. Man has become so
skillful at killing that the tiger is no longer its own master.

What little is known of the Malayan tiger came from
newspaper articles and books written by sportsmen. As recently
as 1968 the tiger was found to be a new sub-species (<u>Panthera
tigris</u> <u>corbetti</u>). From these sources of information, it was
evident that tigers were once abundant in Malaya and Singapore.
As many as forty tigers were reported to be living on or using
the island from the early to mid-1800's. The last tiger in
Singapore was killed in 1932. In the 1950's a British officer,
Colonel A. Locke, estimated that 3,000 tigers inhabited
Peninsular Malaysia. By 1977 the tiger population had declined
to 300 animals.

CAUSES OF MORTALITY

Several factors detrimental to the tiger include: 1) an
increase in the possession of firearms, particularly in rural
areas; 2) the rapid opening-up of forests for agriculture, mining
and human settlements, which has reduced optimum habitats and
lessened available food sources for the tiger, and; 3) increased
accessability into the interior for hunters and poachers, who
kill wild pigs and deer, the main food of the tiger. The tiger
is therefore forced into villages to search for food. In so
doing, it comes across livestock and kills cattle, buffalo and
goats. Livestock depredation is followed by tiger mortality.

The introduction of the steel wire snare by the British
Military Administration in 1945-46 had a very adverse effect on
the tiger population. This "instrument of the damned" killed
large numbers of tigers far away from human habitation, livestock
and cultivation. During a one week period in 1955, five tigers
were killed by this nefarious method in the district of Bentong,
state of Pahang. This meant that some five hundred or more wild
pigs fewer were eaten by tigers in the jungle in that year. The
only people who benefited from this were the pig hunters and
those who set the steel wire snares. The only release from pain
and suffering for the victims of these snares was to kill them.
It is rare for such a victim to be taken alive and nursed back to
health. Thus, tigers which are wounded by gunshot or which
manage to break loose from the noose of a steel-wire snare are
often incapacitated and will become cattle-lifters, and will
attack any unfortunate human beings who cross their paths. Also,
there are a few cases of tigers having died from poisoned baits
and from drowning in the sea after being beaten by fishermen.

LEGISLATION

Until the late 1950's, the tiger was given a status equal to
or lower than that of the wild pig, rat or squirrel. It was to
be destroyed on sight by every possible means. It is ironical
that this beautiful animal was at the same time worthy of
inclusion in the crest of the Federation of Malaya, on postage
stamps and various trade marks.

The tiger was declared a reserved animal in 1957 in the state of Pahang under the Wild Animals and Birds Protection Ordinance No. 2 of 1955.   This was followed by the states of Selangor and Negeri Sembilan in 1959, and Perak in 1965. The state of Kelantan classified the tiger as a big game animal in 1962.   Under the protection of Wildlife Act No. 76 of 1972, the tiger was classified as a game animal and in 1976 it was given total protection.

As a reserved animal, the tiger's position is equal to that of the monitor lizard or the long-tailed macaque, along with such animals as the porcupine, civet and mongoose.   This means that it can be shot, killed, or taken on license.   As a big game animal in the state of Kelantan, it is subject to hunting on a big game license provided for animals like the elephant and gaur.   A game animal may be shot on license, while a totally protected animal may not be shot, killed, or taken except for scientific research or in defense of life.

Admittedly, protection for the Malayan tiger has been very slow in going into effect.   This is largely due to the reputation of the animal that even legislators feared.   Pug marks of the tiger on the fringes of rubber plantations can frighten rubber tappers from going to work.

MORTALITY RECORDS

Information collected by the Department of Wildlife and National Parks indicate that in all of Malaysia, 16 tigers were killed in 1947; 14 in 1949, and for the years 1955-59; four, seven, eight, 21 and six tigers were killed, respectively.   More detailed information is available on mortality records by state for 1960 and 1964; in Kelantan, seven and nine were killed; Terengganu, 23 and 33; Pahang, 14 and five; Johore, 20 and four; Negri Sembilan, two and none; Perak, four and one; and in Kedah, two and five tigers were killed in 1960 and 1964, respectively. In total, 72 and 57 tigers were killed in Malaysia in these two years.   The most recent information on tiger deaths by state indicate that from 1968 to 1977, five, three, one, four, five, two, four, five and 12 tigers, respectively, were killed in Kelantan, that from 1972 to 1977, two, 13, 16, eight and seven tigers were killed in Terengganu; and that from 1977 to 1983, three, two, four, three, one and three tigers were killed in Pahang.   Also, two tigers have been reported killed in Pahang in 1985.

The Department of Wildlife and National Parks is responsible for the protection of human life and livestock against tiger attacks, but not all tigers were killed by wildlife rangers. This was particularly true before the tiger was put on the totally protected species listed in 1976, after which time the incidence of public killing of tigers in defense of livestock or human life became rare.

Of the 46 tigers and two leopards killed in the state of Terengganu between October 1972 and September 1976 (Table 1), only 12 (26%) of the animals were killed by staff of the Department of Wildlife and National Parks.   Tiger sexes were 21 males, 19 females, and four unknown.   The greatest mortality, 14 tigers, occurred in 1974 and the lowest, eight tigers, occurred in 1975. Of the 14 tigers killed in 1974, only three were killed by staff of DWNP.

Of the approximately 36 tigers killed in the state of Kelantan between 1968 and 1976, sexes are known only for animals killed in 1976; these include seven males and five females. Of the 21 tigers killed between 1974 and 1976, only five were accounted for by wildlife rangers; 11 were killed by Rela members and villagers. There are no details for the 15 tigers killed in 1974 (Table 1).

Tigers were killed in all months during the five-year period in the state of Terengganu. The months with the largest numbers of kills were July and September with six and seven animals, respectively. Eight tigers (17.4%) were killed in the monsoon period between November and February. Between March and June, 15 (32.6%) of the tiger mortalities occurred. The highest mortality, 26 tigers (50%), occurred between the months of July and October (Table 1).

In the state of Kelantan, only four (18.2%) of the 22 known mortalities occurred between November and February. Between March and June, 11 tigers (50%) were killed, and seven tigers (31.8%) were killed between July and October. The month with the largest number of kills, five, was April. Tiger mortality occurred in all months except February for the five-year period (Table 1).

The available data indicated that 26 (56.5%) of the tigers killed in the state of Terengganu occurred between 1630 and 2230 hr. Nine (19.6%) of the tiger mortalities occurred between 0430 and 1030 hr, while six (13%) tigers were killed between 2230 and 0430 hr. Five (10.9%) tigers were killed between 1030 and 1630 hr. Tigers were killed during all hours of the day with the exception of 0100, 0600 and 1100 hr (Table 2).

Mortality data for Terengganu and Kelantan (Table 2) appear to indicate that tigers are most active between 1630 to 2230 hr; 60% of all mortalities occurred at this time. The next peak period of activity appeared between the hours of 0430 and 1030, but this only accounted for 17.2% of all mortalities. Tigers were killed in all districts of Terengganu, with 13 in Ulu Terengganu and only two in Kemaman. In the state of Kelantan, tiger mortalities for 1976 were reported only in the districts of Ulu Kelantan and Tanah Merah. Tigers exist in low numbers in most of the southern and western states of Peninsular Malaysia except in a few isolated states with large forest reserves.

BODY DIMENSIONS

Body lengths, heights and weights were measured for 16 and five female tigers (Table 3), killed in the states of Terengganu and Kelantan, respectively, and for 21 males (Table 4) killed in Terengganu.

Table 1. Reported monthly tiger and leopard mortality in west Malaysia (1972-76). Dashes indicate no data and blanks indicate zero mortality.

Year						Month						
	Jan	Feb	Mar	Apr	May	Jun	Jul	Aug	Sep	Oct	Nov	Dec
Terengganu												
1972	-	-	-	-	-	-	-	-	-	1		1
1973		2	1	2	3				5			2
1974	2		1		2	2	2	1		4		2
1975	1		1	1		1	3		1			
1976			1				1	4	1	-	-	-
Total	3	2	4	3	5	3	6	5	7	5	0	3
Kelantan												
1972	-	-	-	1	2	1					1	
1973												1
1974			2				1					
1975										1		
1976	1			4	1		1	3	1		1	
Total	1	0	2	5	3	1	2	3	1	1	2	1

GENERAL

In the states of Terengganu and Kelantan, fewer tigers were killed during the monsoon season than at other times of the year. In the west coast states of Peninsular Malaysia, the opposite was observed.

Wild pigs, which constitute the main diet of the tiger, are more common in the cooler and wetter months when they move from higher to lower elevations. It is believed that this may be the result of seasonal variation in food supplies. Wild pigs did a great deal of damage to crops and remained on the fringes of cultivated areas where hunters and trappers killed them.

Tiger mortality records for the state of Terengganu indicate equal numbers of males and females were killed during the year, except from December to February, when all five animals killed were female. The breeding season occurred during this period. The gestation period is reported to vary from 95-112 days, which coincides with the heaviest reported livestock losses during April, May, and June, when the young are two to four months old.

In many instances, tigers which were killed had old wounds and mutilations inflicted by man. These may have been gunshot wounds or wounds suffered from the steel-wire snare. It is not surprising that these animals, after being weakened by such wounds, preyed on domestic animals instead of their normal, wild prey. Equally surprising is the few head of cattle, numbering two or three, that a tiger killed in the course of a year.

Table 2.   Time of day (24-hour clock) when tigers and leopards were reported killed in west Malaysia (1972-76).

Year	\	\	\	\	\	\	\	\	Hour of Day															
	0	1	2	3	4	5	6	7	8	9	10	11	12	13	14	15	16	17	18	19	20	21	22	23
Trengganu																								
1972																1		1						
1973	1			1	1	1					1						1	1	2	1	1	2		
1974			1			1		1	1	1			1	1	1				2	1	3		2	
1975			1								1							1		2	2			1
1976										2									2	1	1		1	
Total	1	0	2	1	1	2	0	1	1	3	2	0	1	1	1	1	1	3	6	5	7	2	3	1
Kelantan																								
1976						1					1						1		1	3	5			

Cattle-lifters may be perfectly healthy animals. Observations were made of mothers killing cattle together with their young, with obvious results; the young grew up and became cattle-lifters. Equally to blame is the method of cattle management, which may be unique in that cattle are free to roam day and night. No one looks after the cattle, and they become easy prey to the tigers. Cattle and buffalo are easier to kill than wild pig or deer. A wounded or sick tiger may even turn on man because he is easy to kill and is present in abundance.

The three basic requirements for tigers to survive are: 1) food must be adequate in the form of either wild game or domestic animals from nearby villages; 2) water must be present in abundance, and; 3) they must have extensive cover. In the absence of any of these essentials, normal life will not be possible for the tiger.

The average life span of tigers in captivity is 11 years. There are very few authentic age records of tigers in the wild. Twenty-five years is considered the maximum life potential of tigers. Comparatively few tigers survive beyond ten to 15 years.

The ratio of tigers:tigresses in Malaya is three:two. A tiger consequently may have to fight for, and possibly lose, his mate whenever she comes into estrus. In captivity, tigers breed at the age of two, but in the wild it is assumed that few mate before the age of three or four years.

POPULATION AND DISTRIBUTION

In a study carried out in two cattle ranches, it was determined that there were at least eight tigers. Four animals were captured in Darabif and two were killed in Behrang Ulu. The serious problem of cattle-lifting continued. Tigers hitherto

Table 3. Physical characteristics of twenty-one female tigers killed in west Malaysia (1972-76).

Year	Total	Length (in) Body+Head	Tail	Height (in)	Weight (kati[a])
Terengganu					
1972 (N=1)	90.0				
1973 (N=4)					
Mean	99.5	71.3	28.3	36.1	
Range	95.0-103.0	65.0-86.0	17.0-34.0	32.1-41.0	
1974 (N=8)					
Mean	86.9	58.5	28.4	30.7	121.9
Range	70.0-97.0	47.0-67.0	21.0-31.0	23.0-38.0	40.0-150.0
1975 (N=2)					
Mean	84.0	55.5	28.5	30.0	
Range	76.0-92.0	49.0-62.0	27.0-30.0		
1976 (N=1)	91.0	60.0	31.0	29.0	135.0
Kelantan					
1976 (N=5)					
Mean	87.6	57.2	30.2	27.6	164.0
Range	72.0-98.0	48.0-65.0	24.0-34.0	24.0-31.0	90.0-265.0

[a] 1 kati - 1.3 lb.

unknown in the study area appeared, probably from the surrounding forests, and took the places of the captured or dead animals. Experiences in overcoming livestock depredation revealed that at least three or four tigers were killed before the problems were solved or reduced.

It is probable that a minimum of three tigers occurs in each area reported on in the states of Pahang, Terengganu, Kelantan, Perak, Johore, Kedah, Selangor and Negri Sembilan. The distribution maps were drawn from ranger reports who were directed to record the occurrence of tigers while on field work. From the maps and records of occurrence by area, the state of Pahang has an estimated population of 150, Terengganu 84, Kelantan 87, Perak 60, Johore 72, Kedah 21, Selangor 15, Negeri Sembilan 24, and Taman Negara 72 tigers. This is a total of 505 animals, but the records are conservative and do not account for remnant populations not observed. In all probability, there are approximately 600-650 tigers in Peninsular Malaysia.

Table 4.   Physical characteristics of twenty-one male tigers
killed in west Malaysia (1972-76).

| Year | | Length (in) | | Height | Weight |
	Total	Body+Head	Tail	(in)	(kati[a])
Terengganu					
1972 (N=1)					
	94.0	63.0	31.0	38.0	
1973 (N=5)					
Mean	100.0	70.0	30.0	37.8	
Range	79.0-109.0	60.0-73.0	19.0-36.0	32.0-45.0	
1974 (N=6)					
Mean	95.5	65.3	30.1	28.6	154.7
Range	82.0-112.0	53.0-80.0	25.0-35.0	24.0-31.0	80.0-219.0
1975 (N=5)					
Mean	85.8	58.3	27.2	29.4	147.5
Range	75.0-100.0	48.5-65.0	23.0-35.0	27.0-32.0	120.0-180.0
1976 (N=4)					
Mean	95.0	66.4	28.8	34.2	163.3
Range	80.0-108.0	64.0-72.0	16.0-36.0	26.0-42.0	130.0-200.0

[a] 1 kati = 1.3 lb.

PROSPECT FOR THE FUTURE

     Peninsular Malaysia has gone through rapid development since
gaining her independence in 1957.   About 51% of the best land for
agriculture has been cleared for cultivation and human
settlements.   The conflict between man and wildlife was
undoubtedly one of the worst ever to be encountered in the
country.   Countless numbers of animals had to be destroyed or
perished due to loss of habitat.   It was a difficult period and
the duties of wildlife officers were unpleasant to perform.
Presently, approximately 6.54 million ha of the land in
Peninsular Malaysia are still under forest.   Figure 1 shows the
existing and proposed forest reserves, national parks, wild
reserves, and wildlife sanctuaries.   The existing forest reserves
are spread over an area of 2.994 million ha.   Existing national
park wildlife reserves and wildlife sanctuaries occupy 566,066
ha.   The proposal for additional national parks, wildlife
reserves and sanctuaries is for approximately 537,947 ha, of
which 63,489 have been approved for the Endau-Rompin National
Park.   The original proposal for a permanent forest estate of
4.75 million ha will have to come from the same much-reduced
remaining area of 2.98 million ha of forest.   Most of the
existing forests will likely remain under forest simply because
they are not suitable for agriculture; the land is hilly and
includes mountains.

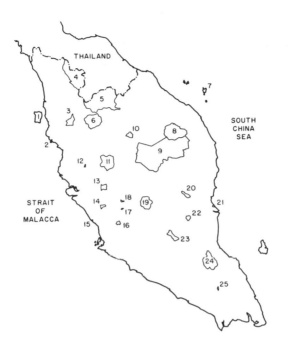

Fig. 1. Conservation areas in Peninsular Malaysia. 1) Batu
Feringgi Wildlife Reserve (WR); 2) Kuala Gula Bird Sanctuary; 3)
Selama WR; 4) Ulu Muda WR; 5) Belum WR; 6) Gerik WR; 7) Pulau WR;
8) Ulu Terengganu WR; 9) Taman Negara; 10) Sg. Nenggeri WR; 11)
Cameron Highland WR; 12) Q. Tampurong WR; 13) Surgkai WR; 14) Sg.
Dusun WR; 15) K. Selangor WR; 16) Temple Park; 17) Bt. Kutu WR;
18) Bt. Frazer WR; 19) Kerau WR; 20) Ulu Lepar WR; 21) Pahang Tua
Bird Sanctuary; 22) Tasek Chiai WR; 23) Tasek Bera WR; 24) Endau-
Rompin National Park; 25) G. Blumut WR.

Tigers are only killed by wildlife officers in extremely
serious cases. One of two man-eaters was killed in 1985. The
other was located deep in the forest where it had killed a
logger. Cattle-lifters are driven away or captured for captive
breeding. This line of action naturally has some adverse
effects. Villagers who are not satisfied may take the law into
their own hands. A few tigers were killed this way by local
police and volunteers. Such cases were later revealed to the
wildlife department and were followed with appropriate action.

Tiger units were set up in the states of Pahang, Terengganu,
Kelantan and Perak. Wildlife rangers were trained to investigate
and report on problems with cattle-lifting or man-eating tigers.
The killing of a tiger is allowed only after permission is
granted by the director general. The units also report on the
existence of tigers in their respective states.

The population was estimated at about 300 tigers in 1976,
and even with all the killing that had taken place years before,
the estimate was conservative. The present population estimate

of 505 tigers is based on work carried out over the years. The
conservation efforts that were made since 1972 appear to have
resulted in an increase in the tiger population.

EFFORTS IN CAPTIVE BREEDING

Cattle-lifting became extremely serious in two cattle
ranches in the states of Pahang and Perak between 1979 and 1982
when a total of 708 head of cattle were killed by tigers. A male
and female were killed during this period in the state of Perak.
Four tigers, two males and two females, were captured from
Darabif in Pahang between 1980 and 1983, another female was
captured later. One male was captured from the state of
Terengganu. A pair about two years of age was taken from a
litter in Ulu Tembeling, Pahang. The total number of captives
was eight; four males and four females. A pair was given to the
National Zoo and 3 pairs are kept in the Malacca Zoo.

Three cubs were born to one of the pairs in the Malacca Zoo.
All three cubs died from wounds inflicted by the male prior to
separation from the female. Another three cubs were later born
to the same pair. The male was separated from the female and her
cubs. Two of the cubs survived and are doing well in captivity.
There are now ten Malayan tigers in captivity in Peninsular
Malaysia.

ACKNOWLEDGEMENTS

I would like to take this opportunity to thank everyone who
assisted in the accumulation of information for this paper. In
particular, I am grateful to the following State Wildlife
Directors: En. Jasmi Abdul, of Pahang; Puan Misliah Basir, of
Terengganu; En. Abdul Razak b. Abdul Majid, of Kelantan; En.
Zainuddin Abdul Shukur, of Perak; En. Sivanathan a/l Elagupillay,
of Kedah; Puan Halimah Muda, of Selangor; En. Ebil Yusof, of
Negeri Sembilan; and En. Saharuddin Anan, Superintendent of Taman
Negara. The efforts of all wildlife rangers in the survey are
gratefully appreciated.

REFERENCES

Blanchard, R.F. 1977. "Tigers in Malaysia: Battered, Beaten and
     On the Edge of Extinction." Unpublished Reports to the
     Department of Wildlife and National Parks: Preliminary
     Analysis of Tiger Mortality and Livestock Depredation in
     Terengganu and Kelantan, West Malaysia.
Locke, A. 1954. The Tigers of Terengganu. London: Museum Press
     Ltd.

9

Tiger Numbers and
Habitat Evaluation in Indonesia

Charles Santiapillai and Widodo Sukohadi Ramono

INTRODUCTION

Prior to the large scale modification of its habitat by man, the tiger (Panthera tigris) thrived in substantial numbers and had a much wider geographical distribution in Indonesia than it does today, being once found in Sumatra, Java, and Bali. Since the turn of the century, a combination of high human population densities and extensive land-use activities, has led to the extirpation of two of the three subspecies of the tiger, viz. Panthera tigris sondaica and P. t. balica from Java and Bali respectively. Today only the Sumatran tiger (P.t. sumatrae) survives (Fig. 1), but even in Sumatra, the disruptive processes that squeezed the tiger out of Java and Bali are beginning to threaten the species there as well (Santiapillai and Widodo 1985).

Nevertheless, the fact that it is still found in Sumatra indicates that the tiger is an exceptionally adaptable predator, able to survive as long as sufficient prey, fresh water, and ample vegetative cover are available to it (Schaller 1967). If the current disruptive processes are not mitigated, the tiger, like the rest of the Sumatran large mammals is likely to have its range and numbers shrink at an accelerating rate.

NUMBERS OF TIGERS IN SUMATRA

At the turn of the century, the tiger was so numerous in Sumatra that it was considered a serious threat to man in several areas, and so rewards were paid by the East India Company for every animal that was killed. Considerable number of people fell victim to tiger depredations annually and there were even instances of entire villages being depopulated as a result (Marsden 1811).

The question of just how many tigers are present in Sumatra today presents many difficulties. The tropical rain-forest

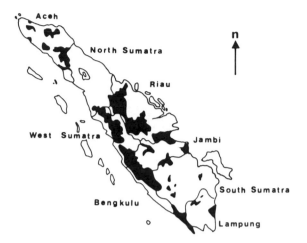

Fig. 1.   The distribution of the Sumatran tiger (<u>Panthera</u> <u>tigris</u> <u>sumatrae</u>).

environment in Sumatra coupled with the secretive nature of the tiger, makes it extremely difficult to estimate their numbers.

Borner (1978) found evidence of the tiger in all eight provinces of Sumatra and estimated their population to be about 1,000 individuals. Since then, disruptive and competitive human land-use patterns have further eroded much of the animal's habitat, especially in the lowlands. Whilst it is impossible to be certain about the total number of tigers remaining in Sumatra today, it's numbers can now be measured in 'hundreds'. In the past, their numbers would have been estimated in the 'thousands' (Santiapillai and Widodo 1985).

HABITATS FOR TIGER IN SUMATRA

Sumatra, with a total land area of 473,606 sq. km, represents one of the largest islands in the world. It's topography is dominated by the chain of volcanic mountains - the Burit Barisan - along the west coast (Fig. 1). The lowland evergreen forest formations occur on the narrow alluvial plains between the Burkit Barisan in the west and the extensive swamps in the east (Seidensticker in press).

Sumatra's human population have always been small compared to those of Java and Bali. By 1930 Sumatra had only about six million people, while there were 42 million in Java (Loeb 1972). During the next five decades, while the population of Java more than doubled to 91 million, that of Sumatra increased by over four fold to 28 million. This was largely due to the annual arrival into Sumatra of thousands of transmigrants (both official as well as spontaneous) from over crowded islands such as Java and Bali.

The largest increase in the human population occurred in the southern province of Lampung, where between 1961 and 1980, the numbers increased from 1.6 to 4.6 million - an increase of 177% (Scholz 1983). The aspirations of these transmigrants for a better life style in Sumatra points to an even greater pressure on the remaining undisturbed wildlife habitats.

The most optimistic estimate of the extent of the protected areas in Sumatra is 92,637 sq. km or 20% of the total land area (IUCN 1985). Only a few of these areas are large enough (Fig. 2) to maintain viable populations of tiger, but even these are not completely immune to the effects of environmental disturbances outside their boundaries.

It is the vulnerability of the tiger to man-induced changes in its environment that explains why it has virtually disappeared from the densely populated province of Sumatra Utara (North Sumatra). Tigers overall numbers in the two provinces of Lampung and Sumatra Barat (West Sumatra) also remain low. The stronghold of the tiger in Sumatra seems to be the province of Riau (94,562 sq. km) where 30% of the island's tiger population is thought to be present (Borner 1978). Nevertheless, the existing protected areas in Riau are unlikely to remain viable and self-sustaining in the face of relentless disruptive land-use activities. The only large reserve, Kerumutan, was established in 1968 and is now almost totally destroyed by activities of illegal settlers (FAO 1982).

The area between Barumun and Rokan Rivers in Riau, recommended by Borner (1978) as a viable reserve for tiger, was found to be 50% peat-swamp forest - a poor habitat for large mammals in general (Seidensticker 1986). Furthermore, not only is the area completely blanketed by timber concessions, it is also being exploited for its oil reserves (Seidensticker 1986) and is earmarked for conversion to oil-palm or sugar cane plantations.

The provinces of Aceh, Jambi, Sumatra, Selatan, and Bengkulu account for almost 50% of the island's tiger population (Borner 1978). Tiger depredations on livestock and human beings are still being reported from Bengkulu and Aceh provinces (Santiapillai and Widodo 1985). The limitations imposed by man on tiger in the provinces of Sumatra Utara, Sumatra Bartat, and Lampung are likely to increase in other provinces as well.

The tiger in Sumatra inhabits a variety of habitats, but it is particularly associated with riverine forests, swamp forests, and grasslands. In the Berbak Game Reserve (Fig. 2), which is renowned for its tigers, the main habitats are freshwater swamp forests and river edge forests (Silvius et al. 1984). However, tiger distribution is not solely determined by the amount of suitable habitats or forest cover left. Other factors, such as the availability of suitable and vulnerable prey populations, and other carnivore competition, chiefly from the red dog (Cuon alpinus), are equally important. The principal prey of the tiger in Sumatra is the wild pig (Sus scrofa), followed by the sambar (Cervus unicolor).

Rain forest habitat in general does not support a high biomass of large ungulates (Eisenberg and Seidensticker 1976). Ground living herbivores such as the wild pig and sambar are generally found in low densities in primary forests (Santiapillai

Fig. 2.   Major conservation areas in Sumatra.

and Widodo 1985).   In view of this and because a large proportion
of the conservation areas in Sumatra are mountainous (Fig. 2),
the tiger occurs in low numbers.

       In the Kerinci-Seblat National Park, the tiger is known from
altitudes of over 1,000 m (Blouch 1984).   As agricultural
pressures and large scale logging operations in the lowland areas
in Sumatra increase, mountainous habitats may offer the only
sanctuary for the tiger in the future.   It is likely that before
the end of this century, tigers will survive in Sumatra only in
discontinuous populations.

       Optimum habitat is provided by sub-climax vegetation.
Reproductive success of tigers in successional habitat can be
high (Smith 1978, Sunquist 1981, Tamang 1983).   Ecotonal areas
are especially favorable to tigers.   The transition zone between
forests and grasslands provides ideal habitats to the tiger's
principal prey species, such as the wild pig and sambar, so these
'edges' enhance the tiger numbers too.

       In the early stages of habitat alteration by man, the tiger
might benefit in so far as it finds such 'edges'.   Selective
logging can promote prey population build-ups and so could
benefit the tiger as well.   Sustained yield exploitations of the
forest, if carried out carefully, is compatible with tiger
conservation (Ashby and Santiapillai this volume).   The threat to
the tiger comes not from the logging per se, but from the human
element that invariably follows such activities.   In planning for
tiger conservation, it is imperative that tigers and new
settlements be kept apart.

       In optimum habitat in India such as the Corbett Reserve,
tiger densities can be as high as 14 per 100 sq. km (Sankhala
1979).   In prime habitat in Nepal, adult tiger densities of 6-7
per 100 sq. km have been recorded, whereas outside such areas,
densities are much less, about 1-2 per 100 sq. km (McDougal 1977,
Smith 1978,   Sunquist 1981).   In Sumatra, tiger densities could

be as high as 3.7 per 100 sq. km in the lowland forests of Bengkulu (Santiapillai and Widodo 1985).

In general, tiger densities are much lower. In the lowland Way Kambas Game Reserve (Fig. 2) in Lampung province, tiger densities range from 2.3 to 3.0 per 100 sq. km (Nash and Nash 1985). In the more mountainous habitat as in the Gunung Leuser National Park in Aceh province, the tiger is known to exist at a density of 1.1 per 100 sq. km (Borner 1978). In summary, the tiger is able to maintain a density of 1 per 100 sq. km in montane habitats and 1-3 per 100 sq. km in more favorable lowland areas in Sumatra.

THREATS TO THE SURVIVAL OF TIGER

The most serious threats to the survival of the tiger in Sumatra today are indiscriminate forest clearance, poaching, and poisoning (Santiapillai and Widodo 1985). If the Sumatran tiger is already under some threat from these, then its status is likely to become even more precarious in the future. The tiger is a range-sensitive species (Seidensticker and Hai 1983) and is therefore susceptible to large scale changes in land-use patterns.

Since 1972, the tiger is fully protected in Sumatra by a Governmental Decree (van der Zon 1979). But poaching still goes on. Today, a tiger skin on the black market is worth over US $3,000, which offers a strong incentive for poaching.

Poison has always been a major threat to the tiger's survival in Indonesia. According to Hoogerwerf (1970) it was perhaps the most important direct cause for the decline of the Javan tiger. Chlorinated hydrocarbons like DDT or organo-chlorides such as Toxaphen (used in the control of cattle ticks) are used as poison in the provisioned baits (Myers 1976).

In the 1800's, one method used to kill tigers in Sumatra, was to leave a vessel of water impregnated with arsenic near the bait (Marsden 1811). Because the tiger has had an adverse press in Indonesia, many people view the animal as vermin that should be destroyed. However, attitudes are slowly changing and villagers are now beginning to understand the role of the tiger in reducing the number of wild boar near agricultural areas. Where the tiger does not occur wild boar have become a serious pest to agriculture, second perhaps to the elephant in the scale of its depredations and damage (Santiapillai and Widodo 1985).

THE OUTLOOK FOR THE TIGER IN SUMATRA

Up to now, the conservation measures have largely been limited to the enactment of legislation for the protection of the tiger in Sumatra and setting aside reserves for its survival. Ideally, the emphasis must be on maintaining the highest number of tigers and widest possible distribution if we are to ensure their long-term survival in Sumatra. This is easy to recommend, but difficult to implement in many of the developing countries where rapid human population growth entails the development of every piece of land to its maximum. Nevertheless, Indonesia has set aside a number of large areas for wildlife and nature conservation, despite its high human population.

The major conservation areas in Sumatra (Fig. 2) are the Kerinci-Seblat National Park (14,846 sq. km), Gunung Leuser National Park (8.025 sq. km), Barisan Selatan National Park (3,568 sq. km), Berbak Game Reserve (1,900 sq. km), and Way Kambas Game Reserve (1,300 sq. km). Because population pressure will continue to increase, the land set aside for conservation must be viewed in this context. Therefore, the best opportunity for the conservation of the tiger in Sumatra lies in some form of multiple use pattern of its peripheral habitats away from the core areas.

REFERENCES

Blouch, R.A. 1984. Current status of the Sumatran rhino and other large mammals in southern Sumatra. IUCN/WWF Report 4, Bogor, Indonesia.

Borner, M. 1978. Status and conservation of the Sumatran tiger. Carnivore 1:97-102.

Eisenberg, J.F. and J. Seidensticker. 1976. Ungulates in southern Asia: a consideration of biomass estimates for selected habitats. Biol. Cons. 10:293-308.

FAO. 1982. A national conservation plan for Indonesia. Vol. II: Sumatra. FAO/INS.78.061. Field Report No. 39. Bogor, Indonesia.

Hoogerwerf, A. 1970. Udjung Kulon: the Land of the Last Javan Rhinoceros. Leiden: E.J. Brill.

IUCN. 1985. Directory of Indomalayan Protected Areas. Indonesia. Cambridge, UK: Protected Areas Data Unit, Conservation Monitoring Centre.

Loeb, E.W. 1972. Sumatra, its History and People. Kuala Lumpur: Oxford Univ. Pr.

Marsden, W. 1811. The History of Sumatra. 3d ed. Reprint. London: Oxford Univ. Pr.

McDougal, C. 1977. The Face of the Tiger. London: Rivingdon Books and Andre Deutsch.

Myers, N. 1976. The Leopard Panthera pardus in Africa. IUCN Monograph No. 5. Morges, Switzerland: IUCN.

Nash, S. and A. Nash. 1985. An evaluation of the tourism potential of the Padang-Sugihan Wildlife reserve. WWF/IUCN Project 3133. Field Report   No. 2.

Sankhala, K. 1979. Tigers in the wild - their distribution and habitat preferences. In International Symposium on the management and breeding of the tiger, pp. 43-59, ed. S. Seifert and P. Muller. Zool. Gart., Leipzig.

Santiapillai, C. and S.R. Widodo. 1985. On the status of the tiger (Panthera tigris sumatrae Pocock, 1829) in Sumatra. Tigerpaper 12(4):23-29.

Schaller, G.B. 1967. The Deer and the Tiger: a Study of Wildlife in India. Chicago: Univ. Chicago Pr.

Scholz, U. 1983. The Natural Regions of Sumatra and Their Agricultural Pattern. A Regional Analysis. Vol. I. Bogor, Indonesia: Central Research Institute for Food Crops (CRIFC).

Seidensticker, J. and A. Hai. 1983. The Sunderbans Wildlife

Management Plan. IUCN/WWF Project No. 1011.
Seidensticker, J. 1986. Large carnivores and the consequences
    of habitat insularization: ecology and conservation of
    tigers in Indonesia and Bangladesh.
Silvius, M.J., H.W. Simons, and W.J.M. Verheught. 1984. Soils,
    Vegetation, Fauna, and Nature Conservation of the Berbak
    Game Reserve, Sumatra, Indonesia. Report 758, Nature
    Conservation Department. Wageningen, Netherlands: Agric.
    Univ.
Smith, J.L.D. 1978. Smithsonian tiger ecology project. Unpubl.
    Report No. 13 Washington, DC: Smithsonian Inst.
Sunquist, M.E.   1981.   The social organization of tigers
    (Panthera tigris) in Royal Chitwan National Park, Nepal.
    Smithson. Contrib. Zool. 336:1-98.
Tamang, K.M. 1983. The status of the tiger (Panthera tigris
    tigris) and its impact on principal prey populations in the
    Royal Chitwan National Park, Nepal. Ph.D. Diss. Michigan
    State Univ., East Lansing, Michigan.
van der Zon, A.P.M. 1979. Mammals of Indonesia. FAO/INS/78/061.
    Bogor, Indonesia: FAO.

10

The Problem Tiger of Bangladesh

Mohammad Ali Reze Khan

INTRODUCTION

The tiger is one animal in Bangladesh which has earned both good and bad names during the last couple of centuries. It has been designated as the national animal of Bangladesh and is used as a water-mark for the nation's currency. During British rule in India everything that attracted Royal attention was termed "Royal" and the Bengal tiger was no exception to this.

The situation of the Bengal tiger in Bangladesh has not changed appreciably in recent years (Khan 1982) except that in the last four years three more tigers were shot and about half a dozen tiger skins were recovered by the law enforcement authorities at Dhaka, Bangladesh. It was not clear whether the tiger skins were smuggled into the country or not.

DISTRIBUTION

I earlier (Khan 1982) reported that about half a century ago tigers were distributed over the entirety of Bangladesh. By 1960 the tiger disappeared from all forest types except the Sundarbans mangrove forest in the southwest, bordering the West Bengal Sundarbans. The Bangladesh Sundarbans cover an area of 5700 sq. km; 4100 sq. km covered with forest and the remaining covered with water. This is the last stronghold for the Bangladesh tiger. Stray tigers occassionally were reported in the Greater Sylket, Mymensingh, Chittagong and Chittagong Hill Tracts districts, but these reports were prior to 1982.

Husain (1981), citing various sources, reported that in 1792 rewards were paid for killing ten tigers in Madaripur (Faridpur district). In 1864 a party of shikaris brought the skulls of 257 tigers and leopards to the (District) Collector at Bogra and obtained about Rs.700/- as reward. In Dinajpur around 1881, the Pranuagar forest in the Birganj Thama (Police Station) was so notorious for tigers that no traveller would pass through at night, or even in the daytime, if alone. In 1905 rewards were paid for killing 11 tigers and 10 leopards in Sylaat. According to a report in 1917, tigers were still well-represented in the Madhupur forest (Mywensingh district).

Husain (1981) also said that even in the early 1920's, tigers, though scarce, were found in Rangpur, Dinajpur, Bogra, Rajshahi, Sylhet, Comilla, Noakhali, Bakerganj, Chittagong and Chittagong Hill Tracts. According to Mitra (1957), tigers were present from 1930 to 1940 in 11 of 17 districts of Bangladesh, including Mymensingh, Barisal (Bakerganj), Rajshahi, Bogra, Dinajpur, Chittagong, Dhaka, Faridpur, and Jessore, Khulna and Noakhali.

The last tiger of Lawachare forest, near Srinongal township, Sylhet district, was shot in 1948. The last tiger of Mainimukh in the Chittagong Hill Tracts was killed in 1958 (Ahamed 1981), although another was also shot there in 1955 (Husain 1981). The last tiger of Banglabanda, Dinajpur district, was shot by Gen. M.A.G. Osmani in 1962 (Khan 1985).

Tigers Killed by Man

There are fascinating accounts of tigers killed by man in the literature. A summary of tiger deaths occurring in the Bangladesh Sundarbans appears in Table 1. Information is unavailable on how many tigers fell prey to hunters in the Bangladesh Sundarbans prior to 1927, but Hendrichs (1975), reporting on the combined Sundarbans (Bangladesh and Indian), indicated that 563 tigers were killed from 1912-26, or 40.2 per year; 41 from 1927-30, or 10.2 per year; 133 from 1930-35, or 26.4 per year (author's estimate based on Hendrichs 1975); 201 from 1935-42, or 28.5 per year; 112 from 1942-47, or 22.3 per year; and 160 tigers were killed from 1947-57, or 16.0 per year for this period. From the figures given by Hendrichs, an estimated 1,395 tigers were killed in the Bangladesh and Indian Sundarbans from 1912 to 1971.

By comparison, Mitra (1957) reported that 329 tigers were killed in government forests of West Bengal and Bangladesh between 1935 and 1941, an estimated 222 from the Bangladesh Sundarbans alone. From 1947 to 1957, he estimates 171 tigers were shot in the Bangladesh Sundarbans and about 153 in the whole of the government forest of West Bengal, India.

In spite of the stringent measures suggested in the 1971 Bangladesh Wildlife Preservation Act against poachers, reports from woodcutters and other people working in the Bangladesh Sundarbans indicate some 20 25 tigers were shot or poached between 1971 and 1973; another seven tigers were killed outside the Sundarbans between 1979 and 1985. Four of these were shot because they were man-eaters/cattle-lifters or they were killed by poachers in the Greater Khulna district, but outside the Sundarbans. One each was killed at Greater Mymensingh (actually in Jamalpur) and in Cox's Bazar forest division of Chittagong district; one was shot in Sylhet in 1985. All these specimens strayed from the Sundarbans or from the neighboring states of India.

Table 1.   Tiger-killing by men and man-killing by tigers in the
Bangladesh Sundarbans from 1925-84.

Years	Tiger Deaths (Husain 1981)	Human Deaths (Habib pers. comm.)
1925-29	33+[a]	---
1930-34	99[a]	---
1935-39	157[b]	---
1940-44	122[b,c]	---
1945-49	90[c]	37+[e]
1950-54	104	92[e]
1955-59	54	100[e]
1960-64	26	117
1965-69	24	97
1970-74	30+[d]	184
1975-79	---	89
1980-84	---	83

[a] Information unavailable for 1925-26, includes author's estimate of 20 for 1933.
[b] Includes author's estimate of 32 per year for 1936, 1940-41.
[c] Includes author's estimate of 19.4 per year for 1942-46.
[d] Includes 20+ reported to author for 1971-73, information unavailable for 1974.
[e] Information unavailable for 1945-47, includes estimate of 18.3 per year for 1948-56 based on Hendrichs (1975).

Man-Killing by Tigers

The earliest known record of man-killing by Sundarbans tigers is that of Bernier, who toured this part of the Mughul Empire during 1665-1666 (Constable 1891).   This was also documented by Jerdan (1874), Anonymous (1908), O'Malley (1914) and in the forest department reports and working plans. Hendrichs (1975) and Habib (pers. comm. 1986) reported on man-killing in the Bangladesh Sundarbans in considerable detail (Table 1).   Including the 15 deaths reported by Habib for 1985, an estimated 814 people have been killed by tigers in the Bangladesh Sundarbans from 1948 to 1986.

Hendrichs' (1975) report   further indicates another 360 people were killed in the combined Sundarbans of Bangladesh and West Bengal between 1912 and 1939.   Mitra (1957) reported that 280 people were killed in the government forest of Bangladesh and

West Bengal between 1930 and 1947, barring four years from 1942 to 1945 for which there were no data.

Mitra (1957) also presented interesting information on the number of people killed by tigers in areas outside government forest in the 1930's: 77 people died in Khulna district, outside the Sundarbans; 36 in Mymensingh; 35 in Barisal; 18 in Rajshahi; eight in Chittagong; four in Bogra; three in Dinajpur; two in Noakhali; and one each in Dhaka, Faridpur and Jessore. Thus, a total of 211 people were killed by tigers in 11 out of 17 districts of Bangladesh. Some people obviously died during this period in the Chittagong Hill Tracts district, but this information was not available separately as it was incorporated in the figures provided for government forests.

Tiger Census

There are no census reports on Sundarbans tigers subsequent to Hendrichs' (1975). Hendrichs worked in the Bangladesh Sundarbans for the first three months in 1971 and estimated the total tiger population at 350 individuals.

I doubt that this number has changed since 1971. The forest department attempted to census tigers from 1978 onwards with no success. The most startling figures were arrived at by a journalist of Reuter's news agency who reported, "the number of tigers in Bangladesh has doubled in only seven years -- to more than 550(!) -- the country's wildlife officials have revealed." (Nunn 1985). Reuter's quoted the figure to be 580(!). By contrast, I (Khan 1985) conjectured the figure to be between 300 and 430 tigers, incorporating Hendrichs (1975) 25% range. Habib (pers. comm. 1986) suggests a conservative estimate of 400 tigers. Although ther has been no recent census of tigers in the Sundarbans, the concensus is that the number of tigers have probably increased since March 1971. But by how many, is a question yet to be answered.

CONCLUSION

I do not believe it matters whether we have precise numbers for the tigers of the Bangladesh Sundarbans or not. We have some figures to work with. The important aspect is that tiger-killing has ceased and that man-killing has been reduced substantially.

How much longer man-killing tigers of the Sundarbans will be tolerated by the people of Bangladesh, and how this interaction can be lessened, are the questions that need immediate answers. I hope cat specialists, in and out of the International Cat Group of the IUCN, will come forward with some solution to this problem.

ACKNOWLEDGEMENTS

I am extremely grateful to Mr. Golam Habib, Divisional Forest Officer, Sundarbans Division, Khulna, Bangladesh, for his response to my letter and providing me with casualty figures from 1971 and tiger estimation. He also helped me to conduct field studies in the Sundarbans back in 1979. Thanks are also due to Mr. Md. Farid Ahsan, Assistant Professor of Zoology, Chittagong University, Bangladesh for his assistance in the field.

REFERENCES

Ahamed, Y.S. 1981. With the wild animals of Bengal. Dhaka: Published by author.
Anonymous. 1908. Bengal District Gazetteers, Khulna. Calcutta: Bengal Secretariat book depot.
Anonymous. 1986. What is a viable tiger population? Cat News 4:3-4.
Constable, A. 1891. Travels to the Megul Empire, A.D.1656-1668. (Original in French by F. Bernier. Paris: Claude Barbin. 1670.) London.
Hendrichs, H. 1975. The status of the tiger (Panthera tigris L. 1758) in the Sundarbans mangrove forest (Bay of Bengal). Saugetirk. Mitt. 23(3):161-99.
Husain, K.Z. 1981. Development activities and their impact on the terrestrial fauna of Bangladesh. Proc. 3rd Nat. Zool. Conf, Dhaka, pp 1-7.
Jerdon, T. 1874. The Mammals of India. London: John Wheldon.
Khan, M.A.R. 1982. The status and distribution of the cats in Bangladesh. Intern. Cat Symp., Kingsville, 1982.
Khan, MA.R. 1985 Mammals of Bangladesh - A Field Guide. Bangladesh: Nazma Reze, House 25, Rd. 1, Dhanmandi, Dhaka-5.
Mitra, S.N. 1957. Banglar shikar prani (In Bengali). Calcutta: Govt. press.
Nunn, W. 1985. New tigers bite back! Khaleej Times Dubai, U.A.E. 26.7.85.
O'Malley, L.S.S. 1914. Bengal District Gazetteers, 24-Parganas. Calcutta: Bengal Secretariat book depot.

11

Female Land Tenure System in Tigers

James L.D. Smith, Charles W. McDougal and Melvin E. Sunquist

INTRODUCTION

The study of the social system of tigers (Panthera tigris) in Royal Chitwan National Park (RCNP), Nepal, revealed that they have a polygnous mating system in which males establish intra-sexual territories from which other males are excluded, but in which one to several females reside (McDougal 1977, Sunquist 1981, Tamang 1982, Smith 1984). Territorial behavior by males is most pronounced when females are in estrus (McDougal, Smith pers. observ.) suggesting that competition for mates is the proximate factor favoring male territoriality (Brown 1964). Females also occupy exclusive ranges, but their ranges are much smaller than those of males, and the critical resource for which they compete appears to be a combination of food, cover, and a secure place to raise young (McDougal 1977, Sunquist 1981, Smith 1984).

The purpose of this paper is to examine the land tenure system of female tigers. First we will attempt to provide answers to the following questions: 1) are female tigers territorial; 2) how and where do tigresses acquire territories; 3) what factors affect territory size; and 4) what is the degree of relatedness of adjacent females?

Based on the answers to these questions, we will discuss the apparent function of territorial behavior and examine the extent to which territoriality may be limiting population density (Brown 1969). We also will examine the genetic implications of the pattern of relatedness among neighboring females.

STUDY AREA

Our study was conducted in Royal Chitwan National Park, Nepal, between January 1974 and June 1985. It was a continuation of research on the ecology and behavior of tigers and their major prey begun in 1973 (Seidensticker 1976, McDougal 1977, Sunquist 1981, Tamang 1982, Mishra 1982, Smith 1984).

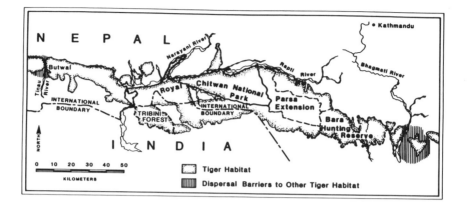

Fig. 1.   Distribution of the tiger population of Royal Chitwan
National Park, Nepal.   Dispersal barriers at Butwal and the
Bhagmati River isolate Chitwan tigers from populations 150 km
east and 250 km west.

The Park lies in the center of an 820 x 40 km lowland belt,
the Terai, which extends the entire length of Nepal (Fig. 1).   To
the north is the Mahabharat Range (2,500-4,000 m), a heavily
settled region, and to the south is the intensively cultivated
Gangetic plain of India.   Running the length of the Terai are low
hills, the Siwalik Range (300-950 m).   In places these hills
merge into the Mahabharat Range; elsewhere they bend away to form
the Dun valleys.   The Park encompasses an 80 km section of the
Siwalik Range and the southern half of the Chitwan Dun Valley.
The Park's northern border is formed by the Narayani and Rapti
rivers (Mishra 1982).

The Park is composed of three ecological zones: the Churia
hills (61%), sloping benches (25%), and floodplain (14%).   The
Churia hill zone includes mature sal forest (<u>Shorea</u> <u>robusta</u>) on
the dry, deeply eroded upper hill slopes. The sloping upland
benches that lie between the hills and the floodplain are
composed of open sal forest with a tall-grass understory. The
floodplain lies along the Rapti, Narayani, and Reu Rivers, and is
composed of a dynamic interspersion of riverine forests, tall
grasses, and broad, sandy riverbanks.   There is an unusually high
degree of vegetational interspersion in the alluvial region of
the Park, and this may account for the high density and diversity
of animals (Mishra 1982).

METHODS

To study the social structure and land tenure system of
female tigers, we captured and radio collared 11 tigresses using
a system described by Smith et al. (1983).   In addition, the
movements of seven uncollared and four radioed females were

monitored based on the recognition of distinctive tracks. The
identity of these animals was confirmed by observing unique
facial markings at bait sites, through photographs taken by
camera trip devices, or by radio locations. Animals that were
radio collared are referred to by numbers; letters are used to
indicate those identified by facial markings and tracks. When
growling was heard, we subsequently examined tracks and other
spoor such as blood at the site in an attempt to reconstruct the
sequence and outcome of an encounter. In several cases the
individuals involved in the agonistic interactions were known
because they were radio collared, whereas in other cases they
were identified from tracks.

    Radio marked animals were tracked on foot, and from elephant
back, vehicle and aircraft. Locations were determined by
triangulation (Mech 1983) and by approaching to within 200-400 m
of the animal and then partially circling it to pinpoint the
position. Tiger tracks, scent marks and other signs were
recorded opportunistically and, in addition, two transects were
monitored regularly for scent marks and tracks (Smith 1984).
Information on reproductive status was based on direct
observation of young or the presence of cub tracks in association
with those of an adult female. The presence of new born-cubs was
established by closely monitoring a female at the predicted
parturition date following a mating association (gestation is
about 102 days). Both radio and track locations were plotted on
1:50,000 aerial photos and transferred to a base map which was
constructed by digitizing a satellite photograph. Home ranges
were defined by a convex polygon subscribed by connecting outer
locations but excluding areas not considered tiger habitat (i.e.,
cultivation; Mohr 1947).

RESULTS

Are Female Tigers Territorial?

    We have four categories of data that indicate tigresses are
territorial: These include 1) essentially non-overlapping
ranges; 2) scent-marking behavior; 3) strong site fidelity and;
4) aggressive behavior between females.

    Non-overlapping ranges: The degree of overlap between
adjacent ranges was small. During the 1977-79 period the ranges
of seven adjacent females had a mean of 7.1% overlap between
neighbors (Fig. 2). Furthermore, from 1979 to 1981 we monitored
the movements of three females which settled in areas adjacent to
their mothers. While each was a subadult they resided within
their mother's range, but as they matured either daughter or
mother or both gradually shifted their ranges to adjacent areas.
By the time daughters were established breeders, the amount of
overlap in their respective ranges was only 3.7%. Thus, even
closely related individuals established mutually exclusive
ranges.

    Scent-marking behavior: The pattern of deposition of urine
and feces and the associated visual marks made by females
indicated that one function of scent-marking was to delineate
territory boundaries. Young females marked extensively when they
began to establish breeding ranges (Sunquist pers. observ., Smith

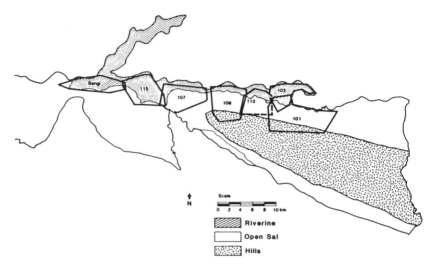

Fig. 2.   The ranges (indicated by convex polygons) of seven adjacent female tigers along the northern border of Royal Chitwan National Park, Nepal.

1984).   Furthermore, females marked in response to adjacent females (Fig. 3, Smith 1984). Marking also was concentrated at home range borders. McDougal (unpublished data) monitored a 7.9 km transect that crossed the entire longitudinal axis of a resident female's territory to determine the distribution of marks in relation to range boundaries. The zones at the eastern and western borders only comprised 21% of the transect, but contained 65% of the marks.   The rate of marking in these border zones was 6.8 times higher than in the central portion of the range.

Site fidelity: There was strong site fidelity throughout the breeding lives of tigresses. No female left her range to establish a breeding territory elsewhere, although several shifted so that part of their range was aquired by their daughters. Two females remained within their territory until they died and one female, who was pushed out of her range, became a transient and never bred again. The one female that left her natal range used an area > 250 sq. km during her dispersal period. Once she settled and began breeding her range was reduced to < 18 sq. km; in April 1986 she was still occupying the same range.

Aggressive Interactions: The strongest indicator of territorial behavior was aggressive interactions between females that resulted in either one female leaving the area or a shift in range boundaries. The following four cases demonstrate how females use aggressive behavior to maintain or obtain exclusive territories.

Case 1: On several occasions over a two month period during F111's dispersal, we located her within the territory of a resident, FB. On one occasion we heard growling from an area where both animals were residing. When we inspected the site, tracks and marks indicated that a fight had occurred. Within two

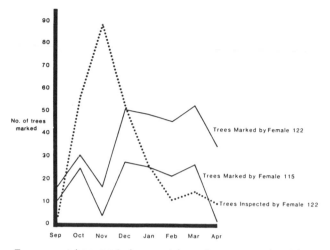

Fig. 3. Tree marking and inspecting frequency by tigresses F115 and F122 along a transect at their common territorial boundaries.

days F111 left the area, shifted across the Narayani River, and established a breeding territory in a different area. The Narayani River then became a natural boundary between the two females.

Case 2: In 1981 F111 attempted to drive her 22- month old daughter from a buffalo kill at a bait site, but was unsuccessful. Several months later another fight occurred between these two females along the sandy river bed of the Reu River. The outcome of this interaction was the establishment of a boundary along the Reu River that the daughter did not cross again to gain access to the bait site.

Case 3: On two occasions fights occurred when the subadult daughter (FN) of F111 left her natal area north of the Narayani River and crossed south into F122's territory. On the second occasion (May 1983) F122's right forefoot was injured. Following this encounter, FN ceased her visits to the south side of the river.

Case 4: F3T settled adjacent to her mother (FB) in 1981 and remained there for two years. Beginning in late 1983, F3T began to encroach on her mother's territory. Finally, in May 1984, when FB was a minimum of 13.5 years old, she abandoned her range and became a transient. She shifted east into and through the area occupied by her other daughter (F122) and became temporarily localized on the eastern border of F122's range. During this time she fought with F122 and eventually had to be removed from the Park.

The seriousness of fights between females was indicated by the fact that we found blood at the site of the fight described in Case 4 and that three females sustained damage to their front paws to the extent that the injured feet became permanently deformed. Thereafter the tracks of these tigresses could easily be identified because their injured paws left unique impressions in the substrate. On another occasion we examined an old female whose tracks were also distinctive and who we suspected was injured before our study began. She had an old canine puncture through a pad on her front paw. During an immobilization of F103 we observed several fresh wounds; she had two punctures on her right shoulder and her right ear was split for 3.5 cm. However, fights that result in permanent noticeable injuries do not appear to occur very often; in 14 immobilizations of five females over a period of four years the case of F103 was the only occasion when fresh wounds were observed.

We believe that these data on non-overlapping home ranges, especially the pattern of daughters settling next to their mothers, scent-marking behavior, long-term site fidelity and examples of female-female aggression constitute a convincing set of evidence that female tigers are territorial.

How and Where Do Females Acquire Territories?

We recorded two settling patterns of subadult tigresses. The typical pattern was for a female to establish a territory adjacent to that of her mother. Animals of both sexes remained within their natal area after the birth of a new litter. When the new litter was approximately two months old the subadult littermates dispersed; at this time the gradual separation of the ranges of a mother and a daughter became apparent. The process took several months and began when females were 22 to 30 months old (Fig. 4, 5 and 6). Typically a mother shifted her territory so that a daughter acquired a portion of the mother's range (see the cases of F122 and FJ in Fig. 2 and 4) or nearly all of the mother's territory as in the case of F103 (Sunquist 1981). It is interesting to note that often there is still some territory overlap when a daughter's first litter is born (Fig. 4c, 5c), but by the time the daughter's offspring are approximately one year old and regularly traveling with her (Fig. 4d, 5d) there is typically very little overlap between the territories of a mother and daughter (Smith 1984).

The second pattern of acquiring a breeding territory was for a female to disperse and become established away from the natal area. During our studies only one radio marked female dispersed; she settled 33 km west of her natal area (Fig. 6).

Because territorial acquisition by females is a gradual process, it is difficult to determine the precise time of its occurrence. We used age at first conception as an indicator of the time when a female had successfully become a territory holder. We were able to monitor four females closely as they established territories and produced their first litters. For these female the age at first conception ranged from 33 – 38 months. Once females became established as territorial breeders they remained territory holders throughout their breeding life (Table 1).

Of 14 females we determined to be territorial, all but one gave birth to at least one litter; the one that did not reproduce probably would have, but she killed a man who was cutting grass within her territory and was removed from the park before she had a chance to breed. As of April 1986 four females have lived over 11 years; each produced 4 to 5 litters (Table 1). The oldest, F115, who is 14.5+ years old, is raising her 5th litter.

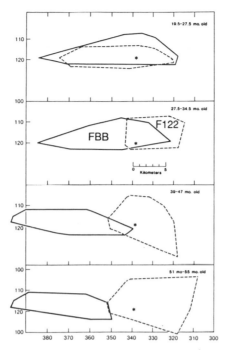

Fig. 4.   Tigress F122's gradual establishment of a territory
adjacent to her mother, FBB.   The latter shifted her territory to
the west.

What Factors Affect Territory Size?

Female territories are, on average, 38% the size of male
territories.   Whereas male territories ranged from 19 to 151 sq.
km (x = 54.4, SD = 35.8), females territories ranged from 10 to
51 sq. km (x = 20.7, SD = 9.2).

To examine the role of prey abundance in determining
territory size, we weighted each habitat type based on a measure
of relative prey abundance (Smith 1984).   Riverine habitat (both
forest and grasslands) had the highest prey density and was
weighted 1.0; other habitat types were given weightings based on
the percentage of prey abundance relative to riverine habitat.
Open sal forest (Shorea robusta) was given the value of 0.48 and
upland sal was weighted as 0.34.   Based on these factors we
recalculated territory size to obtain a measure of relative
territory size based on prey abundance.   The question posed was
whether these weights produced an unusually constant set of
territory ranges.   The coefficient of territory sizes was
compared to a sample of 5,000 coefficients of variation obtained
from a set of randomly chosen prey weighting factors.   The prey
abundance weightings gave a coefficient of variation less than
all but 3% of the samples.   This analysis indicated that prey
abundance was a significant factor (p < 0.03) in determining
territory size.

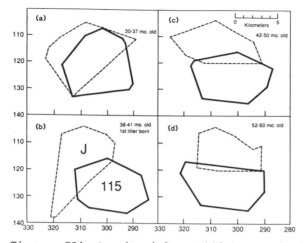

Fig. 5. Tigress FJ's territorial establishment within her natal area and the shift of her mother, F115, from the northern part of her range.

Another factor that influences territory size is the pattern of territory turnover. When a female dies, a young animal often establishes a range in the vacant area. We recorded three cases where the incoming female acquired esentially the same range as that occupied by the former female. Territory expansion was restricted by the presence of adjacent females and no territory holder expanded its range to include a vacancy.

What is the Degree of Relatedness of Adjacent Females?

While we monitored a group of resident, breeding females between 1975 and 1985, the pattern of daughters settling next to their mothers emerged (Fig. 7). The first daughter we observed to establish a range adjacent to her mother was F103 who settled next to F101 in 1975. She mated with her father (M105) and produced a daughter, F111, who was 75% related to M105. By 1980 three more daughters had settled in areas adjacent to their

Table 1. Minimum longevity and reproductive history of six female tigers.

Female	Minimum Age (Yrs)	No. Litters	Remarks
101	12	4	died
Bangi	13.5	4	removed
115	15	5	alive
111	9.5	3	alive
109	13	4	alive
103	5.5	3	died

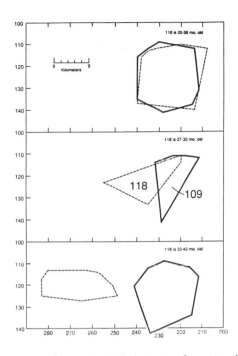

Fig. 6. Tigress F118's establishment of a territory adjacent to her mother, F109. The mother initially shifted eastward, but subsequently the daughter shifted westward entirely outside of her natal area.

mothers. F111 dispersed 33 km west of her natal area and settled to the north of FB who also produced two litters fathered by M105. A daughter from the first of FB's litters (F122) settled to the east of her mother and established a territory bordering F111, her half-sister. A daughter from the second litter (F3T) settled to the west of her mother and F3T's territory partially abuts that of F111 (Fig. 7). These females were 0.375 related to F111. As of spring 1986 we have documented 13 pairs of neighbors who are related to each other (Table 2) and the average degree of relatedness of neighbors was 0.35.

DISCUSSION

Our data on female spacing behavior meet all three of Brown and Orian's (1970) criteria for territoriality; there was very little overlap between adjacent ranges, animals marked their ranges, and females interacted agonistically. It is difficult to weigh the relative importance of aggression and marking in establishing and maintaining territorial boundaries. The high rate of scent marking when territories are first established (Smith 1984) and the infrequency of fights, indicated by the lack of wounds on immobilized females, suggest that marking may play a

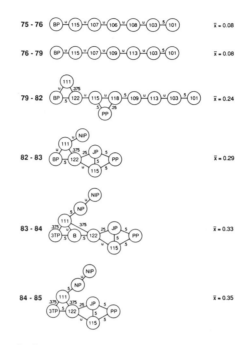

Fig. 7. The minimum mean degree of relatedness of adjacent female tigers. When the degree of relatedness is unknown (U) it is calculated as zero.

significant role in maintaining territorial boundaries. Davies (1980) argues that once a border between neighbors is established there are disadvantages to invading a neighbor's territory (e.g. foraging in an unknown area). Of the aggressive interactions we recorded, none were between established territorial neighbors.

We found a wide range in territory sizes among females and believe this variation reflects habitat quality. When we recalculated territory size based on a prey abundance weighting the coefficient of variation was greatly reduced. A similar inverse relationship between territory size and prey abundance has been reported in a number of other species (e.g. Stenger 1958, Smith 1968, Gill and Wolf 1975, Pyke 1979). Because there is also increased intrusion pressure in areas where prey abundance is high, territory size may not be based on assessment of food abundance alone (Myers 1979). If the availability of prey is a major factor affecting territory size, then territoriality may be a limiting factor determining the density of breeding females. Brown and Orians (1970) suggest that at the highest density (level 3) territorial behavior by the residents prevents non-territorial females from settling. In Royal Chitwan National Park it appears that there are young females looking for territories; on three occasions non-territorial females moved into areas previously occupied by a resident female within a month of the resident's disappearance. For a period of eight months one of these dispersing females (F111) shifted constantly

Table 2.   From 1976-85, 13 pairs of related neighbors were documented.   The degree of relatedness of all neighbors is not known. Projections for several generations of daughters settling adjacent to mothers give a degree of relatedness higher than observed to date.

Mother/Daughter	Sister	Half-Sisters (by Father)
101 - 103	FPP - FJP	111 - 122
109 - 118	122 - F3T	111 - F3T
115 - FPP		118 - FPP
FPP - FJP		122 - FJP
115 - FJP		
FBP - 122		
111 - FNP		

throughout a zone of prime habitat on the northern border of the Park.   She finally settled where the first vacancy became available.

Among polygnous mammals, observations of male dispersal and female philopatry are widespread (Greenwood 1980). However, the degree of relatedness of neighboring females that results from philopatry is not well documented.   Rogers (1976) reported a pattern of daughters settling next to their mothers in black bears (Ursus americanus) similar to what we found in tigers. Yeaton (1972) suggests that mothers may assist daughters in acquiring adjacent territories. The pattern of female philopatry we found in tigers may be the cruicial first step in the evolution of altuistic traits by kin selection (Hamilton 1964, 1972, Sherman 1977).   However, it is also true that the strongest competition is with one's close neighbors, which may be kin. Tigresses allowed daughters to establish territories adjacent to them, but as the aquisition process progressed there appeared to be a gradual increase in the level of aggression between mother and daughter, as was evident in the gradual reduction in territorial overlap between their ranges.   Furthermore, several aggressive interaction were recorded between mothers and daughters, as described above.

We suggest that the function of female territorial behavior is to ensure exclusive access to food, cover, and other resources needed by a female for the survival of herself and her offspring. Of the 14 females that became territorial all but one successfully reproduced; the one that did not produce cubs was removed before she had a chance to breed.   There were no non-territorial breeders.

The genetic implications of the high average degree of relatedness of neighboring females are unclear.   The movement of male genes through the population is restricted because females that males mate with are likely to be related.   Thus, male genes will not spread through the population as rapidly as they would if males mated with a random sample of females from the population.   The genetic structure of the Chitwan population therefore reduces the effective population size and increases the potential for inbreeding depression.

ACKNOWLEDGEMENTS

We thank His Majesty's Government of Nepal for permission to carry out this research and members of the Department of National Parks and Wildlife Conservation for their cooperation. The success of the research depended on the talented field staff of the Tiger Ecology Project (Mahendra Wildlife Trust) and Tiger Tops. F. Cuthbert and C. Wemmer provided critical comments on earlier versions of this manuscript. The Financial support for the project was provided by the Smithsonian Institution and World Wildlife Fund - U.S. Appeal.

REFERENCES

Brown, J.L. 1969. Territorial behavior and population regulation in birds. Wilson Bull. 81:293-329.
Brown, J.L. and G.H. Orians. 1970. Spacing patterns in mobile animals. Ann. Rev. Ecol. Sys. 1:239-62.
Davies, N.B. 1980. The economics of territorial behavior in birds. Ardea 68:63-74.
Gill, F.B. and L.L. Wolf. 1975. Economics of feeding territoriality in the golden-winged sunbird. Ecology 56:333-45.
Greenwood, P.J. 1980. Mating systems, philopatry and dispersal in birds and mammals. Anim. Behav. 28:1140-62.
Hamilton, W.D. 1964. The genetical evolution of social behavior. I,II. J. Theor. Biol. 7:1-52.
Hamilton, W.D. 1972. Altruism and related phenomena, mainly in social insects. Ann. Rev. Ecol. Syst. 3:192-232.
McDougal, C.W. 1977. The Face of the Tiger. London: Rivington Books and Andre Deutsch.
Mech. L.D. 1983. Handbook of Animal Radio-tracking. Minneapolis: Univ. Minn. Pr.
Mishra, H.R. 1982. The ecology and behavior of chital (Axis axis) in the Royal Chitwan National Park, Nepal. Ph.D. diss., Univ. of Edinburgh, Scotland.
Mohr, C.O. 1947. Major fluctuations of some Illinois mammal populations. Trans. Ill. Acad. Sci. 40:197-204.
Myers, J.P. 1979. Territory in wintering sanderlings: the effect of prey abundance and intruder density. Auk 96:551-61.
Pyke, G.H. 1979. Are animals efficient harvesters? Anim. Behav. 26:241-50.
Rogers, L.L. 1976. Movement patterns and social organization of black bears in Minnesota. Ph.D. diss., Univ. of Minnesota, Minneapolis.
Seidensticker, J.C. 1976. On the ecological separation between tigers and leopards. Biotropica 8:225-34.
Sherman, P.W. 1977. Nepotism and the evolution of alarm calls. Science 197:1246-53.
Shields, W.M. 1982. Philopatry, Inbreeding, and the Evolution of sex. Albany: State Univ. N.Y. Pr.
Smith, J.L.D. 1984. Dispersal, communication, and conservation strategies for the tiger (Panthera tigris) in Royal Chitwan

National Park, Nepal. Ph.D. diss., Univ. of Minnesota, St. Paul.

Smith, J.L.D., M.E. Sunquist, K.M. Tamang and P.B. Rai. 1983. A technique for capturing and immobilizing tigers. J. Wildl. Man. 47:255-59.

Smith, C.C. 1968. The adaptive nature of social organization in the genus of tree squirrels Tamiasciurus. Ecol. Mon. 38:31-63.

Stenger, J. 1958. Food habits and available food of ovenbirds in relation to territory size. Auk 75:335-46.

Sunquist, M.E. 1981. The Social Organization of Tigers (Panthera tigris) in Royal Chitawan National Park, Nepal. Smithson. Contrib. Zool. No. 336.

Tamang, K.M. 1982. Population characteristics of the tiger and its prey. Ph.D. diss., Michigan State Univ., East Lansing.

Wasser, P.M. 1985. Does competition drive dispersal? Ecology 66:1170-75.

Yeaton, R.I. 1972. Social behavior and social organization in Richardson's ground squirrel (Spermophilus richardsonii) in Saskatchewan. J. Mam. 53:139-47.

12

Project Tiger: The Reserves, the Tigers and Their Future

H.S. Panwar

INTRODUCTION

India's experience with Project Tiger is now well over a decade old. All along it has been a great process of learning-learning about the tiger's habits and habitats and about the problems of field conservation vis-a-vis a major predator like the tiger requiring large viable home ranges for long-term survival. Many of these problems arise from the inescapable and unrelenting pressures of demography and development. Despite the odds, the project has achieved some reckonable measure of success in not only improving the chances of the tiger's survival, but also in saving some of the prime and unique ecosystems. The tiger reserves have fostered the overall biological diversity and helped a host of other endangered species. The experience has strongly vindicated the comprehensive ecosystem approach of the project. Last but not least, it has provided an insight into how the habitat requirements of the tiger can be compatibly related to the use of the forests by the communities inhabiting the surrounds of the tiger reserves.

BACKGROUND

The tiger is distributed across the bulk of the Indian wilderness, having adapted itself to mountainous, plain, cold, hot, rainy, dry, forested and grassland environments. Less than a century ago, it thrived over extensive belts of wilderness. Paradoxical as it may appear, the interspersion of human habitation through these forested tracts enhanced the habitat productivity for the deer and the antelope, and hence also for the tiger. This was because, traditionally, the people maintained large areas around villages as pastures and open forests in order to meet their fuelwood and pasture needs. Rotating fallow marginal lands with the long-cycle 'slash-and-burn' cultivation practices further enriched these habitats. In a low people:forest ratio that held well at that time, the rural ecosystems complemented the quality and extent of tiger habitats (Panwar 1985).

Demographic pressures, however, started to strain these equations in the terminal decades of the last century. Relatively remote and sparsely populated valleys and the flood plains of the Himalayan _terai_ came under colonization. With the relaxation of the rigid controls characteristic of the colonial-feudal polity, these diversions acquired an accelerated pace in the post-independence period. The private forests rapidly fell to meet the hunger for land or for their timber resources. The village forests and pastures were likewise brought under the plough. Much of these conversions led to accelerated soil erosion for lack of soil conservation measures. Lack of resources for appropriate land development, the urge to provide early relief to the landless as well as to grow more food for the burgeoning population, were all factors which combined to generate a permissive atmosphere that allowed expediency to dictate land use.

The soaring demands of urban population and industry steadily caused more and more remote forests to be opened up and tapped for wood. Rapidly expanding economic activity took a further toll through irrigation, power, industrial and other projects. The resultant abuse and overgrazing depleted both the fodder productivity as well as the regenerative abilities of valuable timber trees to the disadvantage of people and wildlife alike.

Constrained already by a rapidly shrinking and depleting habitat, the wildlife, including the tiger, had to concurrently face the pressures of legal and illegal hunting. Inadequate legislative deterrents and their even weaker enforcement failed to contend with this onslaught. The status of wildlife rapidly plummeted to its lowest-ever ebb as the sixties yielded to the seventies. This decline became abundantly clear in the results of the first country-wide census of the tiger in 1972. As opposed to the turn-of-the-century estimates of the Indian tiger population as being between 20,000 and 40,000, the 1972 census estimated the population to be just a little over 1,800 (Panwar 1983).

THE PROJECT

Project Tiger was conceptualized out of genuine national and international concern for the tiger and the Indian wilderness. A country-wide ban on tiger hunting was effected in 1970 followed by the promulgation of potent legislation for wildlife protection and the setting up of a special task force under the Indian Board for Wildlife in 1972 for formulating the project.

The task force was fully convinced of the connection between the tiger and the wilderness, and therefore formulated the project along an ecosystem approach, giving it the following primary objectives: 1) to ensure the maintenance of a viable population of tigers in India for scientific, economic, aesthetic, cultural and ecological values, and; 2) to preserve for all time, areas of biological importance as a national heritage for the benefit, education and enjoyment of the people.

The task force had visualized that large conservation areas were needed for the maintenance of genetically viable populations of tigers in the wild. The tiger reserves were, therefore, visualized as the breeding nuclei from which surplus animals would emigrate to adjacent forests. However, conforming to the ecosystem approach, the task force cautioned that, "Under no circumstances, however, shall these operations involve holding the tiger population at artificial high levels by such means as large scale, uncontrolled modification of the habitat, artificial feeding of the tiger or its prey, or the introduction of exotic species which can only cause imbalance in the natural ecosystems and can have disastrous results for the habitat and its dependent fauna" (IBW 1972).

The selection of areas for tiger reserves represented, as close as possible with the available resources, the diversity of ecosystems across the tiger's distribution in the country. Nine reserves were set up and the project was launched in 1973. A core area was defined in each reserve to be managed as a 'nature reserve,' with the remaining area to serve as a buffer zone between the surrounding multi-use areas. Management plans were prepared for every reserve along these objectives and approaches. The project provided 100% central funding for the additive elements of the program, and the states were required to continue their current inputs in terms of staff, infrastructure and funds. The states responded adequately by suspending forestry operations and stopping livestock from grazing in the core areas, and also agreed to the relocation of human habitation from the core areas - all of these at considerable financial cost and political discomfiture.

It is indeed remarkable that, in a situation of acute paucity of development resources, funds and commitment were mustered to support the intensive program of habitat protection and rehabilitation under the project. World Wildlife Fund (WWF)-International pledged US $1 million and this sum was greatly expanded. After 14 years of the project, the WWF grant still remains to be fully utilized, and the pooled resources of the center and the states contribute about US $4 million annually. In addition, the states are sustaining a loss of US $14 million annually by having given up forestry operations in the tiger reserves.

With added staff and resources the protection of wildlife and habitat improved. Suspension of stock grazing and forestry operations then promoted early habitat recovery. Village relocation was, however, a delicate operation. It was explained to the people that continued long-term habitation within the core areas was bound to become unsustainable because of unavoidable crop- and cattle-raiding by wild animals. But there was resistance to shifting, mainly because villagers were unsure if adequate facilities and land would be provided for their rehabilitation. However, the provision of alternative facilities, such as adequately prepared agricultural land, new houses and other amenities well before the movement, coupled with perceptive persuasion, gradually broke through the barriers of doubt and resistance. Once the operation was underway, more people volunteered and in some reserves, even more than the targeted number of villages were relocated.

Comprehensive reamelioration of habitat soon became evident. The recovery of the ground and field level vegetation and the revival of the water regime were indeed striking. Improved and enlarged habitats led to rapid increases in prey populations, followed by increases in the populations of predators, including the tiger. A number of other endangered species also benefited, some because of special measures under the project.

The coverage of the project was enlarged in the following years by increasing the areas of existing reserves and by the addition of new reserves. In 1973-74 the initial nine reserves covered a total area of 9,115 sq. km with a 'core' segment of 4,230 sq. km. The present tally of the reserves is 15, with a total area of 24,700 sq. km and a 'core' segment of 9,950 sq. km. There has been more than a three-fold increase in the tiger population of the reserves, the number in 1984 being over 1,100.

Steady enlargement and enhanced management of the protected areas network also contributed significantly. As a result, the country tiger population, as estimated by a country-wide census in 1984, has grown to about 4,000.

TIGER SOCIOLOGY

The revival process spanned all the components, biotic and abiotic, of the ecosystems in the tiger reserves. These changes in the environmental conditions orchestrated the growth and dispersal of prey and, consequently, changes in the spacing patterns of tigers, increases in reproductive success and enlargement of natal areas. The rapidly changing scenario presented exciting opportunities to study tiger sociology.

The author undertook one such study in Kanha from 1972 to 1978. During this period, the first enlargement of the core area took place in 1972, from 318 sq. km to 446 sq. km, by the addition of adjacent area to the south (Extension I). Extension II, in 1974, added 494 sq. km to the core area on the east side of Extension I and also provided a buffer of 1,005 sq. km. These enlargements and concurrent habitat recovery provided the needed habitat for adult tigers to disperse from their natal areas in the core of the original park. This evolving pattern of habitat utilization was studied by intensively monitoring the spatio-temporal use of the original park area and the two extensions by different individuals. Individuals were identified from tracings of their pugmarks, recorded initially on a glass plate (tiger tracer) and then transferred to a sheet of paper with details on location, date, etc. (Panwar 1979a). The seven-year study provided excellent evidence on the social organization and land tenure systems among tigers as related to habitat quality.

Tigers in the best habitat areas, those with high prey densities, tended to have well-defined territorial organization, with a preponderance of prime adults of both sexes. These were also the best natal areas. Until 1972, the only identifiable natal area was the heart of the original park area (Fig. 1). As a consequence of the enlargement and improvement of habitat, a new natal area developed in Extension I in 1974, and another in Extension II in 1978. The findings of this study were documented elsewhere (Panwar 1979b). Relevant details are as follows (Fig. 2 and 3):
   1) All tigers strive to occupy areas with high prey densities, and competition removes the sub-prime (pre- and past-prime) individuals from such areas. Interindividual competition

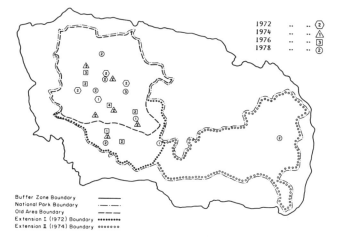

Fig. 1.   Locations of tigresses with cubs in Kanha National Park in 1972, 1974, 1976 and 1978.

also leads to spacing patterns and territorial organization among occupants;

2) Female needs for an assured food supply, in order to raise offspring, and the inclination of males to cohabit with them results in prime adults occupying high prey density areas in a mutually advantageous land tenure system. Dominant-male territories, which are conspecifically recognized and defended, encompass sub-territories of one or more females. Female sub-territories also tend to be well-defined, but may overlap or include transient movement corridors of neighbors;

3) In such a system, territorial males have exclusive conjugal access to sub-territorial females. An additional advantage to females is protection of their cubs from rival males;

4) Areas with high prey densities are also the prime natal areas. The sex ratio favors females, in a 2:1 or 3:1 ratio;

5) Pre- and past-prime adults occupy the peripheries of high prey density areas. These relatively lower prey density areas are the sites of somewhat more loosely established territories. Incidence of breeding is low and the sex ratio favors males;

6) Some old and very young tigers are not able to establish territories in peripheral areas and are forced to disperse into poorer habitats if contiguous forest is available;

7) In extensive forest belts, such dispersing animals, particularly the sub-adults, probably provide the basis of genetic exchange among distant population groups.

CONSERVATION STRATEGY

The social organization of tigers is important to overall conservation strategies in India. It can indicate how to set up a viable and efficacious network of protected areas appropriate to India's demographic and geographic characteristics. The

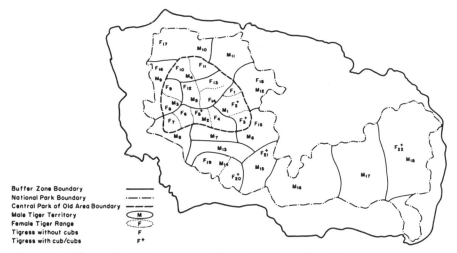

Fig. 2.   Tiger land tenures and population structure.

Indian wilderness is distributed along mountain chains; the Himalayas, the Aravalis, the Vindhyas, the Satpuras, the Eastern and Western Ghats and the Nilgiris. Characteristically and traditionally, human habitation is interspersed through these belts of wilderness. Demographic pressures, as described earlier, have been responsible for the shrinkage and degradation of the wilderness. The sustenance of rural communities living in these areas is mainly provided by marginal-land farming and cattle-raising. However, the lack of soil conservation safeguards and unregulated grazing by burgeoning populations of scrub cattle have caused the productivity of both farmland and pastures to decline. This is the main cause of poverty in these tracts. Wastelands now dominate the landscape where once private and village forests and pastures existed.

On the other hand, in the vaster plains, wide valleys and plateaus, improved irrigation and infrastructural development have resulted in a breakthrough in the 'green revolution,' and the country is now producing a moderate food surplus. The trend is continuing in intensively farmed areas, and estimates indicate that the country's food production is likely to more than double by the year 2000. Even with the present moderate surplus, marginal-land farming has become uneconomical, but the people in the forested regions are compelled to farm on such land for want of employment alternatives.

Diversion of agricultural land to non-agricultural, even to forestry crops like Eucalyptus, are visible in prime agricultural tracts in the plains. Even considerably smaller input and care result in good monetary returns because of the unreasonably high price of wood in these areas. While such diversions are motivated by quick and easy profits, appropriate changes in crop patterns hold the panacea for the communities toiling unreward- ingly in the marginal lands. Yet, such needed diversions of marginal farmland to non-agricultural cash crops, including fuel and fodder plant species, can be brought about only if the livelihood of the people can be ensured during the length of time

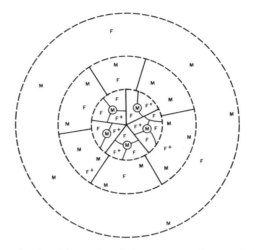

Fig. 3.   Approximate tiger land tenures and population structure in Kanha National Park in 1978.

such crops take to grow to marketable size.   The massive program of wasteland development proposed by the government, aimed at rehabilitating five ha per annum, is another factor which will influence the overall land use, and accordingly, the conservation of wilderness.   Other programs of rural development seek to upgrade the productivity of marginal lands.   In essence, the current scene in India is one of major changes in land use.   This holds crucial portents for the conservation of wilderness.

Experience with Project Tiger has amply demonstrated the efficacy of the core-buffer constitution of wildlife reserves, and has also focused attention on the need to ensure compatibility between such reserves and their multi-use surrounds.   It is in these multi-use areas that the land is to undergo massive restorative changes.   A little care at this stage can enhance the ecological capability of such restored areas and enable them to discharge the role of 'gene corridors' and supplemental habitats, so crucial to the viability and efficacy of the national network of protected areas.

The number and extent of protected areas have greatly increased since the 1970's.   From five national parks and 126 sanctuaries covering 24,000 sq. km in 1975, these areas have grown to 54 national parks and 248 sanctuaries spanning 100,000 sq. km as of March 1986.   The Wildlife Institute of India has evolved a practical biogeographic classification and is at present testing the adequacy of the existing network, and has suggested the enlargement of existing, and the establishment of new protected areas.

It is in this exciting setting that the lessons learned in Project Tiger become relevant:
    1) Core-buffer units with compatible multi-use surrounds, are the only way wildlife reserves can be sustainably managed in India;
    2) A viable and biogeographically adequate network of such

protected areas, managed on an ecosystem approach, will be fully capable of preserving the rich genetic resources of the country's wildlife.
   3) The multi-use surrounds should essentially be developed to meet the needs of the residual rural people.

   If farmlands provide the main sustenance of the rural people in these tracts (not necessarily through cereal crops), and the wastelands are rehabilitated to meet their fuel and fodder needs, the pressures on the protected areas and other forests can be mitigated. Not only will this enhance the efficacy of protected areas, it will also revive the corridor function of other forests and restored wastelands.

   It is no mere coincidence that such a land use stategy is in full accord with the sociological and biological needs of the tiger. The tiger would not have thrived in India if it were otherwise. Doubtless, the tiger can no longer have the extensive habitat it once had, but the people have recognized the value of wilderness. Whatever is to be set apart as wilderness should be managed with this strategy, with harmony ensured through ecologically sound management of multi-use areas. This is the only way conservation of wilderness and genetic reources, essential to long-term human welfare at large, and the well-being of the rural communities in the forested regions, can be ensured.

REFERENCES

Indian Board for Wildlife. 1972. Project Tiger: India - 1972. A planning proposal for preservation of tiger (Panthera t. tigris Linn.) in India. New Dehli: Govt. India Publ.
McDougal, C. 1977. The Face of the Tiger. London: Rivington Books.
Panwar, H.S. 1979a. A note on tiger census technique based on pugmark tracings. Indian Forester, Special Issue (Feb.).
Panwar, H.S. 1979b. Population dynamics and land tenures of tiger in Kanha National Park. Indian Forester, Special Issue (Feb.).
Panwar, H.S. 1983. What to do when you have succeeded: Project Tiger ten years later. Ambio (Jan.-Feb.)
Panwar, H.S. 1985. Protected areas and people: An Indian approach. Proc. 25th Working Session Commission Natl. Parks, Protected Areas. IUCN. Unpublished report.
Sankhala, K. 1978. Tiger: The Story of the Indian Tiger. London: Collins.
Smith, J.L.D. 1984. Dispersal, communication and conservation strategies for the tiger (Panthera tigris) in Royal Chitwan National Park, Nepal. Ph.D. diss. Univ. Minnesota, Minneapolis, MN.
Sunquist, M.E. 1981. The social organization of tigers (Panthera tigris) in Royal Chitwan National Park, Nepal. Smithson. Contrib. Zool. No. 336.

13

Tigers in India:
A Critical Review of Field Censuses

K. Ullas Karanth

INTRODUCTION

The long-term prospect of the tiger surviving as a free ranging species is to a large measure dependent on its chances of doing so in India. There are several reasons for this. Indian tigers constitute about 45% of the estimated total wild population (Mountfort 1981). They are distributed over a large area ranging through a diversity of habitats. Among these, the Terai-Babhar ecotone at the Himalayan foothills and the Peninsular Indian deciduous forests, both with their rich assemblages of ungulate prey, are optimal habitats. Tigers attain greater densities there than anywhere else (McDougal 1977). Also, the scale and scope of the Indian tiger conservation strategy seems to be unmatched elsewhere.

This strategy has two major thrusts. One is the spectacular "Project Tiger" which has created 15 specially designated tiger reserves covering a total area of 23,437 sq. km (Fig. 1). The other less publicized, but perhaps even more effective initiative has been the legislation and implementation of the Wildlife Protection Act of 1972 which has resulted in the creation of many more nature reserves covering an even larger area of tiger habitat. Though the inspiration for these strategies came from International and Indian conservationists, they were essentially formulated and implemented by India's foresters. While doing so they had to overcome many constraints, including public apathy or hostility, paucity of investments, difficult working conditions and a lack of tradition in wildlife management. In spite of these hurdles, Indian foresters have been effective in controlling poaching of tigers and their prey, as well as protecting their habitats from the ravages of excessive fires, stock grazing and wood removal. Sankhala (1978), Mountfort (1981) and Panwar (1984) have described these efforts, which have succeeded far beyond what was expected in a resource-poor, overpopulated developing country like India. Detailed, if somewhat anecdotal accounts by these and several others (e.g., Rathore et al. 1983, Singh 1984) document increased instances of tiger sightings, reproductive success, intraspecific aggression, cattle killing and even manslaughter - all attesting to an increase in tiger densities at many localities.

Fig. 1.   Tiger reserves created as part of India's Project Tiger.

Attempts have been made by Indian wildlife managers to quantity this increase in the tiger population using periodic total counts or censuses.   The censuses show phenomenal increase in tiger numbers over the years in almost all parts of the country.   For example, the official tiger census reports claim that the Indian tiger population has more than doubled: from 1,827 in 1972 to 4,005 in 1984.

TIGER CENSUS IN INDIA

These total counts of tigers have relied solely on the pugmark method which essentially depends on the assumption that each tiger can be identified individually from the tracings or plaster casts of its tracks.   Tracking tigers for hunting them down was a tradition among the Indian Shikaris which often flourished under royal patronage (Sankhala 1978).   Later, Champion (1929) published a study of tiger tracks and their relation to the animal's locomotion.   Many hunters like Corbett (1944) claimed that it is possible to determine the sex, age and even physical condition of a tiger from its tracks.   Often tigers with very distinct pugmarks could also be individually recognized (McDougal 1977).   However, it was S.R. Choudhury who argued that every tiger could be individually identified from its pugmarks. The present system of censusing wild tigers based on pugmarks was

invented by him and has been refined and described by Panwar
(1979a).

The method involves obtaining, on a glass sheet, accurate
tracings of the rear paw imprints of tigers from tracks in the
field. Substrate conditions and probable date when the tiger
made the tracks, stride length of the tiger as well as any
noticeable abnormalities in the tracks are important. The best
season for obtaining tracings should be selected and the field
work carried on intensively for a week or so. It is claimed
that, based on this analysis, experienced persons can segregate
the tracings and identify each individual tiger with a high
degree of reliability (Panwar 1979a). In the past, S.D. Ripley
(as quoted by Sankhala 1978) and Singh (1984) have questioned the
reliability of this method. However, since detailed reasons were
not advanced in support of their skepticism, this census method
is still used exclusively.

RECENT STUDIES OF SOUTH ASIAN CARNIVORES

When the tiger censuses were launched in the early 1970's,
apart from Schaller's (1967) study, scientifically very little
was known about the ecology of large carnivores in south Asia.
Now the situation is qualitatively different. Long-term
ecological studies in Sri Lanka (Muckenhirn 1973), Nepal
(McDougal 1977, Sunquist 1981, Mishra 1982, Tamang 1982, Smith
1984) and India (Rathore et. al. 1983, Joslin 1973, Berwick 1976,
Panwar 1979, Johnsingh 1983, Singh 1984) have considerably
enhanced our knowledge of the tiger, its co-predators and its
prey species.

With this somewhat clearer understanding of the ecology of
large carnivores, especially the tiger, an attempt can be made to
examine the ecological and management implications of Indian
tiger censuses. I believe such a review is urgently needed for
three reasons:
  1) The population boom among India's wild tigers reported
from field censuses seems to have generated some public
complacency that India has already succeeded in ensuring their
long-term, viable conservation;
  2) In the context of increasing man-tiger conflicts around
a few reserves, some public questioning of the need to protect
more habitats when so many tigers are around has begun;
  3) A preliminary analysis of the 1982 census data from
Bandipur tiger reserve, an area I have been familiar with for
over a decade, has shown some disquieting discrepancies (Karanth
1986).

The objective of this review is not to minimize the recent
outstanding efforts at tiger conservation by Indian wildlife
managers. It is only to critically evaluate the census data and
the methodology employed, so that more objective and practicable
systems of monitoring tiger populations can evolve.

OBJECTIVE OF THE REVIEW

Based on current knowledge of the ecology and behavior of tigers and other large carnivores, this analysis evaluates the Indian tiger census data/methods and tries to answer the following questions:
1) Are the numbers of the tiger and its large co-predators (leopard, Panthera pardus and dhole, Cuon alpinus) as reported in the census figures, sustainable on the basis of the published data on populations of their major prey species?
2) Are the numbers of tigers reported from diverse habitats consistent with the recent findings about the ecology and social organization of the tiger?
3) Are the reported increases of wild tiger populations in consonance with observations of population dynamics of tigers at other localities?
4) What are the weaknesses of the present method of monitoring tiger populations, and what improvements can be made in them?

REVIEW OF CENSUS DATA AND METHODS

Here the official population estimates of different species of large mammalian prey (ungulates and monkeys) have been collected from several nature reserves of India. While there can be minor discrepancies in these figures collected from different sources (both published and unpublished) these are unlikely to be of significance.

Analysis of Relative Numbers of Predators and Prey

Tiger, leopard and dhole do not normally kill very large prey like elephant (Elephas maximus) and rhinoceros (Rhinoceros unicornis), and these species have not been taken into account. Conversely, predators like the golden jackal (Canis aureus), hyena (Hyaena hyaena), rock python (Python molurus) and marsh crocodile (Crocodilus palustris) sometimes kill some of the prey species considered in biomass estimates. Although smaller prey species like hare (Lepus nigricollis), porcupine (Hystrix indica) and peafowl (Pavo cristatus) are occasionally killed by the large predators, their overall contribution to the diet is generally low (Schaller 1967, McDougal 1977, Sunquist 1981). Livestock are an important prey item in many parts of the country. However, stock-grazing has been severely curbed in tiger reserves in recent years (Panwar 1984). In the absence of any scat studies it is not possible to estimate the proportion of livestock and smaller prey in the diet of the predators considered in the analysis.

From the studies of Schaller (1967), Sunquist (1981) and Johnsingh (1983), I have derived the average requirement of live weight of prey for the three predators as: tiger - 3000 kg; leopard - 1000 kg; and dhole - 490 kg per year respectively. Several studies of large carnivores (Schaller 1967, 1972, Muckenhirn 1973, Sunquist 1981, Tamang 1982), show that such predators usually do not exercise a decisive limiting influence on the prey populations. Estimated rates of prey biomass removed

annually by predators seldom exceed 10% of the total standing
biomass. In this analysis this rate has been assumed since no
noticeable decline in prey numbers due to predation has been
reported from the ten nature reserves considered. Tables 1 and 2
present the analysis of predator-prey balance using the census
estimates of 1982 for Kanha and Corbett, of 1983 for Sariska,
Bandhavgarh and Periyar and of 1984 for the other reserves. Unit
weights used for predators are: tiger - 113 kg; leopard - 34 kg
and dhole - 13 kg. For prey species these values are within the
range used by different sources cited in Eisenberg and
Seidensticker (1976).

This analysis shows that in all the reserves except Sariska
the biomass of prey available for cropping falls far short of the
prey that would be required if the reported number of predators
were present (Table 1). The estimated discrepancy between prey
biomass required and that available ranges between 46% and 95%!
While it could be argued that small prey and livestock cover a
part of the above deficit, the total shortfalls are unlikely to
be accounted for by this explanation. These can only be
attributed to overestimates of predator populations or under-
estimates of prey populations or a combination of the two.

The predator-prey biomass ratios reported from some other
studies are: Serengeti 1:250, Ngoronoro 1:100 (both in Africa,
Schaller 1972), and from a 20 sq. km study area in Bandipur,
1:124 (Johnsingh 1983). When examined in this light, estimates
of prey and predator populations are likely to be off the mark to
a greater degree in Melghat, Bandhavgarh, Nagarjunasagar,
Bandipur and Periyar.

Estimates available for crude biomass of ungulate prey in
some protected areas of South Asia are: 766 kg/sq. km and 886
kg/sq. km in Wilpattu and Galoya of Sri Lanka and 1,708 kg/sq. km
for Schaller's study area in Kanha (Eisenberg and Seidensticker
1976). Johnsingh (1983) estimated an average prey biomass of
3,320 kg/sq. km in his study area in Bandipur even though the
biomass for the entire reserve is far lower. Estimate of the
biomass of prey in Chitwan, based on line transects, was 2,798
kg/sq. km (Tamang 1982).

In Table 2, I have computed the wild prey biomass that would
have to be present to support the reported predator populations
in these reserves. Within these reserves there are localities
where such high biomass can be present. But it is doubtful
whether the reserve as a whole can support crude prey biomass of
this order. Therefore, it is likely that the imbalance between
prey requirements and availability highlighted in this analysis
for all the reserves except Sariska are likely to be more a
consequence of overestimation of predator population than of
under counting prey numbers.

The methods of counting prey seem to vary widely and do not
seem reliable, at least in some reserves (Karanth 1985).
Therefore, the element of underestimation of prey is also likely
to be present, particularly in Periyar and Bandipur.

Densities and Biomass of Tigers

The overestimation of predator numbers is evident when the
crude densities of tigers reported from different reserves are

Table 1. Prey availability and requirements of predators.

Year	Reserve	Area (km²)	Predator Numbers Tiger	Predator Numbers Leopard	Predator Numbers Dhole	Prey Requirement (kg/Yr)	Prey Biomass (kg)	Prey Availability (kg/Yr)	Prey Availability/ Requirement
1983	Sariska	800	25	25	0	100,000	2,928,000	292,800	293%
1984	Dudwa	613	74	7	a	229,000	1,246,842	124,684	54%
1982	Kanha	940	89	54	a	321,000	1,690,120	169,012	53%
1984	Ranthambore	414	38	45	0	159,000	615,426	64,542	41%
1982	Corbett	521	51	47	a	320,000	1,230,081	123,008	48%
1984	Melghat	1,572	80	50	189	382,610	1,095,684	109,568	29%
1983	Bandhavgarh	105	15	8	6	55,940	156,030	15,603	28%
1984	Nagarjunasagar	3,000	65	200	200	493,940	1,200,000	120,000	24%
1984	Bandipur	690	53	63	130	285,700	412,620	41,262	14%
1983	Periyar	777	41	15	500	383,000	174,840	17,478	5%

a Dhole occurs in Dudwa, Kanha and Corbett Reserves but census estimates were not available. Periyar reported "50 packs" - a pack size of 10 has been assumed here.

Table 2. Biomass of prey and predators.

Year	Reserve	Prey (kg/km²)	Predator Biomass (kg/km) Tiger	Predator Biomass (kg/km) Leopard	Predator Biomass (kg/km) Dhole	Predator Biomass (kg/km) Total	Predator:Prey Biomass Ratio	Prey Biomass Required (kg/km²)
1983	Sariska	3,660	3.53	1.06	0	4.59	1:797	1,250
1984	Dudwa	2,034	13.64	0.39	-	14.03	1:145	3,735
1982	Kanha	1,793	10.70	1.95	-	12.65	1:142	3,415
1984	Ranthambore	1,559	10.37	3.70	0	14.07	1:111	3,841
1982	Corbett	2,361	19.74	3.07	-	22.81	1:103	6,142
1984	Melghat	697	5.75	1.08	1.56	8.39	1:83	2,434
1983	Bandhavgarh	1,486	16.14	2.59	0.74	19.47	1:76	5,327
1984	Nagarjunasagar	400	2.45	2.26	0.87	5.58	1:72	1,643
1984	Bandipur	598	8.68	3.10	2.45	14.23	1:42	4,141
1983	Periyar	225	5.96	0.66	8.37	14.99	1:15	4,929

examined. The relative densities of tigers in different habitats
depend on the following factors:
    1)   The availability and vulnerability of suitable prey
species, which again depend on several environmental parameters.
Generally, tiger densities are likely to be higher in habitats
with greater densities of ungulate prey (Sunquist 1981, Tamang
1982);
    2)   Relative abundance of competing predators like leopard
and dhole. Although several ecological and behavioral mechanisms
result in some ecological separation between predatory species
(Seidensticker 1976, Johnsingh 1983), there is some overlap also.
Therefore, areas with lower numbers of competing predators,
particularly dhole, are likely to support higher densities of
tigers.

    For example, the Terai and Peninsular Indian deciduous
forests are likely to support higher densities of tigers than
evergreen forests or mangrove swamps. Thus, open deciduous
forests, which are good habitats for a coursing predator like
dhole, are likely to be poorer habitats for the tiger when
compared to areas where dhole is not a major predator. However,
the density (expressed in terms of sq. km/tiger) and biomass of
tigers calculated for 18 reserves (Table 3) are not in conformity
with these ecological observations in the case of reserves like
Sundarbans, Periyar and Melghat.

Table 3.   Densities and biomass of tigers in 18 reserves (1984).

Reserve	Area (km^2)	No. Tigers	km^2/Tiger	Biomass (kg/km^2)
Corbett[a]	521	90	5.79	19.52
Bandhavgarh	105	15	7.00	16.14
Dudwa	613	74	8.28	13.65
Sundarbans[a]	2,585	264	9.79	11.54
Kanha[a]	940	89	10.56	10.70
Ranthambore[a]	414	38	10.89	10.38
Bandipur[a]	690	53	13.02	8.68
Palmau[a]	930	62	15.00	7.53
Periyar[a]	777	44	17.66	6.40
Melghat[a]	1,572	80	19.65	5.75
Manas[a]	2,840	123	23.09	4.89
Nagarahole	640	27	23.70	4.77
Sariska[a]	800	26	30.76	3.67
Simlipal[a]	2,750	71	38.73	2.91
Namdapha[a]	1,808	43	42.05	2.69
Nagarjunasagar	3,000	65	46.15	2.40
Buxa[a]	745	15	49.66	2.28
Indravathi[a]	2,084	38	54.84	2.06

[a] Project Tiger Reserves.

    Secondly, these census-based figures can be compared with
the density and biomass figures for tigers derived from studies
in Kanha by Schaller during 1964-65 and in Nepal by the
Smithsonian Tiger Ecology Project during 1974-1985. These study

areas included habitats with very high densities of tiger's prey
and low numbers of leopards and dholes, and can therefore be
considered near optimal habitats for the tiger. Many of the
tigers were individually recognized by their facial markings,
through intensive radio-tracking and year-round monitoring of
tracks (Schaller 1967, Sunquist 1981, Smith 1984, McDougal 1985).

    For comparison with the Indian census data, I have derived
the density and biomass figures from the above studies, using the
estimates of the average numbers of tigers (including immature
ones) in each study site (Table 4).

    Considering these figures it can be expected that really
superior tiger habitats like these can support biomass of 7 to 10
kg of tiger/sq. km, or reach densities of 11-17 sq. km per tiger.
When compared to these figures, the density/biomass figures
derived from the Indian census estimates (Table 3) seem to be too
high even for good tiger habitats like Kanha, Dudwa, Bandhavgarh
and Corbett. The estimates for Sundarbans, Bandipur, Palamau,
Periyar, and Melghat appear to be unreasonably high.
Superficially, the density estimates for other reserves do not
seem unreasonable, though a more detailed examination of prey
biomass estimates would be needed to verify this.

Table 4.   Densities and biomass of tigers derived from scientific
studies.

Reserve	Years	Area (km^2)	No. Tigers	$km^2/$ Tiger	Biomass (kg/km^2)	Source
Kanha	1964-65	318	22	14.45	7.82	Schaller 1967
Chitwan Park	1974-76	544	32	17.00	6.64	Sunquist 1981
Chitwan Park	1980	1,024	90	11.37	9.93	Smith 1984
Chitwan Region	1980	2,700	175	15.43	7.32	Smith 1984

Increase Rates of Tiger Numbers

    At what rates do populations of tigers increase in the wild?
Following his Kanha study, Schaller (1967) suggested that "tiger
population also appears to be self limited, with perhaps a social
spacing mechanism keeping numbers of animals in an area
relatively constant regardless of abundance of prey". The later
data from Chitwan seem to support this.

    Long-term tiger studies in Chitwan since 1974 have yielded
considerable data on the reproduction, mortality, dispersal and
population dynamics of tigers. They show that although tigers
are prolific breeders, mortality among predispersal offspring is
as high as 50% due to various causes (Sunquist 1981, McDougal
1985).

There are only a few "slots" for resident breeding animals in prime habitats. The young dispersers usually linger around as transients along with tigers past their prime, in poorer habitats on the peripheries, until they are able to evict a resident breeder. In this process a large number of transients perish due to conflict with human interests or with resident tigers. Consequently, the tiger population in Chitwan appears to be fairly stable with only minor changes in the number of breeding residents. In spite of an increase in prey availability, Sunquist (1981) did not observe any increase in the number of breeding females in his study area between 1974-77. McDougal (1985), who has monitored the tiger population closely in Chitwan for over a decade, observed an "impressive stability," noting that the number of breeding adults, which was nine in 1975, increased to only 14 by 1985. In Chitwan, as in Indian reserves, the tiger population was responding to vastly improved protection efforts after 1974.

Therefore, when environmental factors improve rapidly, populations of tigers tend to remain stable or increase very slowly. Moreover, because young tigers disperse considerable distances from their natal areas (Smith 1984), densities in prime habitats occupied by resident breeders do not increase.

Are the Indian tiger census data consistent with these findings? Based on census estimates (Tables 5 and 6), there are exponential increase rates of tiger populations for 11 original Project Tiger Reserves and 14 States of India.

Tiger populations in all the reserves, except Simlipal, have recorded net annual increase rates in excess of 6% (Table 5). Even relatively poor tiger habitats like Bandipur show increased rates of 14%, per annum for over 12 years in succession. Since the reserves have not been enlarged in size, these increase rates also imply increases in densities at the same high rates for over 12 years.

Table 5.   Tiger Population Increases in Project Tiger Reserves.

Reserve	Area (km^2)	No. Tigers 1972	No. Tigers 1984	Annual Increase (%)
Bandipur	690	10	53	13.9
Manas	2,840	31	123	11.5
Melghat	1,572	27	80	9.0
Palamau	930	22	62	8.6
Ranthambore	414	14	38	8.3
Kanha	940	43	109	7.8
Corbett	521	44	90	6.0
Sunderbans	2,585	135	264	5.6
Simlipal	2,750	77	71	Decrease
Sariska	800	--	26	---
Periyar	777	--	44	---

Table 6.   Tiger population increases in some states.

State	No. Tigers 1972	No. Tigers 1984	Annual Increase (%)
West Bengal	73	352	13.1
Andhra Pradesh	35	164	12.5
Meghalaya	32	125	11.4
Arunachal Pradesh	69	219	9.6
Tamil Nadu	33	97	9.0
U. P.	262	698	8.2
Assam	147	376	7.8
Karnataka	102	202	5.7
Maharashtra	160	301	5.3
M. P.	457	786	4.5
Bihar	85	138	4.0
Kerala	60	89	3.3
Orissa	142	202	3.1
Rajasthan	74	96	2.2
Nagaland	80	104	2.2
ALL INDIA	1,827	4,005	6.5

In Table 6 we see that from figures reported for larger geographic regions such as States, the increase rates are also very high, in some cases exceeding those of tiger reserves.

As we have already seen, populations of tigers are unlikely to increase at these high rates for over 12 years continuously. Moreover, tiger numbers are regulated by behavioral spacing mechanisms which are density dependent. Therefore, the continuous increases in absolute tiger densities almost everywhere are unexplainable except in terms of wrong population estimates.

The analysis of census data in terms of: 1) relative numbers of larger predators and their prey; 2) densities and biomass of tigers; and 3) long term population dynamics of tigers indicates that the total counts of tigers in India based on the pugmark method are likely to be overestimates.   Therefore, the census method itself merits a critical review.

A Critique of the Census Method

From my examination of tiger census operations in Karnataka State, particularly in Bandipur Tiger Reserve, several weaknesses in the present method of monitoring tiger populations are perceived:

The method assumes that each individual tiger in a given population can be identified from its pugmark tracings by the average reserve manager in India.   This means that given the substrate conditions, locality and the date on which the track was made, the manager is expected to be able to reliably determine: 1) the paw which made the imprint; 2) sex of the tiger; 3) distinguishing features in the tracing that help to identify the individual tiger.   To my knowledge these assumptions have not been validated on a population of known tigers anywhere in India.

I carried out a blind test on wildlife managers who had actively classified pugmarks in the census operations, using 33 pugmarks tracings obtained on two different substrates from four captive tigers. The test results indicate that the above assumptions underlying the census method are unlikely to be valid under field conditions in many cases (Table 7).

However, it is possible that through long-term systematic monitoring of pugmarks a few individual tigers (particularly long-term residents with unusual track patterns) can be identified in a given population. This has been done by McDougal (1985) and his colleagues in Chitwan, Nepal. Moreover, even these identifications have been validated through separate photographic and visual identifications.

Clear identification of the breeding residents in a tiger population are critical to understanding their long-term dynamics. Yet this information is not available in the census records for any tiger population in India. Given the underlying assumption that each tiger (whether resident or transient) is known from its pugmarks, this surprising omission can only mean that the "individual identifications" based on the pugmark tracings are not clear enough to segregate the residents from the transients in any tiger population, even after 14 years of censusing experience.

Given the ease with which tigers are observed in some of the better reserves like Ranthambore, Kanha and Corbett, a few individual tigers and their home ranges have been identified visually, as well as from their tracks (Panwar 1979b, Rathore 1983). However, such exceptional cases cannot be considered as a part of the overall tiger population monitoring in India.

Based on this failure of the census methodology to distinctly identify even resident animals (and their home ranges), the claim that the technique can identify each tiger, including transients, seems highly questionable. Under these circumstance many of the "individual tigers" presumed to exist based on field censuses are likely to be multiple counts of pugmarks of a smaller number of tigers.

Multiple counts of a tiger can occur due to several reasons. Differences in substrate conditions (soil type, moisture content, slope, etc.) can result in large degrees of variations in the pugmarks of even the same paw of a given tiger. The fact that, due to morphological differences, imprints of all four paws can be different from each other - even under identical substrate conditions - further complicates the matter. The pace at which the tiger is moving also causes substantial differences in the shape of the tracings. Tigers can cover large distances in 24 hours, cutting across administrative boundaries of forestry management units. This is particularly true of transient individuals in habitats crisscrossed by forest roads. It is quite likely that these individuals are counted twice or thrice in adjoining reserves or even States. The shape of the tracing can vary depending on who made the tracing. In actual field conditions a complex interplay of several of these factors can result in the failure of the census method, usually leading to overestimates whose magnitude cannot be judged.

Table 7. Ability of wildlife managers to identify tiger pugmarks.

Respondent	Experience w. Tiger Census (Yrs)	No. Correct Choices[a]			No. Tigers Identified Correctly	Total No. Estimated by Pugmarks
		Male/ Female	Hind/ Front	Left/ Right		
A	5	21	19	27	0	24
B	4	22	20	21	0	23
C	5	24	22	28	0	13
D	6	29	22	23	0	7
E	5	27	12	25	0	6
F	12	25	23	23	b	b

[a] 22 or more correct choices are statistically significant using the binomial test.

[b] Respondent did not attempt segregation of pugmarks.

Another flaw of the present censusing method is its exclusive concentration on the total numbers of tigers in an area, neglecting other more important aspects, including a determination of the physical extent of tiger habitat available (as against the legal area of the park or forest range), quality of the tiger habitat in terms of prey abundance and the extent of livestock and wild prey in the diet of tigers.

Perhaps the roots of these problems are partly socio-cultural. Perhaps there was expectation that the Indian Wildlife managers would produce concrete evidence that Project Tiger was succeeding. In the early 1970's, Indian knowledge of modern wildlife biology and management were weak. This may have resulted in managers relying solely on a highly subjective census method to establish total numbers of a thinly distributed, shy, forest dwelling and nocturnal species like the tiger over the entire country.

One result of these overestimates is that the actual population status of wild tigers in India is unknown. Simpler, practical and more objective methods of monitoring tiger populations have not been developed, possibly impairing the ability to manage the Indian tiger populations wisely in the years to come.

FUTURE MONITORING OF WILD TIGERS

Wild tiger populations are faced with increasing possibilities of being forced into isolated gene pools due to habitat fragmentation. Thus, reliable monitoring of their population status is a critical factor for evolving appropriate management strategies. I suggest that the following aspects should be integrated into future population monitoring programs for tigers in India.

The monitoring system should focus on discrete populations of wild tigers rather than on individual nature reserves. Most problems of wildlife management can be tackled using <u>reliable indices</u> <u>of</u> <u>densities</u> of tigers in different areas or at different periods. Questions such as whether tiger numbers have increased, decreased or remained stable, or whether there are more tigers in one site as opposed to another, can be answered adequately with these indices. Estimates of relative densities can perhaps help resolve other management problems. Total counts or censuses have very little practical utility for management.

Objective monitoring of wild tigers, their prey base and the habitat conditions, should be performed by a small number of adequately trained personnel in each reserve. The present practice of deploying large numbers of untrained personnel in counting animals, collecting pugmarks, etc., is likely to defeat the very purpose of objective data collection. The methods used must be simple, repeatable and above all must be within the capabilities of the existing management personnel.

Assessing Habitat Availability

Simple data on presence or absence of tigers in each "forest beat" (about 20 sq. km in area) can be derived from evidence like tracks, scats, kills and sightings of tigers. When added up accurately over larger regions a fairly clear picture of discrete tiger population, and the extent of habitat available to them, will emerge. Over the years extensions of the tiger's range due to improved protection or increased breeding and dispersal can also be assessed from these data.

Assessing the Prey Base

Rough but reliable indices of relative densities of tiger prey species can be obtained from roadside counts and transect counts. These can indicate the trends in prey populations over the years and can reveal which parts of a particular reserve are likely to be better habitats for the tiger. Accurate recording of all livestock killed by tigers can also be very useful for assessing the tiger's prey base and population trends.

Monitoring Tiger Populations

Reliable indices of relative densities of tigers in different parts of a habitat or at different periods can be systematically developed from tiger signs like tracks, scats or scrape marks on roads or trails. The total length of roads, trails and stream beds covered by the investigator can be easily measured. The density index can then be the number of segments (of a suitable length) in which tiger signs were found in proportion to the total length covered. Since tigers habitually use trails and roads these indices can be quite effective in monitoring their densities.

In reserves visited by tourists the frequency of tiger sightings can be another good index of relative density when expressed as number of tiger sightings per year per thousand tourists visiting the reserves.

At a more sophisticated level, year round monitoring of tracks of a few tigers with distinctive pugmarks can also help in identifying some resident breeders, particularly when coupled to identifications based on facial markings. Such identifications can help in determining the home range sizes of residents in habitats of different quality. If the proportion of residents in the population can be estimated, then perhaps absolute densities of tigers can also be worked out and extrapolated to other similar areas.

Most of the above techniques can be fairly easily learned and practiced by the average Indian reserve manager.

ACKNOWLEDGEMENTS

W.A. Rodgers generously allowed me to use the prey biomass calculated by him for some reserves based on census figures. P. Jackson and M.K. Appayya helped me to collect the census data from different sources. Officials of the Mysore Zoo and Karnataka Forest Department helped me in conducting the validation test. A.J.T. Johnsingh, C. Wemmer, and H.S. Panwar offered useful comments on the drafts of this paper. I am grateful to all of them.

REFERENCES

Berwick, S. 1976. The Gir Forest: an endangered ecosystem. Am. Sci. 64:28-40.
Champion, F.W. 1929. Tiger tracks. J. Bombay Nat. Hist. Soc. 33:284-87.
Corbett, J. 1944. Maneaters of Kumaon. London: Oxford Univ. Pr.
Eisenberg, J.F. and J. Seidensticker. 1976. Ungulates in Southern Asia: A consideration of biomass estimates for selected habitats. Biol. Cons. 10:294-308.
Johnsingh, A.J.T. 1983. Large mammalian prey-predators in Bandipur. J. Bombay Nat. Hist. Soc. 80:1-49.
Joslin, P. 1973. Asiatic lion: A study of ecology and behavior. Ph.D. Diss., Univ. Edinburgh, Scotland.
Karanth, K.U. 1985. Analysis of predator-prey balance in Bandipur Tiger Reserve with reference to census reports. J. Bombay Nat. Hist. Soc. (In Press).
McDougal, C. 1977. The Face of the Tiger. London: Rivington Books.
McDougal, C. 1985. Smithsonian Terai Ecology Project, Report No. 5. Washington. Unpublished.
Mishra, H.R. 1982. The ecology and behavior of Chital (Axis axis) in Royal Chitwan National Park, Nepal. Ph.D. Diss., Univ. Edinburgh, Scotland.
Mountfort, G. 1981. Saving the Tiger. London: Michael Joseph.
Muckenhirn, N.A. and J.F. Eisenberg. 1973. In The Worlds Cats, ed. R. Eaton. Vol. 1:142-75.
Panwar, H.S. 1979a. A note on tiger census technique based on pugmark tracings. Indian For. (Special Issue):70-77.

Panwar, H.S. 1979b. Population dynamics and land tenures of tigers in Kanha National Park. _Indian_ _For_. (Special Issue):18-36.

Panwar, H.S. 1984. What to do when you have succeeded: Project Tiger ten years later. In _National_ _Parks,_ _Conservation,_ _and_ _Development_ _-_ _The_ _Role_ _of_ _Protected_ _Areas_ _in_ _Sustaining_ _Society,_ ed. J.A. McNeely and K.R. Miller. Washington, DC: Smithson. Inst. Pr.

Rathore, F.S., T. Singh and V. Thapar. 1983. _With_ _Tigers_ _in_ _the_ _Wild_. New Delhi: Vikas Publishing House.

Sankhala, K. 1978. Tiger: _The_ _Story_ _of_ _the_ _Indian_ _Tiger_. New York: Simon and Schuster.

Schaller, G.B. 1967. _The_ _Deer_ _and_ _the_ _Tiger_. Chicago: Univ. Chicago Pr.

Schaller, G.B. 1972. _The_ _Serengeti_ _Lion_. Chicago: Univ. Chicago Pr.

Seidensticker, J. 1976. On the ecological separation between tigers and leopards. _Biotropica_ 8:225-34.

Singh, A. 1984. _Tiger!_ _Tiger!_ London: Jonathan Cape.

Smith, J.L.D. 1984. Dispersal, communication and conservation strategies for the tiger (_Panthera_ _tigris_) in Royal Chitwan National Park, Nepal. Ph.D. Diss., Univ. Minnesota, St.Paul, MN.

Sunquist, M.E. 1981. The social organization of tigers (_Panthera_ _tigris_) in Royal Chitwan National Park, Nepal. _Smithson._ _Contrib._ _Zool_. 336:1-98.

Tamang, K.M. 1982. The status of the tiger _Panthera_ _tigris_ _tigris_ and its impact on principal prey populations in Royal Chitwan National Park, Nepal. Ph.D. Diss., Michigan State Univ., East Lansing, MI.

PART III

STATUS IN CAPTIVITY

14

Status and Problems of Captive Tigers in China

Tan Bangjie

<u>INTRODUCTION</u>

Of the five existing tiger subspecies (excluding those from Bali, Java and western Asia), the South China tiger, <u>Panthera tigris amoyensis</u> (Hilzhemer), is admittedly the rarest and most endangered. Naturally it is also the rarest in captivity, both in China and abroad. To our knowledge, there were no more than two captive South China tigers outside China as of the early 1980's, both descendants in the Moscow Zoo of a few tigers sent to the USSR in the 1950's. This includes a pair of tigers sent to Moscow by this writer in November 1952. Besides these, China sent two young males to Sudan in 1974. It is not known whether these still exist.

The present status of the wild South China tiger (Fig. 1) is certainly both alarming and disappointing to all of us. Remnants of its present population are so few, and the situation now facing it is so unfavorable, that hope for its natural recovery is scant. To save this subspecies, the only hope lies in developing a well-planned breeding program based on tigers now living in Chinese zoos. Fortunately, the present stock of captive South China tigers is still sufficient for carrying out such a program.

The theme of this paper is to present the status and problems of captive tigers in China, particularly those of the South China subspecies, to serve as a basis for the formulation of that program.

<u>THE GENERAL SITUATION</u>

There are more than 130 zoological gardens and smaller collections attached to municipal parks in China. Among them, nearly 30 are regular zoos, some of which possess fairly good facilities for accommodating tigers.

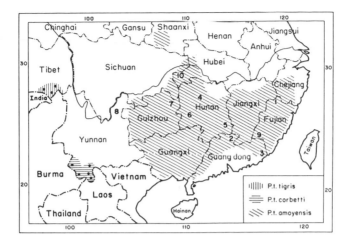

Fig. 1.   Present distribution of <u>Panthera</u> <u>tigris</u> in southern China. Solid circles indicate known localities, numbers indicate specific knowledge of <u>P</u>. <u>t</u>. <u>amoyensis</u>: 1) Pucheng Co., February 1986 - newly caught 30 kg tiger cub seen in cage at bus station, another cub (dead) for sale in street; 2) Shixing Nature Reserve, 1984 - tiger droppings found. Spring 1985 - large male sighted. Four to six tigers believed to be living in and around reserve; 3) Chingxi Commune, Dapu Co., January 1981 - three tiger cubs killed; 4) Anhua Co., May 1984 - two tiger cubs captured by commune militia, tigress fled with third cub; 5) Bamianshan Nature Reserve, 1982-83 - traces of tigers reported; 6) Suining Co., Summer 1984 - tiger seen, traces found at Anyang Mountain; 7) Fenjing Mountain Reserve, a few tigers believed to be living in mountains; 8) Zhenxiong Co., December 1984 - wounded tiger captured, died before arrival of zoo veterinarian; 9) Meihuadong Nature Reserve - newly designated to protect tigers; 10) Lichuan and Tongcheng Co., April 1986 - news reported tigers seen "many times" recently.

      There are nearly 40 Chinese zoos keeping tigers; 17 with over 40 South China tigers (Table 1) and over 30 with more than 100 Siberian tigers (Table 2). A few others have Bengal (<u>P</u>. <u>t</u>. <u>tigris</u>) or Indo-Chinese (<u>P</u>. <u>t</u>. <u>corbetti</u>) tigers (Table 3). All these captive tigers breed regularly. There are also a few hybrids between <u>P</u>. <u>t</u>. <u>altaica</u> and <u>P</u>. <u>t</u>. <u>amoyensis</u>. The pedigree of these hybrids, as well as their descendants, can be traced (Table 4).

      With one exception, all captive tigers in China are zoo-born. A small number are from wild-caught parents, but most are third or fourth generation captive-born. The only exception is a tigress captured near the Jilin-Amur border in the winter of 1984. A considerable number of wild tigers were collected from the provinces of Guizhou, Fujian, Hunan, Guangdong and Heilongjiang in the 1950's, indicating that the populations of wild South China and wild Siberian tigers were at that time very large. However, since a number of large-scale "anti-pest campaigns" have been staged repeatedly in the southern provinces, wild South China tigers have suffered heavily. Meanwhile, wild

Siberian tigers have always suffered from the persecution of deforestation and poaching over the years.

Since the 1960's, it has been difficult for Chinese zoos to collect wild tigers. For instance, the Beijing Zoo collected its last wild South China tiger in 1962. The Guiyang Zoo of Guizhou Province, where tigers once were plentiful and easily captured, collected 12 tigers in the three years of 1957-59, but has not collected one since 1963. In Jiangxi, another well known tiger-infested province, a tiger was killed in the suburbs of Nanchang, the provincial capital, in 1948. Also, as a result of the anti-pest campaigns staged in the 1950's, it was difficult to collect even tiger pelts by the 1960's. The Nanchang Zoo has offered a high price for South China tigers since 1972, but there has been no applicant to date.

All the wild tigers collected by Chinese zoos in the 1950's and early 1960's have since died. There have been only a few wild Siberian tigers collected since 1970. There are no wild-born South China tigers left in any Chinese zoo, and only a single wild-born Siberian tiger living in a small Jilin zoo. However, there are wild Bengal and Indo-Chinese tigers living in the zoos of Beijing, Chengdu and Kunming.

Even though the remnant population of wild Siberian tigers in Jilin and Heilongjiang provinces is very small (Fig. 2), as are the numbers of Bengal and Indo-Chinese tigers in China (Fig.

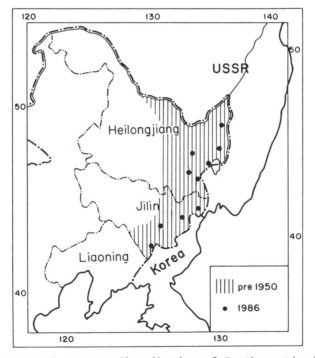

Fig. 2. Past and present distribution of <u>Panthera tigris altaica</u> in northeastern China.

1), their populations outside China are comparatively large and their continued existence is not so questionable. Therefore, the priority of tiger conservation in China must be given to the South China subspecies.

PRESENT STATUS OF SOUTH CHINA TIGERS IN CAPTIVITY

The present condition of the South China tiger in captivity can best be summed up into these few words: lots of problems, many difficulties, but still hopeful.

According to Table 1, 17 Chinese zoos have some 40 certified purebred South China tigers in their collections, including 23 males, 14 females and three cubs. Their ages indicate that, excluding one tiger nearly 18 years of age and seven young tigers less than three years of age, all are of breeding age. Even Wei Wei and Ting Ting of the Chongqing Zoo, now nearly 13 and 12 years old, respectively, still possess good fecundity, having parented a litter of three in 1985. Tigers under three years of age are relatively healthy, and should contribute to the growing captive population in the coming years.

Notable problems include an uneven sex ratio; there are many more males than females in the captive population, and improper pairing; ten out of the 17 zoos have single tigers only, while two others (Guiyang and Wuhan) have more males than females. Such problems can only be solved by implementing a nationwide breeding program.

With the death of Xiao Fu of the Shanghai Zoo two years ago, none of the existing 40 captive South China tigers are wild-born. The last (Wei Ning) of the 12 wild-born South China tigers collected by the Guiyang Zoo in the 1950's died in 1979. Only two of their small-numbered second-generation descendants, Yi Yi of the Shanghai Zoo and a male tiger of the Urumqi Zoo, are still alive. All captive South China tigers now living in Chinese zoos have affiliation to the aforementioned tigers. In other words, all are third or fourth generation descendants of one wild Fujian tigress and five of the 12 Guizhou tigers. As for the 20-odd South China tigers collected by the Beijing Zoo in the 1950's, most were exported abroad and the few remaining did not breed. All died in the 1960's and 1970's. A number of South China tigers were also collected by the zoos of Wuhan, Changsha, Nanchang, Nanning and Guangzhou. They too died some years ago.

Among the 17 Chinese zoos that exhibit South China tigers, only two (Chongqing and Shanghai) have animals which successfully breed regularly. Nanchang Zoo has also bred a few. The zoo at Guiyang was at one time the headquarters of Chinese zoos who kept and bred South China tigers. Their first generation (three of the 12 wild-caught tigers) parented eight cubs (four males, four females) in the 1960's. Five of these were given to other zoos, and the remaining three (one male, two females) produced a total of 24 cubs (Table 5). Most of these were transferred to zoos in other provinces and in foreign countries. The few remaining have not produced offspring in the 1980's. There are two more zoos (Wuhan Zoo and Xiaocyaojin Park of Hefei) which have South China tigers in pairs, and therefore the possibility of reproduction.

Table 1.   Captive South China tigers (P̲. t̲. amoyensis) in Chinese zoos as of March 1986.

Site	Name	Sex	Birthdate		Location	Sire	Dam
Chongqing:	Wei Wei	M	VII	1973	Guiyang	Dagonghu	Weibinghu
	Ting Ting	F	VI	1974	Guiyang	Dagonghu	Ermuhu
	Xiao Hua	M	XII	1978	Chongqing	Wei Wei	Ting Ting
	Xiao Ling	F	IX	1979	Guiyang	Qinggonghu	Ermuhu
	Da Fen	F	VI	1981	Chongqing	Wei Wei	Ting Ting
	Da Yeh	M	VI	1981	Chongqing	Wei Wei	Ting Ting
	Li Li	F	V	1982	Chongqing	Wei Wei	Ting Ting
	Chang Yin	F	VI	1983	Chongqing	Wei Wei	Ting Ting
	Hai Xiao	M	IV	1982	Shanghai	Yi Yi	Xiao Fu
	Cub A	M	VII	1985	Chongqing	Wei Wei	Ting Ting
	Cub B	M	VII	1985	Chongqing	Wei Wei	Ting Ting
	Cub C	F	VII	1985	Chongqing	Wei Wei	Ting Ting
Shanghai:	Yi Yi	M	VIII	1968	Shanghai	Xiao Mao	Xiao Yi
	Ah Fu	F	V	1979	Shanghai	Yi Yi	Xiao Fu
	Xiao Hu	F	IV	1980	Shanghai	Yi Yi	Xiao Fu
	Ah Yi	M	IV	1980	Shanghai	Yi Yi	Xiao Fu
	Qing Qing	F	V	1982	Chongqing	Wei Wei	Ting Ting
	84531 A	F	V	1984	Shanghai	Yi Yi	7758
	84531 B	F	V	1984	Shanghai	Yi Yi	7758
Guiyang:	Qing Gong	M	VIII	1973	Guiyang	Dagonghu	Ermuhu
	Qing Mu	F	VIII	1973	Guiyang	Dagonghu	Ermuhu
Hefei:	Min Min	M	IX	1979	Guiyang	Qing Gong	Ermuhu
	Fang Fang	F	VII	1980	Chongqing	Wei Wei	Ting Ting
Wuhan:	Ah Gui	F		1970	Guiyang		
	Ah Hai	M	IV	1982	Shanghai		
	Xiao Chang	M	II	1984	Nanchang		
	Qian Qian	M	IX	1979	Guiyang	Qing Gong	Ermuhu
	Dong Dong	M	IX	1979	Chongqing	Wei Wei	Ting Ting
Beijing:	Ah Yang	M		1972	Guiyang	Dagonghu	Ermuhu
Baoding:	Lei Mu	M	VI	1981	Chongqing	Wei Wei	Ting Ting
Tianjin:	Qiang Qiang	M	VII	1980	Chongqing	Wei Wei	Ting Ting
Kunming:	Xiao Zhu	M	IX	1979	Chongqing	Wei Wei	Ting Ting
Zunyi:	Xiao Hu	M	X	1972	Guiyang	Dagonghu	Weibinghu
Hangzhou:	Gui Lai	M	VII	1976	Guiyang	Dagonghu	Ermuhu
Urumqi:		M		1972	Shanghai	Xiao Mao	Xiao Fu
Liuzhou:		M		1978	Shanghai	Yi Yi	Xiao Fu
Loyang:	Li Sha	F		1971	Chongqing		
Chichihar:	Hua Gong	M		1975	Guiyang	Dagonghu	Ermuhu
Nanchang:		M			Shanghai	19520	Xiao Fu
		F			Shanghai	19520	?

Nevertheless, the main burden and future hope of breeding South China tigers is placed on the Chongqing and Shanghai Zoos.

One outstanding problem concerning the reproduction of captive South China tigers is that they appear to be too inbred, as practically all are descendants of a single Fujian tigress and a few Guizhou tigers. Therefore, one major goal is to collect one or more wild-born South China tigers for the purpose of improving the bloodline of the present stock.

Table 4 lists hybrids produced by crossing South China tigers with Siberian tigers. Even though the numbers are small, this hybridization has a negative effect that requires serious attention. These hybrids were produced primarily in the 1970's, during the last stage of the "Cultural Revolution." Afterwards, the Shanghai Zoo ceased the practice and there are no more hybrids remaining in this zoo. Nevertheless, some of the hybrids are still living in other zoos. Moreover, as late as the 1980's, more hybrids have been produced by matings between hybrids and normal animals and between the hybrids themselves. To realize the tiger population breeding project, it will be necessary to trace the whereabouts of such hybrids and forbid any further act of hybridization.

To date, there is no comprehensive studbook for captive South China tigers, or any other subspecies of tigers, in Chinese zoos. The establishment of such a studbook is another prerequisite for the successful realization of the breeding project.

A symposium on the conservation of the South China tiger was held in Chongqing in October 1984, and a South China Tiger Conservation Coordination Group set up during this occasion. However, practical works have not begun because problems with organization, official credentials, personnel and budget all remain unsettled. Cries of "South China tigers critically endangered," "South China tigers must be saved promptly," etc., are heard more often, and people showing solicitude are more numerous. To the Chinese, 1986 is the Year of the Tiger. In my first article, written in 1986 (published by the China Youth Daily, 4 January 1986), I voiced my hope like this: "I wish that when the next Year of the Tiger comes (in 1998), our old friend the South China tiger will not only still be alive, but will already be free from its endangered position."

The Chinese Ministry of Urban and Rural Construction and Environmental Protection (MURCEP), in a directive issued in June 1984, allowed the Chongqing Zoological Garden to establish a South China tiger research and breeding center. For this purpose, the Chongqing Zoo has drafted a construction plan that has undergone examination and amendment by scholars and specialists. The big question is, primarily and fundamentally, funding.

Foreign experiences in recent years have shown that, in order to effectively rescue endangered animals (such as the white rhinoceros, Arabian oryx, Bengal tiger, Mountain gorilla, etc.), international cooperation and assistance are important factors. Any rare and precious animal, although endemic to a certain country, is priceless wealth for all mankind. The South China tiger is no exception. To save the South China tiger the Chinese have the primary obligation; but international cooperation is also indispensible.

PRESENT STATUS OF SIBERIAN TIGERS IN CAPTIVITY

From the viewpoint of species preservation, the collection of Siberian tigers for zoos has an entirely different meaning than would that of South China tigers. People all know that the South China tiger is extremely rare, both in the wild and in zoos, and is on the brink of extinction. For Siberian tigers

this is not so. Aside from wild tigers living in the Soviet
Union and Korea, it seems the number living in zoos around the
world already amount to a surplus population. Through the
auspices of the IUCN/WWF, American zoos (including the Bronx and
San Diego zoos) have given ten Siberian tigers to China in the
past two years, an indication of the new situation.

Table 2 lists 96 Siberian tigers kept by 28 Chinese zoos.
This list does not include tigers kept by the zoos of Chengdu,
Chongqing, Jinan, Dezhou, Pengbu, Zhang-jia-kow (Kal gan) and
Huhehot, where Siberian tigers are known to be on display. The
total number of captive Siberian tigers in China is definitely
over 100, more than double that of South China tigers.

A further analysis of Table 2 indicates that: 1) of the 96
tigers in captivity, 43 are males, 50 are females and three cubs
are of undetermined sex. The sex ratio is therefore nearly even,
which is comparatively favorable for breeding; 2) there remains
at least one wild-born tiger, Wan Zhen, the tigress of People's
Park in Yanji. There are also a few born to wild-caught
specimens. For instance, both parents of Da Hu (female) and the
mother of Xiao Lin (male), both of the Dalian Zoo, were wild-
caught tigers. The tigress, Er Hua, of the Xining Zoo, was born
to Beijing Zoo's Hu Lin, a wild tigress captured from
Heilongjiang's Hulin County in 1970. Thus, captive Siberian
tigers in Chinese zoos have more founders represented and more
genetic diversity than captive South China tigers. Furthermore,
the possibility of obtaining wild-caught specimens is definitely
greater for this subspecies than for its South China cousin; 3)
of the present population of about 100 captive tigers,
approximately 10% came from the United States in the early
1980's, and several others came from Canada, the Netherlands and
West Germany. These belong to the Soviet $\underline{P}$. $\underline{t}$. $\underline{altaica}$
bloodline.

In considering the bloodlines of Siberian tigers kept by the
world's zoos, is there any bloodline that originated from Chinese
Siberian tigers? This is quite possible. According to records
kept by the Beijing Zoo (see Table 6; compiled by He Lihua 1980),
a total of 48 Siberian tigers, including many wild-caught
specimens, were exported to the USSR (chiefly the Soviet
Zoocentre) and East European countries from 1953 to 1965.
Presumably some of the Soviet captive tigers inherited this
bloodline.

The Beijing Zoological Garden collected 34 wild-caught
Siberian tigers from northeast China during the period 1950-59;
19 of them passed through the hands of this author from 1952 to
1954. Thirteen young tigers were captured in the early spring of
1953, 21 more (mostly cubs) in the early spring of 1954. No one
knows how many cubs, together with their mothers, died during
this two-year period. No more wild tigers have been collected
since 1959.

In 1970 a three year old, half-tailed tigress, captured by
forest workers at Hulin in eastern Heilongjiang, quite near the
Sino-Soviet border, was sent to Beijing. She died in 1980 after
giving birth to a few cubs. More wild tigers were collected from
the eastern Heilongjiang area and Jilin's Changbaishan mountain
area by the Harbin and Yanji zoos during the 1960's and 1970's.
As a result of forest exploitation, the habitat of the Siberian
tiger in Heilongjiang Province shrank in an easterly direction,

Table 2.   Captive Siberian tigers (P. t. altaica) in Chinese zoos* as of
March, 1986.

Site	Name	Sex	Birthdate		Location	Sire	Dam
Dalian:	Da Hu	F	VI	1972	Dalian	Wild caught	Wild caught
	Xiao Lin	M	X	1977	Dalian	Wild caught	
	Xiao Kwei	M	IX	1980	Dalian	Xiao Lin	Da Hu
	Bin Bin	M	IX	1982	Dalian	Xiao Lin	Da Hu
	Mei Mei	F	IX	1982	Dalian	Xiao Lin	Da Hu
	Feng Feng	F	IX	1982	Dalian	Xiao Lin	Da Hu
	Xi Xi	F	III	1980	W. Germany		
	Ah Xin	F	VI	1983	USA		
	Ah Yuan	M	VI	1983	USA		
	Cub A		VIII	1985	Dalian		Da Hu
	Cub B		VIII	1985	Dalian		Da Hu
	Cub C		VIII	1985	Dalian		Da Hu
Shenyang:	Ke Ke	M	VIII	1981	Harbin		
	Long Long	F	IX	1979	Harbin		
	Da Yang	F	X	1980	Dalian	Xiao Lin	Da Hu
Harbin:	Ermuhu	F		1967	Harbin		
	Sen Sen	M	VI	1974	Harbin		
	763	F	VI	1976	Harbin	Da Gong	Ermuhu
	764	F	VI	1976	Harbin	Da Gong	Ermuhu
	Fang Fang	M	VI	1982	Harbin	Da Gong	763
	Bing Bing	M	VIII	1984	Harbin	741	764
	Jia Jia	M		1980	Canada		
	Mei Mei	F	VI	1983	USA		
Dandong:	Nan Nan	M	VI	1983	Jinan	Shengli	Dong Dong
	Pin Pin	M	VIII	1984	Harbin	741	764
Jilin:	Ha Sheng	M	VI	1975	Harbin		
	Xin Sheng	F	IX	1983	Urumqi		
Yanji:	Fei Fei	M	VI	1976	Dalian	Xiao Lin	Da Hu
	Li Li	F	X	1977	Dalian	Xiao Lin	Da Hu
	Pau Ni	F	VII	1984	Yanji	Fei Fei	Li Li
	Wan Zhen	F	adult		E. Jilin	captured Dec. 1904	
Chichihar:	Bei Mu	F	IX	1981	Shijiazhuang		
	Mao Mao	M	III	1983	Shanghai	7134	69817
Xuzhou:		M			Nanjing	Xiao Shi	Yi Hao
		M	2 years		Nanjing	Xiao Shi	Yi Hao
		F	2 years		Nanjing	Xiao Shi	Yi Hao
Qingdao:		M	III	1983	Jinan		
		F	VI	1984	Nanjing		
Taiyuan:		F	2 years		Shanghai		
		F	4 years		USAian		
	Lian Hua	F	VI	1976	Dalian		
	Lida	F	XI	1980	Netherland		
	Lina	F	XI	1980	Netherland		
	Lin Lin	M	adult		USA		
Tianjin:	Yi Hao	M	VIII	1971	Jinan		
	Guo Guo	M	V	1981	Tianjin	Yi Hao	Er Hao
	Xiao Hua	F	VII	1977	Tianjin	Yi Hao	Er Hao
	Ming Ming	F	V	1978	Tianjin	Yi Hao	Er Hao
	Liya	F	VIII	1982	USA		
Loyang:	Guang Lo	M	IV	1984	Guangzhou		
Nanjing:	Xiao Shi	M	VII	1980	Shijiachuang		
	Yi Hao	F	VI	1974	Nanjing		
	Er Hao	F	IV	1979	Nanjing		
	San Hao	F	XI	1982	Nanjing		
	Xiao Tiaopi	M	VI	1984	Nanjing		

Table 2 Continued.

Site	Name	Sex	Birthdate		Location	Sire	Dam
Shanghai:	7134A	M	III	1971	Shanghai	Ah Er	Hong Hu
	Bin Bin	M		1972	Harbin		
	Tai Yuan	F		1981	Taiyuan		
	Xiao Hua	F				Shanghai	
		M				San Diego	
		F				San Diego	
		F				San Diego	
Nanchang:		F				Harbin	
Ningpo:		F				Chengdu	
Wuxi:	Xiao Hai	M	X	1981	Shanghai	7134	He Ping
	74813	F	VIII	1978	Shanghai		
Changzhou:	Xuanwuhu	M	IV	1984	Nanjing		
Hangzhou:	Jin Guang	F	16 years		Tianjin		
	Hu Hang	M	VI	1982	Shanghai		
Xiamen:	79727	M	VII	1979	Shanghai	74813	7134
Guangzhou:	Wei Di	M	III	1981	Guangzhou	Gong Gong	Li Li
		M		1971	Nanjing		
	Li Li	F		1978	Dalian		
	Mei Nu	F		1981	New York	1303	1297
Kunming:		M	adult				
		F	adult				
Zigong:	Yi Hao	M	VI	1984	Urumqi	Li Li	Xiao Hu
	Er Hao	M	VI	1984	Urumqi	Li Li	Xiao Hu
Urumqi:	Xiao Hu	F	VI	1975	Harbin		
	Li Li	M		1974	Shanghai		
	Xing Xing	F	XI	1982	Urumqi		
	Er Mao	F	VII	1983	Urumqi		
	San Mao	F	VII	1983	Urumqi		
	Mi Sha	M	VI	1984	Urumqi		
Xian:	Shang Xi	M				Shanghai	
	Ning Xi	F	IX	1980	Xining		
Xining:	Mi Mi	F	VI	1973	Harbin		
	Da Hua	M	VI	1975	Shanghai	Heping	7134
	Er Hua	F	VI	1976	Beijingi	Ji Gong	Hu Lin
	Ning Ning	F	IX	1979	Xining	Da Hua	Mi Mi
	Lan Hai	M	V	1978	Shanghai		
	84917A	M	IX	1984	Xining	Lan Hai	Ning Ning
	84917B	F	IX	1984	Xining	Lan Hai	Ning Ning
Inchuan:	Li Na	F	VI	1984	Urumqi	Li Li	Xiao Hu

* No reply from the zoos of Jinan, Dezhou, Bengfu, Kalgan, Chengdu.

and there are no more tigers living in the Xiao-hsing-an-ling mountain areas in northern Heilongjiang. Any remnants are now limited to the eastern Heilongjiang area (Fig. 2).

The Harbin Zoo began to keep Siberian tigers in 1954. From 1959 to 1985, this zoological garden produced 113 tigers; 84 of which (74.3%) survived. A large number of these were given to zoos in more than 40 cities. Another zoological garden with a good breeding record is the Dalian Zoo. A tigress called Da Hu, born there in 1972 of wild-caught parents, has given birth to eight litters totaling 24 cubs (19 survived) from 1976 to 1985. Three of the eight litters consisted of four cubs each. Since

Table 3.   Captive tigers other than P. t. amoyensis and altaica in Chinese Zoos as of March, 1986.

Site	Name	Sex		Birthdate	Location	Sire	Dam
Kunming	Rui Kun	M		1982	W. Yunnan	wild caught	wild caught
		F		1983	W. Yunnan	wild caught	wild caught
		F		1983	W. Yunnan	wild caught	wild caught
		M		1983	W. Yunnan	wild caught	wild caught
		M		1983	W. Yunnan	wild caught	wild caught
		M		cub	W. Yunnan	wild caught	wild caught
Wuzhou		M	X	1973	Kunming	old Rui Kun	Mi Mi
Dan Dong	Ming Ming	M	X	1973	Kunming	old Rui Kun	Mi Mi
Chengdu	Loknow	M	I	1978	Kampuchea	wild caught	wild caught
Beijing	Yao Jin	M	I	1978	Kampuchea	wild caught	wild caught
	Ai Guo	M	VI	1979	Bombay		

she ranks first, both in the birth rate and survival rate among captive Siberian tigers in China, she has won the title of "Hero Tiger Mama." Her mate Xiao Lin was born in 1977 to a wild-caught tigress from Jilin Province. Besides the two zoos mentioned above, the zoos of Nanjing, Shanghai and Urumqi have also bred Siberian tigers from time to time. The Beijing Zoo produced a number of Siberian tigers before the 1970's, but there has been no more breeding since then. The Yanji Zoo, beginning in 1965, has intermittently had five wild tigers from the Changbaishan Mountain area. Now it has only one.

After all, there are few difficulties encountered with the quantity, resource and breeding of captive Siberian tigers in Chinese zoos. The main direction of development hereafter should be the improvement of the standard of scientific management; the gist of which is, of course, the establishment of a comprehensive Chinese studbook, under whose guidance the maintenance of subspecific purity, genetic diversity and strain selection can be achieved. Experimental works should begin at a few selected points and then the experience spread to other places.

Table 4.   Offspring of hybrid tigers in Chinese zoos as of March, 1986.

Site	Name	Sex		Birthdate	Location	Sire	Dam
Xian	Hu Xi	F		1974	Shanghai	Sib. tiger	S. Ch. tiger
Guangzhou	Xiao Hu	M		1974	Shanghai	Sib. tiger	S. Ch. tiger
	Hu Nu	F		1975	Shanghai	Sib. tiger	S. Ch. tiger
	Chi Chi	M	XI	1980	Guangzhou	Xiao Hu	Hu Nu
	Qu Qu	F	XI	1980	Guangzhou	Xiao Hu	Hu Nu
		M	V	1982	Guangzhou	Xiao Hu	Hu Nu
Wuhan	Xiao Guang	F	VI	1984	Guangzhou	Xiao Hu	Hu Nu
Hangzhou	Xiao Hu	M		1968	Shanghai	Sib. tiger	S. Ch. tiger
Baoding	Da Hu	M		1977	Shanghai	Sib. tiger	S. Ch. tiger
Chichihar		M			Chichihar	Hua Gong	Bei Mu

The establishment of a studbook for captive Chinese Siberian tigers may be more difficult and complicated than for South China tigers. Not only are there more Siberian tigers living in Chinese zoos, the origin of these tigers, unlike South China tigers (whose origin is limited to three systems; Guiyang, Chongqing, Shanghai), has a much wider scope. Their origins must be traced back to the wild tigers collected by the zoos of Beijing, Harbin, Dalian, Yanji, Shenyang, Shanghai and Mudanjiang in the 1950's, and then their descendants that were dispersed to Nanjinkg, Jinan, Shijiachuang, Tianjin, Guangzhou, Urumqi and other zoos. In addition, a number of individuals were brought to smaller cities and are unaccounted for.

Information released in December 1985 revealed that the Ministry of Urban and Rural Construction and Environmental Protection has begun planning a Chinese Siberian tiger breeding center in the suburbs of Harbin. This is a promising project, at least for China. The Harbin Zoo, having experienced much success in breeding tigers, may contribute greatly toward the preservation of Heilongjiang's remaining wild tigers.

As for the tigers of Jilin, according to an article entitled "Proposals on the establishment of a Northeast China (=Siberian) tiger breeding and research center" written by Ju Cheng and Jiang Jingsong (1985), Jilin has altogether about 20 tigers distributed in three isolated regions (Fig. 2). Mt. Motianling and Mt. Dalongling each have four to six. The article proposed that such a center be built at Motianling, about 70 km southeast of Huadian County. The proposal added that a group of five breeding tigers would be kept for the first period of five years, during which time ten cubs would be raised. In the second five-year period, the breeding group would increase to 15 and a total of 20 cubs would be raised. Experiments on semi-release and total release will follow in the third period.

PRESENT STATUS AND PROBLEMS FACING OTHER SUBSPECIES

Before the 1950's, Chinese and foreign zoologists knew of only two subspecies of tigers in China, namely P. t. amoyensis and altaica, but were ignorant of other subspecies existing in far off districts on China's southwestern frontier. It was only after animal collectors, sent by Chinese zoos, penetrated into the forests of the Xi-shuang-ban-na District that it became known that there are also tigers in this district, and that they belong to a different subspecies than the South China tiger. Later investigation confirmed that tigers are found in the counties of Mengzhe, Mengla and Dale. Moreover, the tigers are not limited to the Xi-shuang-ban-na District but are also found in counties like Pu-er and Si-mao to the north, and Can-yuan and Ximeng to the northwest of Xi-shuang-ban-na. So the range of this subspecies is fairly wide in the southwestern part of Yunnan (Fig. 1). However, it was not until the beginning of the 1960's that living specimens were obtained from the region. Except for a single specimen collected by the Shanghai Zoo from western Yunnan, the major collector had always been the Uyantong Zoo of Kunming.

One pair of tigers collected by the Yuantong Zoo from Xi-shuang-ban-na lived from the early 1960's until the early

Table 5. A register of South China tigers (P. t. amoyensis) in the Guiyang
Zoo from the 1950's to 1980's. Compiled by Fang Xi and Yang Tongli.

Serial Number	Sex	Birthdate	Location	Sire	Dam	Remarks
5701	F	wildcaught 1957	Tongren			Gave Beijing 1959
5801	M	wildcaught 1958	Chingzhen			Died 1968
5802	F	wildcaught 1958	Changshun			Gave Zhengzhou 1974
5803	M	wildcaught 1958	Chingzhen			Gave Shanghai 1959
5804	F	wildcaught 1958	Chingzhen			Gave Beijing 1959
5901	F	wildcaught 1959	Bijie			Died 1967
5902	F	wildcaught 1959	Changshun			Died 1964
5903	M	wildcaught 1959	Bijie			Gave Shanghai 1962
5904	F	wildcaught 1959	Bijie			Gave Shanghai 1962
5905	F	wildcaught 1959	Weining			Gave Ningbo 1974
6301 (Weibinghu)	F	1963	Guiyang	5801	5901	Gave Beipei 1982
6302	M	1963	Guiyang	5801	5802	Gave Zunyi 1964
6303	F	1963	Guiyang	5801	5802	Gave Zunyi 1973
6401 (Dagonghu)	M	1964	Guiyang	5801	5802	Died 1979
6402	M	1964	Guiyang	5801	5901	Gave Beijing 1967
6403	F	1964	Guiyang	5801	5901	Gave Beijing 1967
6501 (Ermuhu)	F	1965	Guiyang	5801	5802	Died 1980
6502	M	1965	Guiyang	5801	5802	Gave Beijing 1971
7201	M	1972	Guiyang	6401	6301	Gave Beijing 1973
7202	F	1972	Cuiyang	6401	6301	Gave Beijing 1973
7203	M	1972	Guiyang	6401	6501	Gave Loyang 1973
7204	F	1972	Guiyang	6401	6501	Gave Loyang 1973
7205	M	1972	Guiyng	6401	6501	Gave Dalian 1977
7301	M	1973	Guiyang	6401	6301	Gave Zunyi 1973
7302 (Wei Wei)	M	VII 1973	Guiyang	6401	6301	Gave Chongqing 1973
7303 (Qing Gong)	M	VIII 1973	Guiyang	6401	6501	at Guiyang
7304 (Qing Mu)	F	VIII 1973	Guiyang	6401	6501	at Guiyang
7305	M	VIII 1973	Guiyang	6401	6501	Gave Sudau
7401 (Iing Ilug)	F	1974	Guiyang	6401	6501	Gave Chongqing 1974
7402	M	1974	Cuiyang	6401	6501	Gave Sudan 1974
7403	M	1974	Gyiyang	6401	6501	Gave Anshan 1975
7404	F	1974	Guiyang	6401	6501	Gave Anshan 1975
7405	M	1974	Guiyang	6401	6301	Gave Beijing 1975
7501	M	1975	Guiyang	6401	6501	Gave Harbin 1975 Chichihar 1977
7502	M	1975	Guiyang	6401	6501	Cave Yichang 1975
7503	M	1975	Guiyang	6401	6501	Gave Jiamusi 1975
7504	M	1975	Guiyang	6401	6301	Gave Zunyi 1975
7505	F	1975	Guiyang	6401	6301	Gave Zunyi 1975
7601	M	1976	Guiyang	6401	6501	Gave Beijing 1976
7901 (Min Min)	M	IX 1979	Guiyang	7303	6501	Gave Hefei 1985
7902 (Qian Qian)	M	IX 1979	Guiyang	7303	6501	At Guiyang
7903 (Xiao Ling)	F	IX 1979	Guiyang	7303	6501	Gave Chongqing 1980
7904 (Dong Dong)	M	1979	Congqing	7302	7401	At Guiyang

1980's. Zoo staff indicated that Mi Mi, the female tiger, had
lived a total of 22 years and 4 months (August 1960–January
1983). If true, it created a record of longevity. In the autumn
of 1982, her mate Ruikun had already died. Mi Mi was so old that
she had difficulty in walking and feeding. Two adult tigers (her
sons) were living in the same cage with her. She had given birth
to 18 cubs. Three males, born in 1973, were given to the zoos of
Beijing, Tianjin and Wuzhou (Guangxi) in 1975. The one (Ming

Ming) sent to Beijing was transferred to Dandong (Liaoning) in
1980. As for the whereabouts of the other cubs, nothing is known
except that two were sent to the zoos of Maoming (Guangdong) and
Zigong (Sichuan). Table 3 is therefore incomplete.

The Yuantong Zoo now keeps six tigers from Yunnan; four
males and two females. One of the males, collected in western
Yunnan in early 1982 as a tiny cub, has grown to adulthood.
Others that were collected in 1983 in the vicinity of Nankan, on
the Sino-Burmese border, are now sub-adults.

Morphologically, the tigers from western Yunnan look quite
similar to those of Xi-shuang-ban-na. They are larger and paler
than South China tigers, and have a shorter coat. The stripes
are narrow and long. All these points seem to be identical to
the Bengal tiger but differ from the Indo-Chinese tiger. To
which subspecies these tigers belong is a question that requires
further study. If, as indicated by Mazak (1968), the range of P.
t. corbetti includes Thailand, the Malay Peninsula, Indo-Chinese
countries and Burma east of the Irawaddy River, then the tigers
living in Xi-shuang-ban-na and southwestern Yunnan must obviously
be included in this sphere.

The Beijing Zoo received two young male tigers, each
weighing about 27 kg, from Kampuchea in August 1978. They
belonged to a litter of four one-month old cubs captured in
February 1978 in the forests of Pawisha Province in northern
Kampuchea. Beijing retained one of the cubs and sent the other
to Chengdu in 1980. Both are sound and healthy but without
mates.

The Beijing Zoo also has a male Bengal tiger from India.
This was presented by an overseas Chinese, Mr. Cai Shijin, who
bought a pair in Bombay in September 1979. The male's mate died
in May 1983. This is the only Indian Bengal tiger living in
Chinese zoos.

Table 6. Record of Siberian tigers (P. t. altaica) exported from China during
1953-65. Compiled by He Lihua.

	1953	1954	1955	1956	Year 1957	1958	1959	1960	1962	1965	TOTAL
Zoocentre USSR	3	2	3	4	2/1	7		3			25
Leipzig DDR		1/1	1/1								4
Praha Czech	1/0	0/1		1/1							4
Rumania										1/1	2
Bulgaria						1/1					2
Hungary					1/1						2
Poland					1						1
Burma			0/1						1/0		2
Holland					1/1						2
Demmer		1/1		0/1							3
Circus, USSR			1								1

In my opinion, we should not hurry to mate captive tigers from Yunnan with those from either India or Kampuchea before their subspecific positions are finally ascertained.

There are also tigers living in forested areas on the southern slope of the Himalayas in southern Tibet near Assam (Fig. 1). These are obviously Bengal tigers. No living specimen has yet been captured.

CONCLUSIONS

Six conclusions may be drawn:
1) Captive tigers in China represent the subspecies P. t. amoyensis, altaica, tigris, and corbetti. The South China tiger, P. t. amoyensis, a most endangered subspecies endemic to China, is rare both in the wild and in zoos. Its remnant population in the wild is estimated at no more than 30-40 individuals, with "at most eight" in Hunan Province (Gui Xiaojie 1986), "only four to six" in northern Guangdong (Lo Wenjin 1986), "about ten" in Guizhou Province (Fang Xi and Yang Topngli 1985) and a few more believed to exist in other provinces. Nature reserves set up at Shixing, in northern Guangdong; Guidong and Yizhang, in southeastern Hunan; and Fanjingshan, in southeastern Guizhou, may serve some purpose in preserving the South China tiger, but their function must be greatly strengthened.
2) The total number of South China tigers in captivity is around 40. Chinese specialists agree that upon these tigers lies the main hope of survival. To establish a breeding and research center with a "world's central breeding group" in Chongquing, and perhaps one more in Shanghai, seems to be a feasible step in the realization of a South China tiger recovery program. A South China Tiger Conservation Coordination Group has been set up for this purpose, but lacks financial and material support.
3) The wild population of Siberian tigers (P. t. altaica) in the northeastern provinces of Jilin and Heilongjiang is no larger than the South China subspecies. To save the remnant tigers in Chinese territory from doom is an obligation of Chinese zoologists and conservationists. There are already two nature reserves in eastern Heilongjiang (Qixinglazi) and southern Jilin (Changbaishan) whose effectiveness in preserving the tigers is insufficient. The proposal to establish a breeding and research center in the Changbaishan Mountains, with the object of mating wild and captive tigers and finally releasing the offspring back to forest, is worthy of consideration. Captive Siberian tigers in China are much better off than South China tigers.
4) In order to determine the exact size of the wild tiger populations and their whereabouts, both in the northern and southern provinces in China, a large-scale survey is indispensible. The survey may also settle questions concerning the subspecific position and actual population of the tigers in Yunnan. It may even clarify the rumors that there are remnant tigers still persisting in southeastern Xinjiang.
5) The establishment of a comprehensive studbook for all captive tigers in China is an indispensible and prerequisite work for carrying out the recovery program.
6) International cooperation and assistance are important factors in the conservation of top-listed endangered animals. For the recovery of such an important animal like the South China

tiger, close cooperation between Chinese specialists and their foreign colleagues is invaluable to the cause.

ACKNOWLEDGEMENTS

I am very grateful to the following persons who have kindly given me much valuable information concerning the captive tigers in their collections: Director Xiang Peilon of the Chongqing Zoo, Director Fang Xi of the Guiyang Park Zoo, Deputy-director Zhang Cizu of the Shanghai Zoo, Deputy-director Xian Rulun of the Kunming Yuantong Zoo, and Mr. He Lihuya of the Beijing Zoo.

REFERENCES

Fang Xi and Yang Tongli. 1985. The South China tiger of Guizhou Province. (In Chinese.) Unpublished.
Gui Xiaojie. 1986. At most eight South China tigers left in Hunan Province. Hunan Environmental Protection Post, Jan. 23, 1986.
Ju Cheng and Jiang Jingsong. 1985. Proposals on the establishment of a Northeast China tiger breeding and research center. Natural Conservation or Changbai Mountain. No. 2.
Lo Wenjin. 1986. A trip to the "Tiger Mountain." Yangcheng Wanbao (The Canton Evening Paper), Feb. 10, 1986
Mazak, V. 1968. Nouvelle sous-espece de Tigre provenant de l'Asie du Sud-Est. Mammalia 32(1):104-12.
Tan Bangjie. 1983a. Tigers in China. WWF Monthly Report. April, 1983. Gland, Switzerland.
Tan Bangjie. 1983b. The classification and distribution of tigers in China. Paper submitted to the annual meeting of the Chinese Mammalogist Society held at Hefei, October 1983.

15

Nutritional Considerations in Captive Tiger Management

Ellen S. Dierenfeld

INTRODUCTION

Success in maintaining and breeding tigers in captivity is dependent, in large part, on proper feeding practices. Unique aspects of feline nutrition set cats apart from other carnivores. These features have been identified and detailed in the domestic cat, the species currently used as the model for all Felidae, and will be discussed with special reference to the tiger. In addition, problems specific to the captive tiger will be addressed.

NUTRIENT REQUIREMENTS OF THE FELIDAE

Nutrients to be considered, in order of their importance, include water, energy, protein, minerals and vitamins.

Water

Water is the single most important nutrient for the proper function of all living cells, yet is sadly neglected in many nutritional programs. Cats drink very little free water, instead fulfilling their fluid requirements from normal dietary or metabolic sources. Carcass meats contain approximately 70% water; semi-moist (30% moisture) or dry diets (10%) moisture fed to domestic cats often do not contain adequate moisture to maintain physiological function and exacerbate problems of mineral deposition in the urinary tract (MacDonald et al. 1984). Feline Urinary Syndrome (FUS) will be discussed in more detail later in this paper.

One study with domestic cats fed a canned diet of 74% moisture showed daily water intake for young animals (6-8 months) to be about 80 ml per kg body mass; for adult cats, the value was 57 ml per kg (General Foods Corporation 1977). Most of the water was provided by the feed. Only four and 13% of the totals were obtained from drinking by young and older animals, respectively.

Comparative values of water intake (from feed sources only) for a tiger cub averaged 94 ml of water per kg body mass from one to 20 weeks of age (Henry Doorly Zoo unpublished data). Values calculated from intake data with seven adult tigers averaged only 15 ml of water per kg body mass (Minnesota Zoological Garden unpublished data) provided in the feed. More emphasis needs to be placed on defining the fluid intake needs of animals in captivity.

Wild felids have rarely been observed drinking, but tigers appear to prefer habitats containing open water (Schaller 1967), particularly during hot weather. A fresh source of drinking water should be provided continually. Water quality should be examined periodically. Chemical and microbiological monitoring assays are particularly important when water is supplied via free streams or pools in enclosures.

Energy

Basal energy requirements for domestic cats have been estimated at 57 kcal/kg body mass (Macdonald et al. 1984). Maintenance energy requirements (approximated at 1.5 times basal heat production) are thus calculated at 86 kcal/kg.

Digestible energy (DE) requirements calculated from 143 digestibility trials on six adult cats resulted in a mean value of 76 kcal/kg body mass per day (Kendall et al. 1983). Based on zoo feeding trials, DE requirements of 64 and 67 kcal/kg for male and female leopards, respectively, were calculated (Barbiers et al. 1982). These similar values suggest that small domestic cats may, indeed, be reasonable models for larger species although relative energy requirements decrease with increasing body size.

Energy is obtained from three sources: feed carbohydrates, fat and glucogenic amino acids.

Carbohydrates: Experimental and commercial diets containing isolated carbohydrates have shown that cats have the capacity to digest limited amounts of cereal starches, lactose, glucose, sucrose and dextrin, with apparent digestion coefficients near 90% (Morris et al. 1977). However, no dietary requirement for carbohydrates has been demonstrated for the cat (MacDonald et al. 1984). Natural meat diets, while widely used, contain little carbohydrate. Enzyme studies have determined that cats possess low glucokinase activity, and may have a limited ability to effectively utilize high glucose loads (MacDonald et al. 1984).

Lipids: Cats obtain about 60% of their calories from dietary fats, with diets composed of up to 67% fat efficiently digested and utilized (Scott 1968). Apparent digestion coefficients of crude fat (CF) in commercial diets range between 79 and 98% for a variety of lipids (MacDonald et al. 1984). Similarly, the apparent digestion of CF by captive wild felids was 95-99% on a meat-based diet containing 16% crude fat (Barbiers et al. 1982).

Cats can develop steatitis, or yellow fat disease, if fed diets high in polyunsaturated fatty acids or lacking in antioxidants. This condition has been observed on diets of marine fish or fish oils (Scott 1968).

Essential Fatty Acids. Most mammals require linoleic acid as an essential dietary fatty acid which is then converted to arachadonic acid, a precursor or prostaglandins. Felids, however, lack the enzyme 6-desaturase and require a dietary source of arachadonic acid (General Foods Corporation 1977). Lack of this enzyme activity has been shown in the lion as well as domestic felids (Scott 1968), and likely holds true for all of the Felidae.

Diets supplying 4.8% of total kilocalories as linoleic, and 0.0.4% as arachadonic acid provide adequate quantities of essential fatty acids to maintain normal reproduction (MacDonald et al. 1984). High levels of arachadonic acid can spare the linoleic requirement; dietary minimums with interactions have not been specifically determined.

Protein and Amino Acids: Cats, as strict carnivores, possess highly active nitrogen catabolic enzyme systems for metabolizing protein, and are incapable of adapting to very low nitrogen diets (MacDonald et al. 1984). Total protein requirements range between 15 and 30% of the diet (on a dry matter basis) for adult and young animals, respectively. Total nitrogen quantities required range from 0.5-2.0 g/kg body mass (General Foods Corporation 1977).

Approximately one-third of dietary energy for maintenance and growth in young kittens is supplied by protein, but this level drops to about 10% of dietary energy in adults (MacDonald et al. 1984).

Cats require the same essential amino acids as all animal species, but may differ in the quantities needed: arginine (1.05% of the diet), histidine (0.30%), isoleucine (0.50%), leucine (1.20%), lysine (0.80%), methionine (0.75%), phenyalanine (0.90%), threonine (0.70%), tryptophan (0.11%) and valine(0.60%). Each of these estimates is based on experimental diets containing 4.7 kcal/g (MacDonald et al. 1984). These levels are primarily important in the formulation of commercial, cereal-based or experimental diets, and should be adequately met on meat diets.

Protein quality is important in meeting requirements; these essential amino acids must be provided in adequate proportion as well as quantities. Connective tissues or organ meats alone do not contain an adequate amino acid spectrum for proper feline nutrition, whereas muscle meats are properly balanced (Scott 1968).

Several aspects of sulphur amino acid (SAA) nutrition and metabolism are unique in the cat, with various enzyme systems responsible (MacDonald et al. 1984). Methionine requirements are higher than for most mammals but are just met on all-meat diets, and represent the most limiting dietary amino acid. The cat excretes a SAA, felinine, in the urine which may (in part) account for his higher requirement. Amounts excreted are greater for males than females. Felinine may be related to territorial marking, but reasons for its presence are currently unknown. Finally, cats require a dietary supply of taurine (approximately 350 mg/kg diet) to prevent central retinal degeneration (MacDonald et al. 1984).

Minerals

There are few data available on mineral requirements of the Felidae, but needs do not appear to be extraordinary compared with other species. Examples of mineral deficiency diseases have been noted for captive wild felids fed improper diets, and are discussed in the section on nutritional diseases.

Young, growing animals and lactating females have greater mineral needs. Diets should supply a total of 80-100 mg/kg body mass calcium; these amounts should be doubled during lactation and growth (General Foods Corporation 1977). Salt requirements are unknown, as are those of iron. Copper should be supplied at a level of 5 mg/kg diet (General Foods Corporation 1977). Potassium requirements of the growing kitten are influenced by dietary protein, and have been estimated between 0.3 and 0.5% of the diet (General Foods Corporation 1977). Zinc deficiency has only been experimentally produced; however, iodine deficiency has been reported in zoo and domestic felids (Scott 1968). Guidelines for mineral levels in the diet can be found in Table 1.

Vitamins

More information is known about vitamin than mineral requirements of felids. Again, cats possess unique enzyme systems influencing vitamin nutrition.

Fat Soluble Vitamins: Unlike other species, cats lack the enzyme to convert carotene to active vitamin A, and require a dietary source of preformed vitamin A (retinol) (Scott 1968). Thus felids can utilize animal, but not plant sources of vitamin A.   Suitable levels (3-5 ug/g diet) are found in animal fats and organ meats, particularly liver (General Foods Corporation 1977). Vitamin A activity is decreased with storage, exposure to heat and/or oxygen. Deficiency of retinol can interfere with reproduction, vision and/or epithelial tissue integrity.

Vitamin D3 can be synthesized in the skin of the cat via exposure to sun or ultraviolet light. Requirements are probably quite low; rickets due to low dietary vitamin D have only been produced experimentally in animals kept in the dark. Excess vitamin D in the diet (2.5 mg/kg diet) has led to calcification of soft tissues in felids (MacDonald et al. 1984).

Vitamin E requirements are proportional to the level of polyunsaturated fatty acids in the diet. Due to its major function as an antioxidant, deficient cats develop steatitis from fat oxidation. This is not a problem with meat-based diets, but naturally occurring cases have been recorded on marine-fish diets (Scott 1968). Vitamin K is likely synthesized in adequate quantity by intestinal microflora.

Water Soluble Vitamins: Cats require thiamin, riboflavin, pantothenic acid, pyridoxine, biotin, folacin, B12 and niacin (MacDonald et al. 1984, General Foods Corporation 1977). Ascorbic acid is synthesized and is not a dietary essential. Although most species can meet their niacin needs by the enzymatic conversion of dietary tryptophan, in the cat there is almost no synthesis of niacin via similar pathways. Dietary requirements are supplied in animal tissues, but not by plants. Thus, again the cat is constrained, via its metabolism, to an all meat diet.

Table 1.   Recommended nutrient allowances for cats per kg diet, on a dry matter basis.

Nutrient	Unit	Amount
Energy (gross)	kcal	4000
Protein	%	28
Fat	%	10
Linoleic acid	%	1
Arachadonic acid	%	.1
Minerals		
Calcium	%	1
Phosphorus	%	.8
Potassium	%	.3
Sodium chloride	%	.5
Magnesium	%	.1
Iron	mg	100
Copper	mg	5
Manganese	mg	10
Zinc	mg	30
Iodine	mg	1
Selenium	mg	.1
Vitamins:		
Vitamin A	IU	10000
Vitamin D	IU	1000
Vitamin E	IU	80
Vitamin K		Negligible
Thiamin	mg	5
Riboflavin	mg	5
Pantothenate	mg	10
Niacin	mg	45
Pyridoxine	mg	4
Folic acid	mg	1
Biotin	mg	.05
Vitamin B12	mg	.02
Choline	mg	2000
Vitamin C		Negligible

Summarized from MacDonald et al.1984, Scott 1986, Basic Guide to Canine Nutrition 1977, Nutrient Requirements of Cats 1978.

Recommended dietary minimums of vitamins are found in Table 1.

Vitamin needs increase during lactation and infection; it may be wise to externally supplement at these times.

CURRENT FEEDING PRACTICES

A survey distributed at the 5th Annual Dr. Scholl Conference on the Nutrition of Captive Wild Animals (Chicago, IL, December 1985) was returned by personnel from 14 zoos in the U.S. and Canada. Their responses provide the basis of information for this section.

Maintenance

The majority of zoos surveyed (86%) offer either Nebraska Brand Feline Diet (horse meat based) or Nebraska Brand Spectrum Feline Diet (beef based) as the main meat product, with no supplements. Vitamins and minerals are added to both meat mixtures in processing (Central Nebraska Packing, Inc., North Platte, NE).

One zoo alternates feeding chicken, oxtails, boneless beef and Spectrum. One reported the use of different commercial feline diet, whereas another mixes its own diet composed of horse meat, chicken heads, a commercial cat food and dicalcium phosphate. Fourteen percent of the zoos surveyed regularly supplement with vitamin premixes and one adds 2 T. of vegetable oil to the diet.

Regardless of the source, tigers should receive diets containing at least 21% crude protein and 10% crude fat on a dry matter basis. Periodic quality checks should be performed to confirm dietary levels of all nutrients.

Tigers are generally fed once per day, at times ranging from 0700 to 1600. A one-day weekly fast is practiced by about 70% of the zoos in this survey. Amounts fed range from 2.3-6.8 kg of meat per non-reproducing adult animal, with means of 3.7 kg (N=30) and 4.8 kg (N=21) for females and males, respectively. Assuming a dietary energy concentration of 2400 kcal/kg (as fed basis) and a digestibility of 90%, these diets supply an average of 65 kcal/kg body mass. These values are identical to estimated digestible energy requirements of captive felids as discussed previously.

Both males and females show decreased appetites during breeding periods. Hot weather tends to decrease intake whereas cold weather has the opposite effect. Amounts offered should be adjusted to minimize wastage while maintaining relatively constant weight. If intake decreases, it may be necessary to increase the energy, vitamin and mineral concentrations of the diet through the addition of fat and/or vitamin/mineral supplements.

Abrasives are offered by all zoos in the survey with one exception, with no reason given for exclusion in the latter. Abrasives include whole chicken bodies and parts, mammal carcass sections (particularly shanks or tails), bones, and large strips of rawhide fed one to three times per week, biweekly, or on a random basis.

Growth

Surveyed zoos indicated no occasions for hand-rearing cubs over the past several years. If bottle-feeding is necessary, nutritionally complete milk replacers such as KMR or Esbilac (Pet/Vet Products, Borden, Inc., Norfolk, VA) should be used to supply total kilocalories up to 300 kcal/kg body mass. The number of daily feedings necessary can be determined by dividing the total quantity of milk required by the amount that can be consumed in a single feeding with comfortable stomach stretch. Records of bottle-fed cubs showed weight gains between 100 and

200 g per day (Henry Doorly Zoo unpublished data, Binczik this volume).

In the wild, tigers nurse exclusively from four to eight weeks, they are fed regurgitated solid food at six to ten weeks of age, and weaned at five to six months (Schaller 1967). Captive cubs first receive meat supplementation (0.25 kg) at about one month of age. Rapidly growing cubs require higher protein levels in the diet, and can utilize up to 35% protein on a dry matter basis.

Various alterations in captive feeding regimes were noted for young growing cubs including vitamin/mineral supplementation, mixing milk replacer with meat, the exclusion of fasting periods, and feeding two or three times daily. Regardless of the method, the objective is the same: to increase the nutrient supplies to the young animals. Energy needs of young growing felids have been estimated at approximately 200 kcal/kg body mass (Scott 1968).

It is probably a good idea to offer an abrasive during weaning periods to promote normal chewing behavior and dental health.

Pregnancy/Lactation

No zoos altered feeding regimes during pregnancy. Energy, vitamins and minerals were supplemented for the first month of lactation, however, by removing fast days, increasing meat (0.5-1.4 kg) and/or through the addition of multi-vitamin, calcium lactate and milk replacer supplements. Lactating tigresses may need to be fed more than once per day to ensure adequate and continual energy supplies. Energy needs more than double during lactation to up to 250 kcal/kg (Scott 1968).

Geriatrics

Few zoos reported specialized feeding routines for older animals. Appetite decreases with age, as do energy and protein requirements. Conversely, vitamin and mineral requirements increase. Reduced amounts of meat are offered, for longer periods of time, to geriatric tigers. Various methods are used to increase palatability of the diet, including the addition of organ (liver, heart) meats, chicken fat and alternative main meat products (hamburger, horse meat, feline diet). Only one zoo regularly adds a vitamin-mineral premix supplement.

NUTRITIONAL DISORDERS

Obesity

Much of the daily activity of free-ranging tigers, like other animals, is spent in food procurement (Schaller 1967). With normal behavior much reduced, there is a tendency for obesity to become a problem in captive animals. Although a well-fed tiger is by no means slim, care should be taken to avoid overfeeding by following the caloric guidelines outlined earlier.

It is virtually impossible to overfeed growing cubs and lactating females. In fact, females will tend to lose weight during lactation even when fed the best diets ad libitum (Scott 1968).

Practical feeding management is to maintain an average body mass range, allowing 10-20% more calories in winter for animals housed outdoors, and fewer calories in summer.

Reproductive impairment and other health problems are associated with obesity in various species. Most surveyed zoos reported that obesity is seldom or never a problem in their collections. Where it occurs, it is easily corrected by decreasing the amount of food offered.

Some zoos noted thinness, particularly among older animals, to be a frequent occurrence. As discussed previously, both energy requirements and appetite tend to decrease with age. Attempt to maintain a satisfactory body weight range and condition by feeding a more palatable diet if undernutrition is a problem.

Periodontal Disease

Dental problems associated with dietary factors have been reported for captive zoo felids. Focal palatine erosion resulting from infections following molar malocclusion has been related to soft textured commercial diets which do not provide adequate jaw exercise, leading to atrophy of jaw musculature (Fitch and Fagan 1982).

More often, however, buildup of tartar and calculi, leading to gingivitis and periodontal disease, has been documented (Fagan 1980). The disease appears to be related to a lack of abrasive factors in processed meat diets, and justifies the inclusion of some form of roughage as a normal feeding practice.

Calcium/Phosphorus Imbalance

Skeletal problems have been reported in captive feline populations for more than a century (Scott 1968). When only meat is fed, nutritional secondary hyperparathyroidism can result due to lack of calcium in the diet (Scott 1968, Fitch and Fagan 1982). The disease is unrelated to vitamin D status, and most frequently seen in young cubs.

The classical ratio of calcium (Ca) to phosphorus (P) necessary for sound bone growth is 1:1. Muscle tissue alone has a Ca:P ration of 1:20, thus, supplemental $CaCO_3$ (or some other Ca source) must be added to an all-meat diet to prevent problems. Dietary concentrations of both Ca and P should be analyzed, and supplements provided to supply both adequate quantities (from Table 1) and ratios desired.

It is often stated that in a wild state, felids consume whole carcasses, thus obtaining complete mineral nutrition from the skeleton. Conflicting reports from other studies (Schaller 1967) indicate that bones are rejected, rather than eaten. The skin of prey items is consumed, however. It is possible that in the wild minerals may be obtained from soil or dust adhering to the carcass.

Adequate Ca is especially important for young growing animals and lactating females, as both physiological states draw heavily on Ca supplies. In addition to a proper Ca:P ratio of 1:1, these animals should be receiving at least 160-200 mg/kg body mass Ca daily. Excess dietary Ca will be excreted via the feces and/or urine, and does not appear to cause kidney damage or urinary calculi formation.

Current survey results indicate no or very seldom occurrence of bone disorders due to improper mineral nutrition. A majority of zoos sampled supplement lactating females with Ca lactate.

Feline Urinary Syndrome

High sulphur loads from protein metabolism are excreted as ammonium salts, including ammonium sulphate, in the urine. These salts tend to come out of solution in concentrated urine, and can block the ureters and urethra. Thus, feline urinary syndrome (FUS) can become a nutritionally related problem, particularly in male cats.

High magnesium levels in the diet can also predispose cats to FUS through the formation of magnesium ammonium phosphate salts (struvite) (MacDonald et al. 1984).

This disease is exacerbated on diets of dry foods. Increased water intake (either through drinking or mixed with the diet) should be encouraged to minimize the potential for occurrence. FUS has not been a problem for the surveyed tiger populations.

Iodine Deficiency

Iodine deficiency has been observed in both captive zoo felines and domestic cats (Scott 1968). The disease causes an increase in thyroid gland mass due to hypertrophy of cells. Symptoms, when they occur, are most prevalent in growing cubs and kittens and lactating females fed an all-meat diet.

Meat contains little iodine naturally. As well, the high protein content of the diet results in high thyroid activity, particularly during the active metabolic states of growth and lactation. Daily requirements for large zoo felids range between 500-1000 ug. Muscle meat iodine levels average about 50 ug/kg; liver tissue contains up to five times that amount (Scott 1968). No cases of iodine deficiency were reported in the current survey.

BEHAVIORAL FACTORS

Psychological satisfaction is an important nutritional consideration that should not be ignored. Much of tiger feeding behavior in the wild involves stalking, prey capture and consumption by active tearing of the flesh. In addition to oral health as previously discussed, the inclusion of carcass sections may satisfy these instinctual needs of tigers.

Pacing behavior associated with the anticipation of feeding may be alleviated by complete randomization of feeding schedules, a concept that is, at best, impractical to achieve. Occasional supplementation with live prey items may be useful in minimizing displaced feeding behaviors.

Cats' appetites may be sensitive to environmental factors including noise, lighting, the presence or absence of other cats or humans, and presentation of the food itself. It may be possible to stimulate intake in geriatric animals by utilizing environmental cues, although little research has been conducted in that area.

Palatability

Taste does not appear to be a large factor in diet acceptability; cats are neutral to sugar and salt, but respond positively to lipids and certain protein constituents in the diet (MacDonald et al. 1984). Odor has been suggested as especially important in palatability of cat diets, particularly tigers (Schaller 1967); texture and previous experience may be equally valuable.

Cats are generally neophilic, or prefer novelty in the diet (MacDonald et al. 1984). Variation in the diet has been suggested as one mechanism of appetite stimulation in captive tigers. On the other hand, some animals resist dietary changes whether they consist of different commercial feeds or the addition of animal carcasses. These latter behaviors may be stress related, as stressed animals prefer a familiar diet. Necessary diet changes should be introduced gradually.

Aggression

Aggressive behavior directed against other cats or certain employees can result from an infringement of territorial requirements, particularly in confined zoo enclosures. These behaviors can, and do, interfere with diet consumption, prompting many zoos to feed paired animals separately. Large exhibit areas are felt to promote good health and appetites, but may be a luxury item.

DIETARY TOXINS

One peripheral area of nutritional management of captive species is the identification of potential dietary toxins. Benzoic acid, a food preservative, is toxic (both acutely and cumulatively) to cats at levels greater than .45 g/kg body mass (MacDonald et al. 1984). Butylated hydroxytoluene (BHT), a feed antioxidant, has been shown to cause enlarged livers in cats, but does not appear toxic (General Foods Corporation 1977). Neither compound should be present in meat products.

Low doses of estrogens, as from feed additives in beef or chicken diets, are also extremely toxic to cats (General Foods Corporation 1977). These compounds may interfere with reproduction, but evidence is not conclusive.

Poisonous Plants

One final area of concern in the captive management of tigers is that of plant poisoning. Most reports of toxicity due to plant ingestion by zoo animals are understandably associated with herbivores rather than carnivores (Fowler 1981). Although not a common occurrence for tigers, the potential for poisoning should be recognized from the many ornamental exotic plants, shrubs and trees found in or near both indoor and outdoor enclosures. Novel stimuli, combined with boredom often experienced by captive animals, can lead to unusual behaviors such as chewing and even swallowing items that would otherwise be rejected.

Fifty-four species of plants have resulted in suspected or clinical signs of toxicosis in domestic felids over the past two years (Anonymous 1984-1985). Most reports consist of single occurrences; however, repeated poisoning incidents were seen with Dieffenbachia (dumb cane), Dracena (dragon tree), Philodendron and Rhododendron species. These plants warrant consideration due to both their common usage as display vegetation, and specific toxicity to felids.

Dieffenbachia and Philodendron spp. contain oxalic acid in the form of calcium oxalate crystals which, if ingested, can cause irritation of the mouth and gums leading to possible blockage of the respiratory passages. No toxic principle has been identified in Dracena spp., but Rhododendron spp. contain cardiac glycosides which cause a range of clinical signs from nausea, vomiting, diarrhea and slight excitation to irregular heart rate, convulsions, circulatory collapse and coma. If toxicosis from these or other plants is suspected, seek prompt medical attention.

The most prudent management strategy, however, is simply to avoid the problem. Be aware of potentially dangerous plants. Periodically survey all enclosures and remove known or suspected toxic species. Particularly following renovation projects, restrict access to outdoor exhibits until vegetation (particularly weeds) can be properly examined.

SUMMARY

At this time, the domestic cat remains the best model for nutrition of the tiger. Unique nutrient requirements of felids must be recognized in feeding captive tigers, including the need for high protein and fat diets, inclusion of dietary niacin, taurine, arachadonic acid and vitamin A as retinol. Studies need to be conducted to determine if the tiger is, indeed, simply a large cat.

Current feeding practices appear to adequately meet energy and protein requirements of tigers as defined for domestic cats.

Chemical analyses should be performed on a regular basis to ensure diet quality, particularly for minor nutrients such as vitamins and minerals. Research needs include defining the water, vitamin and mineral needs of growing, lactating and geriatric animals.

Behavior aspects of captive feeding must also be addressed. Nutritional management of animals is both a science and an art: one must consider not only the physiological, but also the psychological benefits of any diet offered.

ACKNOWLEDGEMENTS

Information obtained from surveys returned by the following institutions is gratefully appreciated: Minnesota Zoological Garden, Houston Zoological Gardens, Detroit Zoological Park, Lincoln Park Zoological Gardens, Sunset Zoo, Columbus Zoo, New York Zoological Society, Topeka Zoological Park, Potter Park Zoo, Brookfield Zoo, Calgary Zoo, Sedgwick County Zoo, Glen Oak Zoo, Pittsburgh Zoo.

REFERENCES

Anonymous. 1977. The nutritional requirements of cats. In Basic Guide to Canine Nutrition. White plains, NY: General Foods Corp.

Anonymous. National Animal Poison Control Center. Annual Reports. 1984-85. Urbana, IL: Univ. of Illinois.

Barbiers, R.B., L.M. Vosburgh, P.K. Ku and D.E. Ullrey. 1982. Digestive efficiencies and maintenance energy requirements of captive wild felidae: cougar (Felis concolor); leopard (Panthera pardus); lion (Panthera leo); and tiger (Panthera tigris). J. Zoo Anim. Med. 13:32-37.

Fagan, D.A. 1980. Diet consistency and periodontal disease in exotic carnivores. In Proc. Ann. Dr. Scholl Nutr. Conf. pp 241-60. Chicago, IL: Lincoln Park Zoological Gardens.

Fitch, H.M. and D.A. Fagan. 1982. Focal palatine erosion associated with dental malocclusion in captive cheetahs. Zoo Biol. 1:295-310.

Fowler, M.E. 1981. Plant poisoning in captive nondomestic animals. J. Zoo Anim. Med. 12:34-37.

Kendall, P.T., S.E. Blaza and P.M. Smith. 1983. Comparative digestible energy requirements of adult beagles and domestic cats for body weight maintenance. J. Nutr. 113:1946-55.

MacDonald, M.L., Q.R. Rogers and J.G. Morris. 1984. Nutrition of the domestic cat, a mammalian carnivore. Ann. Rev. Nutr. 4:521-62.

Morris, J.G., J. Trudell and T. Pencovic. 1977. Carbohydrate digestion by the domestic cat Felis catus. Brit. J. Nutr. 37:365-73.

National Research Council. 1978. Nutrient Requirements of Cats. Washington, DC: National Academy of Sciences.

Schaller, G.B. 1967. The Deer and the Tiger. Chicago, IL: Univ. Chicago Pr.

Scott, P.P. 1968. The special features of nutrition of cats, with observations on wild felidae nutrition in the London zoo. Symp. Zool. Soc. London 21:21-36.

16

Digestibility and Metabolizable Energy of Diets for Captive Tigers

Michael Kurt Hackenberger, James L. Atkinson, Cheryl Niemuller
and Robert F. Florkiewicz

INTRODUCTION

Tigers (Panthera tigris), the largest of the felids, are consummate carnivores, requiring a high quality diet in large quantity. Conway (1980) noted that $1.2 million are spent annually on feeding captive Siberian tigers (P. t. altaica). If the entire captive population of tigers were considered, this figure would increase fivefold. As zoological institutions expand their role as reservoirs for the world's dwindling genetic material, the nutritional needs of their charges must be elucidated. The determination of specific nutrient requirements and investigation of the economics of appropriate feeding regimes are vital to the maintenance and propagation of captive species.

Information on the feeding and nutritional needs of tigers is limited (Morris et al. 1974, Wittmeyer 1980, Barbiers 1982). Generally, diets for captive wild felids have been developed by trial and error, or from extrapolation of domestic cat requirements. An important aspect of our nutritional knowledge concerns the efficiency with which dietary components are digested and metabolized. The present studies were designed to investigate the digestibility of various diets by adult and juvenile tigers and the metaboliz-able energy value of a diet fed to adults and cubs.

MATERIAL AND METHODS

The initial diet digestibility study utilized the total collection method. The subjects were 18 tigers, comprised of 12 adults and six juveniles, kept at the Hawthorn Corporation, Richmond, Illinois. The animals were identified by sex and subspecies, and as all were captive bred, their ages were well established (Table 1). All animals were individually housed in 3.5 m cages with concrete floors, which allowed complete and separate fecal collection for each tiger. The enclosures were maintained at 12°C and were provided with a sleeping platform.

During the five-day trial period, a known weight of feed was provided daily at 17:00 hr. All food was consumed and water was available ad libitum throughout the experiment. No adjustment period was required, since the diets were identical to those the tigers received prior to the experiment.

The adult tigers received a chunk beef diet stored at -30°C and thawed for two days prior to feeding. The beef was well-trimmed and dusted with a 3:1 ratio of calcium lactate:Pervinal vitamin mix at the rate of 100 g/kg meat. Total intakes ranged from 19.3 to 25.6 kg fresh weight of diet over the course of the study.

Five of the juvenile animals received identical rations of 1.6 kg fresh weight of a commercial frozen feline diet and 320 g kitten milk replacer (KMR) daily. The sixth, a male white tiger, slightly older than the others (Silver, Table 1), received the

Table 1. Details of tigers used in feeding trials.

	Sex	Subspecies	Age	Estimated Weight (kg)
Digestibility Trial				
Adults				
Jack	M	ssp.	4 y	127
Babe	F	ssp.	2 y	90
Overbite	F	ssp.	4 y	95
Nasty	F	ssp.	3 y	110
Indira	F	ssp.	16 m	90
Bandola	M	tigris	22 m	123
Nazar	M	tigris	16 m	104
Saber	M	tigris	13 m	82
Honey	F	tigris	16 m	105
Mindy	F	tigris	22 m	110
Sheba	F	altaica	9 y	120
Roman	M	sumatrae	5 y	130
Juveniles				
Silver	M	ssp.	8 m	50
A	F	ssp.	6 m	23
B	F	ssp.	6 m	23
C	F	ssp.	6 m	23
D	M	ssp.	5.5 m	23
E	M	ssp.	5.5 m	23
Digestibility and metabolizable energy trial				
Adults				
Tito	M	tigris	8 y	225
Ceylon	M	tigris	8 y	250
Cubs				
A	M	ssp.	3 m	6.1
B	F	ssp.	3 m	5.7

Table 2.   Proximate composition and gross energy of tiger diet components.

Diet	DM %	CP %	EE %	Total NFE %	ASH %	GE (kcal/g)
Chunk Beef	29.7	60.8	32.7	2.8	3.7	6.0
Commercial Feline Diet[a]	9.5	50.0	28.5	10.5	11.0	5.8
Kitten Milk Replacer[b]	17.72	41.5	23.0	29.2	6.3	5.8

[a] Guaranteed Analysis:
   Crude Protein...Min. 19.0 %
   Crude Fat.......Min. 12.0 %
   Crude Fiber.....Max. 1.5 %
   Ash.............Max. 4.5 %
   Calcium.........Min. .60%
   Phosphorus......Min. .50%
   Moisture........Max. 62.0 %
   Vit. A Min...... 7,500
   Vit. D_3 Min..... 850 USP u/lb.

[b] Guaranteed Analysis:
   Crude Protein...Min. 7.5%
   Crude Fat.......Min. 4.5%
   Crude Fiber..... 0.0
   Moisture........Max. 82.0%
   Ash.............Max. 1.5%

same ration supplemented with about 1.4 kg/day of the chunk beef diet fed to the adults.

Samples of the various diet components were retained for later analysis (Table 2). Feces were removed from the cages immediately following defecation, weighed and stored frozen.

The fecal samples were dried in a forced-air oven at 43°C and the diet samples were freeze-dried. Dry samples were ground through a 1 mm screen in a Wiley mill and analyzed by standard methods (A.O.A.C. 1980). Nitrogen or crude protein (N x 6.25) was determined by Kjeldahl analysis, crude fat by Soxhlet extraction with anhydrous ether, ash by incineration at 600°C and dry matter by oven-drying. The total carbohydrate fraction was determined by difference. Gross energy was measured with a Parr adiabatic oxygen bomb calorimeter.

For the measurement of dietary digestible and metabolizable energy values, two adult male tigers and two cubs were used (Table 1). The adults were circus animals, housed in individual travelling cages, while the cubs were also kept separately in wire-mesh enclosures. The design was similar to that of the digestibility study, except that an attempt was made to quantify urine production in addition to monitoring food intake and fecal production. The three-day collection for the adults was carried out in Toronto and the five-day collection for the cubs in Cambridge, Ontario. The animals received a chunk meat diet with a gross energy of 7.15 kcal/g dry matter, equivalent to 2.83 kcal/g as fed. Feed samples and feces were frozen for storage, freeze-dried and ground prior to analysis for energy content as described previously. Fifty ml aliquots of urine were freeze-

Table 3.   Apparent Digestibility (%) of energy and proximal components of tiger diets (Mean ± SD).

	Dry Matter	Crude Protein	Crude Fat	Total Carbohydrate	Energy
Adults	97.8 ±0.6	98.2 ±0.5	99.2 ±0.3	79.0 ±6.0	98.0 ±0.5
Juveniles	86.0 ±1.8	91.7 ±0.9	97.8 ±0.2	65.2 ±2.8	91.4 ±1.1

dried and the gross energy of the residue measured by bomb calorimetry.

RESULTS AND DISCUSSIONS

The apparent digestibility values for the energy and approximate diet components were determined (Table 3). Data for males and females were pooled within each age class, since they were not statistically different. Diet digestibilities were uniformly high for all components, with the lowest values being recorded for the total carbohydrate fraction. This is of little significance for the adult tigers, since the total carbohydrate content of the chunk beef diet was less than 3% on a dry matter basis (Table 2). For the juveniles, substantially more carbohydrate was present due to the inclusion of kitten milk replacer and the presumable presence of some cereal in the commercial feline diet (Table 2). Overall the juveniles showed slightly less efficient digestion of the approximate components and energy than the adults. Although this may reflect the different composition of the diets fed to the two groups, it could also imply some immaturity of digestive function in the younger tigers.

In comparison with previously published data for digestibility of commercial carnivore diets (Table 4), the adult tigers gave distinctly higher values for dry matter, protein and energy than others have reported. This is probably due to the low fraction of connective tissue and other indigestible protein in the well-trimmed chunk beef diet compared to diets based on meat and meat by-products. The similarity of the published values and those for the juveniles in this study lends support to this view.

However, some age effect is also evident in the results of the digestible and metabolizable energy study (Table 5). Although data for only two adults and two cubs is presented, the older tigers appear to be slightly more effective in energy digestion than the cubs fed an identical diet. Similarly, the metaboliz-able energy values were higher for the adults, though this could be due to an artifact of the collection process. The total collection from the adult males was compromised by the production of an aerosol spray of urine which may have led to an under-

Table 4.   Published diet digestibility values for captive tigers.

Diet	Digestibility				
	Dry Matter	Crude Protein	Crude Fat	Total Carbohydrate	Gross Energy
Nebraska Brand frozen feline[a]	83.7	88.2	98.9	----	91.8
Frozen commercial meat and meat by-product based[b]	85.2	90.6	98.9	66.3	92.9
Zu/Preem frozen carnivore[b]	79.2	88.8	----	----	----

[a] Barbiers et al. 1982.
[b] Wittmeyer Mills 1980.
[c] Morris et al. 1974.

estimation of the total volume. The urine energy expressed as a percentage of the digestible energy was 3.4 for the adults and 8.2 for the cubs. The latter is closer to the intra-species value of 9.7 reported by Robbins (1983) for fissiped carnivores.

In summary, the digestibility of tiger diets, in common with that of other carnivores, is high but may be affected by the actual composition of the diet fed and the age of the tiger. Such information is valuable in assessing the appropriate feeding

Table 5. Diet digestibility and metabolizable energy values for adult tigers and cubs[a].

Animal	Digestible		Metabolizable	
	Dry Matter (kcal/g)	Gross Energy (%)	Dry Matter (kcal/g)	Gross Energy (%)
kcal/g Dry Matter				
Adult				
Ceylon	7.07	98.9	6.81	95.2
Tito	7.03	98.3	6.79	94.9
(Mean)	7.05	98.6	6.80	95.1
Cub				
Male	6.84	95.7	6.22	87.0
Female	6.82	95.4	6.34	88.7
(Mean)	6.83	95.6	6.28	87.8

[a] Gross energy of feed = 7.15 kcal/g DM; 2.38 kcal/g as fed.

level necessary to achieve optimum growth, maintenance and reproductive performance in captive tigers.

ACKNOWLEDGEMENTS

The authors wish to thank John F. Cuneo, Josip Marcan and Don Dailley for generously making tigers available for use in these studies, Jim Sorenson, General Manager of the Hawthorn Corporation for invaluable assistance during the collection period, Jeff for his generous efforts at Maple Leaf Gardens, and Kathryn M. Dickson for invaluable help with the statistical analysis.

REFERENCES

Association of Official Analytical Chemists. 1980. Official Method of Analysis of the Official Analytical Chemists. 13th. Ed. Washington, DC: AOAC.

Barbiers, R.B., L.M. Vosburgh, P.K. Ku and D.E. Ullrey. 1982. Digestive efficiencies and maintenance energy requirements of captive wild Felidae: cougar (Felis concolor); leopard (Panthera pardus); lion (Panthera leo); and tiger (Panthera tigris). J. Zoo. Anim. Med. 13:32-37.

Conway, W.G. 1980. Where do we go from here. Int. Zoo Yearb. 20:184-89.

Morris, J.G., J. Fujimoto and S.C. Berry. 1974. The comparative digestibility of a zoo diet fed to 13 species of felid and a badger. Int. Zoo Yearb. 14:169-71.

Robbins, C.T. 1983. Wildlife Feeding and Nutrition. New York: Acad. Pr.

Wittmeyer M.A. 1980. A comparative study of the digestibility and economy of three feline diets when fed to lions and tigers in confinement. In The Comparative Pathology of Zoo Animals, ed. R.J. Montali and G. Migaki. Washington, DC: Smithson. Inst. Pr.

17

A Neonatal Growth Model for Captive Amur Tigers

G. Allen Binczik, N.J. Reindl, R. Taylor, U.S. Seal and R.L. Tilson

If, as it appears, captive propagation is destined to play an integral part in ensuring the future of the Amur tiger (Panthera tigris altaica), it behooves zoo workers to find an adequate means of monitoring, and thereby controlling, the development of young of this species in captivity.

Although efforts toward this end are reflected admirably in the sometimes extensive in-house records maintained by zoos on several facets of neonatal tiger development (e.g. growth in weight and height, age at tooth eruption, etc...), little of this information has been published, and formal studies based on large data sets have been few. The growth model described herein is presented to amend this situation.

Mixed-longitudinal weight data on 111 Amur tiger cubs, all born within the past 20 years, were compiled from records made available by 15 North American zoological parks. Only the 50 cubs with weights for five or more of the first 60 days of life were used to construct the model.

Analyses were restricted to the 0-60 day period for three reasons. First, the data were more complete for this period than for any time later. Second, initial curve-fitting for these cubs indicated that the pattern of growth became both more complex and variable soon after 60 days. Finally, this early period was crucial for the cubs; 63% of all neonatal mortality (N=19) occurred during this time (see also Mlikovsky 1985).

The cubs were split into two groups; those which were nurtured by their dams throughout the entire period are referred to as 'mother-reared' (N=12); and those which were removed from their dams soon after birth are referred to as 'hand-reared' (N=38). Individual hand-rearing formulas and techniques reported by 12 zoos varied only slightly and were not perceived by the authors to be cause for differences in cub growth among institutions.

Regression analyses were made with conventional micro-computer software. A P<0.05 level of significance was accepted for all statistical tests.

Table 1.   Comparison of mathematical models for describing neonatal (0-60 days) Amur tiger growth exhibited by 50 cubs.

| Equation | Number | | %[a] | Mean |
	Best Fit	2nd Best		R-squared[a]
$y = a + bx$	36	14	100	.980
$y = a^{bx}$	12	24	72	.965
$y = ax^b$	2	10	24	.968
No fit	0	2	4	---

[a] Based on number of best and second best fits.

The mathematical models tested for correlation to actual neonatal tiger growth included linear, exponential, power and log equations. The linear equation $y=a+bx$, where y is the weight (kg) of a cub at age x (days), provided the best fit in 72% of the trials, and the best or second best fit in all trials (Table 1). This model was also the most highly correlated with the actual data, with a mean R-squared =.980 for all cubs surviving through 60 days. Ranked second and third, respectively, were the exponential and power equations. None of the cubs exhibited a logarithmic pattern of growth.

Of the 12 cubs which did not survive through 60 days, six showed inconsistent or no growth prior to death and were not well represented by any model, four were best represented by linear equations, and one each by exponential and power equations. All but one of the cubs exhibited from one to several depressions or plateaus in their growth curves prior to death. The exceptional cub exhibited steady linear growth until its accidental death at 60 days of age.

The a and b values determined for individual cubs by linear regression analysis showed:
1) The growth of mother- and hand-reared cubs differed significantly throughout the entire 60 days. This difference was reflected in the model intercept values, with mother-reared animals exhibiting an a=1.559 kg (N=11, SD=.113, range=1.425-1.847) versus a=.971 kg (N=36, SD=.364, range=.266-1.913) for hand-reared animals (t=5.250, 45 df, P<0.001). There was no corresponding difference between the two groups with respect to actual birthweights (mother-reared=1.360 kg, N=5, SD=.152, range=1.200-1.600 vs. hand-reared=1.285 kg, N=27, SD=.161, range=.900-1.600; t=.964, 30 df, P>0.20). Hand-reared animals were not, therefore, predisposed to depressed growth due to low birthweights;
2) Differences between model slopes for the two rearing strategies were not significant (mother-reared b=.1053 kg/day, N=11, SD=.0096, range=.0876-.1218 vs. hand-reared b=.1008 kg/day, N=36, SD=.0180, range=.0567-.1383; t=.790, 45 df, P>0.40). Consequently, the fitted growth curve of hand-reared cubs was parallel to but lower than that of mother-reared cubs. A similar

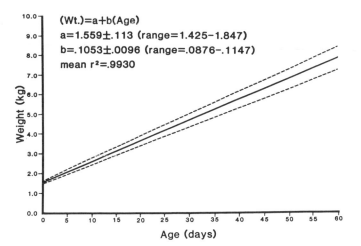

Fig. 1.   Neonatal (0-60 days) Amur tiger growth model based on 11 mother-reared cubs (studbook no. 1853, 1854, 1855, 2308, 2309, 2310, 2311, 2498, 2499, 2657 and 2658; from Seifert and Muller 1985).   Dashed lines delineate 95% confidence interval.

difference in growth between mother- and hand-reared young has been reported for other large felids (De Carvalho 1968).

Also, although not statistically significant, there was some evidence that the depressed growth correlated with hand-rearing was also correlated with higher mortality for this group.   During the first 60 days of life, 22.4% (N=49) of the hand-reared cubs died compared to only 8.3% (N=12) of the mother-reared (with Yate's correction, G=.528, 1 df, P>0.10).   The deleterious effects of hand-rearing may also be long-lasting.   By two years of age, mortality among hand-reared cubs had risen to 42.5% (N=40) while 20.0% (N=10) of the mother-reared had died (with Yate's correction, G=.938, 1 df, P>0.10).

For the above reasons, only data from mother-reared cubs were grouped to determine the parameters of the growth model presented in Fig. 1.

Little information exists in the literature to compare with this model.   Data published on seven mother-reared cubs prompted Veselovsky (1967) to describe young Amur tiger growth as ". . . very rapid and on the whole even," and correspond closely to that presented in this paper.   Hemmer (1979, summarized from literature) reported a lower rate of growth (.086 kg/day) for an unspecified number of cubs.

One shortcoming of the model is that the intercept failed to accurately describe birthweight.   Linear representations of growth consistently overestimated the birthweights of mother-reared animals and underestimated those of hand-reared.   This may have been due to a slight postnatal growth surge common to mother-reared cubs having been replaced in hand-reared animals by a slight depression (unpublished data).   In both cases, the model corresponded very closely to actual cub growth within the first week after birth.

Also, being based on only 11 mother-reared cubs (of four litters), the model may not account for natural variation in cub growth, particularly with respect to the above-mentioned intercept. Slight refinement of the model parameters may be indicated after a greater mother-reared database has been compiled. Nonetheless, the model presents a standard by which an institution may gauge the effectiveness of its rearing regimen.

On the basis of this study, a few recommendations are apparent:

Cubs which fail generally do so after a brief interruption of normal growth. Barring gross pathological or behavioral abnormalities, detection of this pause in development by frequent weighings may offer the first indication that special attention, perhaps even clinical treatment, is warranted. The authors recommend cubs be weighed daily during the first week of life, and as often as once every two to three days thereafter.

Whether the result of institutional policy or circumstance (e.g. abandonment by dam), hand-rearing remains a widespread and common means by which infant tigers are raised in North American collections. Because this technique is correlated with both depressed growth and increased mortality, it would most appropriately be considered only a last resort for rearing tiger cubs.

If an animal must be hand-reared, protocols utilizing milk replacers should be critically re-evaluated. For a species suffering over 30% infant mortality (this study, Mlikovsky 1985), there is clearly merit in working toward the development of new, better hand-rearing regimens.

ACKNOWLEDGEMENTS

The authors wish to acknowledge the generosity shown by the Bronx, Calgary, Detroit, Henry Doorly, Lincoln Park, Metro Boston, Milwaukee, Minnesota, Philadelphia, Potawatomi, Rio Grande, Riverbanks, St. Louis, San Diego and Washington Park zoos in sharing information on animals in their collections.

REFERENCES

De Carvalho, C.T. 1968. Comparative growth rates of hand-reared big cats. Int. Zoo Yearb. 8:56-59.
Hemmer, H. 1979. Gestation period and postnatal development in felids. World's Cats. 3(2):90-100.
Mlikovsky, J. 1985. Sex ratio distribution in the Siberian tiger Panthera tigris altaica (Mammalia: Felidae). Z. Saugetierkunde. 50:47-51.
Seifert, S. and P. Muller. 1985. International Tiger Studbook. Leipzig: Zoologischen Garten Leipzig.
Veselovsky, Z. 1967. The Amur tiger in the wild and in captivity. Int. Zoo Yearb. 7:210-15.

18

Clinical Management of
Captive Tigers

Mitchell Bush, Lyndsay G. Phillips, and Richard J. Montali

The display and care of tigers in zoological parks has a long history. The beauty, power and awe of this large felid has been appreciated by both the staff of zoological gardens and the visiting patrons. There are few zoos that do not display tigers. It is surprising that with this long history and the numerous tigers kept in captive conditions that proportionally little medical data has been published. The majority of data that exist report medical problems that are preventable, given the present state-of-the-art and/or science of zoological medicine, namely dietary related deficiencies and diseases or viral infections.

The medical and surgical care of captive tigers has become easier with advancements in other areas. Major nutritional problems are almost nonexistent due to the increased understanding of dietary needs including proper ratios of vitamins and minerals. The majority of contagious infectious diseases can be prevented by appropriate vaccines. Parasites (external and internal) can be effectively controlled or eliminated with newer drugs.

The development of new anesthetic drugs and refinement of anesthetic techniques allows safer access to the patient and permits aggressive diagnostic evaluations and more sophisticated surgical procedures.

The present challenge to a zoo veterinarian is to establish and maintain a strong preventive medical program. Once this is instituted and functioning, the major problems encountered will be the geriatric problems of an aging collection.

In dealing with tigers, it has been stated repeatedly they are just big domestic cats and share many of the same anatomical, physiological, and medical conditions. Fortunately, this is true in some situations and allows the veterinarians to extrapolate from experiences and expertise dealing with the domestic counterpart. It is best not to be overzealous in utilizing this comparison. Tigers are unique animals; new problems and new techniques should be approached through careful planning and good clinical judgment. Always be ready for the unexpected!

PREVENTIVE MEDICAL PROCEDURES

The scope of preventive medical procedures is extensive and covers the time the animals enter the collection (i.e., born or shipped) to it's disposition to another zoo, or to the complete post-mortem examination when it dies. Specific preventive factors relate to diet, cleaning procedures, pest control and the more medically oriented procedures such as vaccinations, parasite control and quarantine.

Shipment procedures for tigers require good organization and coordination to minimize stress. Prior to shipment the health status of the tiger is evaluated to help insure the animal's safety. The animal should have access to its shipping crate for 2 weeks prior to shipment and preferably be fed in it. The design of the crate should be large enough to meet the USDA requirements and strong enough to safely hold the tiger. The crate should provide adequate ventilation without allowing the tiger access to people or visa versa. If an extended trip is anticipated (>12 hr), provisions should be made to water and feed the animal while it is in the crate. The crate should drain well, and absorbent bedding should be utilized to prevent the tiger from lying in urine. The crate should be strong and large enough for the tiger but the size should allow easy movement and access through doorways. When a tiger is moved, one of its keepers should accompany it to care for and help the tiger adjust to the new environment.

Prior to the introduction of any new tiger to others in an existing population, the newcomer is quarantined for at least 30 days. Ideally the tiger is held in a separate facility and cared for by different keepers than those who care for other felids. Unfortunately, this may not always be possible. In these instances, the tiger is separated from other cats as much as possible and keepers work with it after he/she is done with the existing collection. Personnel working with or near a quarantined cat should wear coveralls and rubber boots designated for the quarantine area. A foot bath going to and from the quarantine may help prevent spread of potential contamination. The quarantine area should have separate drainage which will not cross contaminate other cat areas.

The quarantine period allows observation and testing to monitor the animal for infectious diseases and/or parasites. The 30-day time period is adequate to cover the incubation period of most infectious diseases. Newly captured animals or animals from foreign zoos may require a longer quarantine.

The evaluation of a new tiger begins with a review of its past medical history which should be part of the health certificate. It is unacceptable to send an animal to a new collection without sending its medical history. This data will alert the clinician to previous and potential problems and document past vaccinations, fecal examinations and blood values.

Vaccinations are given during quarantine to allow time for antibody response prior to release of the tiger into the collection. Severely stressed animals may not mount appropriate titers and should be revaccinated if conditions indicate. There is one vaccine that provides good antibody titers to 3 major diseases 1) panleukopenia, 2) rhinotrachitis, 3) calicivirus.

This is a killed product, (Fel-o-vax, Fort Dodge Lab., Inc., Fort Dodge, IA 50501) and provides apparent protection at a 1 ml dose (standard domestic cat), in adult tigers (Bush et al. 1981).

The use of other vaccines is dependent on local situations. In areas endemic for rabies, a killed vaccine (Imrab, Pitman-Moore, Inc., Washington Crossing, NJ 08560) is recommended by the authors for protection of the animal. It is realized this recommendation is at odds with the Compendium of Animal Rabies Vaccines, 1986, which was prepared by the National Association of State Public Health Veterinarians, but we view it as necessary to protect our patients particularly in areas where rabies is enzootic in the wildlife. The use of a killed rabies vaccine at the recommended dose (1 ml) produced, in captive tigers, what is reported to be a protective titer in domestic species (Bush et al. 1985). This vaccine should be repeated yearly in rabies endemic areas.

Leptospirosis is a potential disease in all mammals but has not been reported in tigers. In environments where leptospirosis has been found, vaccination of tigers with a bacterin may be indicated. Two problems arise with this, first, the correct vaccine serotype may not be available, and secondly, the duration of protection for vaccines is usually short lived, 2-3 months.

The recent report of canine distemper encephalitis in a tiger (Blythe et al. 1983) poses the question as to the use of a canine distemper vaccine. At this time, this would not seem to be indicated. Hopefully, most zoos have canine distemper under control in their collection by vaccination of susceptible animals and control of feral animals. A possible exception for indicating vaccination might be if an outbreak of canine distemper occurred near tigers, especially if young tigers were present. If a canine distemper vaccine were to be used, only a killed product should be considered. Unfortunately, no such product is commercially available at this time, but can be obtained through some veterinary schools or laboratories on an experimental basis. We have found that some killed vaccines historically do not produce good antibody titers against canine distemper in a variety of exotic carnivores (Montali et al. 1985).

Vaccination of tigers with feline leukemia vaccine has been done and the animals developed good titers (Citino pers. comm.). The use of this vaccine as a regular procedure, however, requires more consideration since no tigers have been found to be positive for this virus, and neoplastic and immunosuppression syndromes, as occur in domestic cats with feline leukemia virus have not been documented in tigers.

During quarantine a blood sample is obtained to evaluate the animal's health status and provide a baseline for future comparison. An initial blood sample can be taken prior to anesthesia by placing the tiger in a squeeze cage and bleeding it from the lateral tail vein. This will help evaluate the animal's status prior to anesthesia for a complete physical examination. Clinical parameters of primary interest at this time would be total WBC count and differential, hematocrit, BUN, creatinine, liver enzyme values and an examination for red blood cell parasites. A summary of normal blood values for tigers is listed in Table 1 which has been compiled by ISIS. It should be recognized that values might differ between various laboratories and the best set of values is the one that your laboratory compiles. Values such as BUN also may vary from collection to collection, probably in response to diet.

During the quarantine period tigers should also be screened for internal parasites by repeated fecal examinations. If present, parasites should be eliminated, with appropriate anthelmintics, before the tiger is released into an exhibit. This is extremely important in situations where naturalistic exhibits (i.e., dirt and grass) can become contaminated with parasite eggs. Screening for enteric pathogens by stool culture may help identify tigers that are carriers of Salmonella spp.

Usually two to three weeks into the quarantine period a complete physical examination is performed under general anesthesia. A second blood sample is collected for hematologic and clinical chemical screening; at least 20 ml of serum should be frozen for future reference. The skin and ears are examined for external parasites. The feet are examined for ingrown claws and pad lesions. A careful oral examination should particularly assess the teeth for problems which could relate directly to the animal's health and attitude. Un- proven adult males destined as breeders should have semen collected by electroejaculation to evaluate reproductive potential.

When in quarantine, the tiger's diet may be gradually changed to the new diet if different from its original. Any dietary alterations should be gradual to minimize gastro-intestinal upset. It is not unusual to have a newly arrived tiger stop eating because of the environmental change. In some cases it is advantageous to have some of the animal's previous diet accompany it if the food is not available locally. To stimulate appetite, whole carcasses of rabbits or chickens may be offered.

ONGOING PREVENTIVE MEDICINE

Preventive medical programs during quarantine have been stressed in opening sections. It is important to continue this concept as an ongoing program in the maintenance of the tiger in the exhibit units.

An important portion of this preventive approach should include routine observation of the tigers in the collection, not only by the keepers, but also by the veterinary staff. Routine rounds through the cat area should be made to remain familiar with the tigers, to evaluate their overall appearance, activity, and facility conditions, and to talk with the keepers. Occasionally, the veterinarian's experience allows him to detect an abnormality or develop an impression about the tigers' health that may not be obvious to the keeper staff. Familiarity with the normal conditions and tigers will allow a better comparison when the animal is suspected of having a subtle abnormality. As well, a good relationship with the keepers promotes open communications between the animal and veterinary staffs.

A parasite monitoring program is established to provide periodic, regular stool examinations to detect parasitic infections. In many instances, it is realistic to assume that some parasites will not be eradicated from the tiger collection. In these situations the goal is to keep parasite infections at low levels or periodic in nature; a sound monitoring program can accomplish this (Phillips 1980).

Most internal parasites found in stool examinations are relatively common and ubiquitous in captive situations, with some less frequent infections reported (Nagar 1978, Abderasoul 1979, Dasguptz 1979, Lennsink 1979, Gaur 1980). Commonly identified species are from the orders Ascarididae and Strongyloidae (i.e., Toxocara, Toxascaris, Ancylostoma). It is seldom possible to eliminate ascarids totally in tigers, but they are controllable with periodic administration of oral anthelminthics. These agents can be more effective when the full daily recommended dosage is given for more than one day, such as 3 consecutive days, rather then as single treatments. Post-treatment fecal examinations are necessary in assessing efficacy of the initial treatment as are follow-up treatments to remove larval stages not susceptible during the initial treatment.

In our experience, we have found the following anthelminthics to be effective and safe when administered using appropriate dosage regimens:
   1) pyrantel pamoate (Strongid-T, Pfizer Inc., New York, NY 10017) - 3-5 mg/kg per os. Can be given at this level for 3-5 consecutive days.
   2) fenbendazole (Panacur, American Hoescht, Somerville, NJ 08876) - 5-10 mg/kg per os. Most commonly single day treatment, but can be given 3 consecutive days at this level;
   3) ivermectin (Ivomec, Merck & Co., Rahway, NJ 07065) - 0.2 mg/kg, subcutaneous, per os. We have used injectable cattle formulation orally at this dose for 1-3 days. Limited use in tigers with the parenteral route;
   4) piperazine adipate - 88-110 mg/kg per os. Use of this drug orally in either 1 or 2 day treatments. The adipate form has been shown to cause fewer, if any, side effects than has been reported with the citrate form. Citrate should be avoided;
   5) praziquantel (Droncit, Haver-Lockhart, Shawnee, KS 66201) - 5.5-6.6 mg/kg. Either as the oral or parenteral form for cestodes;
   6) sulfadimethoxine (Albon, Roche Chemical Div., Nutley, NJ 07110) - 50 mg/kg, parenteral or per os, as a coccidiostat.

Additional references have listed other anthelminthics used in tigers (Wallach and Boever 1983):
   1) dichlorvos (Task, Atgard, Equigard, E.R. Squibb & Sons Inc., Princeton, NJ 08540) - 24-30 mg of active ingredient/kg in 3-5 divided doses orally. Care should be taken with this chemical to watch for signs of organo-phosphate toxicity;
   2) thiabendazole (Thibenzol, Merck & Co., Rahway, NJ 07065) - 50-100 mg/kg per os;
   3) levamisole (Levasole, Pittmann-Moore, Washington Crossing, NJ 08560) - 10 mg/kg per os;
   4) sodium disophenol (DNP, Professional Veterinary Pharmaceuticals, Wayne, NJ 07470) - 9 mg/kg, subcutaneous;
   5) mebendazole (Telmin, Pittman-Moore, Washington Crossing, NJ 08560) - 15 mg/kg per os, given 2 consecutive days.

Not all eggs or parasitic forms observed in fecal examinations may necessarily be parasitic to the tiger. The tiger may be serving as a paratenic host depending on what it has been fed or what feral animals may be consumed. Coccidia

Table 1. Normal blood values for Panthera tigris (compiled by ISIS).

Hematology

	WBC	RBC	HGB	HCT	NUC RBC	MCV	MCH	MCHC
<1 Yr Male/Female	13.9 13.4	5.6 5.8	10.2 10.7	29.6 31.5	0.0 0.0	52.9 51.2	17.9 17.7	34.3 34.7
>1 Yr Male/Female	13.5 12.3	7.0 6.4	13.8 11.8	38.5 34.4	0.0 0.0	54.6 53.4	19.6 18.6	35.6 34.9
Total Mean, SD	12.8 3.4	6.3 1.0	11.9 2.1	34.2 5.7	0.0 0.0	53.3 5.4	18.6 2.0	34.9 1.5
# Samples, Spec.	305 92	261 70	266 71	305 92	194 51	256 70	240 59	259 69

Differential

	SEGS	BANDS	LYMP	MONO	EOSIN	BASO	RETICS	PLATLETS
<1 Yr Male/Female	5.3 6.2	0.3 0.1	2.4 1.8	254 249	150 191	14.2 6.8	0.3 0.2	494 272
>1 Yr Male/Female	8.0 4.3	0.5 0.5	2.0 1.1	287 133	216 94.9	5.0 2.8	0.1 0.2	271 120
Total Mean, SD	5.2 4.6	0.5 0.7	1.5 1.6	187 231	131 205	4.9 22.5	0.2 0.3	318 120
# Samples, Spec.	298 89	288 83	298 88	286 86	291 88	267 77	41 26	30 20

Chemistry

	CA	PHOS	GLUCOSE	BUN	URIC ACID	CHOLEST	T BIL	D BIL	I BIL	CREAT
<1 Yr Male/Female	10.6 11.1	8.0 7.9	131 122	28.7 25.6	0.5 0.4	238 225	0.2 0.3		0.1 0.1	1.1 1.0
>1 Yr Male/Female	10.3 9.9	5.8 5.5	115 105	31.0 28.2	0.5 0.3	242 186	0.2 0.2		0.1 0.0	2.6 2.3
Total Mean, SD	10.1 0.7	5.9 1.3	110 29.3	28.5 7.3	0.4 0.3	200 57.0	0.2 0.1	0.1 0.1	0.0 0.1	2.1 0.6
# Samples, Spec.	265 70	266 70	255 68	262 69	254 66	261 70	264 70	7	6 7	255 66

Table 1. Continued.

Chemistry (cont.)

	ALK PTASE	NA	K	CL	MG	HCO3	LDA	SGOT	SGPT	CPK
<1 Yr Male/Female	158 122	151 150	4.9 5.0	118 116			208 161	28.0 24.5	39.5	159 153
>1 Yr Male/Female	23.0 22.0	151 153	4.2 4.2	121 123			111 104	28.2 25.3	0.0 60.3	252 266
Total Mean. SD	40.1 52.3	151 3.2	4.3 0.5	121 3.7	15.0 13.8	12.3 3.4	117 65.8	26.0 10.3	0.1 42.3	232 241
# Samples, Spec.	264 69	249 65	250 55	248 65	13	8	259 70	263 70	7 22	11 17

Electrophoresis

	T PROT	GAMMA GLOB	ALBUMIN	OSMOLARITY
<1 Yr Male/Female	6.2 6.3	2.5	3.4 3.0	304
>1 Yr Male/Female	7.4 7.2	3.2 3.4	3.4 3.4	308 307
Total Mean. SD	7.1 0.5	3.1 0.5	3.4 0.5	306 6.1
# Samples, Spec.	280 79	7 6	195 48	36 19

observed may be associated with feeding whole carcass specimens
(e.g., whole rabbits). This emphasizes the need for specific
identification of parasite stages seen in stool and an awareness
of the tiger's diet.

Feral animals serve as sources of additional problems for
tigers. Rodents, birds, domestic cats or dogs, or other pests
that have access to the tiger or its enclosure may serve as a
source of contamination for microorganisms or parasites. An
obvious need for pest control exists in any animal holding
facility.

Well maintained perimeter fencing provides an initial
deterrent to the larger feral animals, particularly dogs.
However, climbing animals, such as cats, can easily defeat such
barriers; therefore, areas around tiger enclosures should be
monitored regularly for feral animal activity. Live trapping
provides a method of removing some feral animals that is
acceptable to the public and humane animal interest groups.
Local animal shelters will usually assist in removal of domestic
animals captured this way. Removal of wildlife trapped in this
manner may be coordinated through state agencies or local
rehabilitation groups. Trapping does not provide a total
eradication of pests; therefore, the design of the tiger
enclosures should be such that it reduces exposure to feral
animals.

Rodent pests must be handled through a well planned,
supervised, continuous pest control program. Safe rodenticides
are available for use around tigers when they are applied
according to their directions. Care must be taken in choosing
compounds that are effective, yet not highly toxic, especially
when considering secondary toxicities. A number of anti-
coagulant rodenticides are available that are effective and have
little or no secondary toxicity potential, e.g., warfarin,
diphacinone, cholecalciferol, brodifacoum. These are the
backbone of most vermin control programs. When rodent
populations become unmanageable or resistent to anti-coagulants,
other more toxic compounds, such as zinc phosphide may be needed,
requiring extra care in their application. It may seem too
obvious, but it should be emphasized that at no time should
tigers have primary access to any rodenticide. In addition, the
program should be designed and instituted to minimize secondary
exposure (i.e., consuming rodents that have been feeding on
poisonous baits).

Insect pests are included in the pest control program. Good
sanitation aids in reducing insect populations, but all
zoological situations experience insect pests, particularly
cockroaches. Insecticide applications can be made around tiger
enclosures with chemicals that are safe when applied in a proper
manner. There are many chemicals available, both primary
insecticides and newer growth regulator compounds, that have low
toxicity potential when used correctly. (Examples of
insecticides include: diazinon, piperonyl butoxide, natural and
synthetic pyrethrins, carbamates, chlorpyrifos; example of growth
inhibitor is Gencor.) Tiger enclosures can be treated by
removing the tigers, applying chemicals that have been deemed
safe to use in primary enclosures, and then cleaning the cage to
avoid exposure to returning tigers. The residual chemicals in
cracks and crevices should have no contact with the tigers but,
if so, exposure levels should be minimal.

All personnel involved with the tigers should participate in the planning stage of the pest control program so everyone is aware of the compounds being used, where and how they are applied, and knowledgeable of the safety of the compounds themselves. Safety of the tigers is utmost in any program.

Inadvertent use or misuse of insecticides can lead to accidental exposure of tigers and possibly fatal results. One author (L. Phillips) has personally experienced loss of several tigers due to incorrect application of insecticides by inexperienced personnel due to a breakdown in communications. The veterinary staff was not aware that an organo-phosphate was being used by keepers. Situations like this must be avoided by carefully planned pest control programs and subsequent correct applications of pesticides.

Besides the esthetic reasons for eliminating pests, the more important reason is eliminating potential diseases found in feral mammals, birds, rodents and insects. The ectoparasites of mammals such as fleas, ticks, and mites, can be transmitted to tigers; as well, internal parasites of these same pests can be acquired by tigers and cause infection. Feral animals also serve as potential sources of pathogens such as the feline viral diseases, rabies, yersiniosis, leptospirosis, salmonellosis, toxoplasmosis, feline infectious peritonitis, and others.

Preventive medical programs also depend on complete post-mortem examination of animals dying in a zoological collection. This service should provide rapid tentative diagnosis from the gross pathological examination to allow immediate medical care of the remaining collection if indicated. Histopathological examination of tissues is mandatory and should be done in a timely manner to make those findings relevant to the health care delivery to the collection. Concurrent cultures may be indicated for bacteria, fungi, and viruses. Appropriate tissues not formalin fixed may be frozen for future studies (i.e., viral, toxicology, genetics). Besides determining the cause of death a complete post-mortem examination allows review of anatomical structure, assessment of nutritional status and the parasitic burden of the animal.

CONTROL OF REPRODUCTION

Long term maintenance of captive tiger populations obviates the need for reproducing them in captivity. Dr. David Wildt addresses the technical aspects of reproductive physiology in his section of this symposium. Concerned zoologists and SSP programs address the genetic components of planned breeding and the desire for maintaining species by eliminating "crossbred" species and maintaining genetic diversity. There are times when animals for display reproduce well and soon reach the carrying capacity of a particular institution or have become over represented in the genetic pool. The need then arises for controlling reproduction either permanently or temporarily.

Reversible control of reproduction has been successfully and easily accomplished through the use of contraceptive implants placed subcutaneously in female tigers. The implant is a medical grade silastic compound impregnated with melengesterol acetate

(Seal 1975, 1976). The slow continuous release of this progestogen-like chemical effectively prevents estrus and suppresses follicular activity on the ovaries. Each implant is usually effective for a two year period after which the implant can be replaced if further contraception is desired.

Removal of the implant will result in a quick return to estrous cyclicity with no adverse effects on subsequent reproductive potential (Black 1979). This method is a simple, dependable, and inexpensive technique for contraception.

Permanent sterilization in tigers can be accomplished surgically by castration or ovariohysterectomy. This approach presents a minor risk due to the need for general anesthesia and surgery, and most importantly, creates physiological changes by removing the hormonal influence provided by the gonad. An alternative is a simple surgical technique of bilateral ductus deferens or oviductal ligation using laparoscopy (Wildt 1977a, 1977b, 1981). This is a minor invasive contraceptive procedure that has been proven safe and effective in several mammalian species with normal sexual behavior maintained. However, converting the animal back to fertility is extremely difficult, if not impossible, requiring microsurgery.

FACILITIES

A major component of tiger management is the facility designed and built to house them. From the outset, the team of people designing the structure, from the architect, curator, zoologist, director, should include input from the veterinary staff. Problems in design and construction lead to unfavorable facilities that may promote health problems that can be prevented by forethought. Many physical aspects of the facility must be considered.

The basic enclosure design is utmost. The size must be adequate for movement and exercise to decrease boredom, stimulate activity, and give the tiger a feeling of security and comfort. Several restraining materials are traditional: 1) bars-metal for strength, relative low maintenance. These are esthetically unpleasant, decrease public visibility, may promote trauma from biting or attacking them, may trap limbs or heads due to inadequate spacing, and may permit trauma from adjacent cats due to improper design barriers; 2) wire-more esthetically pleasing, but not as strong as bars and vulnerable to destruction by the cat. Welded wire material of sufficient gauge can be obtained and is acceptable for tiger enclosures. This material, through improper installation or selection of material, may trap limbs or heads or teeth especially in young animals; 3) glass-esthetically pleasing, better visualization of tiger, but requires more maintenance, expense, and is vulnerable to fracture. No matter what restraining material is used, the composition of the material and the external coatings applied must be non-toxic, non-irritating, or non-trauma inducing.

The newer exhibits have moved toward the use of open air enclosures such as grottos. Many of these are "natural settings" with vegetation and soil. The plants used within an enclosure must be chosen carefully to avoid toxic species. The dirt substrate becomes contaminated over time with microorganisms and parasites thereby exposing the cats to potential concentrations of pathogens. Provisions should be made so that the contaminated substrate can be removed periodically and replaced with clean materials. Reiterating an earlier point, placing animals that have been properly quarantined helps reduce the potential contamination load on the substrate, especially parasitic. The aquatic component of exhibits, the pools and moats, need to be designed for maintaining high water quality, through filtration, ease of cleaning and sanitizing due to the tendency of tigers to defecate in water.

In enclosure areas that have non-dirt substrates, the choice in flooring is extensive. The most common material is concrete, which by itself is not the optimal surface due to its porosity, abrasiveness, and hardness. Coatings over concrete, such as asphalt compounds, epoxy coatings, etc., provide a more acceptable surface by sealing, smoothing, and softening the floor. The important concepts are that the surface be easily cleaned, disinfected, rapidly dried, and non-porous to prevent accumulation of organic debris and contamination. Disinfecting agents should be selected on the basis of effectiveness and low toxicity to tigers; they should not be used in concentrations exceeding the manufacturer's recommended effective dilution. Phenolic compounds should be avoided due to the susceptibility of felids to this chemical. For effective cleaning, hot water and a detergent should be used to remove organic debris followed by or coupled with the disinfectant. In any case, chemicals should be thoroughly rinsed from washed surfaces to prevent exposure. These surfaces must provide good traction for tigers, especially when wet, but not abrasive as to cause foot pad trauma during normal movement or exaggerated pacing. If the surface is too hard, trauma to bony prominences in normal resting or sleeping positions can result. Rubberized flooring, although soft, is easily damaged by tigers providing potential gastrointestinal foreign bodies. The slope of the floor should promote drainage from the tiger's enclosure. The tiger's floor should be above drain level so that a clogged drain will not flood or spill into the enclosure. The design should provide easy shifting during normal daily management routines and especially during manipulative procedures for medical treatment, and to an area where the tiger can be routinely weighed. Good facilities provide the ability to shift tigers from one area of the unit to another without the need of crating or immobilization. Many designs fail in this concept. Under routine conditions the keepers can shift a tiger, but when it does not want to shift (e.g., aware of impending visit or presence of the veterinary staff) it will not. Therefore, the facility should include provisions to force a reluctant animal to a desired location by the use of narrow chutes subdivided by several doors.

The proper cage "furniture" provides the tiger with a variety of sites to stimulate activity, such as different heights, and can be constructed with materials providing a soft and warm place to rest or sleep off the floor. Logs or timbers are provided to allow the natural behavior of scratching for claw wear and maintenance, i.e., to help reduce the ingrown claw and its resultant problems.

Lighting, optimally, should be a combination of natural and artificial illumination. Varying day-night light cycles are beneficial in reproductive cyclicity and health.

In addition to the exhibit enclosure, additional working, holding, and quarantine areas should be present. Off-exhibit holding provides for treatment areas out of the public view and seclusion for a stressed or ill tiger. Within this area, squeeze or restraint cages permit an alternative method of handling for procedures normally necessitating immobilization or anesthesia. A properly designed restraint cage allows simple close examination, collection of samples (e.g., blood, urine, or culture), or drug injection (e.g., antibiotics, vaccinations, anti-parasitic agents, or immobilizing drugs). The use of the cage provides an alternate, less stressful method of drug injection to the remote delivery methods such as darts or pole syringes, especially when large volumes are required.

For female tigers, a maternity unit provides an area where they feel secure to deliver and raise cubs. This area should provide a dry, warm, secluded environment to promote good maternal care.

Irrespective of the cage use, the design must avoid a situation in which an animal cannot be fully seen or reached or shifted for monitoring or potential treatment or immobilization.

Each cage must provide a cleanable water source accessible to both the tigers and keepers that can be shut off and drained. This allows monitoring water intake and water deprivation in certain clinical situations, such as pre- or post-immobilization. The non-reservoir watering systems (such as lab animal design self-waterers) can malfunction and inadvertently deprive the cat of water if not checked daily which may be difficult from outside the enclosure.

Lastly, a simplistic point, the facility must be secure to contain the tiger and protect the public and keepers. The design must provide safe access to animal areas for keepers or the veterinary staff in the event a tiger escapes from its primary enclosure. The situation must be avoided where a tiger is out of his primary cage, cannot be seen, and can only be reached by directly entering the same space it occupies.

DIETS

The common denominator in feeding large felids, such as tigers, is the wide variability in diets offered. There is variability in dietary components such as protein sources, whether they be from meat, fish, poultry, or grains, and variability in formulation, such as percentages of protein, carbohydrates, fiber, ash, etc., in each diet. Without the availability of commercially prepared carnivore diets, it is likely no two institutions would be feeding tigers similarly.

There exist some nutritional studies for exotic felids which have included tigers (Cline 1966, 1969, Vaneysinga 1969, Morris 1974, Grittinger 1977, Mills 1980a, 1980b, Barbiers 1982, Hackenberger in press). Exact nutritional requirements for all nutrients are not known specifically for tigers; therefore, requirements are extrapolated from data on domestic felids (NRC 1978). Diets are formulated, prepared, and fed; some meet dietary needs while others do not and result in tigers with nutritionally related medical problems (e.g., chronic disease, nutritional disorders or poor reproductive performance). Fortunately, most nutritional disorders are of only historical significance due to improved nutritional management (Slusher et al. 1965).

The commercial preparations are formulated from the comparative dietary requirements and received some field testing for varying periods, probably by participants in this symposium. The advantage of the commercial diets is that they are readily available, require little or no labor in preparation, and are assumed to be formulated with a sound nutritional basis. Economics determine the components of these diets as the ingredients vary with the change in cost of producing the diet. Thus, the guaranteed analysis remains the same, but the diet may vary in raw ingredients. The guaranteed analysis label does not guarantee that the ingredients of the diet are actually utilized or available for utilization by the tiger.

Some zoos formulate diets from basic ingredients so the components are relatively constant; however, many times no nutritional analysis on the finished, fed product is conducted. Success with these feeding methods vary with the majority resulting in adequate nutrition. The need still exists for basic research on nutritional requirements for captive tigers.

Food preparation and handling is another area of concern. If the diet is mixed within the institutions, all ingredients should be scrupulously maintained free of contamination from chemicals, pests, or microorganisms. Frozen ingredients should be properly thawed to reduce bacterial growth and diets fed as soon as possible after mixing. Commercial diets should be thawed under clean conditions, free from external contamination, and fed immediately after thawing. Some institutions actually feed the diet while still frozen allowing tigers to eat as it thaws. Avoid allowing raw diets to warm to room temperature for long periods of time prior to feeding. The practice promotes the rapid growth of bacterial organisms.

The food should be weighed and daily records kept as to how much is offered to each individual tiger and how much is consumed. Determination of ration amounts should be a dynamic process to meet changes in metabolic needs, such as in seasonal needs, illness, pregnancy and lactation, and growth. Proper body weight, especially to avoid obesity, should be maintained by dieting alterations. These changes should reflect not only energy needs, but also vitamin and mineral needs. Records of stool consistency assist in determining if the diet is poorly digested or possibly inducing diarrhea indicative of enteric disease. The food should be offered on a non-contaminated surface. In most situations feeding is done on the floor of the enclosure. Feeding stations should optimally be off the floor or substrate, but this is not always practical as tigers often destroy feeding bowls.

Most management programs have found that tigers' appetites and body conditions improve if they are fasted one to two days a week. Either no food is fed on these days or shank or other large bones are fed. Feeding bones has an additional function as discussed in the dental section.

The goal is good diet formulation based on sound nutritional concepts and quality sources of dietary components. In addition, communication between the source of the diets, the veterinary staff, and the keepers will allow monitoring of health status, early evidence of nutritional deficiencies, or potential toxic problems. Only then can dietary inadequacies be assessed. The most common type of poisoning in large felids is from barbiturates used to euthanize feed animals. Felids feeding on this carcass may show varying signs from mild ataxia to general anesthesia that may last for days. The liver from such carcasses are especially high in barbiturate levels and cause more severe signs.

We have also seen toxicity from xylazine in 2 circus tigers who developed ataxia and became recumbent after being fed. Although initially recumbent, the tigers would respond to manipulation and stand up. This type of response is characteristic of xylazine sedation in large felids. To diagnose the toxicity it is best to collect urine for assay. We collected both blood and urine from the circus tigers and only the urine was positive.

These animals are valuable and good diets are most important in the context of preventive medicine. Dietary composition should not be based too heavily on economics or ease of preparation. The responsibility for proper feeding must be assumed when the commitment to exhibit tigers is accepted.

HANDREARING

Ideally all tiger cubs should be raised by their mother for four reasons: 1) she does a better job; 2) the resultant cub usually grows up to be better adjusted behaviorally; 3) it makes a great exhibit; and 4) it saves personnel time.

When this is not possible due to maternal neglect or health reasons the cubs must be hand-raised. Hopefully the cubs can remain with the mother long enough to receive colostrum, but this is not always possible.

When cubs are removed they are given a complete physical examination paying particular attention to possible congenital defects and the status of the umbilicus for signs of infection. A blood sample is collected for baseline values and an accurate weight is obtained. The cub should receive prophylactic antibiotics, usually a long acting penicillin.

To assist with passive immunity the cubs are given subcutaneous and oral serum. This serum is collected sterile from the mother, if she is healthy and not caring for other cubs. If the mother is not available, serum from a healthy adult cat that has been in the collection for at least 1 year can be used as an alternative. The serum is filtered to remove bacteria and given at the rate of 5-8 ml/kg subcutaneous for 2 days and orally 2-5 ml/feeding for 3-5 days.

There are numerous protocols for handraising tigers (Hoff 1960, Husain 1966, Theobald 1970, Kloss and Lang 1976, Hughes 1977) using various products. Certain guidelines are important, initially the cubs should receive 5% dextrose for the first 2 feedings and then started on milk replacer. The choice of milk replacer for tigers seems to be Esbilac (Esbilac, Borden, Inc., Hampshire, IL 60140). We add the enzyme lactase to the milk to break down the lactose and have noted fewer problems with gastrointestinal upsets. The cub should be kept hungry the first day or two and then the diet increased in volume to about 10% of body weight/24 hr. Initially the cub is fed by stomach tube to minimize the risk of inhalation pneumonia, but also to assess residual stomach content by aspiration prior to the next meal. When started on the bottle the first liquid should be 5% dextrose to minimize lung damage if inhalation occurs. The cub is held in a normal feeding position when taking the bottle. When the cub is taking the 5% dextrose well with no coughing, milk can be started. The concentration of the formula is started at 6% and elevated to 12, 15 and 18% as the cub grows to meet the energy requirement without overfilling the stomach. The cub should be receiving the 18% formula at 4-6 weeks of age.

The cubs should be stimulated to urinate and defecate at each feeding by massaging the anogenital area with cotton moistened with warm water. If diarrhea occurs, the formula is diluted with an oral electrolyte solution and total volume decreased by 20-40% for 8-12 hrs. A stool culture prior to antibiotic therapy is obtained to check for pathogenic bacteria. If diarrhea is severe and persistent, all oral intake should be stopped for 12-18 hrs and the cubs supported with subcutaneous fluids, and then started on oral electrolytes followed by dilute formula and returned to normal feeding over the next 12-24 hrs.

Most hand-raised tigers develop hair loss at 6-8 weeks of age (Kloss and Lang 1976, von Seifert 1982). This is thought to be due to some deficiency in the diet. The addition of liver homogenate to the diet has been helpful in preventing and correcting this alopecia. Weaning the cubs to solid food also usually enhances hair coat, growth, and general appearance. This should begin at 5-8 weeks.

Hand-raised cubs should be weighed regularly to monitor weight gain and calculate necessary food intake. A growth chart of these animals can be compared to other published charts (Hemmer 1979). Iron supplements are needed for young growing tigers to prevent anemia. Although many milk replacers have iron added, an additional supplement is beneficial. Iron deficiency is not uncommon in mother-reared cubs that have no exposure to dirt. Iron dextran injections to the cubs may be required in these cases.

Vaccination against feline viral diseases with a trivalent killed product (Fel-O-Vax, Fort Dodge Labs., Fort Dodge, IA 50501) at the recommended dose should be given at 10-12 weeks, 16 weeks, 6 months and 1 year of age. In collections where female tigers have very high titers due to repeated vaccination, passive immunity transferred to the cub can be high enough to delay active immunity induced by the vaccine. This is why additional vaccinations are recommended.

Fecal examinations of mother and cubs should be performed monthly. If hookworms have been a problem in the collection, then the cubs should be prophylactically treated at 6-8 weeks of age.

ANESTHESIA

        Pre-anesthetic preparation of the tiger enhances the success
of the procedure. The tiger should be taken off food for at
least 24 hours or longer if there has been a recent very large
meal. Water should be withheld for at least 12 hours unless
medical concerns or extremely hot weather preclude it. The
patient should be shifted to a small area, preferably a squeeze
cage, for drug administration. Animals that are calmer at this
time usually require less drug and have a smoother induction
period.

        The advances in anesthesiology have markedly improved the
health care delivery to tigers. The history of anesthesia in
tigers prior to phencyclidine was almost nonexistent.

        Phencyclidine did provide safe anesthesia with a low
delivery volume but had adverse side effects; dosages of 0.5-1.2
mg/kg were used (Kloss and Lang 1976). Severe convulsions were
common, especially in Siberian tigers, which required concurrent
tranquilizers and/or anesthetics to control (Seal and Erickson
1969). Excessive salivation and a prolonged recovery period were
also seen. Phencyclidine is no longer commercially available in
this country and there are better choices for anesthesia of
captive tigers as will be discussed.

        The ideal drug for tiger immobilization is CI744 (Zoletil,
Laboratoires READING, Z.A.C. 17, rue des marronniers, 94240
L'Hay-Les-Roses), a 1:1 combination of tiletamine HCL and
zolazapam HCL. The problem is the relative unavailability of
this drug, but it is now commercially available in France, and
hopefully will soon be marketed in the United States. One
advantage of CI744 is its availability as a dry powder which can
be concentrated to 500 mg/ml with a long shelf life. The drug
produces rapid induction requiring only 2-3 minutes for onset of
effects. The drug is remarkably safe in that respiration is not
depressed and the cardiovascular integrity is maintained.
Salivation is minimal and the swallowing reflex is well
preserved. Recovery is rapid and smooth. The disadvantages are
its present limited availability and some minor CNS signs,
usually in the form of mild tremors which are sometimes
encountered. An occasional tiger, especially white ones, may
undergo a re-sedation effect up to 24-36 hr later. These animals
usually show mild sedation with stumbling, which occurs after a
complete initial recovery. As with other anesthetics, the dosage
varies with the activity and status of the patient. In a captive
situation with a calm tiger 0.5 mg/kg is usually adequate while
in the field doses of 3-11 mg/kg are required (Seidensticker et
al. 1974, Smith et al. 1983).

        At present, ketamine (Vetalar, Parke-Davis, Morris Plains,
NJ 07950) is the major anesthetic drug for tigers. With ketamine
alone problems include, poor muscle relaxation and a tendency to
produce CNS signs (tremors and grand mal seizures); also, a large
delivery volume (12-20 ml) is required. Therefore, it is usually
combined with a second drug (i.e., diazepam (Valium, Hoffman
LaRoche, Inc., Nutley, NJ 07110), acepromazine (PromAce, Fort
Dodge Labs., Fort Dodge, IA 50501), promazine (Sparine, Wyeth
Labs., Philadelphia, PA 19101) or xylazine (Rompum, Haver
Lockhart, Shawnee, NS 66201). In a study at the Henry Doorly Zoo
in Omaha, Nebraska, 4 species of cats were anesthetized with

Table 2.   Convulsions resulting from the use of ketamine as the sole anesthesia in four species of big cats (maintained under for ≥1.5 hr.).

Species	No.	No. Anesthetic Episodes	Avg. Dose (mg/kg) Init.	Suppl.	Animals Convulsed (%)
Tiger	12	19	11.2	5.7	89.5
North Chinese Leopard	3	17	14.8	6.6	0.9
Puma	4	10	15.4	6.3	44.4
Cheetah	3	12	14.2	5.5	90.0

ketamine alone; the results are shown on Table 2.   All cats in this study were maintained under anesthesia for at least 1.5 hr. It is noted that about 90% of the cheetahs and tigers had convulsions.   Intravenous diazepam at 0.02 mg/kg was usually administered to control the convulsion after the third seizure or if one of the first two seizures were severe.   We found that some cats only had 1-2 convulsions.   In tigers with CNS signs, 4 of the 17 did not need diazepam.   As noted, the major disadvantage of using ketamine is the large volume of drug required for anesthesia, up to 20 ml in a large tiger.   Even with lyophilization and reconstitution to 200 mg/ml, 10 ml is required.   This volume presents no problems if the animal is restrained in a squeeze cage, but to deliver this by remote injection (i.e., capture dart) requires a large dart or multiple darts.

The addition of other drugs to ketamine not only helps control side effects but decreases the amount of ketamine required for anesthesia.   The "Hellabrun Mixture" combines 125 mg xylazine plus 100 mg ketamine/ml by adding ketamine as a diluent to lyophilized xylazine.   The recommended dose for tigers is 1.5 ml (187 mg xylazine + 150 mg ketamine) for a subadult and 3.0 ml (375 mg xylazine + 300 mg ketamine) + 1 ml ketamine for an adult. The ketamine concentration could probably be increased by using the more concentrated ketamine (200 mg/ml).   This mixture offers the advantage of a small volume for remote delivery.   The high dose of xylazine would seem to cause a prolonged recovery time which can be partially reversed by using yohimbine (0.1-0.15 mg/kg IV), or other alpha-adrenergic blockers (i.e., tolazoline).

Atropine is used with xylazine to prevent bradycardia and help control salivation.   Three to five mg of atropine is given to an adult tiger.   To help decrease the volume of atropine, the more concentrated preparation for large animals is used (15 mg/ml).

There are several reports of using short acting barbiturates to supplement anesthesia (Kloss and Lang 1976).   The longer acting barbiturates are not indicated due to their long recovery time in tigers; some patients require 2-3 days to recover.

At the National Zoological Park we use 25-100 mg xylazine plus 500-1,000 mg ketamine for adult Bengal tiger anesthesia. Complications observed were decreased respiration, vomiting, and convulsions. Many tigers only have 1-2 convulsions during a ketamine anesthesia, therefore, we usually wait until the third convulsion, unless the first one is violent and/or prolonged, before administering intravenous diazapam (3-5 mg IV) to control the seizures.

The use of xylazine alone in tigers is dangerous. The patient may appear asleep, but when stimulated may arouse and react aggressively. If a large enough dose is used to safely handle the tiger, severe respiratory depression occurs.

For prolonged medical treatment or surgical procedures, inhalation anesthesia is used. Following initial anesthesia with injectable drugs and intubation, either methoxyflurane (Metofane, Pitman Moore, Washington Crossing, NJ 08560), halothane (Halothane USP, Holocarbon Labs., Hackensack, NJ 07601) or isoflurane (Forane, Ohio Medical Anesthics, Madison, WI 53713) can be given, depending on the preference of the clinician. Intubation of tigers is not difficult if a long laryngoscope blade is used. Topical anesthetic to the larynx is not usually necessary and it has the disadvantage of blocking gag reflex in case the patient vomits. Most patients do well on spontaneous respiration with occasional assisted respiration but sometimes positive pressure ventilation is indicated.

Physiological monitoring of the anesthetized tiger is an integral part of any anesthetic episode. Once the patient appears anesthetized, the person responsible for the anesthesia should be the first into the cage to evaluate the tiger's status, before allowing others access. Initial observations include: responsiveness to stimuli, respiration rate, color of mucous membranes, pulse rate and intensity, and muscle tone. More sophisticated monitoring can include blood pressure, blood gas measurements and EKG.

A helpful and easy parameter to measure is indirect blood pressure using a regular sphygmomanometer attached to the foreleg. The systolic pressure is read at the point where the needle begins to bounce at the pulse rate as the cuff pressure is released. As the needle continues to fall and the bounce stops at about the level of diastolic pressure. There are commercially available instruments (Dinamap, Critikon, Inc., Tampa, FL 33607) that measure indirect blood pressure. These machines make recordings every minute and transcribe the reading to a printer. Data obtained include systolic, diastolic and mean blood pressure, and heart rate. The machine is also equipped with alarms that can be set for high or low reading.

Indirect blood pressure values may not be as accurate as an intra-arterial transducer, but rather these values are to note trends in the pressures during the anesthesia period. Blood pressure also provides a more functional evaluation of the heart than an EKG recording.

Another parameter to measure is body temperature, especially during prolonged surgical procedures where hypothermia may occur. Elevation of temperature may be seen with convulsions, pre-anesthetic excitement, high environmental temperature, and exposure to direct sunlight. Temperatures greater than 39.4^XDC in a patient should be an indication for cooling with water and air circulation. Severe hyperthermia, >40.6 F, requires more aggressive therapy including water immersion, cold water enemas, IV fluids, corticosteroids, and antibiotics.

SURGERY

Surgical procedures on tigers are similar to other species once the patient is anesthetized. For most prolonged procedures inhalation anesthesia is used.

Cataracts are seen in some young tigers due to nutritional or congenital problems (Magrane and Van De Grift 1975, Benirschke et al. 1976, Kloss and Lang 1976). It is questionable if these cataracts should be removed. A tiger with the hereditary cataract, once confirmed by a veterinary ophthalmologist, should be considered for euthanasia. The nutritional cataracts are usually seen in young cubs receiving a milk replacer. These cataracts appear to be self-limiting and usually require no surgical intervention. A reason for lens removal is traumatic lens luxation (Graham-Jones 1961).

Another congenital problem is cross-eyed animals, especially in white tigers. Again this is a case where euthanasia should be considered and if not, the animal should not be used in a breeding program.

Orthopedic surgery does not require special consideration other than strong rigid fixation due to the patient's strength. There are reports of orthopedic problems in the literature; one involves a fractured mandible (Milton et al. 1980).

Another potential surgical problem is lacerations from tight wounds. Usually lesions are small and are left to drain and granulate in. It is our procedure to give antibiotics orally for 7-10 days after such tights to minimize local infection and bacteremia that may shower to other organs. A common isolate from the mouths of tigers has been Pasteurella multocida (Woolfrey et al. 1985) with Staphylococcus aureus and Streptococcus viridens as potential problems in tiger bites. Pasteurella multocida usually shows good sensitivity to a wide range of antibiotics with cephalosporins being the drug of choice.

DENTAL

A thorough oral examination is an integral part of a physical examination, either planned or done opportunistically whenever immobilization is performed. This examination becomes more important as a tiger ages to prevent dental problems from causing systemic diseases. A common problem reported in exotic felids relates to wear or trauma from fighting between tigers or contact with cage material. The most common dental finding is calculus accumulation, especially along the buccal surface of the upper molars and premolars.

During the examination, not only the teeth, but also the soft tissue structures of the mouth and throat are examined for abnormalities. Foreign bodies lodged between oral structures, such as bone fragments, sticks, etc., can be incidental findings but definite predisposers of oral disease. These should be removed and infections or traumatic lesion treated as indicated. Calculus accumulation is removed from the teeth surfaces with care taken to remove material from the subgingival sulcus. If power equipment is available, the scraped surfaces are polished to smooth dental surfaces which deter future calculus accumulation. At this time, the sub-gingival sulcus, gingiva, and teeth are examined for evidence of gingivitis or periodontal disease. Each tooth is examined, once thoroughly cleaned, for evidence of fractures. Teeth fracture longitudinally or transversely, thus exposing the pulp tissue and periapical structures to infection. The canine teeth are especially prone to fracture or wear because of their location and length. Exposure of the root canal is a common finding in dental fracture or excessive wear. A variety of endodontic procedures provide an option to extraction, which is laborious and disfiguring, especially with canine teeth (Van De Grift 1975, Tinkleman 1979, McDonald 1983). Teeth can be salvaged by performing vital or non-vital pulpotomies and filling the root canals. The remaining crown is left intact. Restorative procedures are available to provide an artificial crown, but this is not always necessary or practical. For subgingival fractures, vital and non-vital root retention is applicable. Radiological examinations of the teeth and bony structures are invaluable in determining the extent of disease or trauma and the course of therapy to pursue. Extractions must still be considered with advanced disease. Veterinarians are becoming increasingly proficient in dealing with dental problems; as well, local dental surgeons can be utilized for consultation.

Sound, regular prophylactic dental care is important in preventing bacteremia of oral origin that can contribute to, or promote systemic disease (Fagen 1980a, 1980b). Good nutrition is needed to maintain healthy oral structures. The feeding of bone twice weekly helps promote good gingival health when tigers are otherwise maintained on a soft diet (Bush and Gray 1975, Haberstroh et al. 1984).

KIDNEY DISEASE

Kidney disease is a recognized problem in aged tigers but is reported to occur at all ages (Benirschke et al. 1976). Chronic interstitial nephritis is the most frequently described histo-pathologic change.

Impaired renal function has been used to explain various clinical syndromes including "tiger's disease". The clinical signs included vomiting, anorexia, depression, increased transit time through the gastrointestinal tract and occasionally convulsions. The concurrent serum chemical values show elevated BUN and creatinine. The theory is the nitrogenous wastes have deleterious effects on other organ systems such as stomach, pancreas and liver (Eulenberger 1981, Straub and Seidel 1983).

The most common kidney lesion seen in tigers is chronic intestinal nephritis (CIN) (Benirschke et al. 1976). A possible etiology for this lesion may be diet related, but may be idiopathic as in other carnivores. The literature also poses long-term or excessive vitamin D supplementation or chronic multivitamin deficiency as possible causes of CIN (Benirschke et al. 1976). Other kidney problems, such as glomerulonephritis, may be due to long term intermittent bacteremia due to bite or scratch wounds that may go unnoticed and untreated. A second cause is dental disease that can provide a ready source of bacteria from the oral cavity. Routine dental examination and treatment plus prophylactic systemic antibiotics after fights are indicated to help minimize bacteremic episodes. Impaired kidney function can be due to kidney lesions other than CIN which can produce the same clinical picture. One such case was a circus tiger with moderate weight loss and a 5 month history of gastro-intestinal upset including vomiting and diarrhea. The meat passed through the gastrointestinal tract almost unchanged even though the fecal trypsin was positive. The tiger also had a mucopurulent nasal discharge. The significant laboratory findings were negative fecal culture for salmonella and other pathogenic bacteria. A mild leukocytosis with neutrophilia and lymphopenia was noted. The BUN and creatinine levels were also elevated. Blood culture was negative and Brucella and leptospira titers were also negative.

The necropsy findings were chronic suppurative odontitis of the four fractured canines which probably led to the acute tubular necrosis and focal nephritis seen in the kidney. Gastritis, purulent bronchitis, cellulitis and mild meningitis were also present.

To diagnose a kidney problem is easier than offering a prognosis for the patient's problem. The clinical signs of vomiting and/or anorexia and/or polydypsia and polyuria and/or convulsions should lead one to a tentative diagnosis of kidney disease. Further diagnostic tests necessitate obtaining blood and urine samples. Elevation of serum BUN and creatinine indicate impaired kidney function, either due to kidney disease and/or dehydration. An improvement in kidney function is difficult to assess on a single sample. Repeated blood samples during therapy will aid in obtaining a prognosis. Urine samples are relatively easy to obtain and provide useful information on the animal's status. Normal tiger urine usually has some protein present-trace to 1+ on a urine dipstick; normal tiger urine also contains fat droplets. The lipiduria originates from lipid accumulation in the proximal tubules of the kidney (Hewer et al. 1948). Urinalysis should be evaluated paying particular attention to specific gravity, glucose, ketones, bile and the microscopic sediment. The evaluation of these factors are the same as in the domestic carnivore.

In some cases, a more aggressive diagnostic approach may be required. The tiger is anesthetized and a complete physical examination including dental examination is performed. Blood samples and cultures are obtained. The patient may be laparoscoped to obtain a urine sample for culture by cystocentesis and a needle biopsy of the kidney for culture and histopathological evaluation. The pathologic evaluation of an end stage kidney may preclude further attempts at therapy where as minimal kidney changes or an infectious component may provide a more favorable prognosis. Treatment for kidney disease is usually supportive, unless a specific infectious etiology can be identified. Supportive care involves systemic fluids for hydration, multivitamins, and prophylactic systemic antibiotics with low propensity for nephrotoxicity. During treatment concurrent stress should be minimal. This is very difficult with the manipulation necessary to medicate a large dangerous patient. There is also a problem in tigers with failing kidneys to adapt to a diet containing less meat which would lessen the production of nitrogenous wastes.

Other specific kidney problems observed include renal papillary necrosis in 2 related tigers of 17 and 18 years of age (McCullagh and Lucke 1978). The gross lesions were pale friable areas in the medulla. The cause was thought to be decreased renal perfusion secondary to severe dehydration and the presence of chronic cortical scarring. Papillary necrosis is a complication also caused by certain types of drugs. One of these drugs, flunixin meglumine (Banamine, Schering, Kenilworth, NJ 07033), is being utilized more in our practice as a non-steroidal anti-inflammatory drug. Caution should be used in administering these drugs to animals with impaired kidney function and/or marked dehydration.

GASTROINTESTINAL DISEASE

Certain gastrointestinal syndromes have been placed under general categories as "general adaptation syndrome" or "tiger disease" (Cocin et al. 1973, Cocin et al. 1976, Benirschke et al. 1976). These problems are historically noted in tigers, especially Siberians. The proposed etiology of "tiger disease" has been reported to be a pancreatic dysfunction (Eulenberger 1981, Straub and Seidel 1983) or disruption of gastrointestinal flora (Kloss and Lang 1976); stress was documented as the cause in one report which correlated the digestive upsets with sudden changes in the tiger's environment (Cocin et al. 1973). Gastrointestinal upsets should be investigated as being caused by diet, infectious agents (Salmonella spp., Clostridium), or concurrent kidney failure. The specific cause is then treated if identified or supportive care instituted.

Gastric ulcers have been seen in conjunction with other disease processes in tigers (Kloss and Lang 1976). The role of these lesions may be difficult to assess but their presence could cause anorexia, abdominal pain, vomiting, blood loss and/or sites for secondary infection (bacterial or mycotic). With this finding of gastric ulcers it seems efficacious to include the new anti-ulcer medications (Zantac, Research Triangle Park, NC 27709), which block the histamine receptor sites, as part of any supportive care protocol for tigers.

Selected reports of specific abnormalities of the gastrointestinal tract include, pyloric stenosis (Fowler and Gourley 1970, Starzynski and Rokicki 1974), gastric impaction with straw (Zenoble et al. 1982), diaphragmatic hernia (Starzynski and Rokicki 1974), and cricopharyngeal achalasia (Kloss and Lang 1976). Gastric hairballs may cause upper gastrointestinal problems and have been successfully treated by pouring a layer of mineral oil over the tiger's drinking water (Theobald 1978).

When investigating the cause of anorexia and/or vomiting in a tiger, a detailed oral examination is included to search for dental problems, pharyngeal foreign bodies, and oral abscesses. Radiographs are used to screen for foreign bodies and barium series help observe transit time and space occupying lesions within or outside the lumen of the gut. The proportional size of the x-ray machine and the patient determine whether a diagnostic radiographic study is feasible.

Liver disease is reported as a problem in tigers. Studies on tissue distribution of enzymes showed alanine aminotransferase (SGPT, old usage) to be liver specific in the tiger and useful as a diagnostic tool in active diseases (Keller et al. 1985). Liver function is evaluated by functional tests such as Bromsulphalein (BSP, Hynson, Westcott & Dunning, Baltimore, MD 21201). Proper evaluation of these liver function tests requires values from control tigers to establish reference values. Diagnosis of liver disease by a liver biopsy either blind transabdominally or by direct observation laparoscopically aids in diagnosis and prognosis. Myelolipomatosis, a focal benign lesion in the liver, has been reported in one tiger and six cheetahs (Lombard et al. 1968). This lesion is usually an incidental finding at necropsy.

BACTERIAL DISEASES

Tuberculosis, caused by Mycobacterium bovis, has been a major disease problem in tigers in some settings (Michalska 1972). It presents as a chronic non-responsive disease with the lungs as target organs. The lesions are exudative and necrotic. Localized tuberculous involving the eyes has also been reported (Michalska et al. 1978). Ante-mortem tests are reported to be unreliable. In collections with severe problems, cubs have been vaccinated with BCG starting at 4 weeks of age, but BCG vaccinations may not be indicated in most situations.

Anthrax has been seen in large felids including tigers (Abdulla et al. 1982). Clinical signs include ataxia, apathy and weakness. Less frequent signs are vomiting, frequent defecation and convulsions. The patient usually dies in 1-4 days. On post-mortem examination there are blood clots reported on the spleen that in the earlier literature were called tumors. Reported treatment includes antisera and antibiotics which have met with limited success.

Anthrax is almost always due to feeding contaminated meat. Other diseases of large cats that have been reported to be caused by this route include glanders in a lion fed a sick horse (Kloss and Lang 1976) and encephalomyocarditis virus (EMCV) that killed 20 lions fed an elephant that died of the disease (Gaskin et al. 1980).

Systemic bacterial diseases have been seen in captive tigers such as bacterial meningitis from Klebsiella and Diplococcus (Wallach and Boever 1983). Colisepticemia (Selbitz et al. 1979, Sathyanarayana et al. 1983), Shigella flexneri (Boro et al. 1980, Zaki 1980), Salmonella sp. (Kloss and Lang 1976, Houck pers. comm.), Corynebacterium pyogenes (Sathyanarayana et al. 1981) and Clostridium perfringens (Pulling 1976) have caused fatal disease in tiger cubs.

Salmonellosis, caused by <u>Salmonella</u> <u>typhimurium</u>, is a recognized medical problem in tigers, occurring either sporadically or as outbreaks (Kloss and Lang 1976). Other species of Salmonella are implicated in enteric disease affecting both young and adult tigers with symptoms ranging from mild gastrointestinal upset to acute death. Clinical signs in adults include lethargy and anorexia initially, with subsequent diarrhea possibly containing varying amounts of blood. Young tigers exhibit a more acute disease and can quickly become moribund. Deaths occur more frequently in young tigers or adults stressed during the clinical course. Stool cultures may or may not be positive for <u>Salmonella</u> <u>spp</u>., namely type C or D (<u>S. enteritidis</u>, <u>S. newport</u>, <u>S. braenderup</u>). Simultaneous culture of food sources, such as commercially prepared diets or beef, are often positive for the same <u>Salmonella</u> type. Therefore, not all outbreaks correlate with both positive stool and food source <u>Salmonella</u> cultures. Most sick tigers respond to supportive care with fluids, bland diets, and a seven day course of antibiotic therapy (based on sensitivity testing). During this therapy diarrhea usually decreases and the animals return to feeding and increased activity.

In two suspected outbreaks of salmonellosis in tigers, the method of food preparation and time from thawing frozen meat to feeding was associated with the incidence of the disease (Houck pers. comm., Campbell pers. comm.). Meat allowed to sit at room temperature for long periods or meat thawed during shipment, and then possibly refrozen, were implicated in disease. Repeated isolations of certain <u>Salmonella</u> <u>spp</u>. from the tigers' food sources and the simultaneous recovery from their stools indicated that certain sources of meat or certain production batches were the cause of the salmonellosis. This suggests that the meat diets may be contaminated at the production source or during handling and shipment. To prevent salmonellosis in captive groups of tigers there must be quality control at the production source, the diet should be maintained frozen prior to feeding, and the thawing process and feeding method must reduce possible contamination.

Another source of salmonellosis may be a tiger that is an asymptomatic carrier of <u>Salmonella</u> <u>spp</u>. These tigers serve as sources of infection for others, especially young tigers, and may break with the disease themselves if stressed.

There is a potential for zoonosis with <u>Salmonella</u> <u>spp</u>., as the organisms found in stool cultures from diseased tigers can also cause salmonellosis in humans (Rettig 1983, Bryson 1986).

VIRAL DISEASE

A viral meningoencephalitis in lions and tigers was reported once in the German literature (Melchior 1973). Clinical signs included hyperactivity, prolapse of the tongue followed by ataxia, convulsions and paralysis. The disease lasted 2-6 weeks and treatment was considered a failure.

Pox disease has been reported in several species of large felid but not tigers (Marennikova et al. 1977). This may be due to lack of exposure. This disease should be considered a potentially fatal disease.

The upper respiratory viral diseases have been reported in tigers (Von et al. 1981) and have signs similar to those reported in the domestic cat and usually have a high morbidity and low mortality. Treatment is aimed at supportive care.

Panleukopenia has been proven by viral isolation in tigers (Woolf and Swart 1974, Zoo. Soc. London 1978, Montali et al. 1986) and presents in similar ways to the disease in domestic cats with signs varying from acute deaths to a more prolonged course of diarrhea and depression with secondary infections. Suspected ocular problems, secondary to this disease, were reported in a tiger with advancing keratitis, uveitis and loss of vision (Graham-Jones 1959).

Feline infectious peritonitis (FIP) has been confirmed in one Sumatran tiger (Worley pers. comm.); it may be an emerging disease problem. FIP has already had a devastating effect on some cheetah populations (Pfeifer et al. 1983).

Serological screens of large felids have found in some species, including the tiger, reactivity with human hepatitis B virus surface antigens (Worley 1983). The significance of this observation is currently uncertain.

FUNGAL DISEASES

Microsporum canis is not an uncommon cause of hair loss in young tigers (Kloss and Lang 1976, Kymhapeb and Cam 1984). Treatment is similar to that in the domestic cat with equally as good results. Griseofulvin is given orally at 20 mg/kg/24 hr or it can be given weekly at 140 mg/kg. A second cutaneous pathogen, Dermatophilosis cargolensis, has been cultured from skin lesions in polar bears and tigers (Kitchen and Dayhuff 1977). The disease has a chronic course but responds to topical and systemic antibiotics.

A nasal granuloma caused by Adiaspiromycosis was a concurrent finding in a tiger with chronic pancreatitis and metastatic mast cell tumor (Jensen et al. 1985).

Coccidioidomycosis was reported in two Bengal tigers with concurrent liver problems living in endemic areas (Hendrickson and Diberstein 1972).

NERVOUS SYSTEM DISORDER

Tigers seem prone to central nervous signs with a wide variety of disease conditions. Many episodes of CNS problems, including blindness, seem to be nonspecific and respond to supportive therapy; others seem to regress with time (von Krzakowski et al. 1984). The London Zoo reports problems in successive litters of Sumatran tigers (Zoo. Soc. London 1976, 1982). Extensive diagnostic evaluations suggested a metabolic or

toxic problem, but viral etiology was also considered.   One tiger
survived with supportive care and vitamin therapy.

    A non-superative meningoencephalitis was seen in Siberian
tigers associated with paramyxovirus nucleocapsids (Gould and
Fenner 1983) which were similar to a later case of canine
distemper encephalitis seen in another tiger (Blythe et al.
1983).   At the NZP a litter of tiger cubs died with pathologic
changes suggestive of canine distemper including broncho-
pneumonia, lymphoid depletion and nonsuppurative brain lesions
(Montali et al. 1986).   Four of six white tigers which died at
the Bristol Zoo in a 7 year period from 1970-1977 had
histological changes of spongiform encephalopathy or gliosis
found in their brains.   The possibility of a "slow virus"
infection or hepatic encephalopathy was considered because of
concurrent liver lesions (Kelly et al. 1980).

CONGENITAL PROBLEMS

    Congenital problems have been reported, and in some
instances, may be related to inbreeding.   Thymic hypoplasia and
lymphopenia which has caused immunodeficiency and the death of
one cub (DeMartini 1974), possibly due to inbreeding, is reported
in Siberian tigers.   Two cardiac defects have been seen in white
tiger cubs, a patent ductus arterosis and atrial septal defect
(Houck pers. comm.).   A survey of congenital defects reported the
following defects in tigers: cleft palate, diaphragmatic hernia
and umbilical hernia (DeMartini 1974, Leipold 1980).   The
presence of congenital cataracts has been mentioned previously.
White tigers have been investigated for the Chediak-Higashi
syndrome, but certain criteria could not be fulfilled to draw a
correlation (Anon. 1978).

MISCELLANEOUS MEDICAL PROBLEMS

    An inflammatory myopathy, nonresponsive to corticosteroids,
was reported in a captive Bengal tiger (Berrier et al. 1975).
The cause of the condition was not identified.   Among other
problems noted in white tigers, there is one report of central
retinal degeneration (Duncan et al. 1982).   The authors recommend
a closer ophthalmic examination of white tigers which may find
this as an emerging problem.

    A possible incidental finding is the occurrence of
heartworms, _Dirofilaria_ spp., in a Bengal tiger kept in an
endemic heartworm area (Kennedy and Patton 1981).   This is one of
7 reported cases of this parasite in non-domestic felids.

NEOPLASTIC CONDITIONS

    The literature contains several case reports of isolated
cases of neoplastic disease in tigers which include adenocarcinoma of the jejunum (Wallach and Boever 1983), metastatic

mast cell tumor (Jensen et al. 1985), metastatic adenocarcinoma
of adrenal and kidney (Hubbard et al. 1983), chondrosarcoma of
the humeral head (Butler et al. 1981), adenoma of lung (Effron et
al. 1977), adenoma of merocrine gland (Effron et al. 1977),
liposarcoma (Effron et al. 1977), adenocarcinoma of the thyroid
(Curson and Thomas 1930), hepatoma (Smith et al. 1971),
leiomyosarcoma (Saunders 1984), vaginal myxoma (Kollias et al.
1985), and sertoli cell tumor (Michalska et al. 1977).

Reports of lymphocytic tumors were not found. One case of
leukemia has been diagnosed in a circus tiger studied at the NZP
as reported as follows. An adult female Bengal tiger, belonging
to a circus, had a partial anorexia, weight loss, and lethargy
for 3 weeks. The tiger was previously treated with antibiotics
and corticosteroids with transient clinical improvement.

The initial hematologic and clinical chemical values are
shown in Table 3. The patient had markedly elevated serum
glutamic-pyruvic transaminase (SGPT) and serum glutamic-
oxaloacetic transaminase (SGOT) with high normal levels for blood
urea nitrogen (BUN) and low packed cell volume (PCV) and white
blood count (WBC). A urine sample, collected form the cage floor
showed a specific gravity of greater than 1.035, a 2+ protein,
and an elevated urobilinogen. The sediment was negative for
cellular components. Supportive therapy continued, again with
intermittent clinical improvement but an overall deterioration of
the tiger's condition. A second blood sample collected after
seven days again showed elevated BUN, SGOT and SGPT, and low WBC
and PCV.

The tiger's clinical condition deteriorated and weight loss
became more pronounced. Four days later she was anesthetized
with CI744 and a diagnostic laparoscopic examination was
conducted. There was nodular hyperplasia and cirrhosis of the
liver, but the kidneys appeared normal on gross examination. The
liver was biopsied for culture and histopathologic examination.

The results of the blood sample showed decreasing kidney
function with a rise in BUN and creatinine while PCV, WBC, SGOT
and SGPT remained static. cultures of the liver, urine and blood
did not grow in aerobic or anaerobic conditions. The tiger's
condition continued to deteriorate; 5 days following surgery she
received a blood transfusion of 1800 ml of fresh whole blood from
another Bengal tiger to combat the anemia. The transaminase
remained elevated and the BUN and creatinine continued to rise
during the course of the disease.

Supportive therapy continued for 4 more days with no
response; clonic seizures occurred on the 5th day after the blood
transfusion requiring general anesthesia for control. A blood
sample showed a markedly elevated BUN, creatinine and potassium.
The transaminase remained elevated and the blood ammonium was 140
ug/dl.

With poor response to therapy, continually debilitating
clinical course of the disease and the violent CNS signs, the
decision was made, reluctantly, to euthanize the tiger.

The histopathologic diagnosis was a monocytic leukemia of
the liver, spleen, lung, lymph nodes, adrenal, meninges and bone
marrow. This was an aleukemic phase of the neoplasm. The tiger
had been negative for feline leukemia virus on two occasions.

Table 3.   April, 1983 blood work from adult female Bengal tiger diagnosed as having leukemia.

	4/1 Initial Sample	4/8	4/12 Biopsy	4/16 Transfusion	4/21 Euthanasia
WBC/10^3ul	3.5	4.7	2.5	6.7	8.0
PMN %	76	76	66	87	82
Bands %	0	1		1	9
Lymph %	16	6	23	9	5
Mono %	8	13	10	3	4
Eosin %		3	1		
Baso %		1			
PCV %	29	23	21	19	27
Reticulocyte % RBC		0.5			
HgB g/dl	10.2	8.4	7.2	6.9	10.6
RBC 10^6/ul	4.9	4.0	3.6	3.2	5.0
Nucleated RBC		5	7	11	2
BUN mg/dl	55	69	137	113	310
Creatinine mg/dl	2.0	2.4	6.3	4.7	14.6
SGOT IU	337	172	191	167	220
SGPT IU	982	568	394	454	432

Another tumor was seen in an 8-year-old male Bengal tiger while on breeding loan at the National Zoological Park. While in quarantine, physical examination of its oral cavity revealed a 4 cm gingival mass on the lateral aspect of the upper left canine; the sulcus contained purulent material. A circular 6 cm lesion was also present on the hard palate. The palatine lesion was biopsied and diagnosed as an eosinophilic granuloma. The patient was given oral systemic antibiotics to treat the perialveolar abscess.

Surgical removal of both masses was attempted due to the lack of response to antibiotics. The base of the palatine eosinophilic granuloma was infiltrated with repository corticosteroids. Histopathological diagnosis of the gingival mass was a hyperplastic gingivitis.

The surgery was repeated one year later due to reoccurrence of both lesions. The base of the eosinophilic granuloma was again infiltrated with corticosteroids. The histopathological diagnosis of the gingival lesion was now a fibrosarcoma with a low grade malignancy. Blood samples for feline leukemia virus were negative on 2 occasions.

Following the diagnosis of fibrosarcoma the canine tooth and surrounding alveolar tissue were removed as a block and allowed to granulate in. Histopathological changes showed invasion of the fibrosarcoma to alveolar bone. The palantine mass was treated on several occasions with cryotherapy with only short term improvement.

The tiger was returned to its original zoo. The eosinophil granuloma remained a problem requiring intermittent therapy. The fibrosarcoma invaded the nasal cavity and was non-responsive to radiation therapy. The tiger lived for 2.5 years after leaving the NZP; a total of 3 years from the initial diagnosis of fibrosarcoma (Machado pers. comm.).

SUMMARY

We feel that the key to successful clinical management of captive tigers is a strong and functioning preventive medical program. When problems are suspected, a more aggressive diagnostic and surgical approach is now possible with the improved anesthetic procedures.

These advances in anesthesiology and clinical medicine now allow better diagnosis of problems and provide a better understanding of disease processes which allows the formulation of rational therapy to minimize mortality and morbidity in our patients.

REFERENCES

Abdelrasoul, K. 1979. Epidemiology of ascarid infection in captive carnivores. Amer. Assn. Zoo Vet. Ann. Proc.
Abdulla, P.K., P.C. James, et al. 1982. Anthrax in a jaguar (Panthera onca). J. Zoo Anim. Med. 13:151.
Anonymous. 1978. Congenital diaphragmatic hernia in the tiger. Help p.35.
Barbiers, R. 1982. Digestive efficiencies and maintenancies energy requirements of captive wild felidae: cougar, leopard, lion, tiger. J. Zoo Anim. Med. 13:32-7.
Benirschke, K., L.A. Griner and S.L. Saltzstein. 1976. Pathological findings in Siberian tigers. ISEZ. 18:263-74.
Berrier, H.H., F.R. Robinson, T.H. Reed and C.W. Gray. 1975. The white tiger enigma. Vet. Med./Sm. Anim. Clin. 467-72.
Black, D. 1979. Uterine biopsy of a lioness and tigress after melengesterol implantation. J. Zoo Anim. Med. 10(2):53-56.
Blythe, L.L., et al. 1983. Chronic encephalomyelitis caused by canine distemper virus in a Bengal tiger. J. Am. Vet. Med. Assn. 183:1159-62.
Boro, B.R., G. Sarma, and A.K. Sarmah. 1980. Bacteriological

investigation of enteric infections in zoo animals. Indian J. Anim. Health 19:39-41.

Bryson, C. 1986. Unravelling a tiger's tale. Queen's Gazette, Kingston, Ontario 18:17.

Bush, M. and C.W. Gray. 1975. Dental prophylaxis in carnivores. Int. Zoo Yearb. 15:223.

Bush, M., R.C. Povey, and H. Koonse. 1981. Antibody response to an inactivated vaccine for rhinotracheitis, caliciviral disease, and panleukopenia in nondomestic felids. J. Am. Vet. Med. Assn. 179:1203-05.

Bush, M., R.J. Montali, et al. 1985. Antibody response in zoo mammals to a killed virus rabies vaccine. Proc. Symp. Immunol. Zoo & Wild Anim., ed. B.T. Baker, and S.K. Stoskopf.

Butler, R., R.H. Wrigley, R. Horsey and R. Reuter. 1981. Chondrosarcoma in a Sumatran tiger (Panthera tigris sumatrae). J. Zoo Anim. Med. 12:80-84.

Cline, K. 1966. Diets for Siberian tigers at Detroit zoo. Int. Zoo Yearb. 6:74-78.

Cline, K. 1969. Diets for Siberian tigers at Detroit Zoo. Int. Zoo Yearb. 9:166-67.

Cociu, M., G. Wagner, N.E. Micu and G. Mihaeschu. 1974. Adaptational gastro-enteritis in Siberian tiger (Panthera tigris altaica) in the Bucharest Zoo. Int. Zoo Yearb. 14:171-4.

Cociu, M., G. Wagner and N. Micu. 1976. The latest observations concerning the general adaptation syndrome in Siberian tigers from the Bucharest Zoo. Zoological Garden of Bucharest, 275-77.

Curson, H.H. and A.D. Thomas. 1930. On certain pathological conditions in a tiger. J. Comp. Pathol. 43:158-60.

Dasguptz, B. 1979. Trypanosomiasis in tigers and leopards in Darjieling. Indian J. Parasit. 3:61-62.

DeMartini, J.C. 1974. Thymic hypoplasia and lymphopenia in a Siberian tiger. J. Am. Vet. Med. Assn. 165:824-26.

Duncan, I.D., J.D. Stewart and S. Carpenter. 1982. Inflammatory myopathy in a captive Bengal tiger. J. Am. Vet. Med. Assn. 181:1237-41.

Effron, M., L. Griner and K. Benirschke. 1977. Nature and rate of neoplasia found in captive wild mammals, birds, and reptiles at necropsy. J. Natl. Cancer Inst. 59:185-98.

Eulenberger, K. 1981. Measurement of creatinine concentration in blood serum- importance to early diagnosis of functional disorders in kidneys of zoo animals, with particular reference of felids. ISEZ. 24:327-35.

Fagen, D.A. 1980a. Diet consistency and periodontal disease in exotic carnivores. Am. Assn. Zoo Vet. Ann. Proc.

Fagen, D.A. 1980b. The pathogenesis of dental disease in carnivores. Am. Assn. Zoo Vet. Ann. Proc.

Fowler, M.E. and I.M. Gourley. 1970. Pyloric stenosis in a Bengal tiger (Panthera tigris). J. Zoo Anim. Med. 1:12-16.

Gaskin, J.M., M.A. Jorge, et al. 1980. The tragedy of encephalomyocarditis virus infection in zoological parks of Florida. Am. Assn. Zoo Vet. Ann. Proc.

Gaur, S.N.S. 1980. Helminth parasites from tiger (Panthera tigris) in India. Indian J. Parasit. 4:71-72.

Gould, D.H. and W.R. Fenner. 1983. Paramyxovirus-like nucleocapsids associated with encephalitis in a captive Siberian tiger. J. Am. Vet. Med. Assn. 183:1319-21.

Graham-Jones, O. 1959. Uveitis in a tigress. Vet. Rec. 71:446.

Graham-Jones, O. 1961. Operation of lens extraction in a tigress (Panthera tigris). Int. Zoo Yearb. 3:107-108.

Grittinger, T. 1977. Fluctuations in food consumption among
    Siberian tigers. Int. Zoo Yearb. 17:196-97.
Haberstroh, L., D.E. Ullrey, et al. 1984. Diet and oral health
    in captive amur tigers (Panthera tigris altaica). J. Zoo
    Anim. Med. 15(4):142-46.
Hackenberger, M. In press. The apparent diet digestibility of
    captive tigers. Proc. 3d Ann. Dr. Scholl Nutr. Conf. Cap.
    Wild Anim.
Hemmer, H. 1979. Gestation period and postnatal development in
    felids. Carnivore 2:90-100.
Henrickson, R.V. and E.L. Biberstein. 1972. Coccidioidomycosis
    accompanying hepatic disease in two Bengal tigers. J. Am.
    Vet. Med. Assn. 161:674-76.
Hewer, T.F., L. Harrison, et al. 1948. Lipuria in tigers. Proc.
    Zool. Soc. London 118:924-28.
Hoff, W. 1960. Handrearing baby cats at Lincoln Park Zoo,
    Chicago. Int. Zoo Yearb. 2:86-9.
Hubbard, G.B., R.E. Schmidt and K.C. Fletcher. 1983. Neoplasia
    in zoo animals. J. Zoo Anim. Med. 14:33-40.
Hughes, F. 1977. Handrearing a Sumatran tiger (Panthera tigris
    sumatrae) at Whipsnade Park. Int. Zoo Yearb. 17:214-18.
Husain, D. 1966. Breeding and handrearing of white tiger cubs
    (Panthera tigris) at Delhi Zoo. Int. Zoo Yearb. 6:187-93.
Jensen, J.M., R.G. Helman and F.M. Chandler. 1985. Nasal
    granuloma in a Bengal tiger. J. Zoo Anim. Med. 16:102-103.
Keller, P., D. Ruedi and A. Gutzwiller. 1985. Tissue
    distribution of diagnostically useful enzymes in zoo
    animals: A comparative study. J. Zoo Anim. Med. 16:28-49.
Kelly, D.F., H. Pearson, et al. 1980. Morbidity in captive
    white tigers. In The Comparative Pathology of Zoo Animals,
    ed. R.J. Montali and G. Migaki. Washington, DC: Smithsonian
    Inst. Pr.
Kennedy, S. and S. Patton. 1981. Heartworms in a Bengal tiger
    (Panthera tigris). J. Zoo Anim. Med. 12(1):20-22.
Kitchen, H. and K. Dayhuff. 1977. Diseases of hair in polar
    bears and Bengal tigers. Am. Assn. Zoo Vet. Ann. Proc.
Kloss, H.G. and E.M. Lang. 1976. Handbook of Zoo Medicine:
    Diseases and Treatment of Wild Animals in Zoos, Game Parks,
    Circuses and Private Collections. New York: Van Nos. Rhein.
    Co.
Kollias, G.V., Jr., et al. 1985. Vaginal myxoma causing
    urethral and colonic obstruction in a tiger. J. Am. Vet.
    Med. Assn. 187:1261-62.
Kymhapeb, B.N. and H.H.H. Cam. 1984. Microsporosis and
    candidiasis in tigers and cheetah. Ippen
Leipold, H.W. 1980. Congenital defects of zoo and wild mammals:
    a review. In The Comparative Pathology of Zoo Animals, ed.
    R.J. Montali and G. Migaki. Washington, DC: Smithsonian
    Inst. Pr.
Lennsink, B. 1979. Ollulanus infections in captive bengal
    tigers. Zool. Gart. N.F. Jena 2, 121-26.
Lombard, L.S., H.M. Fortna, F.M. Garner and G. Brynjolfsson.
    1968. Myelolipomas of the liver in captive wild Felidae.
    Vet. Pathol. 5:127-34.
Magrane, W.G. and E.R. Van De Grift. 1975. Extracapsular
    extraction in a Siberian tiger (Panthera tigris altaica). J.
    Zoo Anim. Med. 6:11-12.
Marennikova, S.S., N.N. Maltseva, V.I. Korneeva and N.M.
    Garanina. 1977. Outbreak of pox disease among Carnivora
    (Felidae) and Edentata. J. Infect. Dis. 135:358-66.
McCullagh, K.G. and V.M. Lucke. 1978. Renal papillary necrosis
    in the tiger. acta Zoologica et Pathologica Antverpiensia,

3-13.

McDonald, S. 1985. Treatment of subgingival fractures in canine teeth. Am. Assn. Zoo Vet. Ann. Proc.

Melchior, G. 1973. Meningo-encephalitis in lion and tiger. ISEZ. 15:245-54.

Michalska, Z. 1972. Pathology of pulmonary tuberculosis in big felides in the Zoological Garden of Wroclaw. ISEZ. 14:351-53.

Michalska, Z., A. Gucwinski and K. Kocula. 1977. Sertoli cell tumor in Bengal tiger. ISEZ. 19:305-07.

Michalska, Z., K. Kocula and A. Gucwinski. 1978. Ocular tuberculosis in a tiger. ISEZ. 20:297-98.

Mills, A.W. 1980a. A comparative study of the digestibility and economy of three feline diets when fed to lions and tigers in confinement. In The Comparative Pathology of Zoo Animals, ed. R.J. Montali and G. Migaki. Washington, DC: Smithsonian Inst. Pr.

Mills, A. 1980b. A comparative study of the digestibility and economy of three feline diets when fed to lions and tigers in confinement. Comp. Pathol. Zoo Anim. Symp., 87-91.

Milton, J.L., M.S. Silberman and G.H. Hankes. 1980. Compound comminuted fracture rami mandible in tiger (Panthera tigris): A case report. J. Zoo Anim. Med. 11(4):108-12.

Montali, R.J., L.G. Phillips, et al. 1985. Canine distemper in exotic carnivores: update on vaccination procedures. Proc. Symp. Immunol. Zoo Wild Anim., ed. B.T. Baker and S.K. Stoskopf.

Montali, R.J., C. Bartz, M. Bush and S. Grate. 1986. In Viral Infections of Vertebrates ed. M.J.G. Appel. Amsterdam: Scientific Publishing Co.

Morris, J. 1974. The comparative digestibility of a zoo diet fed to 13 species of felid and a badger. Int. Zoo Yearb. 14:169-71.

Nagar, S.K. 1978. Studies of small mammals of Delhi Zoological Park as possible source of babesiosis infection among white tigers in the zoo. J. Comp. Dis. 10:175-78.

National Research Committee. 1978. Nutrient requirements of cats. Wash., DC: Academy of Science.

Pfeifer, M.L., et al. 1983. Feline infectious peritonitis in a captive cheetah. J. Am. Vet. Med. Assn. 183:1317-18.

Phillips, L. 1980. The parasitology of primates and carnivores. Am. Assn. Zoo Vet. Ann. Proc.

Pulling, F.B. 1976. Possible enterotoxemia in a Bengal tiger. J. Zoo Anim. Med. 7:26.

Rettig, T. 1983. A brief review of salmonellosis as a veterinary zoonosis and a case history of Salmonella pyometraina in a Bengal tiger performing in a travelling circus. Am. Assn. Zoo Vet. Ann. Proc.

Sathyanarayana Rao, M., R.N. Ramachandra, R. Raghavan, and B.S. Keshavamurthy. 1981. Isolation of Corynebacterium pyogenes from a case of broncho-pneumonia in tiger cubs. Tigerpaper 8:23.

Sathyanarayana Rao, M., R.N. Ramachandra, R. Raghavan, S.J. Seshadri and B.S. Keshavamurthym. 1983. Colisepticaemia in tiger cubs. Tigerpaper 10:22.

Saunders, G. 1984. Disseminated leiomyosarcoma in a Bengal tiger. J. Am. Vet. Med. Assn. 185:1387.

Seal, U.S. and A.W. Erickson. 1969. Immobilization of carnivora and other mammals with phencyclidine and promazine. Fed. Proc. 28:1410-16.

Seal, U. 1975. Long term control of reproduction in female lion with implanted contraceptives. Am. Assn. Zoo Vet. Ann. Proc.

Seal, U. 1976. Hormonal contraception in captive female lions. J. Zoo Anim. Med. 7(4):12-20.

Seidensticker, J., K.M. Tamang and C.W. Gray. 1974. The use of CI744 to immobilize free ranging tigers and leopards. J. Zoo Anim. Med. 5:22.

Selbitz, H.J., K. Elze, K. Eulenberger, A. Voight, A. Bergman, S. Seifert and D. Altmann. 1979. Incidence and inportance of bacterial infections of felids in zoological gardens of Leipzig and Erfurt, between 1965 and 1978. ISEZ. 21:191-200.

Slusher, R., S.I. Bistner and C. Kirchner. 1965. Nutritional secondary hyper-parathyroidism in a tiger. J. Am. Vet. Med. Assn. 147:1109-15.

Smith, R.E., W. Boardman and P.C.B. Turnbull. 1971. Hepatoma in a Bengal tiger. J. Am. Vet. Med. Assn. 159:617-18.

Smith, J.L.D., M.E. Sunquist, et al. 1983. A technique for capturing and immobilizing tigers. J. Wildl. Man. 47:255-59.

Starzynski, W. and J. Rokocki. 1974. Abnormal findings from abdominal cavity of big Felidae in Warsaw Zoo. ISEZ. 16:127-31.

Straub, V.G. and B. Seidel. 1983. Klinische labordiagnostik bei grosskatzen unter besonderer berucksichtigung der tigerkrankheit. Ippen 429-37.

Theobald, J. 1970. Experiences in maintaining an exotic cat collection at the Cincinnati Zoo. J. Zoo Anim. Med. 1:4.

Theobald, J. 1978. Felidae. In Zoo and Wild Animal Medicine, ed. M.E. Fowler. Philadelphia: W.B. Saunders Co.

Tinkleman, C. 1979. Endodontic treatment of Siberian tigers at the Philadelphia Zoological Gardens. Am. Assn. Zoo Vet. Ann. Proc.

Van De Grift, E.R. 1975. Root canal therapy on a Siberian tiger (Panthera tigris altaica). J. Zoo Anim. Med. 6(3):24.

Vaneysinga, C. 1969. The dietary requirements of lions, tigers, and jaguars. Int. Zoo Yearb. 9:164-66.

von Krzakowski, W., et al. 1984. Reversible blindness of a Bengal tiger (Panthera tigris) in Kradow Zoo. Ippen

Von, H., K. Elze, et al. 1981. Enzootic occurrence of FV rhinotracheitis in large feline animals of Zoological Garden of Leipzig. Ippen

von Seifert, S., et al. 1982. Temporary, partial and total alopecia in artificially raised big cat. Ippen.

Wallach, J.D. and W.J. Boever. 1983. Diseases of Exotic Animals, Medical and Surgical Management. Philadelphia: W.B. Saunders Co.

Wildt, D. 1977a. Reproductive control in the dog and cat: an examination and evaluation of current and proposed methods. J. Am. Anim. Hosp. Assn. 13:223-31.

Wildt, D. 1977b. Reproduction control in dogs. Vet. Clin. N. Am. 7:775-87.

Wildt, D.E. 1981. Sterilization of the male dog and cat by laparoscopic occlusion of ductus deferens. Am. J. Vet. Res. 42:1888-97.

Woolf, A. and J. Swart. 1974. An outbreak of panleukopenia. J. Zoo Anim. Med. 5:32-34.

Woolfrey, B.F., C.O. Quall and R.T. Lally. 1985. Pasteurella multocida in an infected tiger bite. Arch. Pathol. Lab. Med. 109:744-46.

Worley, M.B. 1983. The continuing role of a feline hepatitis B-like virus in the pathogenesis of liver disease in exotic felids. Am. Assn. Zoo. Vet. Ann. Proc.

Zaki, S. 1980. Isolation of Shigella flexneri from two tiger cubs. Curr. Sci. 49:288.

Zenoble, R.D., D.L. Rigg, D.H. Riedesel and S.V. McNeel. 1982.
    Straw gastric impaction in a tiger. <u>J</u>. <u>Am</u>. <u>Vet</u>. <u>Med</u>. <u>Assn</u>.
    181:1401-02.
Zoological Society of London 1973-75. 1976. Scientific Report.
    <u>J</u>. <u>Zool</u>. 178:449-53.
Zoological Society of London 1975-77. 1978. Scientific Report.
    <u>J</u>. <u>Zool</u>. 184:323-48.
Zoological Society of London 1979-81. 1982. Scientific Report.
    <u>J</u>. <u>Zool</u>. 197:63.

19

Clinical Diseases of Captive Tigers—European Literature

Bernd Seidel and Jutta Wisser

INTRODUCTION

The following review is based on 18 years of clinical experiences and over 10 years of pathological examination of captive animals, and on published papers - especially those of European zoo veterinarians. By previous agreement, the US-American literature was omitted, as it can be assumed that it is widely known to this audience. The described clinical cases refer to animals in Tierpark Berlin. They are Bengal, Siberian and Sumatran tigers.

CLINIC

Pediatrics - Congenital Defects

As a rule, inborn defects are visible in newborns or can be recognized within the first few weeks after birth because of functional developmental disorders (i.e. hereditary hydrocephalus, cataracts, entropion).

Umbilical hernia: When the behavior of new mothers (especially of primipara) is disturbed, they may sever the newborns' umbilical cord too close to the body or not at all. This results in open umbilical rings and increases the risks of local and ascending infections or complete umbilical hernias (Christensen 1963, Seifert 1970). If detected early such cases can be successfully treated, as we did with six neonate Sumatran, four Bengal and two Siberian tigers, by suturing and by administering local and systemic antibiotics for several days. True umbilical hernias were treated twice in neonate Sumatran tigers, one of which died of septicemia after repeated opening of the wound. One young Siberian tiger, for which anesthesia was not possible due to a severe circulatory disturbance, received successful conservative treatment using a truss for three weeks (Haensel 1972). In these cases hand-rearing of the cubs was necessary to facilitate treatment. A rare case of four neonate

Siberian tigers being firmly tied together by their umbilical cords is described by Seifert (1970).

Palatoschisis, cheiloschisis: Cleft palate and/or harelip as actual lethal factors are also described for tigers (Hill 1957, Nouvel et al. 1972). A female Bengal tiger born with open gums reached an age of eight weeks before dying of weakness due to starvation (in spite of forced feedings) during the preparations for an operation.

Syndactyly: Various congenital defects of the legs are described by Christensen (1963) and Dollinger (1974). We have had in our stock of Bengal tigers only one case of a symmetric syndactyly of the soft parts of both front legs ($Ph_{III}D_{2+3}$) in a non-viable neonate. Its litter mates were reared without problems.

Cataracts, entropion: See under Organ Diseases-Eye.

"Kinked tail": Traumatic alterations in the caudal vertebral column can be cured surgically. Inborn defects (deformation or fusing of the vertebrae, lack of disks) also appear in the tiger. This occurred in one neonatal Bengal tiger which died at 12 days of age due to a bacterial infection. Anatomical pathology revealed no further malformations. Similar to domestic cat breeds, surgical repairs should be undertaken only in the case of extremely valuable animals. These should not, however, be used for breeding.

Diaphragmatic hernia: At present, only the report by Brodey and Ratcliffe (1956) on inborn diaphragmatic hernias exists.

Open cranium: Upon dissection of a young Bengal tiger which had been euthanized due to central nervous system problems which did not respond to treatment, extensive open sutures were found. These occurred on the braincase (including the Dura Mater) between the right os parietal, os temporal, os frontal and os interparietal.

Pediatrics - Nursing Phase and Early Youth

Because veterinary intervention in routine prophylactic wormings and vaccinations is rarely necessary, these data are based exclusively on cases of abandoned kittens. Artificial rearing was necessary for 86 (29%) of the 298 tigers born in our collection; 57 (19%) were Siberian, 155 (51%) Bengal and 86 (29%) Sumatran tigers. Of these animals, 53 (61%) were raised successfully. Removal of newborns or kittens up to one week of age was necessary due to behavioral disorders of the mother (refusal to nurse the kittens, constantly carrying the kittens around), inborn lack of vitality, congenital defects, weight loss, pulmonary defects, or, in two cases, delivery of the cubs by Caesarean section.

Because specific cat-milk substitute preparations were not available, abandoned cubs were fed a combination of a dry (cow) whole milk preparation with condensed milk, rice flour, egg yolk, an oral insertable horse serum (with 7-8% Albumen and Globulin) from human pediatrics, and supplementary multivitamins and minerals. Details of this diet, along with an exact analysis of the nutritive substances it contains, are reported by Haensel

(1972). Our young animals, which clearly could not get any colostrum from their mothers, received 0.1-2.0 ml heterologous (human) gamma globulin (at least 90 rel. & Immunoglobulin) per kg of body weight IM (see also Wagner 1971) the same day they were removed from their mothers and, in the case of more serious disorders, during the rearing phase, according to the degree of illness. No reactions due to incompatibility occurred.

The following sections have been arranged according to symptoms observed in cubs during the rearing phase which necessitated intervention by a veterinarian. Dominant symptoms of the digestive and respiratory tracts, as well as hypothermia and general lack of vitality, appear simultaneously or consecutively, particularly under septisemic conditions. The high percentage of hand-reared kittens among our Bengal and Sumatran tigers were largely the result of abandonment by their mothers (kittens lacking vitality).

Indigestion/dysentery: Indigestion (loss of appetite, diarrhea, tympanites) was most striking in kittens which first nursed from their mothers and were then taken away to be fed by bottle or stomach tube. Among other complicating factors, (temperature changes, primary illnesses, general infections) the animals were stressed by this change. Due to the lack of a commercial milk replacer, this problem may also be caused by hygienic irregularities which occur during the preparation of the four to eight ingredients composing the substitute milk. This problem may also appear during the teething phase.

Extensive works by Elze (1967), Eulenberger et al. (1974), Selbitz et al. (1979) and Schroeder (1984) contain representative analyses and combinations of primary infections and abnormal flora in tigers. Based on our experience, diarrhea often leads to fatal changes in the electrolyte balance, dehydration and circulatory collapse. These symptoms appear before the results of bacteriological feces analysis are available. Therefore, we immediately introduce the following symptomatic treatment; change the formula (reduce fat, increase serum, add astringents and absorbents), possibly force-feed (via stomach tube) broad spectrum antibiotics (streptomycin-penicillin combinations, polymyxin B, Terramycin, oxytetracycline) human gamma globulin, electrolyte solutions and, when required, Spacmolytic parenteral. If there is no response to treatment by two days, different antibiotics are utilized. When resistograms occur, the antibiotic corresponds to the sensitivity of the organism. To date, we have had an 82% success rate using this method of treatment.

General reduced vitality: Due to poor reflexes or low body temperature, approximately 28% of the hand-reared kittens were taken from their mothers on the day of birth. Some were taken via Caesarean section. Stimulation through the use of high caloric electrolyte solutions, exogenous warmth, essentially directed analeptics, and forced feeding via a stomach tube were used as necessary. This must take place before the general antibiotic therapy and the substitution of gamma globulins, multi-vitamins and minerals is begun.

Pneumonia: In 23% of our nursing tigers, defects of the respiratory system appeared clinically as isolated organ infections or as part of a general infection. In 5% of the cases, bronchial pneumonia was the cause of death. Two cases of aspiration pneumonia occurred in young bottle-fed Bengal tigers.

Since that time, our strategy has been to assist weakened cubs immediately with a stomach tube. As a result, pneumonia has now become insignificant as a cause of death. In most cases, infected respiratory organs are the result of hypothermia which occurs post-partum. Very seldom is it caused by protozoa (Ippen et al. 1985). When respiratory symptoms are detected by trained personnel, a course of broad spectrum antibiotics and serum therapy is undertaken. This is commensurate with the primary disease, and is combined with multi-vitamins and support of the heart and circulatory system. Along with this, we optimalize the air and temperature conditions. The response has been extremely promising.

Nervous disorders: The literature contains numerous descriptions of isolated nervous diseases in tigers, particularly stargazing (Gass 1976), which may correspond to "headturning" in nursing infants (Schneider 1933). There are also reports of meningoencephalitis resulting from skull trauma (Frankhauser 1955), hydrocephalus internus acquisitis (Heymann 1963), perinatal tramatogenic encephalopathies (Bogaert and Bignami 1964), epileptoid seizures of unclear origin (Aland 1972), Parkinson's syndrome in white tigers (Rathore and Khera 1981), meningoencephalitis caused by Salmonella dublin in a young animal (Mannl 1983), ataxia and torticollis as an accompanying symptom of retinal disease (Krazakowski et al. 1984). Only 11 of our hand-raised Bengal tigers exhibited nervous and central nervous symptoms, all accompanying other diseases (osteopathies, pneumonia, indigestion). These disappeared with the curing of the primary diseases. Young animals showed a distinct, but temporary, tendency to cramp during teething. This tendency disappeared, without treatment, when the canines appeared.

Osteopathy: The pathogenesis of osteopathy in juvenile tigers is resolved thanks to the works of Brandt (1965), Dammrich und Hildebrand (1970), Elze et al. (1974), Dammrich (1974, 1979), Brahm et al. (1979) and Peters et al. (1981). Mineral imbalances (lack of calcium, excess of phosphorus) can result from feeding only meat, which causes nutritional secondary hyperpara- thyroidism, where serum calcium is maintained to the disadvantage of the skeleton. Secondary hypocalcemia, caused by malabsorption due to continuous diarrhea, can have the same effect by stimulating the production of parathyroid hormones. This results in osteogenesis imperfecta with accompanying degrees of lameness and hyperesthesia. X-ray reveals a demineralization with very thin cortical bone, lordosis of the lumbar vertebral column, clearly marked epiphyseal cartilages and green stick fractures and fissures of various degrees and in various locations. These clinical changes appeared in ten hand-reared Bengal tigers between the seventh week and the end of the fifth month of life.

Treatment depends on the degree of damage; in all cases we injected calcium gluconate and vitamins A, D_3, E and C and began feeding whole animals. In cases of fissures and incomplete fractures it is often sufficient to house the animals in narrow cages to limit their movement. In three animals with one- or two-sided fractures of the femur and dislocated fracture it was necessary to perform osteosynthesis. One Bengal tiger was operated on repeatedly (in its third, fifth and sixth month) because of femur, tibia and fibula fractures on both sides. It developed a pear-like deformed pelvis. This prohibited normal births and made Caesarean sections necessary.

In the last five years we have achieved a desirable calcium-phosphorus ratio of 2:1 by supplying the cubs from the fifth week with minced mice, rats, chicks and guinea pigs, and by adding a phosphorus-free mineral mixture. From the tenth to the eleventh week the small food animals are only roughly cut up. Since implementing this procedure there have been no additional cases of clinical osteopathy in our collection. A young hand-reared Sumatran tiger with epiphysiolysis of the distal tibia and fibula was treated successfully by conservative means by Svobodnik and Slavik (1974).

Alopecia: Temporary alopecia in hand-reared cubs is described by Seifert et al. (1982), and we also experienced severe complications in hand-rearing. Although the milk formula is considered well-balanced (see Haensel et al. 1972), it is not optimal for the development of tiger cubs because it is based on cows' milk containing casein. The clinical symptoms observed in seven Bengal and three Siberian tigers at our zoo were quite uniform: around the twelfth day the animals started losing hair around the ears, and within the next two to three weeks on their whole body with the exception of the face, legs and tip of tail. The general condition of the animals was good and the skin showed no changes. Regardless of treatment (vitamin complexes both orally and parenterally, intensified supply of minerals, and dexpanthenol) the animals started growing hair again during the seventh and eighth weeks and had a complete coat by the fourth month of age.

One possible cause of this disorder is a lack of essential amino and fatty acids and fat-soluble vitamins (Eulenberger 1973). Therefore, for the past ten years, small quantities (about 5% volume) of beef liver juice or meat has been added to the milk. Since then no cases of alopecia have occurred in our tiger cubs.

Anemia: Bohlert (1979) provides a comprehensive bibliography up until 1977 concerning basic values and pathophysiology of the hemograms and enzyme activities, serum electrolytes and blood groups. Three cases of clinically threatening, yet treatable, hypochromic anemia appeared in our collection, with hemoglobin values between 6.8-8.4g% in three week old Bengal tigers born with low vitality. Through treatment with three parenteral applications of an iron-dextran preparation (=100 mg Fe/kg body weight) within one week the Hb-content increased to 10.2-12.2g%. Substantially more problematic was the treatment of genetic anemia in the offspring of a Sumatran tiger which never reared her litters. The perceived iron deficiency could only be altered in two of the tigers during a six week period, from 6.4 and 6.6g% to 10.7 and 11.3g% Hemoglobin. Because a genetic predisposition must be assumed, the mother tiger has since been removed from the breeding program.

Dystocia, Ovarian Cysts

Nonphysiological infertility may be masked for an unusually long time with no evidence of any general disorders. Three days after delivering three dead young over a two-day period, a macerated fetus and remains of the chorion were removed under "Neuroleptanalgesia per viae naturales" from a primiparous Bengal tiger, which had shown no symptoms. In three Bengal tigers, delivery lasted from one to four days before complications appeared such as listlessness and fever and, when only a portion of the fetus was visible, it was decided to perform Caesarean sections. This procedure is similar to the surgery practiced in domestic cats. We refer here to the summaries by Kaal and Schipper (1970), Behlert (1979) and Strauss and Seidel (1984). After a successful operation, one Siberian tigress died due to self-mutilation. In one six-year old Bengal tigress, cystic ovaries with pyometra were evident only because of a purulent vaginal discharge which led to an ovariohysterectomy.

Parasites

Parasitic diseases in tigers are rarely specific. Whenever there is a disturbance within a group of animals, regardless of age, endoparasites should be thoroughly investigated as the cause; particularly in the case of poor nutrition, abnormal coat, loss of appetite, anemia, diarrhea or paralysis. In addition, fecal, hematological and possibly serological examinations are recommended. Many parasites formerly destructive are no longer significant because of improved hygiene and new low-risk chemotherapies. Through regular control of parasite populations, tigers in our collection no longer suffer parasitic infestations except for occasional reoccurrences of round worms and coccidia during the first few weeks of life. Table 1 shows the findings in our own stock, selected recent literature and recommendations for treatment. Recommended treatments are based upon our experiences and the reports from Gass (1976), Theobald (1978), Ashton (1980), Prescott (1981) and Wallach and Boever (1983). We believe that, in view of the specific zooecology, long-term dehelminthization "at any price" is not attainable. For this reason, a controlled balance between host and parasite through routine diagnosis, targeted prophylaxis and treatment should be striven for, in order to avoid developmental and performance setbacks and loss of animals.

Organ Diseases - Eye

Conjunctivitis, keratitis, uveitis: It is not uncommon that serious purulent conjunctival infections develop a few days after the eyes open in both mother-reared and hand-reared tigers. As long as there are no eyelid abnormalities, these conditions end after three to five days of local antibiotic treatment (Terramycin, oxytetracycline, penicillin). Behlert and Behlert (1981) were able to surgically heal heavily vascularized chronic keratitis in a young Sumatran tiger after attempts with scraping and cauterization. Graham-Jones (1962) suspected a viral origin for chronic keratitis, uveitis, Iris bombe and retinal detachment which he treated unsuccessfully with antibiotics in a tigress. According to Michalska et al. (1978), one-sided keratitis in an adult Bengal tiger was an external symptom of an untreatable eye tuberculosis which had destroyed the entire tissue of the bulbus.

Hordeolum: In one young Bengal tiger a purulent infection of the Glandulae tarsales healed within six days without treatment through reabsorption of the secretions.

Entropion: Four Bengal tiger cubs from different litters of the same pair exhibited, in the fifth week of life, typical cases of spastic entropion with purulent conjunctivitis and superficial keratitis. Conservative treatment failed, so surgery was

Table 1. Parasitic infestations of tigers and recommended treatments.

Parasite	Treatment	Dosage (mg/kg)	Number of Applications	References
Trematoda				
Troglotrema spp.				
Paragonimus westermani	Emetine	0.5	10 oral	Mandel and Choudhury 1985
	Tetramisole	10	1 oral	Hilgenfeld 1969,
	Levamisole	11	1 oral	Singh and Somvanshi 1978
Cestoidea				
Taenia pisiformis	Nicosamid	150	1 oral	Patnaik and Acharjyo 1970
Spirometra spp.	Praziquantel	5	1 oral	Tscherner 1974
Nematoda				
Toxocara mystax (cati)				Tscherner 1974, Mandel and Choudhury 1985
T. canis	Piperazine	200	1 oral	Sprent and English 1958
Toxascaris leonina	Mebendazole	100	1 oral	Tscherner 1974, Prescott 1981, Abdel-Rasoul and Fowler 1980
Strongyloides spp.	or	15	2 oral	Tscherner 1974
Ancylostoma spp.	or	3	10 oral	Tscherner 1974
A. duodenale	Fenberdazole	10	3-5 oral	Soulsby 1968
Gnathostoma spinigerum				Mandal and Choudhury 1985
Capillaria spp.				Tscherner 1974
Ollulanus tricuspis				Mandal and Choudhury 1985
Trichinella spiralis				Lang 1955
Sporozoa				
Cystoisospora (Isospora) felis	Mebendazole	100	5 oral	Chaudhury and Choudhury 1980,
	Amprolium	300	5 oral	
	Nitrofurazone	50	10 oral	Strauss 1983
	Sulfadimethoxine	50	10 oral	
	Sulfaguanidine	40	5 oral	Strauss 1983
C. rivolta				
Flagellata				
Trypanosoma evansi	Antrycid	?	? s.c.	Nair et al. 1965
Arachnoidea				
Demodex spp.	Dip w. Amitraz	(250 ppm)		Dollinger 1974
	Ivermectin	200ug	1 s.c.	
Ixodidae spp.	Apply acaricide locally (caution: chlorinated hydrocarbons are tox-c for felids!)			Summarized in Mazak 1965

performed with subsequent antibiotic treatment. Complete functional and cosmetic healing was effected. Since in this case exogenous causes can be dismissed, and since the animals were all related, it cannot be excluded that there may have been a genetic disposition (Seidel 1983). Kuntze (1985) reports successful surgery of a secondary entropion in an old Bengal tiger and the treatment of a leukoma in another animal.

Cataracts: Clouding of the lenses in varying degrees is relatively common in Siberian tigers. The causes are related to nutrition, diabetes or genetics (Hammerton 1933, Elze et al. 1974, Benirschke et al. 1976, Krzakowski and Przepiorkowski 1978). Certain cases have been successfully treated by using a "lense sling" or ultra-sound emulsification (Magrane and Vandegrift 1975, Fisher et al. 1978).

Lens luxation: An anterior lens luxation caused by trauma developed in a one year old tiger in the London Zoo. Surgical removal resulted in a bulbar atrophy (Grahm-Jones 1962). Schroeter (1972) and Beehler et al. (1984) mention ophthal-mological problems in full- or part-albino tigers. Krzakowski et al. (1984) describe inborn, reversible blindness as a result of disturbance of the development of the retina in a Bengal tiger.

Organ Diseases - Ear

Othematoma: The literature contains no information about diseases of the ear. In a five week old Sumatran tiger we observed a fresh hematoma of the left ear resulting from trauma. Because there was no open injury, the young animal displayed no general disorders and, in order not to jeopardize natural rearing, we refrained, at that point, from any treatment in the hope that the discharge would be reabsorbed. This did not occur to the hoped-for degree, and when the animal was removed from its mother in its tenth week the ear was distinctively deformed by connective tissue induration and could not be corrected through cosmetic surgery.

Organ Diseases - Skin

Microsporosis: There are numerous reports by European authors concerning the epidemiological/anthropozoonotic signifi-cance of the microsporum. Microsporum canis and M. gypseum clinically dominate in tigers as opposed to Trichophyton tonsurans, Anixiopsis stercoraria and Pseudoarchiniotus spp. (Stehlik and Kejda 1966, Bohm 1968, Jacob 1971, Schonborn et al. 1971, Gass 1976, Schnurrbusch et al. 1976). Most saprophytic fungi become pathogens as the result of resistance-lowering factors such as feeding disorders (especially during hand-rearing), dietetic changes, inanition, injuries, infections, stress or long term antibiotic treatment. According to our observations of M. canis infections in an adult Bengal tiger, the described clinical picture is almost pathognomonic. Initially, individual round, dry spots of eczema can be recognized dorsally and laterally on the extremities. These spread confluently to the back and the rump and cause no itching. Oral application of griseofulvin is the method of choice: 20 mg/kg for at least five weeks. Gass (1976) also states that 200mg/kg four or five times every ten days proves successful. Along with the treatment of the primary disease, the cleanliness of the quarters, bedding, and all other materials must be ensured.

Injuries, pyoderma: Flesh wounds resulting from trauma occur occasionally during transport of the animal or scuffles with other tigers. They are treated according to principles of general practice for cats. In our collection, one adult Bengal tiger became sick while in an outdoor enclosure during the summer. The cause of this illness, a multiple pyoderma without any general disorders, could not be determined. Initial treatment with antibiotics and corticosteroids was unsuccessful. A high leucocytosis (24,240/cu. mm) indicated parenteral application of a penicillin-streptomycin combination with glucocorticoids and multi-vitamins. This led to clinical healing within 12 days of treatment.

Alopecia: See under Pediatrics - Nursing Phase and Early Youth.

Demodicidosis: See Table 1.

Infectious Diseases - Virus Induced

Panleukopenia: The existence of latent parvovirus carriers under conditions where large numbers of animals of varying ages are kept must constantly be kept in mind. Young animals up to the age of six months are particularly susceptible, despite vaccination. In our experience, actual vaccine breakdowns are also possible, in the case of older animals, when their condition is not ideal (mixed infections, parasites, organ diseases as well as temporary over-crowding). Two typical progressions are characteristic: peracute disturbances of general health, loss of appetite, rapid weight loss, occasional vomiting, rapidly increasing circulatory deficiency as a result of blood coagulation and sudden death. Acute/subacute-vomiting, diarrhea, loss of appetite, dehydration, ptosis, pale mucous membrane, striking pallor of the middle of the tongue with inflamed edges can all be considered pathognomonic (Eulenberger et al. 1974). Bieniek et al. (1968) described an atypical, chronic developing form in tigers of varying ages which resulted from inadequate immunity; loss of appetite, occasional vomiting, extremely rapid emaciation, disorders of the electrolyte balances and exsiccosis. According to Elze et al. (1974), Siberian tigers take a middle position as far as their susceptibility is concerned.

Commensurate with clinical description, treatment develops according to the symptoms (including a wide range of antibiotics or combinations corresponding to a resistogram) and calls for high doses of specific serums, aided by parenteral gamma globulins. We find the question concerning the use of killed or live vaccines to be a decision based on personal experience: We repeatedly vaccinated four approximately one-year old tigers, three Siberian and one Bengal, following a vaccination regime specified by the manufacturer of a killed vaccine for peracute panleucopenia. No more clinically manifested cases of panleucopenia occurred in our collection after using live vaccine. This was administered to healthy, parasite-free, young animals. The first vaccinations were administered when the cubs were four weeks old, repeated at eight weeks of age, and finally, at four months of age. According to Scott (1979) and Ashton (1980), killed vaccines are recommended for basic immunization of

newly imported animals or those in non-optimal health conditions.
However, live vaccines are recommended as boosters.

Feline viral rhinotracheitis: Because this infection has
not occurred thus far in our collection, we refer to the report
by Elze et al. (1974), which also covers clinical diseases of
Panthera tigris. Entire stocks became ill in three to six days
resulting from latent agents of the herpesvirus and picornavirus
groups. This appears as infections of the respiratory tract and
the mucous membranes of the head. Clinical signs include
listlessness and loss of appetite to varying degrees. Minor
cases recover in ten to fourteen days without treatment. When
complications develop because of secondary agents, which can
become chronic, the respiratory symptoms' intensity increases
accompanied by glossitis and stomatitis. The "Ulcerative
glossitis" also found in a circus tiger described by Stehlik
(1968), might be directly related to this group of organisms
based on its clinical and epidemiological appearance. Parenteral
or oral applications of antibiotics or chemotherapy (tylosin,
tetracycline, penicillin, chloramphenicol), multi-vitamins and
gamma globulins are first rate in the control of secondary
invaders. Ashton (1980) discusses vaccination with FVR-
Calicivirus live vaccine in Whipsnade.

Feline infectious peritonitis: Chiefly young animals
contract this specific viral infection of felids, characterized
by a chronic progressive fever, ascites, apathy, weight loss,
often accompanied by icterus, less by vomiting and diarrhea.
Fibrinous discharges in the stomach and/or chest cavity and
pericardium as well as granulomatous reactions of the serosa
dominate pathologically-anatomically. In spite of a rapid
worldwide spread of this disease and proven occurrences in
different species in numerous zoos (Gass 1976, Barlough et al.
1982, Wisser 1984), it has not yet been described in tigers.

Encephalomyelitis: Flir (1973) reports on a fatal enzootic
of unknown origin in tigers in a safari park. Within two weeks
the animals died as a result of central nervous signs, inanition
and exsiccosis in spite of symptomatic treatment. Encephalomyelitis, poliomyelitis, neuronophagia and gliaproliferation have
been histopathologically identified. Because of the pathogenesis
and the histological changes, a virus is suspected as the cause
of the illness.

Rabies: Burton (1950) and Pendit (1951) report casuistic
about this disease in tigers and its epidemiological significance
in India.

Infectious Diseases - Bacterial

Tuberculosis: This disease in tigers, caused thus far
almost exclusively by Mycobacterium bovis, continues to reflect
that as long as tuberculosis cannot be totally eradicated in
cattle, alimentary infections in zoo animals must constantly be
dealt with. Works by Schroeder (1965), Ippen (1969), Gass
(1976), Michalska et al. (1978), Behlert (1979), and Rathore and
Khera (1981) contain comprehensive descriptions of the frequency,
pathological anatomy, diagnosis, problems of prophylaxis and
possible treatment, as well as newer isolated findings. In our
experiences, necessitated by the treatment of two Sumatran, one
Bengal, and one Siberian tiger (all adults), it is extremely

difficult to make a clinical diagnosis. In addition to chronic, reduced general turgor, auscultatory and percussion non-objective respiratory signs, body temperature raised by mechanical (squeeze cage) or chemical immobilization, and gradual emaciation, no pathognomonic changes were recognizable. X-ray examination of the abdomen and lungs is of little value. Hemograms produce no specific results, and allergy test readings (subcutaneous, intradermal, intrapalpebral) cannot be evaluated because the conditions under which they are conducted cause increases in temperature. Improved evidence can be expected through traces of organisms (germs, bacteria) found in secretions and excretions, as well as through hemagglutination. All in all, diagnosis in living animals is still totally unsatisfactory. In spite of this, attempts at objectivity through previously mentioned examinations should be made in all cases of refaractory lung diseases and in additional purchases of clinically symptomless animals. Although the results of treatment with streptomycin, PAS and INH have been reported (Heymann 1959), attempted treatment (as far as the given veterinarian legislature permits) should be limited to the most valuable single breeding animals. For the time being, the safest prophylactic measure for sanitation of stock is, in our experience and along with optimal management, hygiene and feeding regimes following active immunization (see also Ehrentraut 1965): For more than six years all our felids have been vaccinated once with lyophilized BCG-vaccine intracutaneous at ten days of age and boosted after five years. Since introduction of this vaccination program, no cases of tuberculosis have occurred in our cats.

Pseudotuberculosis: According to the report by Schuppel et al. (1984), this disease seldom appears in large cats in zoos. However, because of its anthropozoonotic and epidemiological risks, we refer to a report from Pierce et al. (1973) concerning an infection of Yersinia pseudotuberculosis in Siberian tigers. In peracute progression, rapid weight loss and a comatose state dominated shortly before death.

Anthrax: Evidence of infection by Bacillus anthracis also exists in tigers (Jezic 1929, Goret et al. 1947, Gass 1976). The infection resulted from ingestion of infected meat and in three days led to salivation, vomiting, forced defecation, bleeding from the body openings and death by convulsions. Diagnosis was confirmed through identification of the virus. Broad spectrum antibiotics, cortisone and specific serum are therapeutically promising. In the Kabul Zoo a live vaccine has proven prophylactically effective (Rietschel and Senn 1977).

"Tiger disease": To our knowledge the first occurrence of this disease was in a tiger in the Leipzig Zoo in 1963. Since then the problem has become almost worldwide. According to Seifert and Muller (1978), almost half of all zoos keeping tigers have had this problem. Both in absolute numbers and in percentage, Siberian tigers are most often affected. There is no apparent sex or age correlation (there are more cases from the fourth to the sixth year, however). The duration can be a few weeks to twelve years. Mortality is about 56%. No correlation with the type of feeding (nutritional disturbances or deficiencies) or management has been established (Krische 1978).

In our experience with one Siberian and 12 Bengal tigers and in the literature (Ambrosium 1978, Elze 1978, Eulenberger 1978, Selbitz et al. 1978, Storch 1978), the clinical history of the disease is quite uniform. The first signs of the disease are regular regurgitation of undigested or semidigested food and excretion of foul-smelling and soft feces containing undigested food particles. After a few days the animals show increased appetite, sometimes eating the vomited food again.

These gastrointestinal symptoms suggest an inability to digest protein, fat and carbohydrates. In 21 cases hematology, clinical chemistry and electrolytes were examined. The results of the blood studied did not show conformity. The number of leukocytes varied from between 18,000-52,000/ml in the beginning to 11,180/ml after three to five weeks (during and after treatment). We did not find the significant lymphopenia mentioned by Eulenberger (1978) in every phase, but found it between 7-20% (even in clinically critical conditions), with severe leucocytosis, in almost all phases the neutrophils dominated with 64-88%. The red blood cells were normal. At the onset of the disease there were 7.6-11.13 (X 10^6) red cells/ml with 11.0-16.0g% hemoglobin (Hb) and 41-46 packed cell volumes (PCV). During recovery there were 10.4-13.6 (X 10^6) red cells/ml, 12.4-18.2g% Hb and 33-42 PCV. The serum enzymes support the assumption that these are signs of affected pancreas, whereas functional disturbance of the kidney is not clearly diagnosed in living animals.

Considering the blood examinations, clinical symptoms and reports in the literature (and after some less successful attempts at treatment), we attempted the following:
--   feeding whole animals and beef two or three times daily in small portions,
--   giving high doses of antibiotics (oxytetracycline) parenterally for several days,
--   giving multi-vitamins (with high concentrations of A, E and B) and iron twice weekly,
--   giving glucocorticoids intramuscularly every two or three days during the acute phase,
--   giving standard pancreatin (amylase, lipase, protease) orally during the first treatment days up to several weeks after recuperation,
--   giving steroids weekly until recovery (testosterone to males and estrogen to females).

In the last four years we have treated four animals thus. All recovered and there have been no relapses.

Successes with this treatment, as well as diagnosis by elimination, leads us to conclude that a disturbance of the carbohydrate and fat metabolism is leading to a secondary, severe gastroenteritis with lack of normal bacterial flora. Whether this disturbance is due to infectious-toxic or non-infectious causes, or perhaps to a renal disorder, is not known. Therefore a specific treatment is not possible at present.

Trichobezoars: Fluctuating appetite and emaciation were the only symptoms exhibited by an eight year old Siberian tigress (Anonymous 1985). While under sedation for examination the animal expelled four bezoars as large as a man's fist and then recovered. This source also describes a similar case in another Siberian tiger in the Rotterdam Zoo.

PATHOLOGY

Materials and Methods

This evaluation includes all tigers dissected (N=114) in the past ten years in the Foschungsstelle fur Wirbeltierforschung, Berlin Tierpark. The animals came from different zoos in East Germany and other countries, as well as from circuses. They can be divided as follows:

Subspecies	Juvenile	Subadult	Adult	Total	%
P. t. tigris	22	9	15	46	40.4
P. t. altaica	15	6	10	31	27.2
P. t. sumatrae	21	1	3	25	21.9
P. t. (spp.?)	10	–	2	12	10.5
Total	68	16	30	114	
%	59.7	14.0	26.3		100.0

Juveniles were those from newborn up to weaning age, subadults those from weaning to sexual maturity, and all others were considered to be adults. Along with pathological-anatomical findings, histological and bacteriological examinations were routinely performed on all dissected tigers. Parasitological determinations were made only in cases where infestation was suspected.

Results and Discussion

Over 70% of the animals used to compose the following material were young animals. Losses in the age group of juvenile animals dominate at a rate of 60%. Ippen (1978) had similar results, of 64 dissected tigers, two thirds were young animals. Muller and Seifert (1978), having analyzed causes of death and dissection findings as part of the International Studbook for Siberian tigers, indicate that animal losses are highest in the first year of life. Table 2 shows causes of death for dissected tigers listed by age group. The following discussion of the findings corresponds in sequence to frequency of cause of death.

Trauma: Fatal injuries were shown as the cause of death in 18.3% of all materials, the majority of which were juvenile tigers. Death due to rupture of the liver as well as skull injuries (concussions, skull fractures) were particularly common in the first week of life. Bite injuries from other large cats were prominent in the subadult group. Dissection of one 12 year old Sumatran tiger showed death resulted from a ruptured spleen, the cause of which could not be definitely determined.

Stillbirth: 11.4% of the dissected tigers were stillborn. Bleeding in the abdominal cavity due to liver rupture and organ damage and subdural hematoma of the skull were overwhelmingly in evidence. Three stillborn Siberian tigers from one litter were born two weeks early, commensurate with the breeding date (birth weights were 900g, 925g and 1070g). Two of these showed evidence of high grade hyperplasia of the thyroid gland. The mother exhibited symptoms of the so-called "Tiger disease" during gestation and had been treated with ampicillin, Trimethoprim, sulfamerazine and prednisolone. Ippen and Schroeder (1970)

Table 2.   Causes of death determined upon dissection of 114 tigers.

Cause of Death	Juvenile	Subadult	Adult	Number	%
Trauma	17	3	1	21	18.3
Stillbirth	13	-	-	13	11.4
E. coli infection	13	-	-	13	11.4
"Tiger disease"	-	-	9	9	8.0
Nephropathy	-	1	4	5	4.4
Panleukopenia	-	5	-	5	4.4
Gastroenteritis	3	2	-	5	4.4
Bronchopneumonia	3	1	-	4	3.5
Coccus infection	4	-	-	4	3.5
Salmonellosis	1	-	3	4	3.5
Tuberculosis	-	-	4	4	3.5
Thyropathia	4	-	-	4	3.5
Postoperative (Caesarean section) complications	-	-	4	4	3.5
Anemia	4	-	-	4	3.5
Parasitosis	2	-	1	3	2.6
Unidentified infection	-	2	-	2	1.8
Osteopathy	-	2	-	2	1.8
Congenital defects	2	-	-	2	1.8
Tumor	-	-	2	2	1.8
Old age	-	-	2	2	1.8
Suffocation	-	1	-	1	0.8
Circulatory failure	1	-	-	1	0.8

report an average of 16.5% mortality due to stillbirth and lack of vitality in their dissected material for carnivores. As possible causes for this they indicate genetic factors or changes in metabolism. They also indicate that stillbirths are problematic for pathologists to interpret, since often neither macroscopically nor histologically detectable findings are available for diagnosis. Elze et al. (1984) report that between 5-15% of all zoo animal pregnancies in the Leipzig Zoo end in abortion or stillbirth. They cite hormonal imbalances, narrow pelvis, vitamin A, D, E and Ca deficiencies or obesity as major causes in felids. Prophylactic measures recommended by these authors include keeping the animals at approximately 80% normal body weight, keeping them in breeding condition, ensuring sufficient vitamin support and limiting the time between breedings. In our material, stillborn tigers with ruptures of the liver and damaged organs must likewise be assigned to the complex of causes mentioned above, especially indicated were narrow pelvis and hormonally weak labor. In four cases of difficult births (Table 2) the mothers died as a result of postoperative complications (Caesarean section).

E. coli infection: This caused the death of 13 tigers (11.4%), 11 within their first year of life. Septicemia and gastroenteritis are in the pathomorphological forefront. However, nonspecific organ changes and negligible pleuropneumonia have also been diagnosed. Two tigers, one five and one six weeks old, showed enteritis ranging from catarrhal to fibrinous. One exhibited a fibrinous peritonitis as well, the other liver and spleen necrosis. Bacteriological examinations led predominantly to the isolation of hemolytic coliforms. Serotypes also known as calf pathogens could be isolated in three tigers. In numerous publications (Elze 1967, Schroeder 1967, 1984, Kronberger 1970, Eulenberger et al. 1974, Ippen and Schroeder 1974, Selbitz et al. 1979, 1986), the significance of E. coli infections as cause of death of young predators has been emphasized. According to Elze (1967), two typical progressions appear in carnivores. On the one hand the infections are septicemic, acute to apoplectic progressing disease forms which are difficult to recognize clinically and pathomorphologically. These are found overwhelmingly in cubs in their first three to five days of life. On the other hand he describes dyspeptic or enterotoxemic forms, which frequently infect young animals up to two weeks of age. He characterizes E. coli infections as infectious, multifactored illnesses which result not only from a particular disposition of young animals (inborn lack of vitality, insufficient support with colostrum, low body temperature) but through a reservoir of bacteria as well (mother animal, hospitalization, resistance to antibiotics). Eulenberger et al. (1974) report infections in older juveniles which begin at the age of two to three weeks, but which can frequently be observed at the time of weaning as well, in the form of acute to subacute progression. Acute manifestations result mainly in general health disorders such as listlessness, loss of appetite and constant thin, runny, yellowish to grey-bloody excrements. With regard to disease progression and postmortem findings, our cases of E. coli infection concurred with the basic description of both of the above-named teams. The results of the bacteriological examination also conformed to the determination of E. coli by Selbitz et al. (1979). Selbitz et al. (1966) recommend conducting examinations to determine the direct virulence factors of the serotype, in order to exactly characterize epidemiologically and epizootologically relevant strains.

"Tiger disease": This caused the death of six Bengal and three Siberian tigers in our collection. Four of the animals were between two and four years old, the other five were between nine and 18 years old; three were males and six females. Clinical symptoms have been described above. Postmortems of the emaciated animals revealed acute to chronic gastroenteritis. Lymphoplasmocytic and histiocytic infiltrations of stomach and intestinal mucosa were predominant. The kidneys exhibited glomerulonephritis, thickening of the Bowman's capsule and serous exudate in the capsular space, as well as metastatic calcifications in the medulla. Chronic interstitial nephritis with atrophic glomeruli were also present. The liver commonly showed fatty change with small fat droplets in the hepatocytes and hemosiderosis of Kupffer's cells. The adrenals were macroscopically extremely flat and firm and histologically exhibited a narrow medulla with hemorrhages (in six tigers) and fibrosis of the cortico-medulla junction. The thyroids were active in most cases. The pituitary gland was examined in three tigers and showed no pathological lesions. The pancreas in all nine examined cases showed no abnormalities. Bacteriological examination revealed no pathogenic bacteria in the organs of any of the tigers.

There have been several publications on the "tiger disease". Brach and Kloppel (1963) believe it is caused by a chronic insufficiency of the adrenal cortex. Cociu et al. (1974, 1976) report changes in the adrenal cortex in dissected animals and consider it in context with adaptational difficulties of the tiger. Elze et al. (1974) believe the disease is caused by a disturbance in the hypothalamus, adenohypophysis, and adrenal glands, possibly in connection with stress. Another possible cause of the "tiger disease" is mentioned by Jarofke et al. (1977): In a dissected Siberian tiger they detected a chronic atrophic gastritis with intestinal metaplasia. They consider as a possible cause an auto-immune disease. Elze and Seifert (1978), after conducting an international inquiry about the "tiger disease" and considering the results of research in Leipzig, report clinically a prominent disturbance of the epigastric region, which is initiated by primary or secondary disturbance of the pancreas. Schoon et al. (1980) found in three adult Bengal tigers pathomorphologically a subacute to chronic gastroenteritis as well as an activation of the glandular system, and conclude that during the course of the disease a weakening of the immunological system may occur.

The real cause of "tiger disease" can only be solved by further thorough research. We believe that in our nine dissected tigers the uniform formal pathogenesis was the result of different causative factors. Initially there are specific and considerable changes in the renal tubules and glomeruli, which cannot be detected with the light microscope, and which only during the course of the disease lead to interstitial nephritis or glomerulonephritis. This results in gastroenteritis due to a pre-uremic state accompanied by signs of vomiting, foul-smelling feces and dehydration. This initiates a vicious cycle as damage to the kidneys is sustained or increased by the loss of body fluids and electrolytes, and in turn increases the gastroenteric signs and finally results in emaciation. This theory is supported by the success of our treatment, in which the cycle is broken by "calming" the gastrointestinal tract by means of atropine. Similar histories of tigers with primarily non-inflammatory kidney disturbances are described by Benirschke et al. (1976) and McCullagh and Lucke (1978). We consider the changes in the adrenal (revealed in the histopathology of the examined nine tigers) to be due to treatment with prednisolone. There are several possible pathogens involved in the primary non-inflammatory damages of the tubules and the glomeruli.

The following causes appear probable:
1. Damage to the tubular epithelia due to an excess of phosphorous in the diet, leading to a secondary nutritional hyperparathyroidism. This hypothesis is supported by the curing of a Sumatran tiger by maintaining it on a strict diet (Goltenboth 1978). Possibly this affects more young tigers, like our cases in two to four year old cubs. Furthermore, Eulenberger (1978) detected significantly higher values of serum calcium in some tigers during the acute phase of the disease. In 1981 he concluded that the functioning of the kidneys is impaired. Benirschke et al. (1976) point out that tigers have a very sensitive renal tubular system;
2. Damages to the glomeruli or tubules due to various immunopathological processes. For example one could consider the possibility of concretions of immune-complex deposits due to occult viral infections, as for instance in feline infectious peritonitis or an infection with a non-tumorous forming feline oncorna virus.

Nephropathy: Fatal kidney failure was diagnosed in five tigers (4.4%). Upon dissection an eight year old Bengal tiger was found to be very uremic, had a high degree of chronic glomerulonephritis and a bronchial sarcoma as large as a man's fist in the right half of the lung. The remaining four tigers showed interstitial nephritis, one animal as a result of a pyelonephritis. In the latter, a massive calcification process of the basal membranes of the tubuli and the glands in the stomach mucosa appeared together with chronic interstitial nephritis. This case may be casually related to a disorder of the calcium-phosphorus metabolism, since similar calcification processes are found in other types of young cats (Wisser 1986). Starzynski and Rokicki (1974) report on a uremia resulting from a chronic kidney infection with hyaline cylinders and chronic glomerulonephritis in a four year old male Bengal tiger. Nephrosis and the formation of chalk metastasis in tigers are diagnosed by Ippen (1980) to a much greater extent (12.5%) than we have found. Certainly, changes in the kidney due to inflammation play a decisive role here, as in other felids. However, it can be assumed that they are of a more secondary nature in their etiopathogenesis and primarily affect endogenous metabolic disorders. In this respect, the findings of Hewer et al. (1948), as well as Benirschke et al. (1976) on interstitial nephritis of undetermined etiology and those from McCullagh and Lucke (1978) on papillary necrosis in tigers are particularly worth mentioning.

Panleukopenia: Five subadult tigers with panleukopenia were all in very good nutritional condition; typically acute fibrinous enteritis with edema in the intestinal submucosa were not found. Intranuclear bodies were not detected. A lack of lymphocytes was in evidence in the spleen, lymphnodes and Peyer plates. The myeloid cells in the bone marrow were destroyed; in two cases a panmyelophthisis existed. Basic examinations on the pathology of this virus infection was conducted by Johannsen (1967a,b) as well as Johannsen and Ehrentraut (1967), so that exact determination of panleukopenia compared with other enteritis is histologically possible. The pathomorphological changes in our diagnoses in five tigers concur with the results of these authors and also with the description of the histological findings by Langheinrich and Nielsen (1971). In retrospect a vaccine break was determined, which coincided with the change from killed to live vaccines.

Gastroenteritis, bronchopneumonia: The findings shown in Table 2 were composed primarily from juvenile tigers. Because of antibiotic treatment of the animals they could not be clarified. Dissection statistics from Kronberger (1970), Dollinger (1974), Ippen and Schroeder (1974), Maran et al. (1974) and Goltenboth and Klos (1970) show that gastroenteric and respiratory diseases play a significant role in carnivores. The report on diseases of the digestive organs by Zaki et al. (1980) concerning the isolation of Shigella flexneri in tigers with enteritis, and that by Henrickson and Biberstein (1972) on coccidioidomycosis with ascites, fibrinous perihepatitis and liver necrosis in two Bengal tigers, are particularly worth mentioning. Viral infections, superimposed through secondary organisms, are frequently the cause of pneumonia. Lloyd and Allen (1980) report on the unusual finding of a granulomatous pneumonia with isolation of bacteria of the group EF-4 (Eugenic Germenter 4) in a five month old tiger. Sathyanarayana et al. (1981) isolated Corynebacterium pyogenes from pneumonic lung sections of a young tiger.

Coccus infection: Four tigers ranging from three to 14 days
of age died as a result of coccus infections. Dissection showed
septicemia, peritonitis, enteritis and pneumonia. In all cases
Streptococcus were grown from the abnormal organs, in one animal
a mixed microflora with coliforms. Coccus infections
(superficially Streptococcus) were equal to E. coli infectious
disease factors in the postnatal developmental phase of Zoo
felids. Hamerton (1929), Murer (1939) and Williamson et al.
(1965) report on the manifestation of Streptococcus and
Staphyloccus infections in the respiratory organs of tigers.
Murmann (1982) reports on purulent meningoencephalitis.

Salmonellosis: They are pathomorphologically characterized
by gastroenteritis, serious liver necrosis and secondary
pneumonia. Bacteriological examinations lead to isolation of
Salmonella typhi-murium and S. dublin. Eulenberger et al. (1974)
give prominence to the transmission via food animals, especially
with S. typhi-murium as the primary source of salmonellosis in
felids. Further findings on salmonellosis in tigers are reported
by Khan (S. senftengberg and S. johannesburg; 1970), Benirschke
et al. (S. meningitis; 1976) and Mannl (S. dublin; 1983).

Tuberculosis: All cases in our collection appeared as
chronic organ tuberculosis, acid-fast rods were microscopically
seen in the altered membranes. Cultures from one tiger showed
Mycobacterium avium. There appears to have been a decrease in
the frequency of this disease. Ippen (1978) found evidence of
tuberculosis in 50% of the adult tigers he dissected. These
animals originated from near the same locations as those
considered in our report. The four cases of tuberculosis in our
material correspond to a relative portion of only 13% of the
dissected adult tigers.

Thyropathia: Four juvenile tigers died due to hyperplasias,
with enlargement of the thyroid glands up to ten times normal
size. Histologically these were found to be struma parenchymatosa diffusa with only limited follicular development and
barely visible colloid and papillomatous-epithelial growth into
the follicular lumem. As secondary findings, similar strumen
were diagnosed in 12 young tigers (eight struma parenchymatosa,
three struma colloides diffusa). In an approximately one year
old Bengal tiger, euthanized because of osteopathies, a
follicular adenoma with struma hyperplastica diffusa could be
determined. An 18 year old tiger, euthanized due to senility,
upon histopathological examination showed evidence of
undifferentiated carcinoma (spindle cell type) of the thyroid
gland with metastases in the lung and the suprarenal gland. Both
such thyroid tumors have already been reported by Ippen and
Wildner (1984). In 114 tigers there was a frequency of
thyreopathias of 15.8%. Meister and Ippen (1983) comment on the
pathogenesis of thyroid gland fluctuations.

Anemia: Four Sumatran tiger cubs, from three litters of one
female, died of anemia at the ages of two to four weeks. Aside
from the high grade of anemia there was pathological-anatomical
evidence of ascites and secondary pneumonia. There were no
subcutaneous histological examinations of the bone marrow. In
spite of many reports on the physiology and pathology of blood of
felids, we are not aware of any literature on genetic anemia in
tigers. Demartini (1974) diagnosed a lymphopenia with thymus-
hypoplasia in a four week old Siberian tiger, which he was not
able to interpret etiologically.

Parasitosis: Histological examination revealed parasitic infestations as the causes of death of three tigers. Marked fibrinous enteritis in two tigers were caused by nematodes, the nature of which could not be precisely determined. An eleven week old Bengal tiger showed an abundance of protozoan stages on the blood vessel endothileum of all examined organs. The invasion had led to pleuropneumonia, lymphadenitis and swelling of the liver and kidneys. The nature of the protozoan has not yet been definitely determined (Ippen et al. 1985). Helminthiasis plays a secondary role in tigers, although Goltenboth and Klos (1980) report persistent ascariasis in large cats.

Osteopathy: Two one year old Bengal tigers showed typical skeletal alterations of osteodystrophia fibrosa generalisata with marked fibrosis marrow. Mineralization of the hypoplastic bony tissue development was insufficient in one tiger. A thyroid tumor was found in the other.

Tumor: Tumors were determined to have caused the death of two tigers. A leiomyoma of the right uterine horn in a 19 year old tigress had enclosed the ovary. In a 14 year old Bengal tiger from a circus, a blister in the region around the nose developed. The neoplasm led to necrosis of the left os nasal and part of the left os frontal, os lacrimale and os maxillare as well as the adjacent cartilage and soft tissue. There was evidence of metastasis to the lungs and esophagus. The determination is not yet complete. Other reports of tumors in tigers include: Simone (1973)--adenoma of the tracheal membrane; Cohrs (1928) epithelial plate carcinoma of the lungs; Ratcliffe (1933)--rectal carcinoma; Steinbacher (1941)--lung cancer; Seitz (1958)--malignant hypernephroma or teratoma, mixed tumor; Huber and Maran (1967)--epithelial plate carcinoma of the esophagus; Misdorp and Zwart (1967) in three cases--spindle cell cancer of the lower jaw, carcinoma on the mammaries and the left ovary; Schroeder (1968)--sarcoma of the liver; Smith et al. (1971)--liver cell hepatoma; Dollinger (1974)--kidney carcinoma; Michalska et al. (1977)--diffuse Sertoli's cell tumor of the right testicle; Saunders (1984)--leimyosarcoma; and Prange et al. (1986)--malignant fibrosal histiocytoma in the perineal area.

Suffocation: Dissection of a three month old Siberian tiger revealed all the signs of death by suffocation. A meat bone had become caught in the air passage in the area of the larynx.

CONCLUSIONS

Essentially unresolved are problems primarily in the perinatal period and in the "tiger disease". In the latter, research may most profitably be centered on the parathyroid gland and kidneys, and possible immune-defects.

ACKNOWLEDGEMENTS

Thanks go to Dr. Ulysses Seal and Dr. Ron Tilson of the Minnesota Zoological Garden for their kind invitation to participate in this symposium, to Jim Dolan of San Diego Wild

Animal Park for his "labor of love" in translating the original
German manuscript to English, and to numerous unnamed but
remembered people for their kindnesses and help along the way.

REFERENCES

Note:
Verh.ber. Erkrg. Zootiere = Proc. Int. Symp. Dis. Zoo Animals

Anonymous. 1985. Gefahrliche Haarballen im Tigermagen. Zoo-
    Freund 56:19.
Abdel-Rasoul, K. and M.E. Fowler. 1980. An epidemiologic
    approach to the control of ascariasis in zoo carnivores.
    Verh.berg. Erkrg. Zootiere 22:273-77.
Aland, A. 1972. Epileptische Krampfanfalle bei einem 6 Monate
    alten Tiger. Verh.ber. Erkrg. Zootiere 14:399-400.
Ambrosium, H. 1978. Serumeiweissfraktionen bei gesunden und an
    der soganannten Tigerkrankheit erkrankten Tigern. Congr.
    Rep. 1st Int. Symp. Manage. Breed. Tiger, Leipzig 168-69.
Aston, D.G. 1980. Some diseases in non-domestic cats/discussion
    of the vaccination of non-domestic cats. Brit. Vet. Zool.
    Soc. News 10:14-16.
Barlough, J.E., J.C. Adsit and F.W. Scott. 1982. The worldwide
    occurrence of feline infectious peritonitis. Feline Pract.
    12:26-30.
Beehler, B.Z., C.P. Moore and J.P. Pickett. 1984. Central
    retinal degeneration in a white Bengal tiger. Am. Assn. Zoo
    Vet. Ann. Proc.
Behlert, O. 1979. Immobilisation und Krankheiten der Raubkatzen
    Eine Literaturstudie. Vet. Dis. Berlin West. 433 pp.
Behlert, O. and C. Behlert. 1981. Cirurgische Behandlung einer
    chronischen Keratitis bei einem jungen Sumatra-tiger.
    Kleintierpraxis 26:301-06.
Benirschke, K., L.A. Griner and S.L. Salzstein. 1976.
    Pathological findings in Siberian tigers. Verh.ber. Erkrg.
    Zootiere 18:263-73.
Bieniek, H., W. Encke and R. Gandras. 1968. Eine seuchenhafte
    Erkrankung des Raubkatzenbestandes im Krefelder Tierpark mit
    protrahiertem Verlauf. Verh.ber. Erkrg. Zootiere 10:243-54.
Bohm, K.H. 1968. Dermatomykosen bei Zootiren. Kleintierpraxis
    13:139-41.
Bogaert, L. van, and A. Bignami. 1964. Neurology and neuro-
    pathology of tigers in captivity. III. Perinatal encephalo-
    pathy and probably traumatism during birth of two young
    tigers. Acta neuropath. 3:601-08.
Brack, M. and G. Kloppel. 1963. Chronische Nebennierenrinden-
    Insuffizienz bei einem Thailand-Tiger. Verh.ber. Erkrg.
    Zootiere 5:97-101.
Brahm, R., E. Brahm and W. Bartmann. 1979. Osteodystrophia
    fibrosa bei Feliden. Verh.ber. Erkrg. Zootiere 21:73-76.
Brandt, H.P. 1965. Osteogenesis imperfecta bei jungen
    Raubkatzen. Verh.ber. Erkrg. Zootiere 7:55-59.
Brodey, R.S. and H.L. Ratcliffe. 1956. Congenital diaphragmatic
    hernia in the tiger--two case reports. J. Am. Vet. Med.
    Assn. 129:100-02.
Burton, R.W. 1950. Rabies in tigers--two proved instances. J.
    Bombay Nat. Hist. 49:538-41.
Chaudhury, S.K. and A. Choudhury. 1980. Investigation of
    coccidiosis in "Tigon", a hybrid carnivore kept in the

zoological garden of Calcutta. Verh.ber. Erkrg. Zootiere 22:267-72.

Christensen, N.O. 1963. Einige Beobachtungen uber die Krankheiten der Carnivoren im Kopenhagener Zoo. Verh.ber. Erkrg. Zootiere 5:32-44.

Cociu, M., G. Wagner, N. Micu and G. Mihaescu. 1974. Adaptational gastro-enteritis in Siberian tigers at Bucharest Zoo. Int. Zoo. Yearb. 14:171-74.

Cociu, M., G. Wagner and N. Micu. 1976. The latest observations concerning the General Adaption Syndrome in Siberian tigers from the Bucharest Zoo. Verh.ber. Erkrg. Zootiere 18:275-77.

Cohrs, P. 1928. Paragonimus westermani und primares Plattenepithel-Karzinom in der Lunge, sowie parasitare durch Galoncus pernitiosus verursachte Knoten im Dunndarm eines Konigs-tigers. Beitr. Path. Anat. 81:101-20.

Dammrich, K. and B. Hildebrand. 1970. Hyperparathyreoidismus bei Zootiren. Verh.ber. Erkrg. Zootiere 12:109-11.

Dammrich, K. 1974. Skeletterkrankungen bei jungen Raubkatzen. Verh.ber. Erkrg. Zootiere 16:137-39.

Dammrich, K. 1979. Zur Pathogenese der Skeletterkrankungen bei Zootieren. Verh.ber. Erkrg. Zootiere 21:65-71.

DeMartini, J. 1974. Thymic hypoplasia and lymphopenie in a Siberian Tiger. J Am. Vet. Med. Assn. 165:824-26.

Dollinger, P. 1971. Tod durch Verhalten bei Zootieren. Vet. Diss. Zurich: 229pp.

Dollinger, P. 1974. Analyse der Verluste im Raubtierbestand des Zoologischen Gartens Zurich von 1954 bis 1973. Ver.ber. Erkrg. Zootiere 16:39-43.

Ehrentraut, W. 1965. Gesundheitsuberwachung und medikamentelle Prophylaxe bei der Aufzucht von Feliden. Verh.ber. Erkrg. Zootiere 9:57-64.

Elze, K. 1967. Zur Coliinfektion bei Zootieren. Verh.ber. Erkrg. Zootiere 7:234-37.

Elze, K., K. Eulenberger, S. Seifert, H. Krongerger, K.F. Schuppel and Ute Schnurrbusch. 1974. Auswertung der Krankengeschichten der Felidenpatienten 1958-1973) des Zoos Leipzig. Verh.ber. Erkrg. Zootiere 16:5-18.

Elze, K. 1978. Zur klinik der sogenannten Tigerkrankheit. Congr. Rep. 1st Int. Symp. Manage. Breed. Tiger, Leipzig 172-73.

Elze, K. and S. Seifert. 1978. Abschlussbemerkungen in Auswertung der internationalen Umfrage zur Tigerkrankheit mit besonderer Berucksichtigung der Leipziger Untersuchungen. Congr. Rep. 1st Int. Symp. Manage. Breed. Tiger, Leipzig 172-73.

Elze, K., H.J. Selbitz, K. Eulenberger and S. Seifert. 1984. Aborte, Totgeburten und Erkgankungen von neugeborenen Zootieren. Verh.ber. Erkrg. Zootiere 26:41-46.

Eulenberger, K. 1973. Untersuchungen zur Hamatologie Vitamin-A-Serumgehalt und Hamogramm) und Klinik von Vitamin mangelcrkrankungen bei Siberischen Tigern u.a. Feliden des Leipziger Zoologischen Gartens. Vet. Diss. Leipzig 170+pp

Eulenberger, K., K. Elze, S. Seifert and Ute Schnurrbusch. 1974. Zur Differentialdiagnose, Prophylaxe und Therapie der Panleukopenie, Samonellose und Koliinfektion der jungen Grosskatzen (Pantherini). Verh.ber. Erkrg. Zootiere 16:55-65.

Eulenberger, K. 1978. Hamogramm, ergebnisse blutbiochemischer Untersuchungen und pH-Wertbestimmungen des Magensaftes bei Sibirischen Tigern mit Symptomen der sogenannten Tigerkrankheit. Congr. Rep. 1st Symp. Manage. Breed. Tiger, Leipzig 157-61.

Frankhauser, R. 1955. Neuropathologische Befunde bei Wildtieren. Schweiz. Arch. Tierhilkd. 97:53-64.

Fisher, L.E., S.J. Vainisi and E. Maschgan. 1978. Cataracts in Siberian tigers. Congr. Rep. lst Int. Symp. Manage. Breed. Tiger, Leipzig 192-93.

Flir, K. 1973. Encephalomyelitis bei grosskatzen. Dtsch. Tierarztl. Wochenschr. 80:401-04.

Garlt, Ch. and D. Ritscher. 1977. Kasuistische Beitrage zu Knochenerkrankungern bei Zootieren. Verh.ber. Erkrg. Zootiere 19:314-23.

Gass, H. 1976. Katzen, Schleichkatzen, Marder. In Zootier-krankheiten, ed. H.G. Klos and E.M. Lang. Berlin West, Hamburg: Verlaug Paul Parey.

Goltenboth, R. 1978. Bemerkungen zur Haltung und zud den Krkrankungen der Tiger im Zoo Berlin. Congr. Rep. lst Int. Symp. Manage. Breed. Tiger, Leipzig 199-201.

Goltenbach, R. and H.G. Klos. 1980. Ubersicht uber die Todesursachen und das Krankheitsgeschehen im Raubtierbestand des Zoologischen Gartens Berlin. Verh.ber. Erkrg. Zootiere 22:2203-09.

Goret, J., L. Joubert and L. Chabert. 1947. Fievre Charbonneuse du tigre. Rec. Med. Vet. 123-507-13.

Graham-Jones, O. 1959. Uvitis in a tigress. Vet. Rec. 71:446.

Graham-Jones, O. 1962. The operation of lens extraction in a tigress. Vet. Rec. 74:553-55.

Haensel, Renate and Joachim. 1972. Beobachtungen und Erfahrungen bei kunstlichen Tiger-Aufzuchten. Zool. Garten N.F. 41:97-113.

Hamerton, A.E. 1929. Report on the deaths occurring in the Society's Gardens during 1928. Proc. Zool. Soc. London 49-59.

Hamerton, A.E. 1933. Report on the deaths occurring in the Society's Gardens during 1932. Proc. Zool. Soc. London 451-81.

Henrickson, R.V. and E.L. Biberstein. 1972. Coccidioidomycosis accompanying hepatic disease in two Bengal tigers. J. Am. Vet. Med. Assn. 161:674-77.

Heymann, H. 1959. INH-Anwendung bei Zootieren. Kleintierprax. 4:123-26.

Heymann, H. 1963. Hydrocephalus bei einem Tiger. Verh.ber. Erkrg. Zootiere 5:72-73.

Hilgenfeld, M. 1969. Beitrag zur Lungendistomatose der Grosskatzen. Verh.ber. Erkrg. Zootiere 11:169-73.

Hill, W.C.O. 1957. Report of the Society's Prosector for the years 1955 and 1956. Proc. Zool. Soc. London 129:431-46.

Huber, J. and B. Maran. 1967. Uber einige Falle von Tumoren bei wilden Tieren. Verh.ber. Erkrg. Zootiere 9:225-27.

Ippen, R. 1969. Moglichkeiten und Wege einer Tuberkulose-bekampfung in Zoologischen Garten. Verh.ber. Erkrg. Zootiere 11:25-33.

Ippen, R. and H.D.Schroeder. 1970. Uberr die Verluste bei der Aufzucht von Saugetieren in zoologischen Garten. Verh.ber. Erkrg. Zootiere 12:5-13

Ippen, R. and H.D. Schroeder. 1974. Auswertung der postmortalen Untersuchungsergebnisse bei in Gefangenschaft gehaltenen Landraubitieren. Verh.ber. Erkrg. Zootiere 16:29-37.

Ippen, R. 1978. Die wichtigsten Erkrankungen der Tiger aus der Sicht des Pathologen. Congr. Rep. lst Int. Symp. Manage. Breed. Tiger, Leipzig 202-04.

Ippen, R. and G.P. Wildner. 1984. 24 comparative pathological investigations of thyroid tumors of animals in zoos and in

the wild. In One Medicine, ed. O.A. Ryder and M.L. Byrd.
Berlin West/Heidelberg: Springer Verlag.
Ippen, R., R. Meister and D.Henne. 1985. Zu einigen
Protozoenfunden bei Zoo-Carnivora. Milu, Berlin 6:60-68.
Jacob, K.J. 1971. Uber einen witeren Fall von Mikrosporie bei
jungen Tigern. Verh.ber. Erkrg. Zootiere 13:195-97.
Jarofke, D., K. Dammrich and H.G. Klos. 1977. Atrophierende
Gastritis mit intestinaler Metaplasie bei einem Siberischen
Tiger (Panthera tigris altaica)--ein kasuistischer Beitrag
zum Problem der sogenannten Tigerkrankheit. Verh.ber. Erkrg.
Zootiere 19:353-56.
Jezic, J. 1929. Eine Milzbrandenzootie in einem Zirkus. Jugosl.
Vet. Glasn. 9:177.
Johannsen, U. 1967a. Untersuchungen zur infektiosen Enteritis
Panleukopenie) der Feliden. II. Mitt. Pathologische
Histologie der Erkrankung. Arch. Exp. Vet. Med. 22:383-406.
Johannsen, U. 1967b. Knochenmarkuntersuchungen zur infektiosen
Enteritis Panleukopenie) bei Haous-und Zookatzen. Arch. Exp.
Vet. Med. 22:293-329.
Johannsen, U. and W. Ehrentraut. 1967. Untersuchungen zur
infektiosen Enteritis Panleukopenie) bei Feliden des
Leipziger Zoos. Verh.ber. Erkrg. Zootiere 9.65 70.
Kaal, G.T.F. and K. Schipper. 1970. Sectio caesarea bij een
Koningstiger. Tijdschr. Diergenceskd. 95:833-34.
Kahn, A.Q. 1970. A note on salmonella infections in wild
animals in Khartoum, Sudan. Brit. Vet. J. 126:302-05.
Krische, Gisela. 1978. Zur Futterung und dem Auftreten der
sogenannten Tigerkrankheit. Congr. Rep. 1st Int. Symp.
Manage. Breed. Tiger, Leipzig 150-53.
Kronberger, H. 1970. Betrachtungen zu uber 100 Jahren
Felidensektionen. Zool. Garten N.F. 39:147-51.
Krzakowski, A. and B. Przepiorkowski. 1978. Slepota tygrysa
bengalskiego w krakowskim zoo. Zycie Wet. 53:202-04.
Krzakowski, W., A. Ramisz, J. Skotnicki and B. Przepiorkowski.
1984. Reversible Blindheit bei einem Bengaltiger (Panthera
tigris) im Zoo Krakow. Verh.ber. Erkrg. Zootiere 26:223-25.
Kuntze, A. 1985. Ophthalmologische Probleme bei Zirkustieren.
Verh.ber. Erkrg. Zootiere 27:409-12.
Lang, E. 1955. Ein Beitrag zur Frage der Trichinose. Schweiz.
Arch. Tierheilkd. 97:246-51.
Langheinrich, K.A. and S.W. Nielsen. 1971. Histopathology of
feline panleukopenia. A report of 65 cases. J. Am. Vet. Med.
Assn. 158:863-72.
Lloyyd, J. and J.G. Allen. 1980. The isolation of group EF-4
bacteria from a case of granulamatous pneumonia in a tiger
cub. Austral. Vet. J. 56:399-400.
Magrane, W.G. and E.R. VanDeGrift. 1975. Extracapsular
extraction in a Siberian tiger (Panthera tigris altaica). J.
Zoo Anim. Med. 6:11-12.
Mandal, D. and A. Choudhury. 1985. Helminth parasites of wild
tiger of Sundarbans Forest, West Bengal, India. Verh.ber.
Erkrg. Zootiere 27:499-501.
Mannl, A. 1983. Surch Salmonellen A. dublin Verursachte
Meningoencephalitis bei einem Tigerwelpen. Verh.ber. Erkrg.
Zootiere 25:287-91.
Maran, B., M. Herceg, J. Huber, M. Tadic and S. Cuturic. 1974.
Zur Statistik der Sektionsbefunde bei Raubtieren des
Zoologischen Garten der Stadt Zagreb. Verh.ber. Erkrg.
Zootiere 16:45-53.
Mazak, V. 1965. Der Tiger Panthera tigris Linnaeus, 1975. A.
Ziemsen Verlag, Wittenberg Lutherstadt, Neue Brehm-Bucherei
Nr. 356:162 pp.

McCullagh, K.G. and V.M. Lucke. 1978. Renal papillary necrosis in the tiger. Acta Zool. Path. Antverp. 70:3-13.

Meister, R. and R. Ippen. 1983. Ein Beitrag zu den Thyreopathien der Feliden in Zoologischen Garten. Verh.ber. Erkrg. Zootiere 25:241-48.

Michalska, Z., A. Gucwinski and K. Kocula. 1977. Sertolizelltumor bei einem Bengaltiger (Panthera tigris tigris). Verh.ber. Erkrg. Zootiere 19:305-07.

Michalska, Z., K. Kocula and A. Gucwinski. 1978. Augentuberkulose bei einem Bengaltiger (Panthera tigris tigris). Verh.ber. Erkrg. Zootiere 20:297-98.

Misdorp, W. and P. Zwart. 1967. Geschwulste und geschwulstahnliche Neubildungen bei Zootieren. Ver.ber. Erkrg. Zootiere 9:229-33.

Muller, P. and S. Seifert. 1978. Erstewissenschaftliche Auswertung von Daten uber den Sibirischen Tiger (Panthera tigris altaica) aus dem Internationalen Tigerzuchtbuch. Cong. Rep. 1st Int. Symp. Manage. Breed. Tiger, Leipzig 92-94.

Murer, B. 1939. Pathologisch-anatomische Untersuchungen and gefangen gehaltenen wilden Tieren des Baseler Zoologischen Gartens. Vet. Diss. Bern.

Murmann, W. 1982. Beitrag zur Statistik der Zootierkrankheiten. Vet. Diss. Hannover.

Nair, K.P.D., P.O. George and E.P. Paily. 1965. A case of trypanosomiasis in a tiger (Panthera tigris). Kerala Vet. 4:87-89.

Nouvel, J., G. Chauvier, L. Strazielle and Demontoy, M.C. 1972. Rapport sur la mortalite et al natalitie enregistrees a la menagerie du Jardin des Plantes pendant L'anneea 1970. Bull. Mus. Nat. Hist., Paris, 3e serie, no. 23:17-32.

Pandit, I.R. 1951. Two instances of proceed rabies in a tiger. Indian Med. Gaz. 85:441.

Patnaik, M.M. and Achariyo. 1970. Notes on the helminth parasites of Vertebrates in Baranga Zoo Orissa). Ind. Vet. J. 47:723-30.

Peters, J.C., E.J. Voute and P. Zwart. 1981. Osteogenesis imperfecta im Wurf einer an "tiger disease" erkrankten Sumatra-Tigrin. Kleintierpraxis 26:373-76.

Pierce, R.L., M.W. Vorhies and E.J. Bicknell. 1973. Yersinia pseudotuberculosis infection in a Siberian tiger and a spider monkey. J. Am. Vet. Med. Assn. 163:547.

Prange, H., K. Katenkamp, M. Kirste and V. Zepezauer. 1986. Maligne Fibhrose Histiozytome bei Tiger und Mahnenwolf sowie Schilddrusenkarzinom beim Waschbaren. Verh.ber. Erkrg. Zootiere 28:269-74.

Prescott, C.W. 1981. Fenbendazole in the treatment of intestinal parasites of circus lions and tigers. Vet. Rec. 109:15-16.

Ratcliffe, H.L. 1933. Incidence and nature of tumors in captive wild animals and birds. Am. J. Canc. 17:116-35.

Rathore, B.S. and S.S. Khera. 1981. Causes of mortality in felines in free-living state and captivity in India. Indian Vet. J. 58:171-76.

Rietschel, W. and J. Senn. 1977. Bekampfung von Milzbrand im Zoologischen Garten von Kabul durch Einsatz von Lebendvakzine. Tierarztl. Umsch. 32:36-39.

Sathyanarayana, R.M., R.N. Ramachandra, R. Raghvan and Keshavamurthy. 1981. Isolation of Corynebacterium pyogenes from a case of bronchopneumonia in tiger cubs. Tiger paper. Bangkok 8:23.

Saunders, G. 1984. Disseminated leiomyosarcoma in a Bengal tiger. J. Am. Vet. Med. Assn. 185:1387-88.

Schneider, K.M. 1933. Uber das "drehen" der Grosskatzen. Zool. Garten N.F. 6:173-81.

Schnurrbusch, Ute, C. Schonborn and K. Elze. 1976. Zur Therapie von Dermatomykosen bei Zootieren. Verh.ber. Erkrg. Zootiere 18:187-90.

Schonborn, C., S. Seifert, W. Braun and H. Schmoranzer. 1971. Untersuchungen uber die Hautpilzflora von Zooteiren. Zool. Garten N.F. 41:7-25.

Schoon, H.A., H.P. Brandt and W. Leiboldt. 1980. Ein Beitrag zum Komplex der sogenannten Tigerkrankheit. Berlin. Munchen. Tierarztl. Wochenschr. 94:37.

Schroeder, H.D. 1965. Ein Beitrag zur Diagnostik der Tuberkulose bei Zootieren. Zool. Garten N.F. 30:244-47.

Schroeder, H.D. 1967. Erkankungen der Verdauungsorgane der Feliden. Verh.ber. Erkrg. Zootiere 9:9-14.

Schroeder, H.D. 1968. Erkrankungen der Respirationsorgane der Landraubtiere. Verh.ber. Erkrg. Zootiere 10:17-21.

Schroeder, H.D. 1984. Zu den pra-und postnatalen Infektionen bei Zootieren. Verh.ber. Erkrg. Zooteire 26:33-37.

Schroeter, W.K.H. 1972. Farbanomalien und Streifenreduktion beim Tiger (Panthera tigris L.). Sitzungsber. Ges. Naturfreunde, Berlin, N.F. 12:154-58.

Schuppel, K.F., H.J. Selbitdz and K. Elze. 1984. Zur Pseudotuberkulose bei Grosskatzen. Verh.ber. Erkrg. Zootiere 26:395-98.

Scott, W.A. 1979. Use of vaccines in exotic species. Vet. Rec. 104:199.

Seal, U.S., E.D. Plotka and C.W. Gray. 1978. Baseline hematology, serum chemistry and hormone data for captive tigers. Congr. Rep. 1st Int. Symp. Manage. Breed. Tiger, Leipzig 174-92.

Seidel, B. 1983. Entropium-Operation bei jungen Bengaltigern Panthera t. tigris. Milu, Berlin 5:812-16.

Seifert, S. 1970. Einige Ergebnisse aus dem Zuchtgeschehen bei Grosskatzen im Leipziger Zoo. I. Zum Sibirischen Tiger (Panthera altaica Temminck, 1845). Zool. Garten N.F. 39:260-70.

Seifert, S. and P. Muller. 1978. Auswertung der internationalen Umfrage zur "Tigerkrankheit" mit besonderer Berucksichtigung der Leipziger Untersuchungen. 1. Auswertung der Umfrage. Congr. Rep. 1st Int. Symp. Manage. Breed. Tiger, Leipzig 149-50.

Seifert, S., K. Eulenberger and K. Elze. 1982. Zeitweiliger partieller und totaler Haarausfall bei kunstlich aufgezogenen Grosskatzen. Verh.ber. Erkrg. Zootiere 24:57-59.

Seitz, A. 1958. Bericht uber das rechnungsjahr 1955 1.IV. 1955 bis 31.III.1956). Zool. Garten N.F. 24:128-31.

Selbitz, H.J., K. Elze and A. Voigt. 1978. Ergebnisse bakteriologischer Untersuchungen bei Tigern--ein Beitrag zur Klarung der Atiologie der sogenannten Tigerkrankheit. Congr. Rep. 1st Int. Symp. Manage. Breed. Tiger, Leipzig 161-66.

Selbitz, H.J., K. Elze, K. Eulengerger, A. Voight, A. Bergmann, S. Seifert and D. Altmann. 1979. Zu Problemen des Vorkommens und der Bedeutung bakterieller Infektionen bei Feliden in den Zoologischen Garten Leipzig und Erfurt in den Jahren 1965-1978. Verh.ber. Erkrg. Zootiere 21:191-99.

Selbitz, H.J., K. Elze and L. Liebermann. 1986. Bakteriologische Diagnostik und Epizootiologie von Koliinfektionen bei Zootieren. Verh.ber. Erkrg. Zootiere 28:321-26.

Simon, T. 1873. Angeborene Adenome der Luftrohrenschleimhaut beim Tiger. Virchow's Archiv. 57:537-38.

Smith, R.E., W. Boardman and P.C.B. Turnbull. 1971. Hepatoma in a Bengal Tiger. J. Amer. Vet. Med. Assn. 159:617-19.

Soulsby, E.J.L. 1968. Helminths, Arthropods and Protozoa of Domesticated Animals. Bailliere Tindall London, 6th ed.

Sprent, J.F.A. and P.P. English. 1958. The large round worms of dogs and cats--a public health problem. Austral. Vet. J. 34:161-71.

Starzynski, W. and J. Rikicki. 1974. Abnormal findings from abdominal caviti of big Felidae in Warsaw Zoo. Verh.ber. Erkrg. Zootiere 16:127-31.

Stehlik, M. and J. Kejhda. 1966. Microsporum gypseum-Infektion bei jungen Tigern und ihre Behandlung mit Likuden/Griseofulvin "Hoechst". Kleintierpraxis 11:197-201.

Stehlik, M. 1968. Ulcerative Glossitis bei Leoparden und Tigern. Kleintierpraxis 13:112-13.

Steinbacher, G. 1941. Nachrichten aus Zoologischen Garten. Zool. Garten N.F. 13:364-70.

Storch, W. 1978. Immunhistologische Untersuchungen bei der sogenannten Tigerkrankheit. Congr. Rep. 1st Int. Symp. Manage. Breed. Tiger, Leipzig 170-71.

Strauss, G. 1983. Untersuchungen zum Auftreten von Kokzidienoozsten und--sporozysten in den Fazes fleischverzehrender Tiere aus dem Tierpark Berlin. Vet. Diss. Berlin: 100 + XXVIIpp.

Strauss, G. and B. Seidel. 1984. Analyse tierarztlicher Eingriffe im geburtsnahen Zeitraum bei verschiedenen Saugetieren 1970-1983). Verh.ber. Erkg. Zootiere 26:65-68.

Svobodnik, J. and M. Slavik. 1974. Traumatische Epiphysiolyse bei einem Sumatra-Tiger. Zool. Garten N.F. 44:290-94.

Theobald, J. 1978. Carnivores (Carnivora). In J. Zoo and Wild Animal Medicine, ed. M.E. Fowler, Philadelphia: W.B. Saunders Co. 611-706.

Tscherner, W. 1974. Ergenbnisse koprologischer Untersuchungen bei Raubtieren des Tierparks Berlin. Verh.ber. Erkrg. Zootiere 16:77-89.

Wagner, G., M. Cociu, N. Micu, A. Rizeanu and M. Cociu. 1971. Uber die Anwendung und Wirkung von Human-Gammaglobulin bei der Tigeraufzucht. Verh.ber. Erkrg. Zootiere 13:321-25.

Wallach, J.D. and W.J. Boever. 1983. Diseases of exotic animals. Medical and surgical management. Philadelphia: W.B. Saunders Co. 12:1159.

Williamson, W.M., E.B. Tilden and R. Getty. 1965. Pathogenic bacteria recovered at the animal hospital at the Chicago Zoological Park since 1954. Verh.ber. Erkrg. Zootiere 12:105-122.

Wisser, J. 1984. Feline Infektiose Peritonitis-Sektionsbild bei einer Lowin (Panthera leo). Verh.ber. Erkrg. Zootiere 26:341-48.

Wisser, J. 1986. Haufiges Vorkommen von Kalzinosen bei Nordluchsen (Lynx lynx). Verh.ber. Erkrg. Zootiere 28:255-60.

Zaki, S., T.S. Nalini, M.S. Rao, and B.S. Keshavamurthy. 1980. Isolation of Shigella flexneri from two tiger cubs. Current Science 49:288.

20

Comments on the "Tiger Disease"

Siegfried Seifert and Peter Muller

For a long time the frequent incidence of the so-called tiger disease has been thought to be a serious obstacle to the establishment of a stable population.

Since in the Leipzig Zoo experiments for the purpose of solving this problem were successful, we would like to add to our paper on development of the tiger stock in man's world a brief contribution on how to avoid and to treat this disease typical to the tiger.

As we already reported on the occasion of the First International Tiger Symposium in Leipzig 1978 during a period of 15 years in the Leipzig tiger stock 14 tigers (8, 6) had fallen ill with the so-called "tiger disease" to a different degree.

Findings of Clinical, Laboratory and Diagnostic Analyses

The following symptoms occurred in animals suffering from this disease:
1) Constant vomiting of intaken meat, undigested;
2) Smearing, pulpy faeces with a high percentage of undigested meat;
3) Vaulted, partly unrelaxed abdomen;
4) Increased number of calf-pathogen coli bacillus and aerobic endospore producing bacteria in the vomit and the corresponding gastric juice and faeces;
5) Increased presence of aerobic endospore producing bacteria in the duodenum and jejunum identified as Clostridium perfringens (toxin proof with mice positive, pathogen for guinea-pigs);
6) Increased intake of fluids;
7) Leucocytosis, shifting to the left, lymphopenia;
8) Increase of gamma globulin content with advancing duration of illness.

Main findings of the postmortem examinations were:
1) Chronic gastritis;
2) Erosions of the gastric mucous membrane;
3) Numerous bean- to one mark-sized ulcera in the area of the smaller curvature and in the zone of fundus at the gastric

231

entrance;
    4) Ulcera in the small intestine;
    5) Chronic interstitial nephritis.

Analyzing these clinical and pathological findings in 1978, Prof. Elze and his team already deduced that in all probability it is primarily a matter of a conditional infectious-toxic event with dysbactery.

The bacteriological findings especially focused attention on opportunistic viruses such as E. coli types and clostridium, which had been found in all examinations of stomach and small intestine contents and faeces of animals which had fallen ill. This hypothesis was confirmed by medical successes before 1978 when six animals were treated with oral tetracycline or furazolidone.

In the following once again coliform bacteria and clostridium perfringens were found in the vomit and in faeces immediately after the beginning of the disease. Thus, the suspicion of the presence of a mixed, especially an E. coli or perfringens enterotoxemia, infection increased. The lymphopenia also indicated a striking toxin effect.

Potential Causes of the Disease

At present we suppose that all C. perfringens types might cause enterotoxemias. From the clinical-epizootiological point of view, the "tiger disease" is an enterotoxemia without infectious character. These only occur at high concentrations of contamination, and especially in connection with supporting factors. The most important of these factors for meat-eaters are:
    1) Meat enriched with viruses and toxins;
    2) Intake of too large rations at once;
    3) Change in quality of meat.

This leads to disturbances in the secretion of digestive ferments, an increase in undigested protein and carbohydrates in the small intestine, and again favors the increase of C. perfringens. Since this analysis was made we have increased the quality of our tigers' diet, decreased the rations and given the animals a high percentage of bones on the base of this model of etiopathogenesis. Nevertheless, eight animals (three males, five females) aged one to 16 have fallen ill with the "tiger disease", i.e. vomiting and defecation with undigested pieces of meat. Immediately after the first vomiting occurred the animals were given low diet for some days (rabbit or a half ration of good pork or beef).

Recommended Treatment

Medicine given to the animals consisted of 500 mg three times daily, 1,000 mg three times daily, 1,500 mg two times daily, or 2,000 mg two times daily Ampicillin, according to body weight (preparation: Penstabil/spofa/CSSR) with a small piece of meat or in meat-balls given orally for five to 12 days.

After two to at latest seven days, none of the eight animals showed vomiting or unphysiological faeces anymore. Also, after stopping the medication none of the animals suffered a relapse.

Our veterinary team, led by Prof. Elze, chose "Ampicillin," since clostridias primarily are highly sensitive to penicillin. Oral application is also very important, the frequently used parenteral method is far less effective.

Furthermore, beginning treatment at the first symptoms of the illness is essential, since by this point a heavy increase in bacteria has not yet taken place. Only the bacteria of not yet-released toxins are subject to this treatment.

Epicrisis: In the diagnostic work and in consistent applications of an oral Ampicillin treatment continued on the basis of the data, theses and hypotheses of 1978, we were able to avoid and to overcome the vehement progress of the disease and deaths due to the so-called "tiger disease."

From the investigations and the results of treatment concerning the tiger disease from 1963 to the present, we may conclude that this disease is an enterotoxemia due to a mixed infection with different E. coli types and C. perfringens sp. At present we are in the process of identifying the mentioned bacteria groups, and of determining optimal dosages of medication.

21

Antibiotic Therapy in the Bengal Tiger

Andrew Teare

A major problem facing the clinical veterinarian in the zoological setting is the lack of information regarding the safety and effectiveness of drugs and vaccines in non-domestic species. Clinical studies have been completed in recent years that have addressed the problems of safety and effectiveness of some vaccines in non-domestic species (Bush et al. 1981, Montali et al. 1983) and some data exists on dosages of antibiotics for a few avian and reptilian species. However, completing the necessary experimental work to gather this knowledge for all drugs and vaccines used in zoological practice is simply not feasible for the vast range of species currently kept in zoological gardens. One common solution to this lack of data has been to extrapolate from knowledge gained in domestic species. Thus, antelope are often viewed as modified sheep or cattle, the horse serves as a model for wild equids, and domestic poultry provide much of the data used to estimate drug dosages in other avian species.

Unfortunately, different species can react in unexpected and often dramatic ways to the drugs and vaccines administered. The literature contains numerous examples of vaccine failures; particularly canine distemper disease caused by the use of modified-live vaccines (Kazacos et al. 1981). The red panda is a species that is extremely sensitive to vaccine-induced canine distemper and disease will occur even with vaccines shown to be safe and effective in a variety of non-domestic canids (Montali et al. 1983). Unexpected drug reactions have also occurred; one example is the extreme sensitivity to ivermectin toxicity exhibited by turtles and tortoises, despite the safe usage of this drug in some snake and lizard species (Teare and Bush 1983). It can be expected that antibiotics will also show marked differences in dosages over the wide range of species exhibited in zoological collections.

While the susceptibility of a disease-causing microorganism to various antibiotics is a major factor in the selection of an antibiotic regimen, therapy failure can still occur if host-drug and host-microorganism interactions are not also considered. An appropriate treatment regimen results in the drug reaching the site of infection at a sufficient concentration to prevent growth of the microorganism, continues for a period of time that allows elimination of the microorganism from the host and should have no permanent, adverse effects on the host. Choosing an appropriate antibiotic regimen then, is dependent upon a knowledge of the pharmacokinetics, pharmacodynamics and toxicity of the antibiotic in the host as well as the sensitivity of the microorganism to the antibiotic.

For zoological veterinarians, the domestic cat usually serves as the model for all non-domestic felids. Certainly there are great morphological similarities for the range of species in Felidae, but this is probably not a good basis for believing that a 350 pound tiger will metabolize a drug in the same manner as a ten pound house cat. However, the use of dosage regimens developed for the domestic cat has been the best solution for the veterinarian treating nondomestic felids. Indeed, for the vast majority of cases, extrapolation from data on other species will continue to be the best initial method for arriving at a dosage regimen for non-domestic animals.

Recently, however, I completed a series of studies in the Bengal tiger (Panthera tigris tigris) that measured the serum concentrations of three antibiotics that are commonly used in zoological medicine. The results allow, at least for these three antibiotics, a much better estimation of the correct antibiotic dosage in the tiger. In addition, the data from this study can be directly compared to the results of previous studies completed in the domestic cat. The direct comparison of equivalent data leads to a better understanding of the differences and similarities between these two members of the Felidae and, hopefully, will allow for the development of better antibiotic regimens in all Felidae.

INTRAMUSCULAR GENTAMICIN

Seven Bengal tigers were given an intramuscular injection of gentamicin at a dosage of 5 mg/kg. Blood samples were obtained just prior to injection and at 0.5, 1, 2, 3, 4, 6, and 8 hours after injection. Serum concentrations of gentamicin were measured using a commercially available radioimmunoassay (Rianen[tm], gentamicin I[125] radioimmunoassay kit, New England Nuclear). The results showed that gentamicin is rapidly absorbed by the Bengal tiger with mean serum levels reaching 17.3 mcg/ml by the 0.5 hour sample and peaking at the 1 hour sample with a mean concentration of 19.5 mcg/ml. Serum concentrations of gentamicin declined rapidly in all subsequent samples to reach a mean level of 2.9 mcg/ml at the 8 hour sample (Fig. 1).

Jacobson et al. (1985) have reported on the results of a similar trial in the cat. A 5 mg/kg intramuscular dose of gentamicin in the cat resulted in peak serum concentrations of 23.1 mcg/ml at 0.5 hours after injection. The decrease in serum gentamicin following the peak level had two distinct phases in the cat; the initial very rapid decline reduced mean serum concentrations to 11.8 mcg/ml by 1 hour after injection while the final slower phase maintained detectable serum levels of gentamicin for as long as 24 hours after injection. Consideration of the recommended minimum, maximum and therapeutic serum concentrations for gentamicin led Jacobson et al. (1985) to propose a dose of 2.5 mg/kg every 12 hours for the cat.

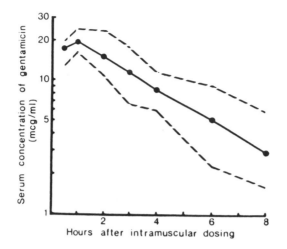

Fig. 1. Mean serum concentration of gentamicin in tigers following intramuscular dosage at 5 mg/kg. Broken lines indicate minimum and maximum serum values at each sampling time.

Comparing these two felids, peak concentrations of gentamicin in the cat occur sooner and are slightly higher than is produced by the same dose in the tiger. However as soon as 1 hour after injection, the initial rapid decline in the cat has reduced serum gentamicin levels to below those found in the tiger. Finally though, the slow disposition phase in the cat causes serum gentamicin concentrations to be maintained longer than in the tiger. Mean gentamicin serum concentrations in the tiger have fallen to 2.9 mcg/ml by 8 hours after injection while the mean levels in the cat are still 3.6 mcg/ml at 12 hours after injection. The results from this study would indicate that appropriate therapy with gentamicin in the tiger requires either a higher dosage or more frequent administration than is currently recommended for the cat.

INTRAMUSCULAR CHLORAMPHENICOL

Blood concentrations of chloramphenicol were measured in six Bengal tigers given intramuscular chloramphenicol sodium succinate (100 mg/ml) at 20 mg/kg. Blood samples were obtained just prior to dosing and at 0.5, 1, 2, 3, 4, 6, 8, and 12 hours after injection and serum concentrations of chloramphenicol were determined using a colorimetric analysis method (Watson 1979a). It was shown that chloramphenicol was rapidly absorbed in the tiger; the peak concentration was found in the 0.5 hour sample in all six animals. Mean serum chloramphenicol levels declined in subsequent samples, but chloramphenicol was still detectable in all the 12 hour samples (Fig. 2). In the majority of tigers, the serum concentrations of chloramphenicol remained above the recommended therapeutic level of 4 mcg/ml for at least 8 hours after intramuscular dosing.

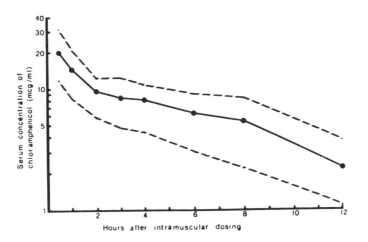

Fig. 2. Mean serum concentration of chloramphenicol in tigers
following intramuscular dosage at 20 mg/kg. Broken lines
indicate minimum and maximum serum values at each sampling time.

        In the cat, plasma concentrations of chloramphenicol have
been reported for the sodium succinate ester given by the
intramuscular route at 20 mg/kg (Watson 1979b). In another study
of chloramphenicol sodium succinate in cats, Watson (1980a)
reported on the plasma concentrations following intramuscular
dosing at 30 mg/kg. While both studies reported plasma
concentrations (rather than serum concentrations) and blood
samples were only collected for the first 8 hours after dosing,
it is still useful to compare this published information on
chloramphenicol in cats with the tiger data.

        The concentration-time curves obtained with intramuscular
chloramphenicol sodium succinate in the tiger and the cat have
striking similarities. Both species show rapid absorption of the
antibiotic with peak measured concentrations occurring at the 0.5
hour sample. For the 20 mg/kg dosage, the initial mean chloramphenicol concentrations are also quite close in value (18.4
mcg/ml for the cat and 20.1 mcg/ml in the tiger). However, over
time, it is the tiger that has the slower decline in chloramphenicol concentrations. At 8 hours after dosing, the mean
serum chloramphenicol concentration in the tiger is 5.3 mcg/ml,
while the comparable sample in the cat has dropped to a mean
level of 3.4 mcg/ml (Watson 1979b). Mean plasma chloramphenicol
concentrations of 5.6 mcg/ml can be obtained in the cat at 8
hours after dosing by increasing the chloramphenicol dosage to 30
mg/kg (Watson 1980a). The results for intramuscular chloramphenicol in the tiger suggests that a slightly lower dosage or
frequency of administration will maintain therapeutic levels in
the tiger than is required for the cat. This is in contrast to
the findings for intramuscular gentamicin, which indicated that
the requirements for appropriate therapy in the tiger was
slightly higher than for the cat.

ORAL CHLORAMPHENICOL

For the veterinary clinician in zoological practice, oral drug administration is often easier and less stressful than the parenteral route. When antibiotic treatment is indicated then, oral therapy is often the treatment route of choice for many species. In the cat, it has been shown that chloramphenicol is an antibiotic that is well absorbed when given orally and that absorption is not significantly effected by the presence of food (Watson 1979a). Seven Bengal tigers were used to investigate the effectiveness of oral chloramphenicol in this species.

Each tiger received sufficient 250 mg chloramphenicol capsules to produce an approximate dosage of 20 mg/kg (actual dosages ranged from 18.8 to 21.2 mg/kg). Blood samples were collected just prior to dosing and at 1, 2, 3, 4, 6, 8, and 12 hours after dosing. Serum chloramphenicol was determined using the same colorimetric assay technique that was used in the intramuscular chloramphenicol study.

In all tigers, serum chloramphenicol concentrations exceeded 4 mcg/ml at the 1 hour sample. Mean serum concentrations continued to rise slowly for the next 2 hours before beginning to decline (Fig. 3). Absorption of chloramphenicol from the intestine varied for different tigers causing the time of individual peak serum concentrations to range from 2 to 6 hours after dosing. Peak levels were lower than those obtained in the tiger following intramuscular administration of chloramphenicol sodium succinate. Prolonged absorption of chloramphenicol from the intestine also resulted in a slower decline in serum concentrations than was seen following intramuscular chloramphenicol, producing mean serum levels of 6.0 mcg/ml even at 12 hours after dosing.

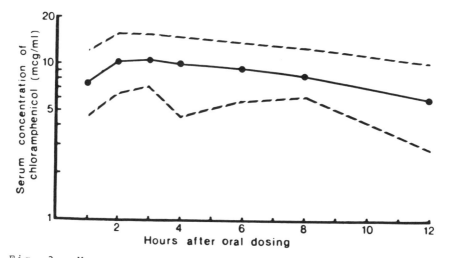

Fig. 3. Mean serum concentration of chloramphenicol in tigers following oral dosage at 20 mg/kg. Broken lines indicate minimum and maximum serum values at each sampling time.

In the cat, Watson (1979b) has reported the plasma concentrations following oral administration of chloramphenicol capsules at a dosage of 20 mg/kg. The mean peak chloramphenicol concentration was 10 mcg/ml, which was lower than those obtained in cats dosed by the intravenous, intramuscular or subcutaneous routes. However, compared to intramuscular chloramphenicol administration, plasma levels declined more slowly in cats dosed by the oral route, so that plasma concentrations at 8 hours were higher for the oral route. In the cat, as was seen with the tiger, there was variation in the absorption of chloramphenicol so that peak plasma concentrations for individuals occurred from 0.5 to 8 hours after dosing.

In general, the results for oral chloramphenicol in the tiger were very similar to those reported by Watson (1979b) for the cat. Mean peak levels in the cat (10 mcg/ml) were nearly identical to those produced by the same dose in the tiger (10.6 mcg/ml). In the tiger the mean peak level occurred slightly later than in the cat, but there was overlap in the time of peak chloramphenicol concentration shown by individuals of the two species. Both species showed a slower decline in chloramphenicol concentrations when the oral route was compared to intramuscular administration; this is presumed to be due to prolonged intestinal absorption in both species. However, in the tiger, serum chloramphenicol concentrations are maintained at higher levels than in the cat; 8.5 mcg/ml at 8 hours in the tiger versus 5.7 mcg/ml for the cat at the same time. This agrees with the results obtained for intramuscular chloramphenicol in the tiger and confirms that tigers require a lower dose of chloramphenicol than the cat. Based on extensive studies, Watson (1980b) has recommended an oral regimen of 25-40 mg/kg twice a day for the cat. Other sources recommend a dosage as high as 50 mg/kg twice a day (Johnson 1983). As the dosage of 20 mg/kg produced mean serum concentrations in the tiger that exceeded the recommended minimum therapeutic level of 4 mcg/ml for 12 hours, it is clear that an appropriate oral dosage regimen for chloramphenicol in the tiger is less than 20 mg/kg twice a day.

ORAL TETRACYCLINE

It has been stated that the tetracyclines are readily absorbed following oral administration, with peak plasma concentrations occurring 2 to 4 hours after administration (Huber 1982). However, studies in the dog have indicated that while tetracycline is absorbed rapidly from the stomach and small intestine, only a small fraction of the dose administered is actually available for absorption. Direct instillation of 20 mg/kg of tetracycline into the duodenum of dogs resulted in peak serum concentrations of only 4.0 - 4.5 mcg/ml (Pindell et al. 1959). In another study, an oral tetracycline dosage of 12.5 mg/kg resulted in maximal serum tetracycline concentrations of less than 0.5 mcg/ml (Tisch et al. 1955). Oxytetracycline appears to be even less available in the dog; oral administration at an average dosage of 44 mg/kg produced mean peak serum concentrations of less than 0.8 mcg/ml with all 4 of the commercial preparations tested (Cooke et al. 1981). No reports of similar studies in the cat were found in the literature, but an oral tetracycline dosage of 20 mg/kg every 8 hours has been suggested for the cat (Johnson 1983).

Serum tetracycline concentrations were measured in six Bengal tigers given oral tetracycline capsules at an approximate dosage of 20 mg/kg. As commercially available 250 mg capsules were used, the actual oral dosage varied from 19.0 to 21.5 mg/kg. Blood samples were collected at the same times as for the oral chloramphenicol study and serum tetracycline was assayed using a previously described agar-well diffusion assay (Teare et al. 1985)

Tetracycline was detected in all post-dosing serum samples from the tigers, but concentrations were much lower than had been anticipated. Of the 42 serum samples collected after dosing with tetracycline, only 5 had tetracycline concentrations of 1 mcg/ml or greater and 4 of these samples were from one animal. Mean serum concentrations peaked at 0.7 mcg/ml, 2 to 3 hours after administration (Fig. 4). Serum tetracycline concentrations did decline slowly, with mean levels of 0.3 mcg/ml even after 12 hours.

A tetracycline dosage of 42 mg/kg was administered orally to one tiger and blood samples were collected at 2, 3, 6 and 8 hours after dosing. Serum tetracycline concentrations in this animal ranged from 1.1 to 1.9 mcg/ml, with the maximum concentration found in the 2 hour sample. While the minimum concentration of tetracycline required to inhibit growth will vary with the microorganism, it is safe to state that the blood levels obtained in the tigers given 20mg/kg are only likely to be therapeutically effective against the most sensitive of microorganisms. It would appear that the dosage regimen recommended for the cat is not appropriate for use in the tiger.

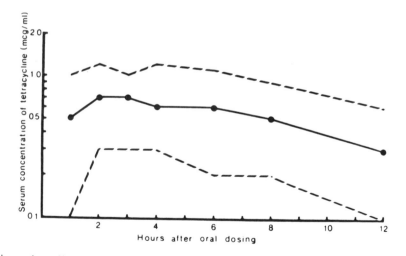

Fig. 4. Mean serum concentration of tetracycline in tigers following oral dosage at 20 mg/kg. Broken lines indicate minimum and maximum serum values at each sampling time.

SUMMARY

   The pharmacokinetics of intramuscular gentamicin, intramuscular chloramphenicol, oral chloramphenicol and oral tetracycline were studied in the Bengal tiger. The resulting serum concentration-time curves for the tiger show considerable similarity to published data for the cat. However, sufficient differences do exist, in the pharmacokinetics of these antibiotics, to indicate that the dosage regimens recommended for the cat require modification to be appropriate in the tiger. Specifically, to maintain therapeutic levels of gentamicin in the tiger appears to require a slightly higher overall daily dosage. A similar situation exists for oral tetracycline; the drug does not appear to be well absorbed in the tiger as dosages of even 42 mg/kg only produced moderate serum concentrations. For chloramphenicol though, the data shows that in the tiger a dosage of 20 mg/kg (intramuscularly or orally) will maintain higher serum concentrations than the same dose in the cat. Chloramphenicol is well absorbed when given orally, and a dosage of slightly less than 20 mg/kg twice a day should maintain therapeutic levels in the tiger.

   Obviously this work is only the beginning step towards understanding the pharmacological similarities and differences within the Felidae. Indeed, the fact that differences were found between tigers and cats raises many more questions than are answered. For example, studying one or more of the other aminoglycoside antibiotics in the tiger could indicate whether the findings for gentamicin apply to the other antibiotics in this class. It would also be of interest to repeat these pharmacokinetic trials in lions or leopards to determine whether the findings in the tiger, relative to the cat, apply to other members of the genus Panthera. Continuing this type of research is critical if we are ever to extrapolate drug dosages between species with any more accuracy than is presently possible.

ACKNOWLEDGMENTS

   This research was funded by Conservation Project Grant IC-40083-84 from the Institute of Museum Services, with matching funds from the Omaha Zoological Society. The author is very grateful to Steve, Alan, Carla, Dave and the other members of the "cat crew" at the Omaha Zoo. Without the continuous assistance of these dedicated personnel, the project could not have been completed.

REFERENCES

Bush, M., R.C. Povey and M.S. Koonse. 1981. Antibody response to an inactivated vaccine for rhinotracheitis, caliciviral disease and panleukopenia in nondomestic felids. J. Am. Vet. Med. Assn. 179:1203-05.
Cooke, R.G., A. Knifton, D.B. Murdock and I.S. Yacoub. 1981. Bioavailability of oxytetraycline dihydrate tablets in dogs. J. Vet. Pharm. Therap. 4:11-13.

Huber, W.G.   1982.   Tetracyclines In _Veterinary Pharmacology and Therapeutics_. 5th ed., ed. N.H. Booth and L.E. McDonald. Ames: Iowa State Univ. Pr.

Jacobson, E.R., J.M. Groff, R.R. Gronwall, A.F. Moreland and M. Chung.   1985.   Serum concentrations of gentamicin in cats. _Am. J. Vet. Res._ 46:1356-58.

Johnson, R.   1983.   Table of common drugs: Approximate doses. In _Current Veterinary Therapy VIII_, ed. R.W. Kirk. Philadelphia: W.B. Saunders Co.

Kazacos, K.R., H.L. Thacker and H.L. Shivaprasad.   1981. Vaccination-induced distemper in kinkajous. _J. Am. Vet. Med. Assn._ 179:1166-69.

Montali, R.J., C.R. Bartz, J.A. Teare, J.T. Allen, M.J.G. Appel and M. Bush.   1983.   Clinical trials with canine distemper vaccines in exotic Carnivores. _J. Am. Vet. Med. Assn._ 183:1163-67.

Pindell, M.H., K.M. Cull, K.M. Doran and H.L. Dickison.   1959. Absorption and excretion studies on tetracycline. _J. Pharm. Exp. Therap._ 125:287-94.

Teare, J.A. and M. Bush.   1983.   Toxicity and efficacy of ivermectin in chelonians. _J. Am. Vet. Med. Assn._ 183:1195-97.

Teare, J.A., W.S. Schwark, S.J. Shin and D.L. Graham. 1985. Pharmacokinetics of a long-acting oxytetracycline preparation in ring-necked pheasants, great horned owls and Amazon parrots. _Am. J. Vet. Res._ 46:2639-43.

Tisch, D.E., K.M. Cull and H.L. Dickison.   1955. Pharmacological studies with tetracycline hydrochloride. In _Antibiotics Annual 1954-1955_, ed. H. Welch and F. Marti-Ibanez. New York: Medical Encyclopedia, Inc.

Watson, A.D.J.   1979a.   Effect of ingesta on systemic availability of chloramphenicol from two oral preparations in cats. _J. Vet. Pharm. Therap._ 2:117-21.

Watson, A.D.J.   1979b.   Plasma chloramphenicol concentrations in cats after parenteral administration of chloramphenicol sodium succinate. _J. Vet. Pharm. Therap._ 2:123-27.

Watson, A.D.J.   1980a.   Plasma chloramphenicol concentrations in cats following intramuscular administration of three different chloramphenicol preparations. _J. Vet. Pharm. Therap._ 3:107-10.

Watson, A.D.J.   1980b.   Oral chloramphenicol dosage regimens in cats. _J. Vet. Pharm. Therap._ 3:145-49.

PART IV
REPRODUCTIVE BIOLOGY

22

Behavioral Indicators and Endocrine Correlates of Estrus and Anestrus in Siberian Tigers

Ulysses S. Seal, Ronald L. Tilson, Edward D. Plotka, Nicholas J. Reindl
and Marialice F. Seal

INTRODUCTION

The large felid species - cheetahs, jaguars, leopards, lions, pumas, snow leopards and tigers - are declining in numbers in the wild and many forms are either endangered or threatened (Goodwin and Holloway 1972, Foose and Seal 1986). The need for careful genetic management of small captive populations (Seal and Foose 1983) has stimulated interest in elective control of reproduction including contraception (Seal et al. 1976, 1978) and artificial insemination (Bonney et al. 1981, Dresser et al. 1982, Wildt et al. 1981a). However, numerous unsuccessful attempts to artificially inseminate big cats have focused attention on the paucity of information concerning the endocrine aspects of the reproductive cycles of these species.

Behavioral aspects of the estrous cycle of Bengal tigers have been described (Sadleir 1966, Kleiman 1974). The present study was initiated to develop a quantitative behavioral profile, based upon daily observations of female tigers during the breeding season, that might be used to identify the stage of estrus in individual animals as a prelude to induction of ovulation and artificial insemination. Blood samples were collected and physical examinations conducted at least once each week to provide physiological and endocrine correlates of the estrous cycle for comparison with the behavioral data.

METHODS

Animals

Five mature female Siberian tigers, Minnesota Zoo numbers 180, 368, 711, 712 and 732 (studbook no. 1295, 1422, 1432, 1328 and 1427, respectively; Seifert and Muller 1985), were used in this study. Females 180 and 732 had produced litters. Females 368, 711 and 712 were nulliparous; details of their endocrine cycles during one previous year have been reported elsewhere (Seal et al. 1985). For the behavioral studies of estrus, the females were maintained in a off-exhibit holding facility. Housed separately, each animal had free access during the day to an outside area (37 sq. m) and was locked in a connecting indoor area (7 sq. m) at night. An adult male, 1159 (studbook no. 1729), was housed in an adjoining cage. The females could see, smell and hear the male.

Water was available ad libitum. The tigers were fed daily, at about 1530 hr, 2-3 kg of a prepared carnivore diet (Nebraska Brand Feline Diet) to maintain an approximately constant body weight. Weighings were periodically conducted with an overhead spring balance which was calibrated with known weights. Female tiger body weights ranged from 108-130 kg during the study.

Behavioral Observations

Three of the females (368, 711 and 712) used for establishing the behavioral estrous cycle were observed daily for four months (February to May). The frequency of occurrence of a set of behaviors was recorded by the same person in five one-minute segments prior to the start of each day's routine maintenance activities. The behavioral indicators of estrus (Kleiman 1974) chosen for inclusion in this study were vocalizing (calling or moaning; Schaller 1967), prustening (a greeting call that sounds like air expelled softly through the nostrils), rubbing the cheek, forehead or flank against the walls/bars of the enclosure, and rolling over and writhing on the back. If a female exhibited lordosis or semilordosis (postures assumed just prior to copulation; see Leyhausen 1956), a score of 1 was given for every 10 seconds the posture was maintained. Each of the above behaviors were weighted equally, in contrast to Kleiman's (1974) method of establishing a symbolic score for estrous behavior by weighting particular patterns more than others (e.g. lordosis, rolling and flank rubbing were scored higher than cheek rubbing, prustening or calling). Other estrus behaviors noted by Kleiman for female Bengal tigers included urine-spraying, exhibiting flehmen (for definition see Leyhausen 1956) and pacing.

Although the frequencies of occurrence of some behaviors showed significant correlations with the endocrine profiles of the females (see Seal et al. 1985), total scores with no weighting of individual behaviors were most indicative of the females' estrous cycles. Like Kleiman's (1974) observations with Bengal tigers, urine-spraying among female Siberian tigers was negatively correlated with the occurrence of estrus and thus was not included in the analysis presented here.

Endocrine Samples

Endocrine events were established by immobilizing the tigers each week with 600-900 mg ketamine, 30-50 mg xylazine and 10-20 mg diazepam administered by blow dart (see Seal et al. 1985). Additional ketamine was used as needed to maintain anesthesia. A baseline sample of 60 ml of venous blood was collected 35-60 min after the first injection of the immobilizing drugs. A portion of the blood was placed into tubes containing EDTA for hematological analysis, and into glass tubes with no additives for serum preparation. The serum was separated on the day of collection and frozen until used for assay. A second blood

sample was collected 15-30 min after the first and was processed
for serum. Weekly hematological analysis indicated that white
cell counts, hemoglobin, red blood cells and hematocrit values
remained stable during the study. Serum levels of estradiol,
testosterone, progesterone and luteinizing hormone (LH) were
measured by specific radioimmunoassays (Seal et al. 1985).

RESULTS

Birth Season

The seasonal frequency of parturition (530 litters, 1,239
young) in Siberian tigers held in collections throughout the
Northern Hemisphere (Fig. 1) was highest in April-June (P<0.001),
suggesting that Siberian tigers are seasonal breeders.

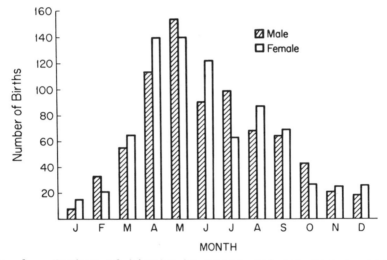

Fig. 1. Number of births by month and sex for registered
Siberian tigers in Northern Hemisphere zoos. Data from the
International Tiger Studbooks 1976-82 (Seifert and Muller 1985)
and ISIS (1983).

Estradiol

Peaks of estradiol concentration occurred from February to
June with low values from June through January under natural
photoperiod. Baseline serum immunoreactive estradiol-17B values
(Fig. 2 and 3) ranged from <5 to 115 pg/ml in 520 weekly baseline
samples collected over 42 months during natural cycles for three
tigers, and 18 months for two tigers. Values greater than 20
pg/ml of immunoreactive estradiol were more than three standard
deviations greater than the mean of the remaining values or the
anestrous values and were considered indicative of a peak and of
an active ovarian follicular phase. During anestrus, estradiol
levels were 4.2 ± 0.5 pg/ml (N=70), ranging from 0.5 to 9.3

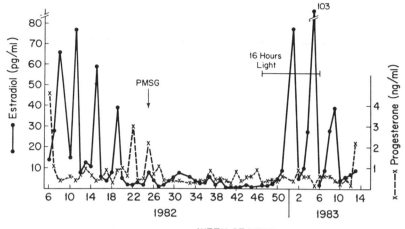

Fig. 2. Baseline serum concentrations of estradiol and
progesterone in weekly samples during two breeding seasons from
female tiger 711. The photoperiod in the second year (1983) was
extended to 16L:8D by use of a 250 watt floodlight mounted in the
ceiling of the indoor cage. The animal was shut into the cage at
night to ensure exposure to the light.

pg/ml. Peak estradiol concentrations were 47.6 ± 6.0 pg/ml
(N=17), ranging from 21 to 115 pg/ml. Interestrous levels were
8.7 ± 0.7 pg/ml (N=28), ranging from 1.7 to 15.1 pg/ml.

The duration of elevated estradiol values was 6-10 days.
Data were collected on a total of 56 cycles in the five animals.
Excluding two outliers and the data from female 180, the interval
between peaks was 24.9 ± 1.3 days for the animals which did not
ovulate spontaneously.

Progesterone

Serum progesterone concentrations ranged from 0.5 to 12
ng/ml in the set of 440 weekly baseline samples (excluding female
180 and experimentally induced ovulatory cycles). Progesterone
values were less than 1 ng/ml in 145 of the baseline samples, and
18 values were between 1 and 2 ng/ml (Fig. 2). Excluding values
greater than 2 ng/ml, serum progesterone concentration was 1.2 ±
0.15 ng/ml (range 0.2-1.8 ng/ml) in the samples collected
February through June (except for female 180). Values greater
than 2 ng/ml were observed in 17 of the 56 baseline samples; ten
from female 712, the most excitable animal, three from female 368
with none above 3 ng/ml, and four from female 711 with one value
(4.6 ng/ml) above 3 ng/ml. Eleven of 17 estradiol peaks were not
associated with elevations of progesterone and none of the
elevated progesterone levels persisted more than two weeks. Six
of the values above 3 ng/ml in female 712 occurred during the
apparent anestrous interval with no hormonal or behavioral
indications of ovarian cycling. This animal was sick when the
highest value, 11.5 ng/ml on January 20, 1983, was obtained.

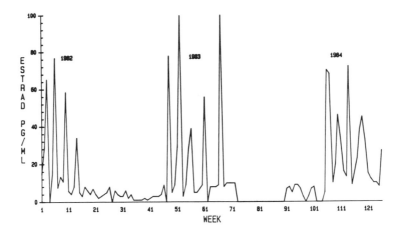

Fig. 3.   Baseline serum immunoreactive estradiol concentrations
in female tiger 711 over three seasons.

Female 180 appears to be a spontaneous ovulator (Fig. 4).
Weekly blood collections were begun in January 1985. She gave no
clear behavioral indications of estrus for several months, then
an endocrine and behavioral estrus was followed by an increase in
serum progesterone to >70 ng/ml. Serum progesterone values
indicated about a five week luteal phase ending in early June.
She was not handled again until early 1986. Progesterone
concentrations in weekly blood samples with no other
manipulations indicated that she ovulated spontaneously three
times in 1986 (Fig. 4). Each luteal phase appears to be five or
more weeks in duration.

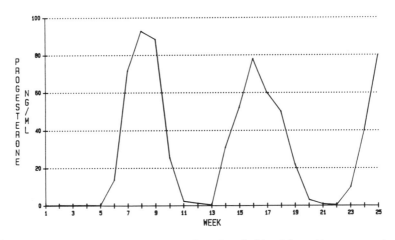

Fig. 4.   Serum progesterone levels indicating three spontaneous
ovulations in an isolated multiparous tiger (female 180).

Testosterone, LH and Androstenedione

Serum testosterone concentrations in the baseline samples
(Fig. 2) ranged from 10 to 100 ng/dl. During anestrus,
testosterone concentrations were 23.4 $\pm$ 1.1 ng/dl (N=70,
range=10-46 ng/dl). Peak testosterone levels were 73.9 $\pm$ 4.2
ng/dl (N=17, range=52-100 ng/dl). All of the 54 estradiol peaks
greater than 20 pg/ml were accompanied by testosterone peaks
(more than three SD greater than the anestrous mean) greater than
50 ng/dl. LH values varied between 0.1 and 2.0 ng/ml (one
outlier of 3.4 ng/ml) with no peaks consistently associated with
either estradiol or progesterone peaks. Androstenedione
concentrations ranging from 0.20 to 5.60 ng/ml were found in
three of these tigers (Seal et al. 1985). Androstenedione
concentrations were correlated with estradiol (r=0.75, N=46,
P<0.001) during the estrous season as were testosterone values
(r=0.74, P<0.001). The correlation of androstenedione with
testosterone was r=0.74 (P<0.001). The correlation of serum
testosterone and androstenedione levels with estradiol, their
lack of correlation with progesterone (r = 0.06, P>0.2), and
their increase after PMSG treatment suggest an ovarian rather
than adrenal origin for these hormones in these tigers.

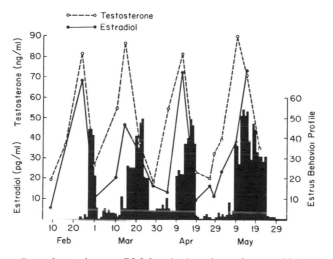

Fig. 5. Female tiger 711's behavioral profile and serum
estradiol and testosterone levels during four cycles in the 1985
breeding season.

Behavior in Estrus

Estrous behavioral profiles were significantly correlated
(P<0.001) with the endocrine profiles of estradiol and testosterone concentrations (Fig. 5). Clear peaks, showing a steady
increase and subsequent decrease in expression, were apparent in
each cycle the tigers exhibited. Also, the peak expression of
each female's estrus profile was relatively constant, showing
more variation between females than among the various cycles of
any individual female. For example, female 711 showed three

estrous behavioral peaks in the 50-60 unit range, as compared to females 368 and 712, which showed behavioral peaks in the 90-120 and 130-155 unit ranges respectively. By contrast, a subsequent study of estrous behavior during the 1985/1986 season showed two females (732 and 180) with differing peaks (unpublished data).

The interval in days that female 711 showed estrous behavior increased from the first to the fourth (or last) cycle she underwent for the season (see Fig. 5). The first estrous cycle was about four days in length, the second and third were close to ten days in length and the last cycle was about 15 days in length.

On average, the interestrous interval for the five females was 25.0 ± 1.3 days (N=10), an interval in agreement with the interestrous endocrine peaks (Seal et al. 1985). This 25 day interval has been confirmed from studies based upon both the behavioral and hormonal data derived from nine animals of different genetic lineages over four seasons of observations.

One behavioral feature of the interestrous period is the complete apathy females display during anestrus. In contrast to the constant vocalizing, pacing, rubbing and rolling that is indicative of estrus, anestrus is characterized by resting quietly, even though measurable (but low) concentrations of estradiol and testosterone are evident (Fig. 5).

The value of using behavioral indicators of an estrous cycle is that they allow a relatively accurate prediction of when the next cycle will occur (approximately 25 days in Siberian tigers) and thus allow for more accurate timing for the induction of ovulation and artificial insemination.

DISCUSSION

Five of the Siberian tigers were seasonal in their estradiol cycles. This correlates with the observed pattern of births in the Northern Hemisphere. The occurrence of postovulatory, postpartum and postlactational estrous cycles, based upon behavioral observations on other tigers at this zoo, could account in part for births at other times of the year. Experiences with one animal exposed to long day photoperiods in the late fall (Seal et al. 1985) suggest that seasonal cycles in these animals may be synchronized by photoperiod, as has been established for the domestic cat (Scott and Lloyd-Jacob 1959). This might also account for some of the off-season births in zoo animals that are exposed to artificially extended photoperiods as a part of exhibit or management practices.

The usual anovulatory cycle length of 25 days based upon both the behavioral observations and the hormone data was derived from nine animals of different genetic lineages over four seasons of observations. A mean cycle length of 49 days has been reported for the Bengal tiger in four institutions (Rowlands and Sadleir 1968, Kleiman 1974). There are no reported hormone data for Bengal tigers. We observed two cycles of extended length in both the hormonal and behavioral studies. These approximately double-length cycles did not appear to be the result of ovulations in the hormone study since there was no significant elevation of progesterone during the interval between estradiol peaks in the two 42 day cycles in female 712. There was evidence of possible follicular activity based upon an increase in testosterone. The time between ovulatory cycles in female 180 was about 60 days, which is significantly longer than the interestrous interval reported for the Bengal tigers.

Four of the five intensively examined tigers in this study did not appear to have spontaneous ovulations based upon the lack of a consistent or sustained elevation of serum progesterone following any of the estradiol peaks. Elevations in progesterone observed in these four tigers were small and of short duration relative to the ovulatory and pregnancy values reported for lions (Schmidt et al. 1979), one jaguar (Wildt et al. 1979), pumas (Bonney et al. 1981) and domestic cats (Verhage et al. 1976, Stabenfeldt and Shille 1977, Wildt et al. 1981b).

The small amounts of label excreted in the urine following administration of tritiated estradiol or progesterone suggest that measurement of these steroids in urine for detection of the reproductive status of tigers would be difficult and lack the necessary sensitivity for delineation of cycles (Seal et al. 1985). These observations are in accordance with reports for the domestic cat (Karim and Taylor 1970, Shille et al. 1984), indicating that the primary route of estradiol excretion in felids is through the gut.

The synthesis of testosterone by ovarian follicles from the ovary of the domestic cat has been described (YoungLai et al. 1976). It appears possible that the pathway of estradiol synthesis in the tiger is by aromatization of testosterone as has been indicated for mares (Silberzahn et al. 1983), cattle and sheep (Herriman et al. 1979, Wise et al. 1982) and rats (Bogovich and Richards 1982). The cyclic elevations in testosterone may be responsible for some of the behavioral aspects of estrus (Martenos and Everitt 1982).

SUMMARY

Seasonal analysis of 1,239 captive births of Siberian tigers (Panthera tigris altaica) indicated a peak in April-June (P<0.001). Studies on seven animals in the Minnesota Zoo indicated that behavioral estrous cycles and ovarian follicular phase cycles began in late January and ceased in early June. Behavioral observation of 10 estrous cycles in three tigers yielded an estrous length of 5.3 ± 0.2 days and an interestrous interval of 24.9 ± 1.3 days. Hormonal assays on weekly blood samples from four female tigers included 46 cycles in four breeding seasons. Peak estradiol-17B levels were 46.7 ± 6.0 pg/ml and interestrous concentrations were 8.7 ± 0.66 pg/ml during the breeding season. Anestrous estradiol levels were 4.2 ± 0.5 pg/ml (N=70). The interestrous interval between estradiol peaks was 24.9 ± 1.3 days with two outliers of 42 days. Serum progesterone concentrations in four animals from February to June were 1.2 ± 0.15 ng/ml (N=32) providing no evidence for ovulation or corpus luteum formation. Luteinizing hormone (LH) levels were 0.56 ± 0.04 ng/ml (N=180). Serum testosterone (r = 0.71, P<0.001) and androstenedione levels (r = 0.75, P<0.001) were correlated with estradiol during the breeding season. The

duration of anestrus was seven to eight months in five of these
tigers. The interval was shortened in one tiger in one season by
exposure to a 16L:8D photoperiod in the fall. The Siberian tiger
appears to be a polyestrous seasonal breeder and an induced or
spontaneous ovulator whose breeding season may be synchronized by
photoperiod.

ACKNOWLEDGEMENTS

    We thank N. Manning, M.K. Twite, M.D. Lewis, J.H. Champlin,
M.J. Heindel and G.C. Schmoller for sustained excellence in
performing the laboratory analytical procedures. We appreciate
the assistance of the many volunteers and zoo staff who assisted
us in working with the tigers including G. Binczik, T. Kreeger
and G. Post. This study was supported in part by the VA Medical
Research Service, the Minnesota Zoological Garden and the
Marshfield Medical Foundation.

REFERENCES

Bogovich, K. and J.S. Richards. 1982. Androgen biosynthesis in
    developing ovarian follicles: evidence that luteinizing
    hormone regulates thecal 17a-hydroxylase and C_{7-20} lyase
    activities. Endocrinology 111:1201-08.
Bonney, R.C., H.D.M. Moore and J.M. Jones. 1981. Plasma
    concentrations of oestradiol-17b and progesterone, and
    laparoscopic observations of the ovary in the puma (Felis
    concolor) during oestrus, pseudopregnancy and pregnancy. J.
    Reprod. Fert. 63:523-31.
Dresser, B.L., L. Kramer, B. Reece and P.T. Russell. 1982.
    Induction of ovulation and successful artificial insemin-
    ation in a Persian leopard (Panthera pardus saxicolor). Zoo
    Biol. 1:55-57.
Foose, T. and U.S. Seal. 1986. Species survival plans for large
    cats in North American zoos. Proc. Inter. Cat Symp., ed D.
    Miller. Washington, DC: Natl. Wildlf. Fed.
Goodwin, H.A. and C.W. Holloway. 1972. Red Data Book. Morges,
    Switzerland: International Union for the Conservation of
    Nature and Natural Resources.
International Species Inventory System (ISIS). 1983. Species
    Distribution Report. Apple Valley, MN: ISIS.
Herriman, I.D., D.J. Harwood, C.B. Mallinson and R.J. Heitzman.
    1979. Plasma concentrations of ovarian hormones during the
    oestrous cycle of the sheep and cow. J. Endocr. 81:61-64.
Karim, M.F. and W. Taylor. 1970. Steroid metabolism in the cat.
    Biliary and urinary excretion of metabolites of [4-]-
    oestradiol. Biochem. J. 117:267-70.
Kleiman, D.G. 1974. The estrous cycle in the tiger (Panthera
    tigris). In The World's Cats, Vol. 2. Biology, Behavior and
    Management of Reproduction, ed. R.L. Eaton. Seattle, WA:
    Feline Research Group.
Leyhausen, P. 1956. Verhaltensstudien au Katzen. Berlin: Paul
    Parley.
Martensz, N.D. and B.J. Everitt. 1982. Effects of passive
    immunization against testosterone on the sexual activity of
    female rhesus monkeys. J. Endocr. 94:271-82.

Rowlands, I.W. and R.M.F.S. Sadleir. 1968. Induction of ovulation in the lion (Panthera leo). J. Reprod. Fert. 16:105-11.

Sadleir, R.M.F.S. 1966. Notes on the reproduction in the larger Felidae. Int. Zoo Yearb. 6:184-87.

Schaller, G.B. 1967. The Deer and the Tiger. Chicago: Univ. Chicago Pr.

Schmidt, A.M., L.A. Nadal, M.J. Schmidt and N.B. Beamer. 1979. Serum concentrations of estradiol and progesterone during the estrous cycle and early pregnancy in the lion (Panthera leo). J. Reprod. Fert. 57:267-72.

Scott, P.P. and M.A. Lloyd-Jacob. 1959. Reduction in the anoestrous period of laboratory cats by increased illumination. Nature 184:2022.

Seal, U.S., R. Barton, L. Mather, K. Olberding, E.D. Plotka and C.W. Gray. 1976. Hormonal contraception in captive female lions (Panthera leo). J. Zoo Animal Med. 7:12-20.

Seal, U.S., E.D. Plotka and C.W. Gray. 1978. Baseline hematology, serum chemistry, and hormone data for captive tigers (Panthera tigris spp) and lions (P. leo). In International Tiger Studbook: Congress Report on 1st International Symposium on the Management and Breeding of the Tiger. Leipzig, DDR: Zoologischer Garten Leipzig.

Seal, U.S. and T. Foose. 1983. Development of a masterplan for captive propagation of Siberian tigers in North American zoos. Zoo Biol. 2:241-44.

Seal, U.S., E.D. Plotka, J.D. Smith, F.H. Wright, N.J. Reindl, R.S. Taylor and M.F. Seal. 1985. Immunoreactive luteinizing hormone, estradiol, progesterone, testosterone, and androstenedione levels during the breeding season and anestrus in Siberian tigers. Biol. Reprod. 32:361-68.

Seifert, S. and P. Muller. 1985. International Tiger Studbook. Leipzig, DDR: Zoologischer Garten Leipzig.

Shille, V.M., A.E. Wing, B.L. Lasley and J.A. Banks. 1984. Excretion of radiolabeled estradiol in the cat. (Felis catus L): A preliminary report. Zoo Biol. 3:201-10.

Silberzahn, P., L. Dehennin, H. Zwain and P. Leymarie. 1983. Identification and measurement of testosterone in plasma and follicular fluid of the mare, using gas chromatography-mass spectrometry associated with isotope dilution. J. Endocr. 97:51-56.

Stabenfeldt, G.H. and V.M. Shille. 1977. Reproduction in the dog and cat. In Reproduction in Domestic Animals, ed. H.H. Cole and P.T. Cupps, 3rd Ed. New York, NY: Academic Pr.

Verhage, H.G., N.B. Beamer and R.M. Brenner. 1976. Plasma levels of estradiol and progesterone in the cat during polyestrus, pregnancy and pseudopregnancy. Biol. Reprod. 14:579-85.

Wildt, D.E., C.C. Platz, R.K. Chakraborty and S.W.J. Seager. 1979. Estrous and ovarian activity in a female jaguar (Panthera onca). J. Reprod. Fert. 56:555-58.

Wildt, D.E., C.C. Platz, S.W.J. Seager and M. Bush. 1981a. Induction of ovarian activity in the cheetah (Acinonyx jubatus). Biol. Reprod. 24:217-22.

Wildt, D.E., S.Y.W. Chan, S.W.J. Seager and P.K. Chakraborty. 1981b. Ovarian activity, circulating hormones, and sexual behavior in the cat. I. Relationships during the coitus-induced luteal phase and the estrous period without mating. Biol. Reprod. 25:15-28.

Wise, T.H., D. Caton, W.W. Thatcher, A.R. Lehrer and M.J. Fields. 1982. Androstenedione, dehydroepiandrosterone and testosterone in ovarian vein plasma and androstenedione in peripheral arterial plasma during the bovine estrous cycle. J. Reprod. Fert. 66:513-18.

YoungLai, E.V., L.W. Belbeck, P. Dimond and P. Singh. 1976. Testosterone production by ovarian follicles of the domestic cat (Felis catus). Hormone Res. 7:91-98.

23

Seminal-Endocrine Characteristics of the Tiger and the Potential for Artificial Breeding

David E. Wildt, Lyndsay G. Phillips, Lee G. Simmons, Karen L. Goodrowe, JoGayle Howard, Janine L. Brown and Mitchell Bush

INTRODUCTION

Species within the genus Panthera generally reproduce well in captivity. The need for selective genetic management, however, has generated interest in controlling reproduction either for the purposes of contraception or artificial propagation (Seal and Foose 1983, Seal et al. 1985). Wild felids could benefit from artificial breeding through the more efficient distribution of genotypes from selected individuals. Theoretically, the genetic component of males under-represented in the captive population could be more effectively utilized through the distribution of semen to other zoos. This concept could be expanded further through semen collection from outbred animals free-ranging in native habitats. The judicious use of sperm cryopreservation could allow long-range banking of gene pools as well as protect existing populations from catastrophe including epizootic diseases.

Many valid reasons exist for establishing artificial breeding programs in wildlife collections similar to those developed in the domestic livestock industry. Economic incentives spurred tremendous financial support of basic research into the reproduction and endocrinology of domestic species, particularly cattle. Buoyed by a comprehensive understanding of these processes and unlimited research animals, it was inevitable that routine, artificial breeding of most farm species would follow. Without similar financial incentives or animal resources, the time-table for success in other species, including carnivores, has been less than spectacular. Domestic cats have produced young after insemination with fresh or frozen-thawed spermatozoa (Sojka et al. 1970, Platz et al. 1978). However, few births have resulted from inseminating wild felids, with the exception of single reports involving a puma (Bonney et al. 1981) and Persian leopard (Dresser et al. 1982) in which single young were born. These two cases probably fail to reflect the actual number of unsuccessful artificial inseminations performed in zoos which go unreported. In a recent review (Wildt et al. 1986), we provided data illustrating the difficulties of artificially breeding wild felids such as the cheetah. Although 23 of 30 cheetahs responded to hormonal treatment by ovulating, none of the 23 artificially inseminated females became pregnant. Similar information is likely available for other felids.

Establishing a successful artificial breeding program, however, is limited by an insufficient reproductive-endocrine data base for practically all Felidae including the tiger. The family as a whole demonstrates a fascinating array of novel reproductive and endocrine characteristics which preclude the direct adaptation of technologies available for other species. An organized, methodical approach to ascertaining physiological norms is the logical prerequisite to implementing large-scale artificial breeding programs. For the tiger, these basic efforts have focused predominantly on studies of ovarian morphology before and after gonadotropin therapy (Phillips et al. 1982) and natural estrous cyclicity as determined by fluctuations in circulating hormones (Seal et al. 1985).

This chapter details the known physiological norms for the male tiger including information on ejaculate characteristics and endocrine interrelationships as well as the potential influence of stress on reproductive function. Related information concerns methodologies for promoting spermatozoal viability in vitro and cryopreserving spermatozoa as well as discussing the possible mechanisms influencing the artificial insemination process. Because our research focuses on many species of Felidae, comparative data are occasionally provided to illustrate remarkable specificities or to provide potential directions for future research.

METHODOLOGY

Animals and Maintenance

Tigers maintained in four U.S. zoological parks contributed data on ejaculate and testis characteristics. As male tigers breed estrual females throughout the year (Nowak and Paradiso 1983), there was no attempt to control for season when evaluating seminal data. Ejaculates were collected during 10 different months. Endocrine data were obtained from tigers maintained at the Henry Doorly Zoo in Omaha, Nebraska. All blood samples were collected from February through April, a period in Omaha coinciding with peak sexual activity of the females. Ovulation-induction and artificial breeding attempts were performed in February and March.

Males and females ranged from 1.5 to 15 years and 2 to 9 years of age, respectively, and generally were maintained in comparable conditions. Standard diets were provided and all animals were housed with access to indoor-outdoor enclosures. Approximately half the males were maintained with one or two females but all had visual, aural and olfactory contact with con-specifics of both sexes. None of the males were observed copulating during the 3 week period before evaluation. Because there were no significant effects of season or presence of a mate on ejaculate-endocrine characteristics, the data were pooled for presentation.

Anesthesia

Males were immobilized with ketamine hydrochloride (Ketaset, Bristol Labs., Syracuse, NY, 11.2 mg/kg, intramuscularly) and maintained in a surgical plane of anesthesia with supplemental ketamine HCl injections (5.7 mg/kg, intravenously). It is common for tigers to experience brief (15-60 sec) but episodic catatonic seizures while under ketamine HCl anesthesia. Although controllable by the use of sedatives, these drugs relax musculature around the urethra causing urine contamination of electroejaculates. A low dosage of diazepam (Valium, Hoffmann-La Roche, Inc., Nutley, NJ, 0.02 mg/kg) was used to minimize severe seizures which occurred in 89.5% of the immobilizations.

Ejaculate-Testis Characteristics

Semen was collected by a standardized electroejaculation technique (Wildt et al. 1983, Wildt et al. in press, Howard et al. in press) using an electrostimulator and 4.5 cm diameter rectal probe. The regimented sequence consisted of 80 incremental stimuli given in an on-off pattern in three series (series 1 and 2 = 30 stimuli each; series 3 = 20 stimuli). Semen was evaluated for volume and spermatozoal concentration, % motility and status (speed of forward progression based on a scale of 0, no movement to 5, steady rapid forward movement). Gross morphological assessments were made by fixing a seminal aliquot in 1% glutaraldehyde and later evaluating 300 spermatozoa/ejaculate under phase contrast microscopy (Pursel and Johnson 1974, Wildt et al. in press, Howard et al. in press). The length and width of each testis was measured with laboratory calipers and these values converted to volume using a previously described formula (Howard et al. 1983).

Twenty-seven electroejaculates were collected from 18 males. Four individuals produced seminal fluid void of any spermatozoa. These tigers were 1.5, 2.0, 2.2 and 3.1 years of age, and although of mature size, were likely prepubertal. The youngest tiger producing spermatic seminal fluid was 2.7 years of age. Ejaculate-testis characteristics for adult versus prepubertal tigers are summarized in Table 1. Seminal fluid and testicular volume were less (P<0.05) in younger compared to sexually mature tigers. Left and right testis volumes were comparable within age groups and there was a 2.5 fold variation in combined testes volume among tigers producing spermatic ejaculates. Marked individual variations were also observed for most ejaculate traits.

Average ejaculate volume and total number of motile spermatozoa/ejaculate (MS/E) were comparable to recent data for free-ranging lions (Wildt et al. unpublished data). However, the MS/E value for tigers was considerably greater than that reported for the cheetah (Wildt et al. 1983), clouded leopard (Wildt et al. in press) or domestic cat (Platz et al. 1978, Wildt et al. 1983). On the average, tiger ejaculates contained a similar proportion of pleiomorphic spermatozoa to those of free-ranging lions (mean, 32.8%) (Wildt et al. unpublished), captive clouded leopards (mean, 38.9%) (Wildt et at. in press) and domestic cats (mean, 29.1%) (Wildt et al. 1983), but considerably less than the high percentage (71%) observed in cheetahs (Wildt et al. 1983).

Table 1.   Comparison of electroejaculate-testis characteristics
between adult and prepubertal tigers.

	Adult		Prepubertal	
No. Males	14		4	
No. Ejaculates	23		4	
	Mean ± SEM	Range	Mean ± SEM	Range
Ejaculate vol. (ml)	5.7±0.9	0.8-14.0	1.7±0.5	0.4-2.8
No. sperm/ml ejaculate ($\times 10^6$)	38.1±10.2	1.0-215.0	-----	-----
Spermatozoal motility (%)	77.7±2.9	50-90	-----	-----
No. motile sperm/ ejaculate ($\times 10^6$)	175.4±46.1	2.2-644.0	-----	-----
Spermatozoal status (0-5)	4.1±0.1	3.0-5.0	-----	-----
Abnormal sperm forms (%)				
Primary defects	14.9±3.4	1.0-62.9	-----	-----
Secondary defects	25.0±3.4	6.7-73.0	-----	-----
Total	39.9±5.5	8.0-91.0	-----	-----
Testis vol. (cm^3)				
Right	36.4±2.6	22.8-60.1	18.7±3.3	12.1-22.4
Left	35.3±2.5	21.7-60.1	19.6±3.7	12.4-24.4
Combined	71.7±5.0	45.3-116.8	38.3±7.0	24.5-46.8

Structurally deformed spermatozoa originate as the result of
spermatogenic dysfunction (primary defects; head, acrosome or
midpiece abnormalities; tightly coiled flagellum) or anomalous
transport and/or maturation of the gamete within the excurrent
duct system (secondary defects; bent neck/flagellum or presence
of a cytoplasmic droplet). The fertilization potential of a
spermatozoon with either aberration may be impaired; however,
primary defects generally are considered more serious due to
severely compromised motility, acrosomal integrity or perhaps
even altered DNA content in instances of micro- or macrocephaly.
When deformities were present in tigers, secondary defects
predominated usually in the form of spermatozoa containing
proximal or distal cytoplasmic droplets (mean, 13.7%) (Table 2,
Fig. 1d). The most prevalent primary abnormality was a tightly
coiled flagellum (mean, 9.4%, Fig. 1b), a pleiomorphic form
frequently observed in other felid species (Wildt et al. 1983,
Wildt et al. in press).

Inexplicably, ejaculates from certain individuals which were
proven breeders, contained relatively high proportions of
spermatozoa with severe deformities. More than 23% (range,
18.7-26.3%) of all cells in three tigers were afflicted with

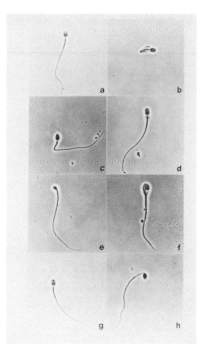

Fig. 1. Spermatozoal forms detected in the tiger electro-
ejaculate   a) normal; b) coiled flagellum; c) bent
midpiece; d) proximal cytoplasmic droplet; e) microcephalic
defect; f) macrocephalic defect; g) acrosomal defect; h) bent
neck and missing mitochondrial sheath.

irregular, loosened or missing acrosomes (Fig. 1g). In 14 of 23
electro-ejaculates, 0.3 to 5.0% of all spermatozoa had major
damage to the midpiece. This defect was characterized by a
partially or completely missing mitochondrial sheath (Fig. 1h), a
critical component of the motor apparatus, the absence of which
renders the cell immotile.

It was of interest that adult tigers varied so widely in
ejaculate-testis characteristics. It is tempting to search for
trait correlates which might provide clues for diagnosing
fertility potential within or among individuals. In cattle,
scrotal circumference is a practical and reportedly effective
index of sperm production (Hahn et al. 1969).In adult tigers,
however, combined testes volume was not highly correlated
(P>0.05) to any electroejaculate characteristic: volume
(r=0.29); spermatozoal concentration (r=0.28); motility (r=0.29);
progressive status (r=0.31); or structural abnormalities
(r=-0.29).   It also was difficult to interrelate any particular
trait with past reproductive performance. Eight of the 14 tigers
producing spermic ejaculates and contributing to the composite
data of Tables 1 and 2 were proven breeders. None of the 14
males was experiencing known fertility problems at the time of
evaluation. Therefore, this information can be considered as the

Table 2. Spermatozoal abnormalities in tiger electroejaculates
(N=23).

Morphological Abnormalities	Mean % (+SEM) of total sperm/ejac.
Primary defect	
Biflagellate	0.01 ± 0.01
Macroephalic	0.04 ± 0.02
Bicephalic	0.08 ± 0.06
Microcephalic	0.52 ± 0.30
Missing mitochondrial sheath	0.71 ± 0.24
Abnormal acrosome	4.10 ± 1.70
Tightly coiled flagellum	9.40 ± 1.80
Secondary defect	
Bent neck	1.10 ± 0.30
Bent midpiece without cytoplasmic droplet	1.40 ± 0.20
Bent flagellum	3.50 ± 1.30
Bent midpiece with cytoplasmic droplet	5.30 ± 1.50
Cytoplasmic droplet	13.70 ± 2.00

best available data base of expected norms for reproductively
competent tigers.

Based on recent studies of the clouded leopard (Wildt et al.
in press), a standardized electroejaculation sequence used on
multiple occasions is helpful in establishing a hierarchy of
reproductive potential within a specific male cohort. The tiger
also could benefit by repeated comparative evaluations of males
with known histories of breeding success versus those
experiencing infertility. This strategy is needed to
definitively establish the relative importance of spermatozoal
concentration, motility and structural characteristics to
fertility.

Endocrine Characteristics of the Male Tiger

Comparative endocrine data may provide new avenues for
discovering or refining taxonomic/ evolutionary relatedness among
species. Certain species also exhibit rather uncommon
reproductive characteristics, such as an increased production of
pleiomorphic spermatozoa. Although such physiological phenomena
may be the consequence of a compromised genotype (O'Brien et al.
1983, O'Brien et al. 1985, Wildt et al. unpublished), it is also
possible that hyperadrenal activity related to stress-
susceptibility could adversely influence reproductive capacity
(Wildt et al. in press). Increased adrenal function has been
shown to have detrimental effects on hypothalamic, pituitary or
gonadal function in a variety of domestic species and man. At
least two species of Felidae (clouded leopard and North Chinese
leopard) exhibit extraordinarily elevated circulating gluco-

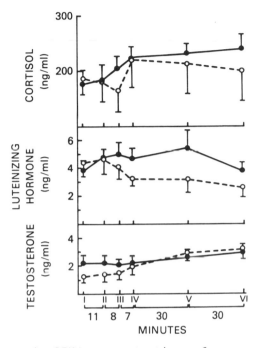

Fig. 2. Mean ($\pm$ SEM) concentrations of serum cortisol, LH and testosterone in control (o-o) and electroejaculated (o-o) adult, male tigers. Sampling times were pre-electroejaculation (I), after each electroejaculation series (II-IV) and 30 (V) and 60 (VI) min after the last electroejaculation stimulus.

corticoid concentrations in the presence of relatively high proportions of structurally defective spermatozoa (Wildt et al. in press). Certainly endocrine studies to precisely relate the significance of adrenal activity to male reproductive capacity are warranted. Monitoring hormonal activity of males also has potential for improving medical care and animal management. Although circulating levels of luteinizing hormone (LH) or testosterone have not yet provided an accurate index of reproductive potential among proven breeder males, such analyses are useful for diagnosing or confirming hypothalamic, pituitary or gonadal dysfunction. More importantly, it is critical to determine if the captive environment and the stresses associated with elective manipulations (i.e., anesthesia, electro-ejaculation, laparoscopy) significantly influence reproduction and, if so, which species are affected. Delineating species-specific hormonal responses provides directive for further studies to improve general management strategies, anesthesia procedures for health care and artificial breeding techniques.

Cortisol, LH and testosterone were measured in the peripheral circulation of anesthetized male tigers to determine adrenal, pituitary and testicular activity. Blood samples were obtained by venipuncture immediately before beginning semen collection, after each electroejaculation series and at 30 and

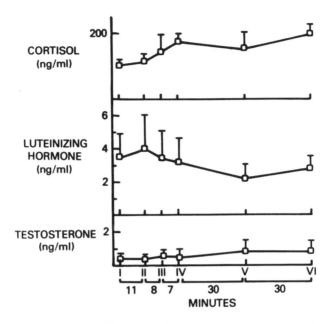

Fig. 3. Mean (± SEM) concentrations of serum cortisol, LH and testosterone in electroejaculated, prepubertal, male tigers. Sampling times were pre-electroejaculation (I), after each electroejaculation series (II-IV) and 30 (V) and 60 (VI) min after the last electroejaculation stimulus.

60 min after the last electrical stimulus. Because the duration of electroejaculation averaged 26.1 ± 0.9 min, the bleeding interval spanned 86 min. Control bleedings (immobilization but no electroejaculation) were conducted over a similar time interval. To evaluate adrenal responsiveness and the potential influence of glucocorticoid release on acute LH and testosterone secretion, a series of tigers also were treated with synthetic adrenocorticotropic hormone (ACTH, Cortrosyn, Organon, Inc., W. Orange, NJ). After taking a control blood sample, a single bolus of ACTH (250 ug) was injected intramuscularly and additional samples taken at 15 min intervals for 2 hours. Specific information on radioimmunoassay techniques are available in earlier publications (Chakrabority et al. 1979, Carter et al. 1984, Wildt et al. 1984, Wildt et al. 1984, Goodrowe et al. 1985).

The average and ranges in serum concentrations of cortisol, LH and testosterone in adult, control (anesthetized only) and adult and prepubertal tigers subjected to anesthesia/electro-ejaculation are illustrated in Table 3. Profiles of mean cortisol, LH and testosterone concentrations over time in control and electroejaculated adult (Fig. 2) and prepubertal tigers (Fig. 3) also are provided. Mean concentrations of each hormone were similar (P>0.05) between adult males subjected to anesthesia only

Table 3.   Serum cortisol (C), lutenizing hormone (LH) and testosterone(T) concentration[a] of tigers subjected to anesthesia/ electroejaculation or anesthesia alone.

	C	LH	T
Adult			
Control (N=7)			
Mean $\pm$ SEM	195.0+45.2	3.7+0.7	2.1+0.6
Range	74.6-394.0	1.7-5.7	0.1-4.0
Electroejaculated (N=13)			
Mean $\pm$ SEM	211.2+19.6	4.6+0.6	2.3+0.5
Range	117.6-331.8	1.5-7.6	0.4-5.2
Prepubertal			
Electroejaculated (N=4)			
Mean $\pm$ SEM	150.5+18.3	3.2+1.3	0.6+0.3
Range	114.4-173.3	1.6-5.7	0.1-1.2

[a] Values are ng/ml of serum.

compared to those anesthetized and electroejaculated. Mean cortisol and LH levels between both adult groups and prepubertal tigers were comparable (P>0.05), however, young males were characterized by comparatively low testosterone concentrations (Table 3, Fig. 3). Average temporal profiles of each hormone during the bleeding interval were not different (P>0.05) between the two adult groups. Immobilization and serial bleeding alone resulted in a slight insignificant rise in serum cortisol which was not influenced markedly by superimposing the electro-ejaculation procedure (Fig. 2). Mean serum LH levels within each group varied by <2 ng/ml over time. Similarly, average testosterone concentrations were rather static in electro-ejaculated males and gradually and inexplicably rose (2 fold) in control males.

Hormonal profiles from individual, adult males provided a more discriminating illustration of endocrine dynamics (Fig. 4). From these analyses it appeared that serum cortisol almost always rose in electroejaculated tigers (Fig. 4a-c) with occasional, more modest elevations in nonstimulated, but anesthetized controls (Fig. 4e,f). A detectable cortisol rise similar to that observed in Fig. 4e was observed in three of seven control episodes, usually occurring about 30 min into the bleeding period. The electroejaculated tiger expressed a relatively narrow range in adrenal responsiveness (range, 1.1 to 2.7 fold increase) as determined by analysis of serum cortisol. Only three of 13 electroejaculated and two of seven control animals produced maximal serum cortisol levels greater than 300 ng/ml. The peak values detected among all electroejaculated and all control males were 421.9 and 462.2 ng/ml, respectively. Based on the ACTH challenges, the peak cortisol levels measured in electroejaculated and control tigers may have been maximal. Individual males receiving a single bolus of ACTH had 1.4 to 3.0

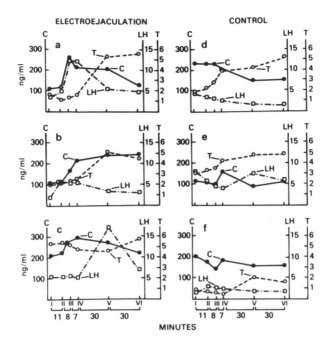

Fig. 4.   Serum cortisol (C), LH and testosterone (T) profiles in
individual, adult male tigers subjected to anesthesia/electro-
ejaculation (a-c) or aesthesia only (d-f).

concentrations no different from that of electroejaculated males.
The magnitude of this induced response was less than that
observed in ACTH-treated cheetahs (Wildt et al. 1984) and clouded
leopards (Wildt et al. in press).   One tiger (Fig. 5c), electroejaculated after the 2 hour ACTH challenge, was capable of
eliciting a cortisol response similar to that observed in other
electroejaculated males (Fig. 4).   Therefore, even after an ACTH-
induced rise in cortisol, the adrenal glands of the tiger appear
immediately responsive to a second manipulatory stimulus.

     Individual profiles demonstrated significant serum LH
concentration changes over time suggesting a possible episodic
release pattern.   Occasionally, a sharp rise in LH was followed
by a subsequent increase in testosterone (Fig. 4a).   Thus,
pituitary and gonadal hormone release in adult male tigers
appears dynamic.   However, to establish true LH/testosterone-
pulsatility, a longer and more frequent blood sampling protocol
would be required.   During the 86 min bleeding interval, LH and
testosterone in individual males varied by as much as 11.8 and
6.18 ng/ml, respectively.   Therefore, a multiple blood sampling
regimen is required in tigers to accurately determine basal
secretion of either hormone.   Overall, mean LH concentration was
highly correlated to mean testosterone (r=0.71, P<0.01) among
individual, electroejaculated tigers.   However, neither mean LH
nor testosterone was correlated (P>0.05) to any ejaculate
characteristic including spermatozoal concentration (r=0.16,

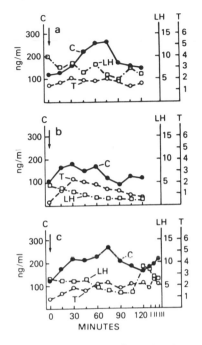

Fig. 5. Serum cortisol (C), LH and testosterone (T) profiles in three anesthetized, adult, male tigers injected with 250 ug ACTH immediately after collection of a 0-min blood sample. Arrow denotes time of ACTH injection. Tiger in panel c was electro-ejaculated after the ACTH challenge and blood samples obtained after the first (I), second (II) and third (III) electro-ejaculation series.

0.09, respectively), motility (r=-0.27, -0.11, respectively), progressive status (r=-0.26, -0.01, respectively) or structural abnormalities (r=-0.11, -0.11, respectively).

The tendency for average testosterone concentration to rise over time was reflected in individual profiles of many of the electroejaculated and control males (Fig. 4). It was difficult, based on the available data, to relate this observation to either a coincidental increase of an adrenal induced response. Data from the ACTH study were of little value in understanding this observation. Although some ACTH-treated males had acute (Fig. 5b) or subtle (Fig. 5c) increases in testosterone which paralleled cortisol release, others demonstrated little, if any, change in testosterone production (Fig. 5a). Increased stress-induced adrenal activity demonstrably alters pituitary and gonadal release of LH and testosterone in other species (see reviews, Ramaley 1981, Rose and Sachar 1981, Collu et al. 1984, Moberg 1984). Interestingly, this effect is not always mediated via the pituitary and there is evidence that elevated gluco-corticoids have a direct effect at the gonadal level, causing either decreases or increases in testosterone release. At present, our data can only be interpreted to suggest that increased adrenal activity in the tiger does not have an appreciable influence on acute testosterone release.

From a comparative aspect, species of Felidae exhibit marked differences in endocrine patterns and function, especially with respect to adrenal activity. Figure 6 illustrates the striking differences among the tiger, cheetah, puma and North Chinese leopard maintained in the same zoological environment and subjected to a regimented electroejaculation-bleeding protocol. The temporal cortisol profile in the tiger was less acute and more prolonged than that of the cheetah but more attenuated and of shorter duration than that of the puma. Basal, pre-stimulation cortisol concentration was extremely elevated in the North Chinese leopard which was increased only modestly by electrostimulation, perhaps as a consequence of an already existing adrenal hyper- function. Although species temperament was not specifically quantitated in this study, serum cortisol appeared to be an accurate index of general behavioral patterns. In this particular colony of felids, the tigers responded less aggressively to human interference (i.e., darting for immobilization) than North Chinese leopards. In contrast, under the same manipulative conditions, tigers reacted in a more excitable fashion than cheetahs. The duration of response was of interest because tigers, North Chinese leopards and cheetahs expressed either a stabilized or acute rise and fall in circulating cortisol after anesthesia/electroejaculation. In contrast, the marked and continuous cortisol rise in the puma suggested an inherent difference in the mechanism governing adrenal control. The circulating concentrations of other hormones appeared less species specific, although in this study the similar LH and testosterone values detected among tigers, pumas and North Chinese leopards were about 2-fold greater than that of cheetahs.

Semen Handling

A prerequisite to the routine use of tiger semen is proper preparation of inseminates as well as maintenance of spermatozoa in a viable condition for indeterminant intervals in vitro. Delays in preparing females for artificial breeding or the need to inseminate multiple females from a single ejaculate require a short-term, maintenance environment for promoting and sustaining spermatozoal viability. Freshly collected felid ejaculates contain a high proportion of accessory gland fluids and are contaminated with cellular debris including dead and structurally deformed spermatozoa. In addition, several components of seminal fluid, such as degradative enzymes, can severely impair spermatozoal viability. Therefore, successful artificial breeding may be optimized by removal of spermatozoa from seminal fluids and concentrating spermatozoa from the dilute electro-ejaculate into a smaller inseminate volume. Yet there have been few efforts to study the biochemistry of felid semen, the maintenance requirements for sustaining in vitro viability or the susceptibility of these spermatozoa to handling stressors including low speed centrifugation.

Our preliminary studies have emphasized improving the duration of sperm motility in vitro by evaluating factors associated with dilution, centrifugation, resuspension and maintenance temperature. Fresh ejaculates from each of four tigers were aliquoted into each of three treatment groups: 1)

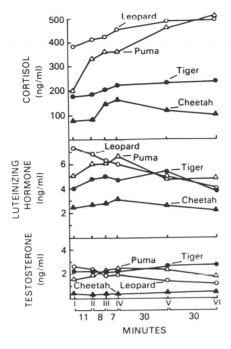

Fig. 6. Mean concentrations of serum cortisol, LH and testosterone in the tiger (n=13), cheetah (n=15), puma (n=7) and North Chinese leopard (n=10). All males were anesthetized with ketamine HCl and subjected to a standardized electroejaculation procedure. Sampling times were pre-electroejaculation (I), after each electroejaculation series (II-IV) and 30 (V) and 60 (VI) min after the last electroejaculation stimulus.

undiluted, raw semen; 2) diluted (v/v) with Hams F-10 medium (Gibco Labs., Chagrin Falls, OH); and 3) diluted (v/v) with Biggers, Sterne and Whittingham's (BSW:Biggers et al. 1971) medium. Aliquots from treatments 2 and 3 were centrifuged at 300 xg for 10 min, the supernatant removed and the sperm pellet gently resuspended with the original dilution medium to give a final concentration of 10×10^6 spermatozoa/ml. Aliquots from each of these treatment groups then were maintained at either 37 $^{\circ}$C (in a water bath) or 23 $^{\circ}$C (room temperature). Spermatozoal motility assessments were made at 20 min intervals.

The dilution/centrifugation/resuspension process markedly increased the duration of spermatozoal motility in vitro in all four test males, on one occasion by as much as 12 fold (Fig. 7). It also was evident that tiger spermatozoa prefer a cooler holding temperature since the duration of sperm motility frequently was enhanced 2 fold (up to 12 hours) by maintenance at 23 $^{\circ}$C rather than 37 $^{\circ}$C. This effect likely was mediated by a reduction in metabolic rate suggesting that even lower temperatures may be more advantageous for sustaining spermatozoal viability.

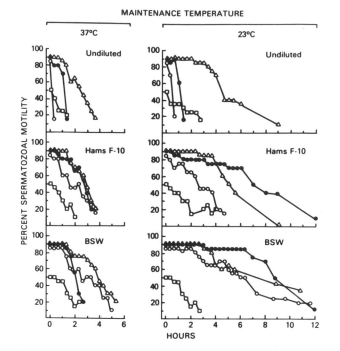

Fig. 7. Spermatozoal motility profiles over time from four individual ejaculates untreated or diluted with Hams F-10 or BSW medium, centrifuged and resuspended. Aliquots from each ejaculate were maintained at 37° C or 23° C.

Semen Cryopreservation

The efficacy of freeze-storing carnivore semen including that of the tiger has not been studied extensively. Our preliminary efforts have been comparative in nature, evaluating the influence of different cryoprotective diluents in protecting post-thaw viability. Following evaluation of spermatozoal motility and progressive status, freshly collected semen was aliquoted (v/v) into each of three diluents (PDV-62, Test, Tris), all with an egg yolk-glycerol base (Howard et al. in press). After equilibration at 5 °C for 30 min, each aliquot was supplemented with additional cooled diluent (2:1) and maintained at 5 °C for an additional 15 min. Each aliquot then was pipetted in pellet form onto a dry ice block (Nagase and Niwa 1964, Platz et al. 1978), left for 3 min and then plunged into liquid nitrogen (LN$_2$). After packaging in individual vials and storing in LN$_2$, three pellets from each aliquot were thawed rapidly (37 °C) in physiological saline (0.9%) or a tissue culture solution (TCS, Tyrode, Difco Lab., Detroit, MI). Thawed samples were examined at 15 min intervals to determine maximal post-thaw viabilities.

Table 4 depicts the comparative results of freeze-thawing electroejaculates from three individual tigers. Because of the small sample size, results were not analyzed statistically,

Table 4. Results of freeze-thawing tiger spermatozoa using three cryoprotective diluents and two thawing media.

Tiger	Spermatozoal Motility (%)/Status[a]						
	Pre-freeze	PDV-62[b]		Post-thaw Test[b]		Tris[b]	
		saline[c]	TCS[c]	saline[c]	TCS[c]	saline[c]	TCS[c]
1	75:4.0	40:4.0	50:4.5	40:3.0	50:4.0	30:3.5	40:4.0
2	70:4.0	35:4.0	50:4.5	30:3.5	40:4.0	25:4.0	45:4.5
3	85:5.0	75:4.5	80:5.0	70:4.5	75:4.0	70:4.0	75:4.0

[a] Progressive sperm status (0, no movement to 5, rapid forward progression)
[b] Cryoprotective diluent
[c] Thawing medium

however, several important observations were noted. First, tiger spermatozoa maintained some degree of progressive motility after experiencing a cryostress. The % recovery of the original pre-freeze motility ranged from 40 to 67, 36 to 71 and 82 to 94% in males #1, #2 and #3, respectively. This variance appeared only modestly related to the cryoprotectant used, but may have been influenced by thawing medium. Slightly greater motility was consistently observed for spermatozoa thawed in TCS compared to the saline solution. Most importantly, the greater motility recovery in tiger #3 compared to #1 and #2 suggested that variation among individual males may dictate the success rates of freezing tiger semen. Similar findings have been reported for other species (Tischner 1979, Howard et al. in press).

Recent data from our laboratory (Howard et al. in press) as well as others (Pursel et al. 1972, Berndston et al. 1981) indicate that post-thaw motility is inadequate as the sole criterion of sperm freezability. Ultrastructural damage to thawed spermatozoa, particularly the acrosome, is common in many species. Therefore, acrosomal integrity is a valuable, ancillary measure for assessing freeze damage. Unfortunately, the acrosome of the felid spermatozoon, including that of the tiger, is extremely narrow in structure and difficult to routinely identify even by high-powered phase contrast microscopy (Fig. 1). Therefore, considerable basic research is required for developing freezing technology for tiger semen. Certainly, preemptive efforts to any large-scale cryobanking program should include establishing the biochemical milieu of seminal fluid as well as further analyses of cryoprotective diluents, cooling-equilibration intervals, freeze-storage containers (pellets versus ampules or straws), post-thaw ultrastructural integrity and in vitro testing of fertilizing capacity of thawed spermatozoa.

Ovulation Induction and Artificial Insemination

Some females fail to exhibit overt estrous behavior which helps identify impending ovulation and optimal breeding time.

Administering gonadotropic hormones to the female permits the effective control of the reproductive cycle, thereby allowing the scheduling of artificial insemination attempts. Hormonal therapy can induce ovulation, increasing the chances of achieving a successful artificial insemination. Our preliminary efforts have focused on adapting a gonadotropin and artificial insemination regimen originally developed in the domestic cat (Platz et al. 1978, Wildt et al. 1970) to other felids (Wildt et al. 1981, Phillips et al. 1982, Wildt et al. 1986) including the tiger.

Hormonal stimulation and artificial breeding studies were conducted in consecutive years using five adult tigresses. Before beginning gonadotropin treatment, each female was immobilized with ketamine HCl (4 mg/kg, intramuscular injection), intubated and then maintained in a surgical plane of anesthesia with halothane gas. Laparoscopy was used to assess the number of immature follicles, mature follicles, corpora hemorrhagica and corpora lutea present (Wildt et al. 1981, Phillips et al. 1982, Wildt et al. 1986). Gonadotropin treatment began 4 days later. Each female received 10 mg of follicle stimulating hormone (FSH-P, Burns Biotec Labs, Inc., Omaha, NE)/day for 5 to 7 consecutive days (Days 1-7, Table 5). On Days 5, 6, 7, 8 and/or 9 another laparoscopy was performed to evaluate treatment effect on ovarian response. Artificial insemination times were determined by subjective estimates of ovarian activity including impending or recent ovulation. Generally, each female was artificially inseminated once immediately following laparoscopy and while still under anesthesia. Ejaculates either were untreated or diluted with tissue culture solution (TSC, Tyrode) and centrifuged (300 xg, 10 minutes) with subsequent removal of seminal fluid and resuspension of the sperm pellet in TCS. Spermatozoa were deposited either intracervically or trans-cervically into the uterine body immediately adjacent to the internal, cervical os. In one female, laparoscopy was used to transabdominally cannulate the uterine horns for insemination (Wildt et al. 1986).

The 5 day regimen of FSH-P (10 mg/day) was shown previously to be either moderately or very effective in inducing ovulation in cheetahs, North Chinese leopards and lions with or without the concomitant use of human chorionic gonadotropin (hCG, Pregnyl, Organon, Inc., W. Orange, NJ) (Wildt et al. 1981, Phillips et al. 1982, Wildt et al. 1986). Compared to these species, the tigress was relatively refractory with mature follicular development and/or ovulation being observed in only five and four of 11 attempts, respectively (Table 5). The additional use of 500 iu of hCG given intramuscularly to two of the anovulatory tigresses was not beneficial in stimulating ovulation. On all of the anovulatory occasions, ovaries appeared swollen, containing numerous immature follicles 3 to 5 mm in diameter.

In 1982, estrous behavior was observed in two females beginning on the afternoon of Days 3 and 5, respectively. One of these tigresses (#1 with 14 ovulation sites) mated over a 36 hour interval. The second estrual female (#2 with three ovulation sites) was artificially inseminated intracervically. A laparotomy was performed in each of these females and the uterine horns and oviducts flushed with a phosphate buffered saline solution in an attempt to recover embryos. Neither unfertilized nor fertilized ova were recovered 5 days after tigress #1 began mating. A bilateral oviductal flush on Day 3 in tigress #2 contained three oocytes with no morphological evidence of fertilization.

Table 5.   Ovarian activity in tigers before and after daily treatment with FSH-P Gonadotropin (10mg/day).

Tiger	Yr.	Treatment Duration (days)	Days Laparoscopy Performed	Results[a,b] Pre-treatment	Post-treatment	Ovulation by Last Laparoscopy
1	1981	5	5	NA	1CH,6MF	+
	1982	5	9	1MF	6CH,8CL,6MF	+
	1983	7	5,7,8	>10TF	>10TF	−
2	1981	7	5,6,7	NA	>10TF	−
	1982	5	6	5MF	3CH	+
	1983	7	5,7,9	4TF	6MF,5TF	−
3	1981	5	5	5ACL	5ACL,>10TF	−
	1982	7	6,9	NA	>10TF	−
4	1982	5	6	NA	1CH	+
	1983	7	5,7	>10TF	>10TF	−
5	1983	7	5,7	2MF,>6TF	>10TF	−

[a] Day 1 = First day of treatment.
[b] NA=no activity;   TF=tertiary follicle;   CH=corpus hemorrhagicum; CL=corpus luteum;   ACL=aged corpus luteum.

Based on the retrieval of the unfertilized oocytes, we suspected that spermatozoa were not being transported to the oviducts. Therefore, on three occasions in 1983, females were laparotomized one hour after insemination. The oviducts and proximal/distal aspects of each uterine horn were flushed separately with physiological saline. The flush medium was centrifuged (1000 xg for 10 min) and the pellet examined for the presence of spermatozoa. No spermatozoa were detected in the proximal aspect of the uterine horn after intracervical or transcervical inseminations. Spermatozoa were absent from the oviducts of all tigresses and were present in the proximal horn only after the laparoscopic, transabdominal insemination.

In domestic species, the movement of spermatozoa to the oviduct is rapid after intravaginal insemination, beginning instantaneously and usually requiring 5 min or less (see review, Overstreet 1983). This dynamic process is accomplished entirely through passive conveyance by the female tract. The accelerated transport of spermatozoa into the oviducts appears mediated by uterine contractions initiated either by seminal fluid constituents (including prostaglandins) or neuroendocrine controls (associated with copulation). Clearly in the present study, sperm transport appeared compromised in gonadotropin-treated, artificially inseminated, anesthetized tigers. If this was the consequence of impaired, passive uterine activity, then at least three potential problems must be addressed before considering the practical application of artificial insemination in tigers.
    1. Tigers are presumed to be copulatory-induced ovulators (Seal et al. 1985). Such stimulation may be critical in initiating or sustaining uterine contractility. A single artificial insemination may simply be insufficient to compensate for the necessity of intense and frequent mating stimuli;
    2. All artificial breedings were performed during surgical

anesthesia and immediately after laparoscopic examination. These manipulatory events could not only compromise uterine activity and thus sperm transport, but also influence other endocrine processes potentially detrimental to the artificial breeding attempt. Other simultaneous observations support this possibility. For each tigress in Table 5, times for each manipulatory procedure (from initial ketamine HCl injection to surgical closure or artificial insemination) were recorded. Blood samples were obtained from each anesthetized female at 15 min intervals throughout all procedures and sera analyzed for cortisol concentrations. Adrenal function appeared particularly sensitive to invasive surgery. When no surgery was superimposed on general anesthesia, cortisol levels over time among females immobilized only with ketamine HCl were comparable to those induced with ketamine HCl and maintained on halothane gas (Fig. 8a,b). In contrast, surgical manipulation, even the rather minor technique of laparoscopy, induced a marked and chronic rise in serum cortisol (Fig. 8c). When the manipulatory procedures were prolonged (in the event of an artificial insemination or laparotomy), cortisol concentrations remained at levels 3-fold greater than basal (Fig. 8d). This increase in cortisol may have the potential of adversely influencing endogenous events responsible for facilitating sperm transport or subsequent fertilization.

3. The sensitivity of the reproductive tract to elements influencing contractility is largely determined by the endocrine status of the female. The exogenous gonadotropin therapy used in tigers affected ovarian morphology in most individuals. It is likely that circulating concentrations of estrogens and progesterone, two critical factors affecting uterine environment

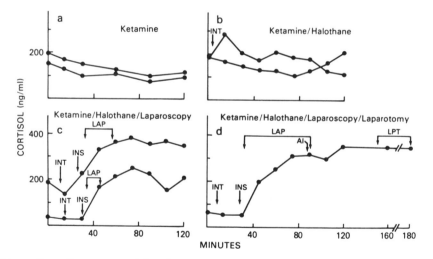

Fig. 8. Serum cortisol concentrations in individual tigresses subjected to: a) ketamine HCl anesthesia only, n=2; b) ketamine HCl induction followed by intubation (INT) and halothane anesthesia, n=2; c) same anesthesia as (b) but with abdominal insufflation (INS) and laparoscopy (LAP), n=2; d) same anesthesia and laparoscopy procedures as (c) but with an intracervical artificial insemination (AI) and a laparotomy (LPT) for uterine/oviductal flushing, n=1.

and activity, were altered, even in females with modest ovarian responses. Recent data in domestic cats strongly suggest that gonadotropin treatment compromises reproductive-endocrine function. Domestic cats treated with 2 mg FSH-P/day for 5 days and then mated produce a greater proportion of unfertilized oocytes than domestic queens copulating while in natural estrus (Goodrowe et al. 1986). Furthermore, serum estradiol-17B levels in gonadotropin-treated queens never achieve the concentrations observed in natural estrus counterparts and, in fact, decline coincident with the onset of mating behavior (Fig. 9). Additionally, circulating progesterone, which normally begins to increase on the fourth day of estrus, rises prematurely in FSH-P treated cats (Fig. 9). There is little doubt that the processes of sperm transport, fertilization and embryo translocation/ implantation could be extremely vulnerable to such perturbations in endocrine status.

Together, the rather fragmentary data now available would suggest the need for further research in several areas. First, considerable improvements are required to pharmacologically control ovarian activity in wild carnivores. Successful artificial breeding is predominantly a consequence of accurate timing of the insemination and ovulatory events. More species specific or purer gonadotropin preparations as well as delivery systems are required which additionally may reduce the need for a multiple injection sequence, thereby minimizing stress. Optimally

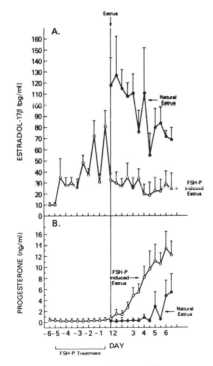

Fig. 9. Mean (+ SEM) concentrations of serum estradiol-17^XB and progesterone in domestic female cats in natural (o) or FSH-P (2 mg/day for 5 days,(o)-induced estrus).

these hormonal treatments should induce ovarian activity and an endogenous endocrine environment simulating that of a naturally estrual female.   The effects of psychogenic or manipulatory stressors on acute or chronic reproductive function require evaluation.   In particular, our preliminary observations indicate poor sperm transport in artificially inseminated tigers under a surgical plane of anesthesia immediately following a laparoscopic examination.   Resorting to pharmacologic treatments (i.e., oxytocin, prostaglandin) to induce uterine contractility, avoiding simultaneous laparoscopy or attempting insemination during sedation rather than anesthesia are all alternative strategies warranting investigation.

Potential of Embryo Transfer and In Vitro Fertilization

Other biotechnologies of proven importance to domestic animal and human reproduction must be considered for future application to wild felids, including the tiger.   However, until effectively developed in animal models such as the domestic cat, there is little justification for attempting similar studies in rare species.   Progress in our laboratory as well as others suggest that eventually embryo transfer or in vitro fertilization may be a viable alternative for propagating wild species of Felidae.   Published data for embryo transfer in exotic felids is very limited.   In one study, two embryos were recovered from one of three gonadotropin-treated tigresses, but, when transferred to an African lion, failed to result in term offspring (Reed et al. 1981).   Another study reported the successful collection of single embryos on three occasions from a naturally estrous lioness treated with hCG (Bowen et al. 1982).   Two laboratories, however, have reported the successful transfer of domestic cat embryos between donors and recipients (Kraemer et al. 1979, Goodrowe et al. 1986).   Recovery of high quality embryos and the pregnancy rates after transfer appear compromised by the use of exogenous gonadotropins in both donor and recipient cats (Goodrowe et al. 1986, Wildt et al. 1986).   Using an FSH-P gonadotropin regimen results in ovarian hyperstimulation, altered endocrine status, reduced fertilization rate and an undesirable, accelerated rate of embryo transport through the reproductive tract.   Therefore, until options for hormonal therapy improve, recovery of viable felid embryos likely will be difficult and success haphazard.

Little information is available regarding in vitro fertilization (IVF) of felid oocytes.   This process involves the extracorporal fusion of mature oocytes and spermatozoa under controlled laboratory conditions.   The major impediment to this approach is the need for sophisticated culture systems, not only to ensure fertilization, but also to promote development of the highly vulnerable, early-stage embryo.   Nevertheless, IVF has been reported for oocytes flushed from the oviducts of naturally ovulating or hormonally-induced domestic cats and fertilized with epididymal or ductus deferens spermatozoa (Hammer et al. 1970, Bowen 1977, Niwa et al. 1985).   In one study (Bowen 1977), embryos advanced in culture to the blastocyst stage; however, no embryo transfers were reported.

If IVF is to have application in rare, wild felids then oocytes must be recovered atraumatically, without repeated laparotomies and oviductal flushing.   The oviduct is particularly sensitive to handling and predisposed to adhesion formation with

adjacent tissue thereby potentially decreasing future reproductive potential. Our preliminary studies have circumvented
this problem by aspirating follicular oocytes using transabdominal laparoscopy, much like the procedure used in recovering
human ova for IVF. Ovaries of gonadotropin-treated queens were
secured using an ancillary probe. Under direct laparoscopic
observation, mature follicles were aspirated using a 22 gauge
needle attached by a series of tubing to a collection apparatus
and vacuum pump. Mature oocytes were washed and placed in a
modified Kreb's Ringer Bicarbonate (KRB) medium in a 5% CO_2 in
air atmosphere (37 ^{o}C). Spermatozoa, collected by electroejaculation were centrifuged and the resulting pellet layered
with KRB medium. After a 1 hour swim-up interval and appropriate
dilutions, a 100 ul drop containing $2X10^4$ sperm cells was added
to culture dishes containing oocytes. Presently, approximately
90% of all punctured follicles result in recovered oocytes. Of
these, about 90% are quality graded as suitable for an IVF
attempt. Based on observations of pronuclear formation (as
determined by differential interference contrast and fluorescence
microscopy) or blastomere formation and cleavage, approximately
34% of all cultured ova are judged fertilized. Although
preliminary, these data present some exciting, potentially
realistic possibilities. Presuming the prerequisites for IVF
among species of felids are even remotely similar, then the
overall transabdominal approach to oocyte retrieval provides a
less invasive and manipulatory approach than conventional methods
of embryo transfer. Theoretically, the oocyte donor need only be
subjected to hormonal treatment and a single laparoscopy for
gamete recovery. The steps involving artificial insemination and
embryo recovery via laparotomy are eliminated, instead
fertilization accomplished in a controlled laboratory
environment.

CONCLUSIONS

     In summary, considerable data now are available concerning
ejaculate-endocrine characteristics of the male tiger. High
quality semen can be collected by electroejaculation from most
anesthetized males greater than 2.5 years of age. Details on
electro-ejaculate characteristics are provided and are found to
vary considerably among males, even in those with positive
breeding histories. On the average, the tiger ejaculate contains
approximately 40% pleio-morphic sperm forms, less than the
proportion observed in another closely studied felid, the
cheetah. No morphometric correlates are indicative of reproductive potential; testicular size is not related to either
electroejaculate volume, spermatozoal concentration or motility
characteristics.

     Based on circulating concentrations of cortisol in electro-
ejaculated and ACTH challenged males, the tiger produces a rather
modest adrenal response compared to other wild felids. Cortisol
levels also occasionally rise even in control males subjected to
anesthesia only. Blood levels of LH and testosterone are
comparable to values measured in the North Chinese leopard and
puma but generally greater than concentrations detected in the
cheetah. Neither LH nor testosterone level is related to any
electro-ejaculate trait measured. Significant fluctuations in
both hormones over the 86 min bleeding interval suggest that a

multiple blood sampling regimen is necessary to assess the endocrine status of any given male.

The in vitro viability of tiger spermatozoa is improved markedly by laboratory processing which involves dilution, centrifugation, resuspension and maintenance at ambient temperature. Spermatozoa of this species also survive pellet freezing on dry ice and storage in liquid nitrogen. However, considerable additional research is required before the massive banking of tiger spermatozoa can be recommended. Variations in post-thaw sperm survival among and within males, observed using different cryoprotective diluents, suggest the need for examining all levels of technology, with emphasis on acrosomal integrity as a means of determining spermatozoal viability.

The female tiger responds to exogenous, daily FSH-P therapy, although much less effectively than the cheetah, North Chinese leopard and lion. Artificial insemination of hormonally treated tigresses has not been successful to date and likely is attributable, in part, to compromised sperm transport. Whether poor movement of sperm through the reproductive tract is related to the lack of copulatory stimuli, the anesthesia condition necessary for artificial insemination or altered endocrine status as a consequence of gonadotropin treatment remains to be determined.

Together, these observations provide directions for more detailed studies which eventually may include considering embryo transfer and in vitro fertilization for assisting propagative efforts. The ultimate potential of any of these advanced technologies is limited only by the imagination and dedication of those responsible for the species. The primary limitation now is an adequate data base of physiological knowledge, a resource critical to future success.

ACKNOWLEDGEMENTS

These data could not have been collected without the generous cooperation and assistance of the following individuals and organizations: Drs. S.J. O'Brien and J. Martenson, Genetics Section, National Cancer Institute; Dr. A. Teare and R. Rockwell, Henry Doorly Zoo; Dr. B. Raphael, Dallas Zoo; Dr. K. Fletcher and L. DiSabato, San Antonio Zoological Gardens; and J. Maynard, Exotic Feline Breeding Compound.

Support was provided, in part, by Friends of the National Zoo (FONZ) and a grant from the Bay Foundation administered by the American Association of Zoo Veterinarians. K.L.G. and J.G.H. are partly supported by a grant from the Womens' Committee of the Smithsonian Associates, the Smithsonian Institution, Washington, DC.

REFERENCES

Berndtson, W.E., T.T. Olar and B.W. Pickett. 1981. Correlation between post thaw motility and acrosomal integrity of bovine sperm. J. Dairy Sci. 64:346-49.

Biggers, J.D., W.K. Whitten and D.G. Whittingham. 1971. In Methods in Mammalian Embryology, ed. J.C. Daniel, Jr. San Francisco: W.H. Freeman and Co.

Bonney, R.C., H.D.M. Moore and J.M. Jones. 1981. Plasma concentrations of oestradiol-17^XB and progesterone and laparoscopic observations of the ovary in the puma (Felis concolor) during oestrus, pseudopregnancy and pregnancy. J. Reprod. Fert. 63:523-31.

Bowen, M.J., C.C. Platz, C.D. Brown and D.C. Kraemer. 1982. Successful artificial insemination and embryo collection in the African lion (Panthera leo). Am. Assn. Zoo Vet. Ann. Proc.

Bowen, R.A. 1977. Fertilization in vitro of feline ova by spermatozoa from the ductus deferens. Biol. Reprod. 17:144-47.

Carter, K.K., P.K. Chakraborty, M. Bush and D.E. Wildt. 1984. Effects of electroejaculation and ketamine-HCl on serum cortisol, progesterone and testosterone in the male cat. J. Androl. 5:431-37.

Chakraborty, P.K., D.E. Wildt and S.W.J. Seager. 1979. Serum luteinizing hormone and ovulatory response to luteinizing hormone-releasing hormone in the estrous and anestrous domestic cat. Lab. Anim. Sci. 29:338-44.

Collu, R., W. Gibb and J.R. Ducharme. 1984. Effects of stress on gonadal function. J. Endocrin. Invest. 7:529-37.

Dresser, B.L., L. Kramer, B. Reece and P.T. Russel. Induction of ovulation and successful artificial insemination in a Persian leopard (Panthera pardus saxicolor). Zoo Biol. 1:55-57.

Goodrowe, K.L., P.K. Chakraborty and D.E. Wildt. 1985. Pituitary and gonadal response to exogenous LH-releasing hormone in the male domestic cat. J. Endocrin. 105:175-81.

Goodrowe, K.L., J.G. Howard and D.E. Wildt. 1986. Embryo recovery and quality in the domestic cat: Natural versus induced estrus. Theriogenology 25:156.

Hahn, J., R.H. Foote and G.E. Seidel, Jr. 1969. Testicular growth and related sperm output in dairy bulls. J. Anim. Sci. 29:41-7.

Hamner, C.E., L.L. Jennings and N.J. Sojka. 1970. Cat (Felis catus L.) spermatozoa require capacitation. J. Reprod. Fert. 23:477-80.

Howard, J.G., D.E. Wildt, P.K. Chakraborty and M. Bush. 1983. Reproductive traits including seasonal observations on semen quality and serum hormone concentrations in the Dorcas gazelle. Theriogenology 20:221-34.

Howard, J.G., M. Bush and D.E. Wildt. In press 1986. In Current Therapy in Theriogenology, ed; D. Morrow. Philadelphia: W.B. Saunders Co.

Howard, J.G., M. Bush, V. de Vos, M.C. Schiewe, V.G. Pursel and D.E. Wildt. In press. Influence of cryoprotective diluent on post-thaw viability and acrosomal integrity of spermatozoa of the African elephant (Loxodonta africana). J. Reprod. Fert.

Kraemer, D.C., B.L. Flow, M.D. Schriver, G.M. Kinney and J.W. Pennycook. 1979. Embryo transfer in the nonhuman primate, feline and canine. Theriogenology 11:51-62.

Moberg, G.P. 1984. Adrenal-pituitary interactions: Effects on reproduction. Proc. 10th Int. Cong. Anim. Reprod. Art. Insem. 4:129-36.

Nagase, H. and T. Niwa. 1964. Deep freezing bull semen in concentrated pellet form. Proc. 5th Int. Cong. Anim.

Reprod. Art. Insem. 4:410.
Niwa, K., K. Ohara, Y. Hosoi and A. Iritani. 1985. Early events of in vitro fertilization of cat eggs. J. Reprod. Fert. 74:657-60.
Nowak, R.M. and J.L. Paradiso. In: Walker's Mammals of the World, 4th ed., p. 1088. The Johns Hopkins Univ. Pr. Baltimore: 1983.
O'Brien, S.J., D.E. Wildt, D. Goldman, C.R. Merril and M. Bush. 1983. The cheetah is depauperate in genetic variation. Science 221:459-62.
O'Brien, S.J., M.E. Roelke, L. Marker, A. Newman, C.W. Winkler, D. Meltzer, L. Colly, J. Everman, M. Bush and D.E. Wildt. 1985. Genetic basis for species vulnerability in the cheetah. Science 227:1428-34.
Overstreet, J.W. 1983. In Mechanism and Control of Animal Fertilization, ed. J.F. Hartmann. New York: Academic Press.
Phillips, L.G., L.G. Simmons, M. Bush, J.G. Howard and D.E. Wildt. 1983. Gonadotropin regimen for inducing ovarian activity in captive-wild felids. J. Am. Vet. Med. Ass. 181:1246-50.
Platz, C.C., D.E. Wildt and S.W.J. Seager. 1978. Pregnancies in the domestic cat using artificial insemination with previously frozen spermatozoa. J. Reprod. Fert. 52:279-82.
Pursel, V.G. and L.A. Johnson. 1974. Glutaraldehyde fixation of boar spermatozoa for acrosomal evaluation. Theriogenology 1:63-68.
Pursel, V.G., L.A. Johnson and G.B. Rampacek. 1972. Acrosomal morphology of boar spermatozoa incubated before cold shock. J. Anim. Sci. 34:279-83.
Ramaley, J.A. 1981. In Environmental Factors in Mammal Reproduction, ed; D. Gilmore and B. Cook. Baltimore: University Park Pr.
Reed, G., B. Dresser, B. Reece, L. Kramer, P. Russell, K. Pindell and P. Berringer. 1981. Superovulation and artificial insemination of Bengal tigers (Panthera tigris) and an interspecies embryo transfer to the African lion (Panthera leo). Am. Assn. Zoo Vet. Ann. Proc.
Rose, R.M. and E. Sachar. 1984. In Textbook of Endocrinology, ed. R.M. Williams. Philadelphia: W.B. Saunders Co.
Seal, U.S., E.D. Plotka, J.D. Smith, F.H. Wright, N.J. Reindl, R.S. Taylor and M.F. Seal. 1985. Immunoreactive luteinizing hormone, estradiol, progesterone, testosterone and androstenedione levels during the breeding season and anestrus in Siberian tigers. Biol. Reprod. 32:361-68.
Seal, U.S. and T. Foose. 1983. Development of a masterplan for captive propagation of Siberian tigers in North American zoos. Zoo Biol. 2:241-44.
Sojka, N.J., L.L. Jennings and C.E. Hamner. 1970. Artificial insemination in the cat (Felis catus L.). Lab. Anim. Care 20:198-204.
Tischner, M. 1979. Evaluation of deep-frozen semen in stallions. J. Reprod. Fert., Suppl. 27:53-59.
Wildt, D.E., G.M. Kinney and S.W.J. Seager. 1978. Gonadotropin-induced reproductive cyclicity in the domestic cat. Lab. Anim. Sci. 28:301-07.
Wildt, D.E., C.C. Platz, S.W.J. Seager and M. Bush. 1981. Induction of ovarian activity in the cheetah (Acinonyx jubatus). Biol. Reprod. 24:217-22.
Wildt, D.E., M. Bush, J.G. Howard, S.J. O'Brien, D. Meltzer, A. van Dyk, F. Ebedes and D.J. Brand. 1983. Unique seminal quality in the South African cheetah and a comparative evaluation of the domestic cat. Biol. Reprod. 29:1019-25.

Wildt, D.E., D. Meltzer, P.K. Chakraborty and M. Bush. 1984. Adrenal-testicular-pituitary relationships in the cheetah subjected to anesthesia/electroejaculation. Biol. Reprod. 30:665-72.

Wildt, D.E., P.K. Chakraborty, D. Meltzer and M. Bush. 1984. Pituitary and gonadal response to LH releasing hormone administration in the female and male cheetah. J. Endocrin. 101:51-56.

Wildt, D.E., M.C. Schiewe, P.M. Schmidt, K.L. Goodrowe, J.G. Howard, L.G. Phillips, S.J. O'Brien and M. Bush. 1986. Developing animal model systems for embryo technologies in rare and endangered wildlife. Theriogenology 25:33-51.

Wildt, D.E., J.G. Howard, L.L. Hall and M. Bush. In press. The reproductive physiology of the clouded leopard. I. Electro-ejaculates contain high proportions of pleiomorphic spermatozoa throughout the year. Biol. Reprod.

Wildt, D.E., J.G. Howard, P.K. Chakraborty and M. Bush. In press. The reproductive physiology of the clouded leopard. II. A circannual analysis of adrenal-pituitary-testicular relationships during electroejaculation or after an adreno-corticotropic hormone challenge. Biol. Reprod.

24

Evaluation of Tiger Fertility by the Sperm Penetration Assay

Hugh C. Hensleigh, Ann P. Byers, Gerald S. Post, Nicolas Reindl, Ulysses S. Seal, and Ronald L. Tilson

INTRODUCTION

The sperm penetration assay (SPA) is a diagnostic test that measures the ability of sperm to fertilize ova. Described by Yanagamachi et al. in 1976, the SPA has been used extensively in humans and has been shown to be correlated with fertility (Zausner-Guelman et al. 1981, Overstreet et al. 1980)and with fertilization rates observed in human in vitro fertilization (Wolf et al. 1983). The SPA has also been used to assess fertility in the bull and stallion (Brackett et al. 1982). During the 1985 breeding season we tested Siberian tiger sperm using the SPA in collaboration with the Minnesota Zoo (Post et al. 1986).

The goal of the present work is to determine the optimal conditions for testing Siberian tiger sperm as part of a captive breeding program. The semen analysis is the basis of evaluation of male fertility; however, use of a functional test like the SPA greatly increases one's ability to differentiate between normal and sub-fertile individuals (Aitken et al. 1984). Complete assessment of male fertility is especially important in a breeding program where choices are being made between males. If cryopreservation of sperm is being used, the SPA is useful to evaluate the sperm before freezing and after thawing.

METHODS

Three Siberian tigers housed at the Minnesota Zoo were used in this study (Table 1). Two were proven breeders, #862 and #2832; the third animal, #1159, has not been used in the breeding program and thus was more available for semen collection. Semen samples were collected from anesthetized animals by electroejaculation. Each fraction was collected with 20 pulses at five volt intervals between 25 and 40 volts. After a five minute rest interval another fraction was collected; two or three fractions were collected on each collection day. Semen samples were transported to the laboratory in a styrofoam container of water at 37 to 32 $^\circ$C.

Table 1.   Breeding record of the three Siberian tigers used in this study.

Tiger ID	Age (yr)	Litters Sired
1159	9	untested
862	13	5 (15 cubs)
2832	11	3 (10 cubs)

A routine semen analysis was done for each fraction. Estimates of percent motile sperm were done using a phase microscope at 100x magnification. Blom stained specimens were counted for percent viable sperm and percent sperm with normal morphology. The SPA was done as described by Yanagamachi et al. (1976) with the following modifications.

Sperm Preparation

Fractions were combined for testing (initial tests indicated fractions gave similar fertilization rates). If samples were washed, an equal volume of culture medium (BWW, Biggers et al. 1971) was added to the sample, the sample centrifuged at 300xG for 10 minutes, and the pellet of sperm resuspended in BWW to the volume of the original sample. Control samples were centrifuged and resuspended in seminal plasma. Incubations were done in a 5% CO_2 incubator at $37^{\circ}C$ or at $23^{\circ}C$ in air.

Collection of Hamster Ova

Hamsters were super-ovulated with pregnant mare's serum gonadotropin (PMS) and human chorionic gonadotropin (HCG) to increase ova yield. Forty I. U. of PMS was given intra-peritoneally on the morning of post-estrus (day of sticky vaginal mucus), 25 I.U. HCG was given 60 hours later, and ova were collected from the oviducts the following morning. Hamsters were killed by ether anesthesia, oviducts were removed, placed in a culture dish containing BWW, and incised to release the cumulus masses and ova. Cumulus masses were dispersed from the ova by incubation in 0.5% hyaluronidase; zona pellucidae were dissolved by incubation in 0.5% trypsin. Ova were washed three times in BWW following enzyme incubations.

Insemination and Scoring of Test

Ten to 20 ova were placed in 20 microliter droplets of BWW under mineral oil; 20 microliters of the sperm preparation were added to the droplet giving a final concentration of 25 million motile sperm per ml. Following a three hour incubation, ova were washed three times in BWW, placed on a microscope slide in 20 microliters of medium, and flattened by placing a cover slip on the droplet so that the structures in the cytoplasm were easily visible. Ova were examined for the presence of decondensed sperm

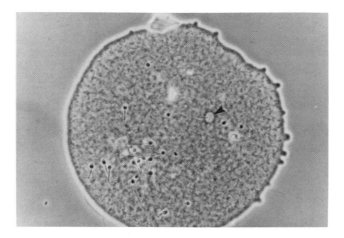

Fig. 1.   Hamster ova penetrated by Siberian tiger sperm (arrow).
The penetrating sperm has undergone decondensation of the DNA and
is several times bigger than the sperm attached to the surface of
the ova.

in the cytoplasm. Decondensation of the sperm head is the first
change in the sperm following penetration of the ova. The
resulting swollen sperm head is about five times the diameter of
sperm seen on the surface of the ova (Fig. 1). Penetration is
reported as a percentage of ova penetrated by one or more sperm.

RESULTS

     Semen analysis of the three animals used in this study for
the 1985 and 1986 breeding season is summarized in Table 2;
values given are the mean values for each fraction. The volume,
concentration and motility was remarkably consistent within
animals.

     The average volume of each fraction ranged between 3 and 6
ml, concentration ranged from 3 to 49 million per ml, and
motility ranged from 40 to 75%. Tiger # 1159 was not being used
in the current breeding program and was therefore more available
for collection (Table 2c); the discrepancy in sperm
concentrations between 1985 and 1986 may be due in part to
repeated collections during 1986. The data for this animal for
1986 is summarized by fraction in Table 3. There was a tendency
toward greater volume in fraction 2 and a higher count in
fraction 1 although the differences in these values were not
statistically different. The motility, viability, morphology,
and number of white blood cells (WBC) was very consistent between
fraction one and two. Only one fraction three was collected in
this series and it had a lower volume and concentration but a
higher motility and viability than the fraction one or fraction
two samples. The number of WBC was consistently low (range=0 to
$400/mm^3$).

Table 2.   Average semen analysis data for the 1985 and 1986 breeding season.

Year	Collections	Fractions	Volume (ml)	No. Sperm $(x10^6/ml)$	Motility (%)
2a. Tiger #862					
1985	1	2	6.0	15	75
1986	1	2	5.5	21	62
2b. Tiger #2832					
1985	1	2	3.3	2.9	40
1986	1	2	3.2	5.5	48
2c. Tiger #1159					
1985	1	3	5.0	6	70
1986	5	11	5.9	49.1	70

Sperm must undergo capacitation prior to fertilization (Austin 1951, Chang 1951). This change normally occurs in the female reproductive tract but can also occur in vitro (Bedford et al. 1978). Human sperm will capacitate if washed and incubated in culture medium for three hours (Sher et al. 1984).

To determine the optimal preincubation time for tiger sperm, samples were incubated for 0, 2, and 4 h prior to testing in the SPA (Table 4). None of the ova were penetrated by sperm that was not preincubated, while 31% of the ova were penetrated by sperm preincubated for 2 h. After 4 h preincubation the penetration rate was 4%; the motility of this sample was also greatly reduced and may be indicative of the functional condition of the sperm.

In human sperm, motility can be preserved by incubation at room temperature (23°C). Tiger sperm preincubated at 23°C for 2h

Table 3.   Semen analysis data for tiger #1159 by fraction for the 1986 breeding season (Mean ± SD).

Fraction	N	Volume (ml)	Conc. $(x10^6/ml)$	Motility (%)	Viability (%)	Morphology (%)	WBC $(\#/mm^3)$
1	5	3.9±0.8	64±24	64±3	64±8	80±2	170±83
2	5	8.6±3.7	39±29	72±8	65±8	84±4	80±56
3	1	2.5	24	90	79	84	0
Total/Means	11	5.9±1.6	49±16	70±4	66±4	82±2	125±44

Table 4.   Effect of preincubation of sperm at 37°C prior to sperm penetration assay.

Preincubation (h)	No. SPA's	No. Ova	Ova Penetrated No.	%
0	3	76	0	0
2	3	96	30	31
4	1	22	1	4

were tested to determine if these conditions are adequate for capacitation. None of the ova inseminated with sperm preincubated at 23°C were fertilized while 24% of the ova inseminated with sperm preincubated at 37°C were fertilized (Table 5).

In most species, capacitation occurs only after washing of the semen sample to remove the seminal plasma (Van der Ven et al. 1982, Kanwar et al. 1979). Samples of tiger sperm were washed in BWW and resuspended in BWW prior to the SPA to determine if fertility would be enhanced by increased capacitation. Control samples were also centrifuged but were resuspended in seminal plasma. Washed and unwashed samples were preincubated at 37°C. Penetration rates for the two groups were virtually identical, about 30% (Table 6).

CONCLUSIONS

Siberian tiger sperm collected by electroejaculation is capable of fertilizing zona pellucida-free hamster ova. From the semen analysis data, little difference was seen in the quality of the first and second fractions and at least two fractions should be collected when an animal is put down for electroejaculation. In the two cases when a third fraction was collected it was comparable to the first two; a bit lower in concentration but higher in motility and viability.

Table 5.   Effect of temperature during two hour preincubation.

Temperature (°C)	No. SPA's	No. Ova	Ova Penetrated No.	%
23	1	21	0	0
37	2	54	13	24

Table 6. Effect of washing sperm prior to a 2h preincubation and sperm penetration assay.

Treatment	No. SPA's	No. Ova	Ova Penetrated No.	%
Not washed	2	66	21	32
Washed	1	30	9	30

The optimal conditions for capacitation of tiger sperm as indicated by the SPA was a preincubation of 2h at $37^{\circ}C$. Preincubation at $23^{\circ}C$ was not sufficient for capacitation although motility was maintained at a higher rate at this temperature. The presence of seminal plasma did not inhibit capacitation as indicated by the SPA.

This work suggests more questions than it answers. Can capacitation take place at $23^{\circ}C$ with longer incubation? Is there a difference in fertility of different fractions from one collection? Can fertilization itself take place in the presence of tiger seminal plasma? And the big question: does the SPA predict tiger fertility? To answer this last question sperm samples from animals known to be infertile are needed. The other area that will be of interest is use of the SPA with frozen sperm samples because this will enable samples frozen at any location to be shipped to a central laboratory where testing can be done to establish the fertility of males for an internationally coordinated breeding program. The SPA has been reported to work well with human samples that had been cryopreserved (Urry et al. 1983). In our preliminary work, tiger sperm was capable of penetrating hamster ova following cryopreservation. Further work is needed to determine the comparative fertility of fresh and cryopreserved tiger spermatozoa.

REFERENCES

Aitken, R.J., F.S. Best, P. Warner and A. Templeton. 1984. A prospective study of the relationship between semen quality and fertility in cases of unexplained infertility. J. Androl. 5:297-303.
Austin, C.R. 1951. Observations of the penetration of sperm into the mammalian egg. Austr. J. Sci. Res. B4:581-89.
Bedford, J. and G. Cooper. 1978. Membrane fusion events with fertilization of vertebrate eggs. Cell Surface Reviews. 3.
Biggers, J., W. Whitten and D. Whittingham. 1971. The culture of mouse embryos in vitro. In Methods in Mammalian Embryology, ed. J. Daniel, San Francisco: W. H Freeman and Co.
Brackett, B.G., M.A. Cofone, M.L. Boice and D. Bousquet. 1982. Use of zona-free hamster ova to assess sperm fertilizing ability of bull and stallion. Gamete Res. 5:217-27.
Chang, M.C. 1951. Fertilizing capacity of spermatozoa deposited into the fallopian tubes. Nature 168:697.

Kanwar, K., R. Yanagimachi and A. Lopata. 1979. Effects of human seminal plasma on fertilizing capacity of human spermatozoa. Fertil. Steril. 31:321.

Post, G.S., H.C. Hensleigh, U.S. Seal, T.J. Kreeger, N.J. Reindl and R.L. Tilson. In Press. Penetration of zona-free hamster ova by Siberian tiger sperm: a first for the felidae. Zoo Biol.

Overstreet, J.W., R. Yanagamachi, D.F. Katz, K. Hayashi and F.W. Hanson. 1980. Penetration of human spermatozoa into the human zona pellucida and the zona-free hamster egg: a study of fertile donors and infertile patients. Fertil. Steril. 33:534-42.

Urry, R.L., D.T. Carrell, D.B. Hull, R.G. Middleton and M.C. Wiltbank. 1983. Penetration of zona-free hamster ova and bovine cervical mucus by fresh and frozen human spermatozoa. Fertil. Steril. 39:690-94.

Van der Ven, H., A. Bhattacharyya, Z. Binor, S. Leto and L. Zaneveld. 1982. Inhibition of human sperm capacitation by a high-molecular-weight factor from human seminal plasma. Fertil. Steril. 38:753-55.

Wolf, D.P., J.E. Sokoloski and M.M. Quigley. 1983. Correlation of human in vitro fertilization with the hamster egg bioassay. Fertil. Steril. 40:53-59.

Yanagimachi, R., H. Yanagimachi and B.J. Rogers. 1976. The use of zona-free animal ova as a test-system for the assessment of the fertilizing capacity of human spermatozoa. Biol. Reprod. 15:471-76.

Zausner-Guelman, B., L. Blasco and D.P. Wolf. 1981. Zona-free hamster eggs and human sperm penetration capacity: a comparative study of proven fertile donors and infertile patients. Fertil. Steril. 36:771-77.

25

Artificial Insemination and Embryo Transfer in the Felidae

B.L. Dresser, C.S. Sehlhorst, G. Keller, L.W. Kramer and B. Reece

INTRODUCTION

Although reproduction in animals normally progresses without artificial intervention, there has become an urgent need for the development of techniques designed to address specific aspects of reproduction in wildlife species, particularly in those which are endangered. Techniques developed for use in domestic animals are gradually being applied to zoo animals but with limited success.

The potential for techniques such as artificial insemination becomes apparent when the risks of having to transport large, dangerous or rare animals are considered. Likewise, embryo transfer, a technique in which a fertilized ovum is transferred to a recipient mother for the tenure of gestation and ultimate birth, has great potential for ensuring the continuity of particular genetic lines and bringing isolated animals into a breeding program. Through freezing, this technique also allows for the preservation of genes for future use in different populations. Aged females, no longer contributors to a breeding group, may still remain a valuable source of eggs or embryos for transplantation. Inferior females, whose genetic qualities are undesirable, are still able to nurture a transplanted embryo through gestation and function in a surrogate role.

Methods also need to be developed whereby the number of offspring from an endangered species can increase by the use of a more common member of another closely related species through the determination of surrogate mothers (interspecies embryo transfer). Since conservation of endangered species is no longer a prerogative but a responsibility, research must be designed to insure the propagation and survival of wildlife.

DOMESTIC FELIDAE

Artificial Insemination

Artificial insemination is possible in domestic felidae with

both fresh (Sojka et al. 1970) and frozen semen (Platz et al.
1978). Domestic cats serve as good experimental animal models
for studying methods of artificially controlling reproduction in
exotic felids since their physiology and anatomy are very
similar. It is a prerequisite of success that the queen be in
estrus, which can be partly verified by examination of a vaginal
smear and partly by the behavior when the neck and pelvic region
are touched (Christiansen 1984).

Since domestic cats and most exotic cats (Wildt et al. 1980)
are induced ovulators, induction of ovulation must be achieved by
a copulation stimulus or through an exogenous source of
luteinizing hormone. Thus, ovulation may be induced by mating
with a vasectomized male or by injection of human chorionic
gonadotropin (HCG) on days 1 and 2 of estrus (Platz et al. 1978).
Ovulation occurs within 25-26.6 hours of an intramuscular or
intravenous injection of HCG (Sojka et al. 1970).

Semen from domestic cats for artificial insemination can be
collected by electroejaculation (Platz and Seager 1978, Ball
1976, Dooley et al. 1983, Dooley and Pineda 1986, Pineda and
Dooley 1984, Pineda et al. 1984, Seager 1977) or with an
artificial vagina (Sojka et al. 1970).

Fresh Semen

Fresh semen for artificial insemination in cats can be
diluted (extended) immediately after collection with an isotonic
saline solution to a volume of 1 ml, and 0.1 ml of this dilution
used for insemination. Undiluted semen may also be used if the
samples collected are very low in sperm concentration. Even
though pregnancy may be achieved after insemination with 1.25 x
10^6 sperm, the number of sperm per insemination dose should be at
least 5 x 10^6 for routine work. The semen is deposited deep into
the vagina or into the cervix using a 9 cm long 20-gauge cannula
tipped with polyethylene tubing (Sojka et al. 1970) or with a 22
gauge lavage needle bulb tipped with silver solder as determined
and employed by the present authors in ongoing artificial
insemination projects with cats. The insemination should be
synchronized with hormonal treatment for induction of ovulation
or within 27 hours of its administration. A conception rate of
50% has been achieved after synchronized insemination, but the
rate increases to 75% if the insemination is repeated 24 hours
later simultaneously with a supplementary injection of 50 i.u.
HCG (Sojka et al. 1970). Fertilization can take place when the
sperm are introduced into the vagina up to 49 hours after an
ovulation-inducing injection of HCG, whereas no fertilization
results after matings that occur 50 hours or more after this
injection, that is approximately 24 hours after ovulation (Hamner
et al. 1970).

Frozen Semen

The diluent that has been used successfully for freezing
sperm which resulted in the birth of live kittens consists of
deionized water containing 20% (v/v) egg yolk, 11% (w/v) lactose
and 4% (v/v) glycerol, together with streptomycin sulphate 1000
ug and penicillin G 1000 i.u./ml (Platz et al. 1978, Platz et al.
1976). Immediately after collection the semen is mixed with 200
ul sterile 0.9% saline and 200 ul diluent at room temperature

($22-23^{\circ}C$). Motility and sperm content must be judged before further dilution to a proportion of at least 1:1 (semen to diluent, v/v), and after an equilibration period of 20 minutes at 5° an additional 200 ul diluent plus an amount equivalent to the collected ejaculate volume are added. Then the sample is frozen in pellets, transferred to, and stored in vials in liquid nitrogen.

A high recovery rate can be obtained from rapid thawing pellets in 0.154 M NaCl at 37° (Christiansen 1984), or in Tyrode's solution at 37° (Sehlhorst et al. 1985).

Freezing of domestic cat frozen semen in straws in liquid nitrogen vapor has been reported by these authors. When dealing with very small ejaculate volumes such as in the domestic cat, extended semen was frozen in as little as 0.05 ml aliquots by separating it via air bubbles from the upper and lower portions of the 1/4 ml straw which were filled with extender alone. This technique allows the small sample to be aliquoted into several straws and, thus, permits future evaluation of a representative sample without sacrifice of the entire frozen ejaculate, as well as greater experimentation in freezing/thawing procedures per collection. To date, no significant difference between thawed samples frozen in straws or by pelleting methods has been found by these authors (Sehlhorst et al. 1985).

For artificial insemination with frozen semen, the number of spermatozoa should be $50-100 \times 10^{6}$ motile spermatozoa, so two or more ejaculates are required. As the insemination dose is only 0.1 ml, concentration by centrifugation of the thawed semen is needed, and removal of some of the diluent. The female cat is anesthetized and placed in dorsal recumbency with the hindquarters elevated during, and until 20 minutes after the insemination (Christiansen 1984).

The freezing and thawing procedures do not seem to significantly lower the viability and motility of the spermatozoa. In ejaculates collected into an artificial vagina, and by use of electroejaculation, the motility was found to be 83 and 70% respectively immediately after collection, and after freezing and thawing 71 and 54% (Platz et al. 1978). In spite of these rates, only six pregnancies resulted after artificial insemination of 56 cats, probably due to a lack of exact synchronization of insemination and ovulation. The litter sizes in these pregnancies were 1, 1, 2, 2, 2, 2, and 4, and all kittens were normally developed with normal birth weights.

Embryo Transfer

Embryo transfer, surgically performed in the natural cycle, has succeeded in cats. The synchronization was with $\pm$ one day and ovulation was induced by either injection of HCG or by natural mating. In two studies (Kraemer et al. 1979, Kraemer et al. 1980), 47 embryos were recovered in 17 collections. In nine transfer experiments, all recovered embryos were transferred, resulting in four pregnancies diagnosed and four offspring. In another study (Goodrowe et al. 1986), 43 embryos were recovered from cats in natural estrus resulting in one litter of live kittens.

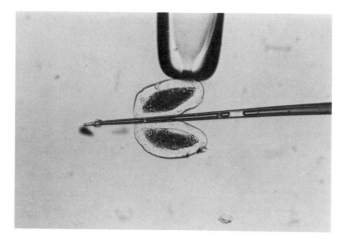

Fig. 1.   Domestic cat embryo being split by micromanipulation for the production of identical twins.

Embryo transfer, surgically performed in cats with a hormonally induced cycle using FSH-P, has also been successful. One study (Goodrowe et al. 1986) did not use HCG to induce ovulation and a total of 82 embryos were recovered with one litter live born.   An ongoing study in the laboratories of the present authors using both FSH-P and HCG has resulted in at least 600 embryos which are being used in cryopreservation and micro-manipulation experiments.   At least 40 of the fresh embryos have been transferred resulting in three litters.   We have recovered as many as 75 embryos from one cat during one collection using FSH-P and HCG.

To date, no kittens have been produced from frozen embryos although attempts are now being made in several laboratories in the United States. The present authors have evaluated frozen-thawed embryos that, when put into culture, continued to develop to hatched blastocysts.   Failures at transfer are probably due to improper synchronization of recipients and attempts are being made to improve this aspect.

Embryo Splitting

Micromanipulation of cat embryos for splitting, in efforts to produce identical twins, is also being attempted by these authors.   Demi-embryos have successfully been produced in culture and transferred.   However, these transfers have not yet resulted in pregnancies.

In Vitro Fertilization

In vitro fertilization of cat eggs has been successful both by sperm capacitated in vivo (Hamner et al. 1970) and in vitro (Bowen 1977) but no transfers of these embryos have been made.

NON-DOMESTIC FELIDAE

Artificial Insemination

Artificial insemination has not been very successful in nondomestic species although many attempts have been made. Only 19 different nondomestic mammalian species have been reported in the literature or by personal communication to date to have successfully resulted in offspring due to artificial insemination (Dresser 1983). In most cases, the procedure was not repeated so that only one animal or one litter was actually produced within the species. Of these 19 species, only two are felidae. Numerous attempts have been made to inseminate cheetah, Bengal and Siberian tigers, North Chinese leopards, African lions, margay, ocelots, clouded leopards, snow leopards, puma (Phillips 1981, Bowen et al. 1982, Dresser et al. 1981, Watson 1978).

In 1981, the Zoological Society of London reported the birth of a puma (Felis concolor) as the first wild cat to be produced by artificially induced ovulation and artificial insemination (Moore et al. 1981a, Moore et al. 1981b). Induction of ovulation was achieved using pregnant mare's serum gonadotropin (PMSG) and HCG. Two intrauterine inseminations were made in three pumas by laparotomy resulting in one conception and a single live birth, thus making this a surgical procedure.

In 1982, the Cincinnati Wildlife Research Federation team was successful in the first nonsurgical approach to artificial insemination in exotic felidae when they inseminated a Persian leopard (Panthera pardus saxicolor) by depositing semen in the uterus through a vagina (Dresser et al. 1982). This leopard cycled naturally but ovulation was induced with HCG. One offspring resulted but the birth was breech and the cub died during parturition. However, the procedure was successful and has served as a model for nonsurgical artificial insemination in nondomestic felidae.

Artificial insemination has also been used for the production of embryos to transfer between nondomestic felidae. This procedure has resulted in embryos from African lions (Panthera leo) (Bowen et al. 1982) and Bengal tigers (Panthera tigris) (Dresser et al. 1981, Reed et al. 1981).

Fresh Semen

Semen has been collected by electroejaculation, artificial vagina and manual stimulation in over 100 species of zoo animals (Seager et al. 1980, Seager 1981). Electroejaculation is the method of choice, due to the need for restraint of most all nondomestic species, even though it does not always allow the best sample. The collection of fresh semen from nondomestic felids has been mostly for fertility evaluation (Phillips 1981, Seager 1974, Seager 1976, Ott et al. 1981, Velhankar et al. 1967, Wildt et al. 1983). Fresh semen has been used in most all artificial insemination in the puma and Persian leopard, fresh undiluted semen was used. There have been only a few comments in the literature as to the best type of semen extender to use for non-domestic felidae but none of these extenders have yet been used in successful pregnancies. Most extenders used with nondomestic cat semen are some variation of the one used for domestic cat semen (Platz et al. 1978). Extended semen was used by these authors to prolong Bengal tiger sperm viability when samples were transported long distance in a joint project between the Knoxville Zoo and Kings Island Wild Animal Habitat in Ohio (Reed et al. 1981). This extended semen resulted in embryos after being used for artificial insemination. The semen extender was a commercially available product (Laiciphas, L'Aigle, France) that contained fresh egg yolk. Other reports in the literature where felid semen is mentioned, do not discuss extenders unless samples are to be frozen.

Frozen Semen

Nondomestic felid semen has been frozen in pellets, straws and ampules. Although many samples have been frozen and thawed, followed by artificial insemination attempts, there have never been any reported pregnancies from frozen semen in nondomestic felids. Most nondomestic cat semen is frozen with glycerol as the cryoprotectant of choice. Studies by these authors (Sehlhorst et al. 1985) as well as personal communications from others have indicated that sperm from cheetah, snow leopard, Siberian and Bengal tigers, African and Asian lions, Jungle cats and Persian leopards have demonstrated motility at post-thaw but most of these results have not yet been published. "Banked" semen stored at locations throughout North American zoos and laboratories needs to be thawed and evaluated by insemination so that methods for semen freezing can be improved.

Embryo Transfer

There have been no reports of successful embryo transfer in nondomestic felids from fresh or frozen embryos. However, members of the Cincinnati Wildlife Research Federation team successfully recovered embryos from a Bengal tigress by inducing estrus with FSH-P and ovulation with HCG. At laparotomy, two viable embryos in the morula stage were recovered and transferred surgically to the uterus of an African lioness in an attempt to use interspecies embryo transfer as a method to increase the numbers of rarer tigers by using common lions as surrogate mothers. This one transfer did not result in any pregnancies although techniques developed in this project, following refinement, could help to make interspecies embryo transfer possible between closely related felidae (Reed et al. 1981, Reece et al. 1981).

African lion embryos have been collected from one lioness after five attempts (Bowen et al. 1982). Midventral laparotomy was performed following natural estrus and included ovulation using HCG or gonadotropin releasing hormone (GnRH). Three embryos were recovered and frozen for later embryo transfer. As of this writing, there are no reports of these embryos being successfully transferred.

Embryo Splitting

Embryo splitting of nondomestic felid eggs has not been reported.

In Vitro Fertilization

Follicular eggs from African lions have been aspirated by the present authors for technique development prior to in vitro fertilization experiments but further progress has not yet been made. Personal communication has provided information to the present authors on attempts by other investigators to aspirate follicular eggs from felid ovaries at necropsy followed by freezing attempts but no reports have been published on this work and no results have ensued.

ACKNOWLEDGEMENTS

The authors would like to acknowledge and thank Emilie Gelwicks, Karen Wachs, Joyce Turner, Dr. Stanley Leibo and Naida Loskutoff for their significant contributions to portions of the projects mentioned and to Vickie Stidham for her preparation of the manuscript. Projects mentioned were supported in part by the National Museum Act, Institute of Museum Services and the Nixon-Griffis Fund.

REFERENCES

Ball, L. 1976. Electroejaculation. In Applied Electronics for Veterinary Medicine and Animal Physiology, ed. W.R. Klemm. Springfield: C.C. Thomas.

Bowen, R.A. 1977. Fertilization in vitro of feline ova by spermatozoa from the ductus deferens. Biol. Reprod. 17:144-47.

Bowen, M.J., C.C. Platz, Jr., C.D. Brown and D.C. Kraemer. 1982. Successful artificial insemination and embryo collection in the African lion (Panthera leo). Am. Assn. Zoo Vet. Ann. Proc.

Christiansen, I.J. 1984. Reproduction in the Dog and Cat. London: Bailliere Tindall.

Dooley, M.P., K. Murase and M.H. Pineda. 1983. An electroejaculator for the collection of semen from the domestic cat. Theriogenology 20:297-310.

Dooley, M.P. and M.H. Pineda. 1986. Effect of method of collection on seminal characteristics of the domestic cat. Am. J. Vet. Res. 47:286-92.

Dresser, B., L. Kramer, P. Russell, G. Reed and B. Reece. 1981. Superovulation and artificial insemination in Bengal tigers (Panthera tigris), African lions (Panthera leo) and a Persian leopard (Panthera pardus saxicolor). Amer. Assn. Zool. Parks Aqua. Ann. Proc.

Dresser, B.L., L. Kramer, B. Reece and P.T. Russell. 1982. Induction of ovulation and successful artificial insemination in a Persian leopard (Panthera pardus saxicolor). Zoo Biol. 1:55-57.

Dresser, B.L. 1983. Reproductive strategies for endangered wildlife in the eighties: A conference summary. Am. Assn. Zool. Parks Aqua. Ann. Proc.

Goodrowe, K.L., J.G. Howard and D.E. Wildt. 1986. Embryo recovery and quality in the domestic cat: Natural versus induced estrus. Theriogenology 25:156.

Hamner, C.E., L.L. Jennings and N.J. Sojka. 1970. Cat (Felis catus) spermatozoa require capacitation. J. Reprod. Fert. 23:477-80.

Kraemer, D.C., B.L. Flow, M.D. Schriver, G.M. Kinney and J.W. Pennycook. 1979. Embryo transfer in the nonhuman primate, feline and canine. Theriogenology 11:51-62.

Kraemer, D.C., G.M. Kinney and M.D. Schriver. 1980. Embryo transfer in the domestic canine and feline. Arch. Androl. 5:111.

Moore, H.D.M., R.C. Bonney and D.M. Jones. 1981a. Successful induced ovulation and artificial insemination in the puma (Felis concolor). Vet. Rec. 108:282-83.

Moore, H.D.M., R.C. Bonney and D.M. Jones. 1981b. Induction of oestrus and successful artificial insemination in the cougar, (Felis concolor). Am. Assn. Zoo Vet. Ann. Proc.

Ott, J., N. Schaffer, A. Prowten, D.E. Wildt, C. Platz, S. McDonald and S. Seager. 1981. Semen collection (by electroejaculation) and fertility examination in the snow leopard (Uncia uncia). Am. Assn. Zoo Vet. Ann. Proc.

Phillips, L.G. 1981. Artificial insemination team project in exotic large felidae. Am. Assn. Zoo Vet. Ann. Proc.

Pineda, M.H. and M.P. Dooley. 1984. Effects of voltage and order of voltage application on seminal characteristics of electroejaculates of the domestic cat. Am. J. Vet. Res. 45:1520-25.

Pineda, M.H., M.P. Dooley and P.A. Martin. 1984. Long-term study on the effects of electroejaculation on seminal characteristics of the domestic cat. Am. J. Vet. Res. 45:1039-41.

Platz, C., T. Follis, N. Domorest and S. Seager. 1976. Semen collection, freezing and insemination in the domestic cat. Proc. 7th Int. Cong. Anim. Reprod. Art. Insem. Cracow, Poland. 4:1053-56.

Platz, C.C., Jr. and S.W.J. Seager. 1978. Semen collection by electoejaculation in the domestic cat. J. Am. Vet. Med. Assn. 173:1353-55.

Platz, C.C. Jr., D.E. Wildt and S.J.W. Seager. 1978. Pregnancy in the domestic cat after artificial insemination with previously frozen spermatozoa. J. Reprod. Fert. 52:279-82.

Reece, B., B. Dresser, G. Reed, P. Russell, L. Kramer, K. Pindell and P. Berringer. 1981. An interspecies embryo transfer from Bengal tiger (Panthera tigris) to African lion (Panthera leo). Am. Assn. Zool. Parks Aqua. Ann. Proc.

Reed, G., B. Dresser, B. Reece, L. Kramer, P. Russel, K. Pindell and P. Berringer. 1981. Superovulation and artificial insemination of Bengal tigers (Panthera tigris) and an interspecies embryo transfer to the African lion (Panthera leo). Am. Assn. Zoo Vet. Ann. Proc.

Seager, S.W.J. 1974. Semen collection and artificial insemination in captive cats, wolves, and bears. Am. Assn. Zoo Vet. Ann. Proc.

Seager, S.W.J. 1976. Electroejaculation of cats (domestic and captive wild felidae). In Applied Electronics for Veterinary Medicine and Animal Physiology, ed. W.R. Klemm. Springfield: C.C. Thomas.

Seager, S.W.J. 1977. Semen collection, evaluation and artificial insemination of the domestic cat. Curr. Vet. Ther. pp 1252-54.

Seager, S.W.J., D.E. Wildt and C.C. Platz. 1980. Semen collection by electroejaculation and artificial vagina in over 100 species of zoo animals. Proc. 10th Intl. Cong. Anim. Reprod. Art. Insemin, Madrid, Spain. pp 571-79.

Seager, S.W.J. 1981. A review of artificial methods of breeding
    in captive wild species. <u>Dodo</u> (<u>J. Jersey Wildl. Preserv.</u>
    <u>Trust</u>) 18:79-83.
Sojka, N.J., L.L. Jennings and C.E. Hamner. 1970. Artificial
    insemination in the cat (<u>Felis catus</u>). <u>Lab. Anim. Care</u>
    20:198-204.
Sehlhorst, C.S., B.L. Dresser and L. Kramer. 1985. Comparison of
    extenders with glycerol in cryopreservation of semen from
    exotic and domestic felidae. <u>Ohio J. Sci.</u> 23:190.
Velhankar, D.P., C.R. Sane and M.P. Kulkarni. 1967. Spermatozoa
    of tiger--some observations on their morphology. <u>Indian Vet.</u>
    <u>J.</u> 44:315-19.
Wildt, D.E., C.C. Platz, S.W.J. Seager and M. Bush. 1980. Use of
    gonadotropic hormones to induce ovarian activity in domestic
    and wild felids. <u>Am. Assn. Zoo Vets</u>. Ann. Proc.
Wildt, D.E., M. Bush, J.G. Howard, S.J. O'Brien, D. Meltzer, A.
    Van Dyk, H. Ebedes and D.J. Brand. 1983. Unique seminal
    quality in the South African cheetah and a comparative
    evaluation in the domestic cat. <u>Biol. Reprod.</u> 29:1019-25.

26

Chemical Communication in the Tiger and Leopard

R.L. Brahmachary and J. Dutta

"Tiger, Tiger burning bright in the forest of the night"

Not only does the tiger burn bright with night-adapted eyes and ears that hear even the faintest sounds, it also makes good use of its olfactory senses (disputed as they are). One aspect of this is the use of semio-chemicals, or pheromones, in communicating with conspecifics.

The Mysteries of Smell

The chemical basis of smell has been a challenging mystery to unravel. Theoretically, even a single molecule (or very few molecules) can generate the sense of smell. As such, it is difficult to obtain enough material for chemical analysis. Recently developed techniques of gas chromatography, mass spectrophotometry and high performance liquid chromatography allow identification of minute quantities of chemical substances. Again, the binding of odor molecules to receptors is also now being studied. Models such as erythrocyte membranes or algal cell membranes are being used.

The Tiger as a Subject

Added to the difficulty of studying the "molecules of smell" is the formidable problem of investigating the tiger. Until recently, this nocturnal, asocial, cover-loving predator had mostly been studied through the rifle sight. Schaller (1967) initiated a scientific study which is now reaping rich dividends, especially in Royal Chitwan National Park, Nepal, where the Smithsonian group carried on a long-term study (see Mishra this volume, Smith et al. this volume). Today in India these animals can be observed for comparatively long periods of time and sightings can be frequent.

But for obvious reasons, it may not be possible to collect samples for analysis from wild or even zoo tigers, or to present compounds from these samples or their synthetic counterparts to the animals, and to observe their responses. We collected some material from one zoo and from Khairi, a tigress raised as a

house pet in Jashipur, Orissa, by the late S.R. Choudhury and his
cousin Nihar Nalini. We believe that the proper study of tiger
ethology in general, and tiger pheromone in particular, needs a
two pronged attack of observing such "pet" tigers and wild
tigers. Singh (1983) raised another tigress and these two
tigresses, Tara and Khairi, observed closely, helped us
understand flehmen and olfactory behavior.

OLFACTION IN THE TIGER

The olfactory abilities of the tiger, leopard and lion have
been hotly debated. Brahmachary and Dutta (1984) discussed this
subject, and in the present context we summarize what has
emerged. Since Jim Corbett (1944) wrote that "tigers have no
sense of smell," numerous persons repeated or believed in this
statement. Even for hunting prey the tiger may occasionally make
use of olfactory senses, especially in areas of thick cover, (see
McDougal 1978) and the two pet tigresses mentioned above
sometimes followed the trail of prey by smelling. Over and above
the data, which are mostly necessarily anecdotal and not
rigorous, we may now mention a significant fact. One hunter
practically revolutionized tiger shooting in Indochina by
offering rotting carcasses as bait instead of live bait or fresh
carcasses (Perry 1964).

One might be tempted to believe that the cat family, unlike
the dog family, has a blunt sense of smell. But this need be no
more than our ignorance of the cat, which is less trainable or
more independent than the dog. Interestingly, the keen sense of
smell of the domestic cat was employed in the first world war in
order to smell out poison gas long before human soldiers could
detect it (Bedicheck 1960). It may also be mentioned that the
Sanskrit word for tiger (Vvaghra) is derived from "keen-smelling
sense". Whether a tiger hunts by smell or not, it certainly
becomes aware of the scent of another tiger or its own, as
evident from the "flehmen" (Leyhausen 1964). This characteristic
gesture was seen to ontogenetically evolve, i.e. to first appear,
on the first day of the fourth month in a litter of tiger cubs
kept under almost daily observation from the age of one month
onwards. In two leopard cubs we detected this gesture at the age
of three and a half months.

Tigers and leopards "flehmen" at their own scent (urine,
marking fluid), at scent of other tigers/leopards, and many other
odorous substances. It is thus a general gesture which indicates
perception of a smell, but they may also perceive other scent
codes which do not elicit flehmen gesture. This is indicated by
certain findings with leopards (Brahmachary and Dutta 1984). In
spite of this, "flehmen" is perhaps one of the very few
indicators which might be employed in the study of pheromones of
tigers.

Urine/Anal Gland/Marking Fluid

Numerous animals are known to mark their territories. The
territorial markings or signatures of the tiger/leopard/lion are
also part of a compelling instinct. Confusion has been created
while describing these markings. Urine, anal gland secretion and

"marking fluid" have been mentioned as the agents for effecting this marking. We have used the rather noncommittal term-marking fluid, as "urine" and "anal gland secretion" do not convey the right meaning and may also lead to contradiction.

Schaller (1967) wrote the following account on the spraying of a smelly fluid by tigers:
> "This liquid shoots upwards at an angle of thirty degrees, hitting the vegetation three to four feet above ground. Tigresses eject a rather wide spray...On two occasions a tiger stopped and sniffed the scent, grimacing afterward with nose wrinkled and tongue hanging out, a gesture described as "flehmen" by Leyhausen. It was twice possible to examine the fluid closely after a tigress had sprayed some leaves, much of it consisted of a clear, pale yellow liquid apparently urine. Several clumps of a granular, whitish precipitate were in it...Tiger urine by itself, however, did not have a particularly strong odor, whereas this fluid smelled very musky, readily discernible to the human nose at a distance of 10-15 feet. The white precipitate was apparently a secretion from the anal glands (see later)...The scent adhered to the vegetation for a long time. One tuft of grass still smelled pungent to me after one week. On tree trunks the scent persisted for three to four weeks except when it rained heavily."

What Schaller observed twice in wild tigers is a familiar occurrence to the observers of Khairi and Tara. Confusion still exists. For example, Albone (1984) stated that it is not clear from the published account of Brahmachary and Dutta (1979), "whether anal sac secretions, urine, or a mixture of the two was analyzed". Again, Sunquist (1984) writes that tigers spray "urine mixed with anal gland secretions" on trees and other substrates.

The odorous, musky fluid is ejected through the urinary channel, whereas products of the anal gland can be secreted through the anus or anal region only, as there is no connecting channel between the urinary tract and the anal gland (Hashimotoy et al. 1963). If so, the anal gland secretion can mark the faeces only. As, in the tiger, the most prominent and frequent marking is made by spraying from or through the urinary channel, perhaps this should not be referred to as anal gland secretion. The musky odor is always present in the urine but more markedly so in the marking fluid. Two-month old tiger cubs did not generate any such characteristic smell in the urine.

Territoriality and Marking

Numerous mammals have by now been observed to mark their territories/home ranges, and there are different degrees of territoriality in various mammalian species (Ewer 1969). Regarding tigers, it may be mentioned that in earlier times, naturalists and hunters talked of tigers' beats, which indicates a degree of "exclusiveness." Even though the territory of one may overlap with that of another, the animals seek no encounter except on special occasions (Smith et al. this volume).

It seems tigers mark a number of places in their territory and periodically visit and renew the markings. These (and urine) would serve as signals for members of the opposite sex. During the estrous or pre-estrous stage marking by the female seems to increase.

Chemistry of the Marking Fluid

Lipids: Matthews (1969) pointed out that tigers daily excrete lipids (through urine and marking fluid) equivalent to one week's butter ration for the British soldier in the second world war. Our first studies were carried out with the idea that lipids might be the carriers of pheromone, imparting stability to the latter. Also, it might be possible that certain lipids or short chain fatty acids might themselves be the pheromones, for fatty acids are important constituents of pheromones ranging from the insect world to mammals. We thus kept an open mind and at first concentrated only on the separation of the lipids with the help of thin layer chromatography.

The MF has no 10% TCA - precipitable component. Thus it does not contain any soluble protein, and probably no lipid bound protein, either.

Lipids were separated on TLC plates with the help of two solvent systems: 1) chloroform:methanol:water (65:25:4) and; 2) petroleum ether:acetic acid:water (80:20:1.5 or 84:16:1.5). Solvent system 2 was used most often. Detection of spots was made with the help of iodine vapor and also from the fluorescence in the plate put under UV lamp after spraying with 6% rhodamine solution. Generally, five to six spots were clearly visible. There may be quantitative variation in the patterns depending on the season, but so far no oestrous-specific (or any season-specific) component has been detected. With higher quantities (50 ul or more) applied, all the spots are visible in samples from all seasons. It was also noticed that no smell was perceptible on TLC plates after these were dried. It has also been confirmed that the chloroform-methanol (2:1) extract of the marking fluid smells strongly after the solvent evaporates, but that the smell is no longer perceptible after the dry residue is treated with petroleum ether and the solvent allowed to evaporate off.

The volatile part: The urinary and/or marking volatiles are apparently similar. Theoretically, the pheromones should be sought at this level (Brahmachary and Dutta 1979, 1984). Some of the urinary volatiles are amines and aldehydes. Phenylethylamine, cadaverine and putrescine were found in both tiger and leopard. The first occurred in comparatively large quantities, for it was not detectable in comparable volumes of steam distillates of urine of some other cats (see below). Marking fluid/urine and their steam distillates elicit flehmen reaction from tigers and leopards.

Leopards: A pair of leopard cubs was raised from the age of three months. They were trained to urinate on the lawn at nearly fixed hours. From about the age of 3 1/2 months they were observed to "flehmen" and their urine had the characteristic musky smell.   Part of the urine could be collected in a clean glass beaker before it fell on the soil.   Urine samples up to the age of one year were collected.

Amine fraction: It was established (Brahmachary and Dutta 1981) that the volatile scent-bearing part of the steam distillate of tiger marking fluid/urine consists of several amines of which phenylethylamine is a very prominent member, although this is absent, that is, undetectably small, in the domestic cat, the golden cat and the lion using comparable quantities of urine.   So, the first attempt was to search for this amine by following the identical procedure, namely converting the amines in the steam distillate into the corresponding hydrochlorides and separating these with the help of paper chromatography.   In this manner, and using the standard reference, B - phenylethylamine has been found to prominently feature in leopard pheromones.   Putrescine, too, could be similarly identified.

Non-amine fraction: Fresh steam distillates were acidified (so that the alkaline medium just reached the acidic PH) and extracted with diethyl ether.   The upper ether layer containing the non-amine fraction (NAF), on drying, was found to be strong smelling.   Thus, in the leopard a considerable part of the odor (and possibly pheromonal activity) resides in the non-amine fraction.

2,4-DNP hydrazones from the NAF were prepared and purified by standard methods.   Following Ellis et al. (1958), paper chromatography of the mixed hydrazones was carried out on propylene:glycol impregnated Whatman paper.   The solvent was light petrol (40_f^o-50°):methanol (96:4).   Of the two clearly visible spots, R_f of one coincided with that of the standard acetaldehyde.   The spots also turned pink after spraying the chromatogram with 10% KOH.

The mixed hydrazones, when subjected to preparative TLC on silica gel plate using n-hexane:diethyl ether (70:30) as the solvent, revealed seven bands of which five were too faint to work with.   The diethyl ether extract of the most prominent band, (second from top) was gas-chromatographed on 5% SE30 column on diatomite C and revealed two peaks of the same retention time as those of the hydrazones of acetaldehyde and propionaldehyde. These two components of the second band could be separated on a silicic acid:celite (2:1) column using n-hexane as the extracting solvent, the column fractions being monitored by the GLC.   Mass spectrograms of the two components showed the presence of molecular ions of values 224 and 238, respectively.   These and the mass fragment values indicate 2,4-DNP-hydrazones of acetaldehyde and propionaldehyde, respectively.

The sixth band produced a single peak of retention time inbetween the hydrazones of hexanaldehyde and heptanaldehyde.   At present we cannot identify it, but the possibility that this is an iso-heptanaldehyde cannot be excluded.

Are These Compounds Pheromones?

Despite the present sophistication of chemical techniques for separating molecules, we cannot say that we have isolated the pheromones.   We have only identified some of the urinary and marking fluid volatiles which are likely to function as pheromones. Theoretically, the distinctive body odor of every individual, comparable to fingerprint or histocompatibility, can have its molecular basis in the c̲h̲a̲r̲a̲c̲t̲e̲r̲i̲s̲t̲i̲c̲ r̲a̲t̲i̲o̲s̲ of different odor molecules which must be largely independent of diet. Both Gorman et al. (1974) and we started with this conceptual framework. Gorman et al. (1974, 1976) established this basis with 24 mongoose, each of which had a distinctive ratio of the different carboxylic acids of the anal gland which are utilized for marking, and the mongoose were trained to distinguish the scent marks of individuals. Such work is impossible with tigers or leopards but differences in the amine spots were found in the chromatographs for different tigers (Brahmachary and Dutta 1981). Also, in the case of Khairi, the ratio of two amines was determined to be three:two in two samples separated by a gap of many months. Evidence available from four tigers indicates differences characterizing individual tigers.

Vomer and Non-volatile Molecules

The vomeronasal, or Jacobson's, organ is found in many reptiles and mammals, ranging from the now extinct therapsidans to living elephants. It is absent in the chiroptera, cetacea and primates, but persists in human embryos. This organ was discovered by Ruysch in 1703 (Schilling, 1970) and studied thoroughly by Jacobson in 1811.

The function of this organ long remained a mystery, however. Following the line of thought expressed by Estes (1972) who used the antelope as a model (it can probably be used for other mammals too), the gesture of flehmen forces some of the odor molecules into the vomeronasal organ. The animal, especially the male, then tests the urine with the tongue. Schilling's X-ray film and fluorescent stain (Wysocki et. al. 1980) show that the tongue can transfer non-volatile molecules into the vomeronasal organ. Interestingly, this organ has a nervous connection with that area of the brain which is known to be associated with sexual behavior. It is probable that the function of this organ is to test the sexual status of members of the opposite sex. As early as 1845, Gratiolet (Estes 1972) had a clever hunch regarding this issue.

Since tigers have well developed vomeronasal organs and frequently indulge in licking urine, it is likely that certain non-volatile pheromones are collected and tested through this agency.

Marking Fluid (MF) and Rain

It is known that the smell of MF clings to vegetation for days. Territory markings or sex scents can thus remain functional for a long time. To simulate the conditions of the rainy season a clump of mango leaves was sprayed with MF and left to dry. After the clump was kept under running tap water for 13-16 hours, the lower leaves, dripping wet, still carried a perceptible odor. A very strong direct shower of the tap eliminated the smell from the upper leaves) within four hours. Thus, in the rainy season, the scent mark on the lower leaves of a clump would be preserved for some time.

The MF was sprayed on a leaf, which after drying, was kept overnight in water. Part of the smell was now transferred to the water and part clung to the leaf. The water was concentrated and spotted on a Whatman paper and allowed to run in the solvent mentioned above. The ninhydrin spots were visible on the paper. Such experiments were undertaken to explore the possibilities of identifying individual tigers in the field.

In connection with the problem of odor being washed off by rains, we may note a slightly different case. In 1935 W.S. Thom (Brahmachary pers. data) described that in the wet rainy season of Burma, all animals, the tiger as well as its prey, emit a stronger smell. This may be Nature's way of compensating for the loss of odor (being washed away).

REFERENCES

Albone, E.S. 1983. Mammalian Semiochemistry. New York: Wiley.
Bedichek, R. 1960. The Sense of Smell. Michael Joseph: John Wiley.
Brahmachary, R.L., and J. Dutta. 1979. Phenylethylamine as a biochemical marker in the tiger. Zeit. Naturforsch. 34C:632-33.
Brahmachary, R.L., and J. Dutta. 1981. Pheromones of tigers: experiments and theory. Am. Nat. 118:561-67.
Brahmachary, R.L., and J. Dutta. 1984. Pheromones of leopards: facts and theory. Tigerpaper 11(3):18-23.
Corbett, J. 1944. Man-Eaters of Kumaon. London: Oxford Univ.Pr.
Ellis, R., A.M. Gaddis and G.T. Currie. 1958. Anal. Chem. 30:475-76.
Estes, R.D. 1972. The role of vomeronasal organ in mammalian reproduction. Mammalia 36:315-41.
Ewer, R.F. 1969. Ethology of Mammals. New York: Plenum.
Gorman, M.L., D.B. Nedwell and R.M. Smith. 1974. An analysis of contents of anal scent pockets of Herpestes auropunctatus. J. Zool. 72:389-99.
Gorman, M.L., D.B. Nedwell and R.M. Smith. 1976. A mechanism for individual recognition of odour in H. accropunctatus. Amim. Behav. 24:141-45.
Hashimoto, Y., Y. Eguchi and J.A. Arakawa. 1963. Historical observation of the anal sac and its glands of a tiger. Jap. J. Vet. Sci. 25:29-32.
Leyhausen, P. 1964. Symp. Zool. Soc. London. No. 14.
Matthews, L.H. 1969. The Life of Mammals. Vol. 1. London: Wedenfeld and Nicolson.
McDougall, C. 1978. The Face of the Tiger. London: Rivington Books and Andre Deutsch.
Perry, R. 1964. The World of the Tiger. New York: Atheneum.
Schaller, G.B. 1976. The Deer and the Tiger. Chicago: Univ. Chicago Pr.
Schilling, A. 1970. Memoirer du Museum National d'histoire Naturelle, Serie A. Zoologie 61:206-80.
Sunquist, M. 1985. Tiger. In Encyclopedia of Mammals, ed. D.W. MacDonald. Oxford: Equinox (Oxford), Ltd.
Wysocki, C.J., J.L. Wellington, and G.K. Beauchamp. 1980. Access of urinary nonvolatiles to the mammalian vomeronasal organ. Science 207:781-83.

PART V

CAPTIVE MANAGEMENT

27

Species Survival Plans and Overall Management Strategies

Thomas J. Foose

SPECIES SURVIVAL PLANS

Traditionally, Action Plans of the IUCN's Species Survival Commission have concentrated on protection of habitats and animals. Such actions are necessary but may not be sufficient. In contrast, species survival plans as being proposed in this paper are predicated on the biology and management of populations. The objective of species survival plans is the preservation of wildlife both as species and as components of ecosystems.

These two approaches may be equivalent if it is possible to preserve natural habitats large enough and well enough. But this objective is difficult for large terrestrial vertebrates, especially those, like the tiger, at the top of ecological trophic complexes. Such species often have significantly greater requirements for environmental resources, especially space, and may frequently disappear before their habitat or other elements of their ecosystem. Population management becomes particularly important for these so-called "charismatic megavertebrates" (Boecklen and Gotelli 1984).

POPULATION MANAGEMENT

Population management is vital because wild populations are becoming small and fragmented. Gene pools are becoming gene puddles. Small populations are vulnerable to extinction from stochastic causes (Shaffer 1981). Four categories of stochastic causes have been distinguished, but it seems possible to further combine them into two types: genetic and demographic.

Genetically, small populations tend to lose variation important for species at both the population and individual level (Frankel and Soule 1981, Soule and Wilcox 1980) (Fig. 1). At the population level, genetic variation is required to permit continual adaptation to ever-changing environments. At the individual level, loss of genetic variation frequently produces inbreeding depression manifested by reduced survival and fecundity (Ballou and Ralls 1982, Soule 1980).

In general, the smaller the population is, the faster genetic diversity is lost due to stochastic changes, which are known as random genetic drift. Actually, there are four major factors that produce genetic changes in populations; drift, selection, mutation, and migration (Crow and Kimura 1970). But in small, closed populations, genetic drift increasingly dominates.

Assessment of the loss of genetic variation due to drift must consider several variables (Lande and Barrowclough 1986):
  1) What kind of genetic diversity is of concern, e.g. average heterozygosity versus allelic diversity?
  2) Are migration, mutation, or especially selection strongly operative?
  3) How long is the absolute period of time being measured?
  4) How large is the genetically effective size of the population? The N_e is not necessarily nor usually the same as the total census N. Rather, the N_e depends on the dynamics of propagation in the population, e.g. the sex ratio and family sizes of reproducing animals (Crow and Kimura 1970). Theoretically, the N_e can vary from a small fraction, e.g. one-tenth, to twice the census number N.
  5) What is the generation time of the species? Standard algorithms measure loss of genetic diversity generation by generation. Hence, assuming equal N_e's, a species with a generation time of seven to eight years (like the tiger) would be expected to lose more genetic diversity due to drift over a period of 200 years than would a species (like the rhino) with a generation time of 15 years.

For a large vertebrate like the tiger, effective populations smaller than several hundred individuals may lose appreciable amounts of genetic diversity due to drift over a period of 200 to 1000 years. Assuming the objective of conservation strategies will be to preserve 90% or more of at least average hetero-zygosity for 200 to 1000 years, populations of several hundreds seem the minimum (Franklin 1980, Lande and Barrowclough 1986).

For very small populations, less than 50, demographic stochasticity may cause greater risks than genetic problems (Goodman 1986, Samson et al. 1985, Shaffer and Samson 1985). Included among these risks are; devastation by natural catastrophes, decimation by disease epidemics, predator or competitor eruptions; stochastic events in the survival or fertility of a small number of individuals, i.e. all the offspring produced by the few adults in the population could be of the same sex.

Small populations must therefore be managed genetically and demographically if they are to survive as evolutionarily viable entities over the long term (Schonewald-Cox et al. 1983). The temporal implications of this point are important. It has been argued that the persistence, even expansion, of small populations over a period of a few decades demonstrate that concerns about genetic diversity are unfounded or unsubstantiated (Hemmer this volume). In evolutionary terms, such contentions are short-sighted and fail to appreciate the time scale over which evolutionary processes and events occur. Success of conservation programs can ultimately be measured only by the behavior of

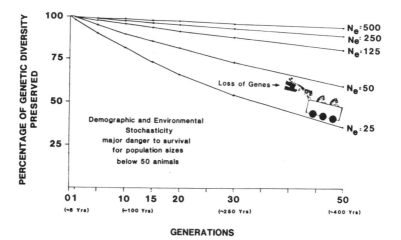

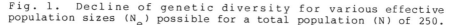

Fig. 1. Decline of genetic diversity for various effective
population sizes (N_e) possible for a total population (N) of 250.

populations over much longer periods of time. In the meantime,
the types of genetic management that have recently been proposed
for small populations are formulated to preserve options for the
future. Lack of genetic management may foreclose many options.

There will be no attempt in this paper to discuss
extensively the types of genetic and demographic management that
may be appropriate, but a few salient points seem pertinent.

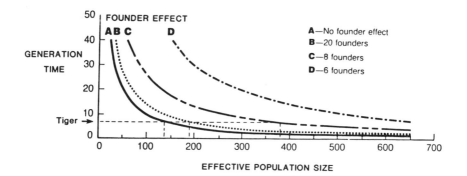

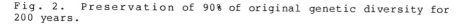

Fig. 2. Preservation of 90% of original genetic diversity for
200 years.

Because of the relationship between loss of diversity and population size, genetic management will entail determination of the current effective population size and estimation of what the minimum viable population (MVP) size should be. The MVP will depend on the genetic objectives of the program and will include consideration of the five variables cited earlier (Soule et al. 1986, Lande and Barrowclough 1986) (Fig. 2):

1) How much, and what kind of, genetic diversity is to be preserved? The optimal answer is, of course, all of it. This may not be feasible. Preserving 95% of the average heterozygosity requires an N_e twice as large as maintaining 90%. Preserving rarer alleles will require larger populations than will merely maintaining average heterozygosity;

2) How long must the diversity be preserved? Again, the optimal answer is forever, but a more realistic objective may be until the "demographic winter" (Soule et al. 1986) has passed and human population and development are stabilized or reduced, so that natural habitat may actually be expanded. This eventuality would seem to be at least 300 to perhaps 1000 years in the future. More immediate relief may be provided by advances in reproductive technology (Soule et al. 1986);

3) Are mutation, migration, or particularly selection operating to mitigate losses due to drift? For small remnant populations over a period of several centuries, mutation is not likely to be a significant force counteracting drift. Moreover, such populations will be closed in the sense that migration from other gene pools will not be available to oppose drift. However, natural selection may be a significant counter force to drift;

4) What is the N_e of the population? Normally, the social structure and dynamics of wild populations are such that the N_e is significantly less (0.5 or less) than N, the census number in the population. Management can theoretically enlarge the N_e/N ratio;

5) What is the generation time of the population or species? Generation will vary with the value of demographic parameters that will be regulated by ecological conditions and/or by population management;

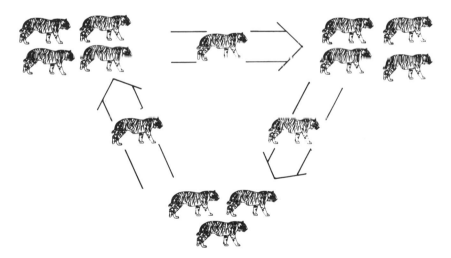

Fig. 3. Managed migration among subpopulations of tigers.

6) How subdivided or fragmented is the population? Depending on the amount of gene flow, natural or managed, subdivision may increase or decrease the N_e. Genetically, management of small and fragmented wild populations will normally require periodic movement of genetic material among the separate demes or subpopulations (Fig. 3). How much managed migration for genetic reasons will depend on whether or not the species or subspecies is to be managed as a single large or several small populations (Shaffer and Samson 1985, Wilcox and Murphy 1985). There are tradeoffs (Lacy 1986). If the N_e of the combined subpopulations is much greater than the MVP, then it may be best to manage as a single large interbreeding population. This strategy will require at least one migrant among each subpopulation per generation. However, if the N_e is less than the MVP for the intended objectives, it is possible to use the fragmentation of the population to advantage (Lande and Barrowclough 1986, Foose et al. 1986). Subdivision of small populations will actually increase the amount of allelic diversity preserved in the entire population. Each subdivision loses genetic diversity faster than a single large population of the same total size as all the subdivisions combined, but different alleles are lost from the various demes and overall allelic diversity is greater. To exploit this strategy, less than one migrant per population per generation is indicated.

Movement of animals among subpopulations may also be necessary for demographic reasons to reinforce or reestablish resident populations that have been decimated. Conversely, it may also be necessary to regulate the sizes of populations by removing animals to prevent overpopulation and consequent damage to the environment (Goodman 1980). Thus, demographic as well as genetic management will be in order.

Tigers clearly represent a species whose populations are small and fragmented. Table 1 documents the data that could be located to perform a survey of smallness and fragmentation. Conservation strategies and species survival plans must be cognizant of this smallness and fragmentation.

CAPTIVE AND WILD POPULATIONS

Preservation of gene pools need not, and perhaps cannot depend solely on wild populations (Foose 1983). Captive propagation can assist. Indeed, the captive population of one and perhaps two subspecies of tigers is larger than the number in the wild. However, it must be emphasized that the primary purpose of captive propagation is to reinforce, not replace, wild populations. Both captive propagation and wild populations may be necessary for the survival of the species or subspecies.

Captive propagation can serve as a reservoir of genetic material that can be periodically reinfused as "new blood" into wild populations. Moreover, captive propagation can be particularly effective in this function because zoo populations can in principle be managed genetically to maximize the effective size of the number of animals being maintained.

Wild populations are vital to subject the species to natural selection and thus maintain better wild traits, which will

Table 1. Sizes of wild populations of tigers (sources from this volume).

Subspecies	Animals	Number	Sizes	> 100
P. t. amoyensis	30-40	?	?	0
P. t. altaica	400-500	?	?	1?
P. t. corbetti	2000	?	?	?
P. t. sumatrae	600-800	?	?	?
P. t. tigris	4000+	21+	15-264	3

eventually erode in captivity despite the best attempts to arrest genetic change occuring over the long term.

Thus, optimal conservation strategies for large vertebrates like the tiger should incorporate both captive and wild populations that are interactively managed for mutual support (Foose et al. 1985) (Fig. 5). Moreover, much the same principles and methods of genetic and demographic management should be applied to small populations in the wild as well as in captivity. For example, the intensive management possible in captivity depends upon individual identification and data management. With species like the tiger, whose individuals are so distinct and whose populations are so manageably small, it may be feasible and desirable to consider studbook types of management. Indeed, such programs are already underway in several areas. As Sullivan and Shaffer (1975) among others have observed, wildlife management in the wild, as well as in captivity, is becoming not unlike a "megazoo."

CAPTIVE POPULATIONS   **WILD POPULATIONS**

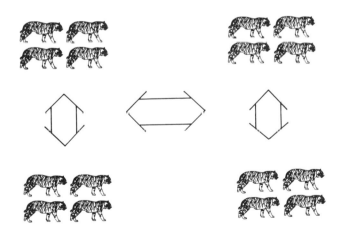

Fig. 4. Interactive management of captive and wild tiger populations.

This vision of cooperating captive and wild programs is not without its critics, many of whom frequently express skepticism about the feasibility of reintroducing captive stock into wild situations. Two of the major concerns are that reintroductions will be disruptive to remnant resident wild populations or that animals produced in captivity will not be able to adapt to natural conditions. These concerns and criticisms are especially prominent for species like the tiger, with whom much cultural transmission of predatory and other survival skills may occur. It is still not known if these problems are real. But if they are, reproductive technology may provide a solution. It is entirely possible that genetic material can be transferred from captive to wild populations via artificially transplanted sperm or embryos (Fig. 5). The technology for such operations is not yet available, but research is vigorously in progress (Dresser et al. this volume, Seal et al. this volume).

Reproductive technology may assist in yet another way. The "frozen zoo" may permit effective populations to be expanded to literally thousands or millions of animals, most of whom are preserved as cryogenically stored gametes or embryos. This prospect is not to propose that living animals be replaced, but once again, merely reinforced. Populations of living animals in both natural habitats and zoos must continue to be maintained, but augmented where technology permits, by stored germinal tissue.

ZOO PROGRAMS

Zoos are organizing to perform this integral role in conservation strategies. In North America, the Species Survival Plan (SSP) has been established (AAZPA 1983, Foose 1983). The SSP encompasses all the zoos maintaining a particular species in

CAPTIVE POPULATIONS   **WILD POPULATIONS**

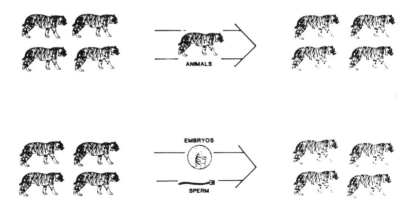

Fig. 5. Possible contributions of captive populations to wild populations of tigers.

Table 2.   Taxa designated for the AAZPA Species Survival Plan
(SSP).

Puerto Rican Crested Toad	Ruffed Lemur
	Black Lemur
Chinese Alligator	Golden Lion Tamarin
Orinoco Crocodile	Lion-Tailed Macaque
Radiated Tortoise	Gorilla
Aruba Island Rattlesnake	Orang-Utan
Madagascan Ground Boa	Asian Small-Clawed Otter
	Maned Wolf
Bali Mynah	Red Wolf
White Naped Crane	Red Panda
Andean Condor	Siberian Tiger
Humboldt's Penguin	Asian Lion
	Snow Leopard
Asian Elephant	Chacoan Peccary
Indian Rhino	Barasingha
Sumatran Rhino	Okapi
Black Rhino	Gaur
White Rhino	Arabian Oryx
Asian Wild Horse	Scimitar-Horned Oryx
Grevy's Zebra	

---

the United States and Canada. Thirty-seven species, including
the tiger, have been designated so far for the SSP (Table 2).
Eventually it is expected that there will be SSP programs for at
least a thousand taxa of mammals, birds, reptiles and amphibians.

For each species designated, an intensive program of
demographic and genetic management is developed (Seal and Foose
1983a, 1983b, Ballou and Seidensticker this volume). The details
of this program are documented in a masterplan which provides
institution-by-institution and animal-by-animal recommendations
for management. Each SSP is administered by a Species
Coordinator assisted by a management committee elected by and
from participating institutions.

Europe is also rapidly developing a species survival plan or
population management program (Europaisches Erhaltungszucht-
Program or EEP). Indeed, the British Federation of Zoos has
already been operating SSP-type programs for several species,
including the tiger. Ultimately, there will be a global program
for management and propagation of tiger populations in captivity
to assist preservation of the species gene pool (Fig. 6).

Capacity of Zoos

Unfortunately, the space and resources available in the
"captive habitat" of zoos are limited (Foose 1983, Foose and Seal
1986, Conway 1986). Because of the sizes of populations of each
taxa that must be maintained for gene preservation, it is very
likely it will be necessary to limit the number of different
populations that are accommodated (Table 3).

Subspecies present particular difficulties. Recently, the
SSP has been re-examining the biological validity of subspecies.

NORTH AMERICAN SSP

EUROPEAN EEP

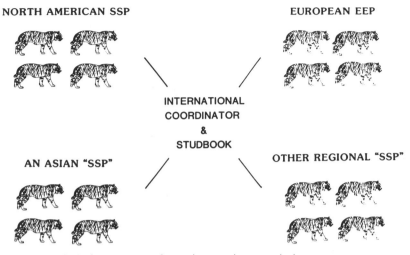

AN ASIAN "SSP"

**INTERNATIONAL
COORDINATOR
&
STUDBOOK**

OTHER REGIONAL "SSP"

Fig. 6. A global program for tigers in captivity.

The tentative approach is to determine which described subspecies are evolutionarily significant units (ESU's) based on investigation of genetic, morphometric, and biogeographic data. Results of such studies should assist the SSP, and the zoo community in general, in assigning priorities and allocating resources in the captive programs. The SSP realizes there may also be biopolitical considerations in subspecies questions, but

Table 3. Capacity of ISIS and studbook captive facilities for large felids.

Species	Extant Sub- species	Red Data Book Subspecies	Captive Population	# Subspecies if Population 100	250	500
Panthera leo	11	1	1079	10	4	2
Panthera tigris	8	8	1429	14	6	3
Lions & Tigers	19	9	2508	25	10	5
Panthera onca	8	8	179	2	0	0
Panthera pardus	15	15	503	5	2	1
Panthera uncia	1	1	312	3	1	0
Felis concolor	29	2	280	3	1	0
Neofelis nebulosa	4	4	202	2	1	0
Acinonyx jubatus	6	6	454	4	2	1
"Other Large Felids"	63	36	1930	19	8	4
Total	82	45	4438	44	18	9

Table 4. Captive populations and capacity for tigers (Panthera tigris).

Subspecies	Population		
P. t. amoyensis	24/16	=	40
P. t. altaica	266/301	=	567
P. t. balica	0/0	=	0
P. t. corbetti	13/12	=	25
P. t. sondaica	0/0	=	0
P. t. sumatrae	69/88	=	157
P. t. tigris	230/290	=	520
P. t. virgata	0/0	=	0
P. t. (generic)	46/74	=	120
Total			1429

ultimately the limitations of captive habitat may require hard decisions based solely on biological criteria.

Pending further elucidation of esu's, it does appear zoos of the world may possess enough captive habitat to maintain acceptably large populations of four or five tiger taxa, esu's, geographical varieties, etc. (Table 4). Acceptability is basically defined in terms of the capacity to preserve at least 90% of average heterozygosity for 200 years. This situation in large part reflects the great exhibition popularity and interest in these animals. The tentative proposal formulated by the IUCN Captive Breeding Specialist Group (CBSG) in collaboration with the International Tiger Studbook Keeper has been for zoos outside China to develop populations for P. t. altaica, sumatrae, tigris, and corbetti. Zoos in China are encouraged to continue their efforts with P. t. amoyensis. For some subspecies, especially P. t. tigris, there will almost certainly be the need to obtain some more pure founder stock.

The phenomenon or dilemma of the "white tiger" needs to be considered in this context (Foose and Seal 1986). The number of white tigers in zoos has been growing rapidly (Roychoudhury and Acharjyo this volume). Currently, about 10% of the tigers in North American Zoos are white. White tigers therefore are occupying much captive habitat that would otherwise presumably be available for other types of tigers. There has been much controversy about white tigers. Some contend that white tigers are a biological oddity; others maintain that the animals can and do contribute to the preservation of P. t. tigris. It does appear to be the case that white tigers have much visitor attraction and hence exhibit, and revenue-generating, value for zoos (Simmons this volume, Maruska this volume).

In trying to resolve how to treat the problem of white tigers, there are several further facts that seem of relevance. Much of the current stock does not represent purely one of the described subspecies. Moreover, even where and if it does, the great proliferation for its exhibit value has expanded the representation of this morph far beyond the frequency with which it occurs in natural populations. Such over representation of the white morph seems inappropriate in a captive population being

maintained presumably and primarily as a back-up for the species
or subspecies in the wild.

Consequently, it would seem to be in the interest of the
species to contain only the number of white tigers in captivity
that will not appreciably reduce the "captive habitat" available
for the populations being managed for gene pool conservation,
while still providing sufficient stock for the exhibit needs and
objectives of zoos.

"Value" of Zoo Programs

Two final observations about the "value" of captive programs
seem appropriate. First, there often has not been an
appreciation of the biological value of captive propagation as
reinforcement for wild populations. Even where its significance
has been recognized, it frequently has not been invoked until the
wild populations are critically low. However, a population
biology perspective suggests that populations in the several
hundreds or even few thousands may not insure survival over the
long-term. In relation to tigers, this point suggests that it
should not be concluded that the existing wild populations, even
if well protected, obviate the need or desirability of a well
managed back-up captive population.

Finally, there has also been some criticism that captive
propagation is not a cost effective way to conserve species. It
has been suggested that if the millions of dollars expended to
maintain a population of hundreds of tigers for a century or two
in captivity (about $5 million/year for the 1500 tigers in zoos
(Conway 1980)) were spent instead on field conservation, much
more could be accomplished.

The problem is that much of the money involved in supporting
zoos is not and probably cannot be available for field
conservation. Zoos traditionally have evolved to serve local
constituencies and communities. Even though zoos are now
changing their orientation so that conservation is their highest
priority, the bases of support are still local and the need to
provide tangible gratification to these constituencies and
communities in the form of the zoo is great. However,
conservation has the opportunity to exploit the considerable
resources of zoos to develop captive propagation that can be
significant and perhaps vital to the survival of species.

CONCLUSIONS AND RECOMMENDATIONS

The next century or two are going to be catastrophic times
for wildlife. Species like the tiger will survive only if
conservation strategies are predicated on principles of
population biology, incorporate both in situ and ex situ methods,
and opportunistically exploit and coordinate all resources
available for gene pool preservation.

To achieve these objectives, it is proposed species survival
plans must be formulated for taxa like the subspecies of tigers.
These species survival plans should incorporate both captive and
wild populations and programs. The plans should provide

subspecies-by-subspecies, population-by-population, sanctuary-by-sanctuary and region-by-region recommendations. It is further proposed that a species survival plan task force led by a species coordinator, much like the management committees for the captive programs, be organized to formulate and implement the species survival plan for tigers. This task force should include representatives of wildlife managers in the countries of origin of wild populations, and of the zoo groups operating propagation programs for gene pool preservation.

REFERENCES

American Association of Zoological Parks and Aquariums. 1983. Species Survival Plan. Wheeling, WV: AAZPA.

Ballou, J. and K. Ralls. 1982. Inbreeding and juvenile mortality in small populations of ungulates: a detailed analysis. Biol. Cons. 24:239-72.

Boecklen, W.J. and N.J. Gotelli. 1984. Island biogeographic theory and conservation practice: species-area or specious-area relationships. Biol. Cons. 29:63-80.

Conway, W.G. 1980. Where do we go from here? Int. Zoo Yearb. 20:183-89.

Conway, W.G. 1986. The practical difficulties and financial implications of endangered species breeding programs. Int. Zoo Yearb. 24.

Crow, J.F. and M. Kimura. 1970. An Introduction to Population Genetics. New York: Harper and Row.

Foose, T.J. 1983. The relevance of captive populations to the strategies for conservation of biotic diversity. In Genetics and Conservation, ed. C. Schonewald-Cox, S. Chambers, B. MacBryde, and W.L. Thomas. Reading, MA: Addison-Wesley.

Foose, T.J., N.R. Flesness and U.S. Seal. 1985. Conserving animal genetic resources. IUCN Bull. 16(1-3):20-21

Foose, T.J., R. Lande, N.R. Flesness, B. Read and G. Rabb. 1986. Propagation plans. Zoo Biol. 5(2):139-46.

Foose, T.J. and U.S. Seal. 1986. Species survival plans for large cats in North American zoos. In Proceedings of the International Cat Congress, ed. D. Miller. Washington, DC: Natl. Wldlf. Fed.

Frankel, O.H. and M.E. Soule. 1981. Conservation and Evolution, New York: Cambridge Univ. Pr.

Franklin, I.R. 1980. Evolutionary change in small populations. In Conservation Biology, ed. M.E. Soule and B.A. Wilcox. Sunderland, MA: Sinauer Associates.

Goodman, D. 1980. Demographic intervention for closely managed populations. In Conservation Biology, ed. M. Soule and B. Wilcox. Sunderland, MA: Sinauer Associates.

Goodman, D. 1986. The minimum viable population problem, I. the demography of chance extinction. In Viable populations, ed. M.E. Soule. Oxford: Blackwell Pub.

Lacy, R.C. 1986. Loss of genetic diversity from captive populations, interacting effects of drift, mutation, immigration, selection, and population subdivision. Unpublished.

Lande, R. and G. Barrowclough. 1986. Effective population size, genetic variation, and their use in population management. In Viable Populations, ed. M.E. Soule. Oxford: Blackwell Pub.

Samson, F.B., F. Perez-Trejo, H. Salwasser, L.F. Ruggiero and
   M.L. Shaffer. 1985. On determining and managing minimum
   population size. Wildl. Soc. Bull. 13:424-33.
Schonewald-Cox, C., S.M. Chambers, B. MacBryde and W.L. Thomas.
   1983. Genetics and Conservation. Reading, MA: Addison-
   Wesley.
Seal, U.S. and T.J. Foose. 1983a. Development of a masterplan
   for captive propagation of Siberian tigers in North American
   zoos. Zoo Biol. 2:241-44.
Seal, U.S. and T.J. Foose. 1983b. Siberian tiger species
   survival plan: a strategy for survival. J. Minn. Acad. Sci.
   49(3):3-9.
Shaffer, M.L. 1981. Minimum population sizes for species
   conservation. BioScience 31(2):131-34.
Shaffer, M.L. and F.B. Samson. 1985. Population size and
   extinction: a note on determining critical population sizes.
   Am. Nat. 125(1):144-52.
Soule, M.E. 1980. Thresholds for survival: maintaining fitness
   and evolutionary potential. In Conservation Biology, ed.
   M.E. Soule and B.A. Wilcox. Sunderland, MA: Sinauer
   Associates.
Soule, M.E. and B.A. Wilcox. 1980. Biological Conservation,
   Sunderland, MA: Sinauer Associates.
Soule, M.E., M. Gilpin, W.G. Conway, and T.J. Foose. 1986. The
   millennium ark: how long the voyage, how many staterooms,
   how many passengers? Zoo Biol. 5(2):101-14.
Sullivan, A.L. and M.L. Shaffer. 1975. Biogeography of the
   megazoo. Science 189(4196):13-17.
Wilcox, B.A. and D.D. Murphy. 1985. Conservation strategy: the
   effects of fragmentation on extinction. Am. Nat. 125:879-87.

28

Captive Management of South China Tigers *(Panthera tigris amoyensis)* in China

Xiang Peilon

INTRODUCTION

The South China tiger (P. tigris amoyensis) is one of the eight subspecies of contemporary tigers, and a special race of tigers in China. This subspecies lives in the central region of tiger distributions in the world, and once dispersed over vast areas of east, central, south, and southwest China. However, because of dense human population and large-scale deforestation, the range of this animal has been largely reduced, and its numbers have sharply decreased during the last twenty years.

At present, the most optimistic estimates indicate there are about 70 South China tigers, but by conservative estimates only about 40 exist. It would be impossible to rejuvenate the wild population naturally because its small number is scattered among separate forest areas. Thus, hope for salvation of this subspecies is contingent upon increasing reproduction by captive breeding, and then perhaps the wild populations can be expanded.

Since 1978, South China tigers in our zoo have bred successfully many times. In seven litters, 19 cubs have been born altogether, and 13 of them have survived the last seven years or more. Of these, all but four have been loaned to other zoos, and three were exchanged with others; we now have nine breeding tigers. Our pedigree for these tigers is shown here.

```
Qingzhen(M)         Zhenhua(F)
   +      ------>      +   -------->  Weiwei(M)
Weining(F)          Zhenfeng(M)         + ----> 7 litters
                       +   -------->  Tingting(F)
                   Weining(F)
```

Qingzhen was wild-caught in 1958 in Qingzhen County, Guizhow Province. The female Weining was wild-caught in 1959 in Weining County, Guizhow Province. Weiwei (July 1973) and Tingting (June 1974) were born in Qianling Zoological Garden, Guiyang City and sent to Chongqing Zoological Garden in October 1973 and December 1974, respectively.

The animals in our zoo are housed together, in outdoor cages providing them 12 sq. m of exercise space and 9 sq. m indoor retreats suitable for resting or birthing. Each animal receives a diet of 4-5 kg of meat and eggs daily, consisting mainly of beef, lean pork and fresh chickens. Each is fasted one day a week. A variety of meats should be fed, otherwise symptoms of vitamin and amino acid deficiency may be induced. Internal organs, such as the heart and liver, should be added when meat is used as the main food. It is best to use meat with cartilage, and it is important to feed tigers bone at regular intervals. During the seasons short of sunshine, cod-liver oil and multi-vitamins should be supplemented.

REPRODUCTION

Sexual maturity: Weiwei (male) and Tingting (female) first mated in March 1978, at the ages of 4.5 and 3.5 years, respectively. Both were considered to be sexually mature. After 102 days of pregnancy, Tingting gave birth to four cubs.

Estrus and mating: Tingting became cyclic after sexual maturity, showing estrus behavior more obviously during spring and autumn (Table 1).

During estrus, the female was more active and had less of an appetite. She usually lay on her back, stretched and growled. After the door separating the two tigers was opened, no immediate mating happened, but they smelt each other. Then the tigress came to the front of the male and lay on her stomach with her tail sidelong. The male mounted her back and bit her nape. Both animals then gave an excited roar. Mating lasted no more than a minute. The tigress then rolled on the ground after copulation. The estrus period of the tigress could be divided into early and late stages. Usually she showed strong sexual behavior in the first four days, then lost interest.

Table 1. Estrus and matings of all parities of a captive South China tigress.

Parity	Date in Heat	Days in: Estrus	Days in: Copulation
1	03/11/78 to 03/16/78	6	6
2	09/04/78 to 09/09/78	6	6
3	05/24/79 to 06/01/79	8	8
4	03/25/80 to 03/29/80	5	3
5	02/26/81 to 03/02/81	6	6
6	01/16/82 to 01/23/82	8	5
7	02/19/83 to 02/26/83	8	6

Pregnancy duration: The tigress usually showed no further estrus once she became pregnant. Rolling on the ground was

observed during the first month of gestation, but this behavior
disappeared quickly. After she was pregnant, the tigress'
appetite and body weight increased. From the second month of
gestation, the mother's median abdomen and udders became swollen,
and her teats were bulgy and pink. When she walked, her abdomen
looked solid, heavy, and drooped. The length of gestation was
generally influenced by maternal health and fetal number and sex
ratio, and ranged from 100 to 105 days (Table 2).

Table 2. The gestation length and the date of partum of all
litters of a captive South China tigress.

Litter	F/M	Gestation	Birth Time/Date
4	2/2	102	
2	1/1	100	4:50 Dec. 17, 1978
3	3/0	102	4:45 Sep. 10, 1979
3	2/1	103	4:20 July 8, 1980
3	2/1	102	21:30 June 12, 1981
2	0/2	103	7:30 May 4, 1982
1	0/1	105	21:22 June 10, 1983

Parturition: In the wild, the tigress usually looks for a
hidden and quiet place before parturition, but has no habit of
nest-building with litter hay in general. Under artificial
conditions, the cubbing den and bed were ready one month before
estimated parturition. The cubbing bed was 200 cm long, 85 cm
wide, and 30 cm deep. It is necessary to keep the cubbing den
dim, comfortable, and quiet, and to adapt the tigress to it as
quickly as possible.

There was an hour interval between the first cub birth and
the second in a litter. The tigress had most of her births in
the early morning (Table 2). Commonly, the tigress began to walk
restlessly about seven hours before parturition. Movement of the
unborn cubs could be observed when the tigress lay on her side.
The tigress ate the placenta immediately after giving birth, then
licked herself and her cub, and lapped fetal fluid up from the
padding board.

Nursing: An hour after parturition, the mother lay on her
side on the cubbing bed, licked the cubs' forelegs, and gently
pushed with her nose in order to induce her cubs to suckle. The
primiparous tigress in captivity is often short of maternal
ability. Nursing behavior and its experience may be related to
endocrine functions. Mammogenesis and milk secretion are all
influenced directly by endogenous hormone. Prolactin in mammals
promotes milk secretion, and in cases where the mother is lacking
in milk secretion, she would have less ability to care for her
cubs.

From Tingting's second litter, we began to feed her
mammotropic tablets one week before parturition to enhance milk
secretion and maternal behavior. Natural nursing may advance the
bonding of the mother with her cubs. Successful nursing may also

be useful to improve maternal behavior and subsequent reproduction in mother tigers.

Nursing lasts about half a year under wild conditions. After this, the young follow their mother for about two years, until they learn how to catch their own prey, and then disperse. In captivity, tiger cubs generally can be weaned by about three months of age, followed by hand-rearing.

HAND-REARING OF TIGER CUBS

Tiger cubs have to be reared by hand when the tigress fails to nurse, or when her health fails.

Tingting gave birth to 19 cubs in seven litters. Among them, 13 cubs lived and six died (Table 3). Thirteen cubs were reared by hand from 3 - 93 days after birth. These practices have led to great success.

Table 3. Duration of lactation periods of a captive South China tigress in successive litters.

Number Born	Born Alive	Lactation period (days)
4	0	0
2	1	3
3	3	10
3	3	68
3	3	70
2	2	93
2	1	76

Hand-rearing Chamber

Ventilation: Fresh air was advantageous to the spiritual stabilization and increased appetite and metabolism of cubs. The cubs suffered from respiratory tract infections if ventilation was poor.

Temperature and humidity: The optimal ambient temperature for cubs at one month of age was 20-22°C. If it was decreased to below 18°C, the cubs shivered. Suitable humidity was 60-80%; if too high, the cubs showed shortness of breath, if too low, the mouth and tongue of the cubs felt dry. When the temperature in winter was increased with a wood stove, a basin filled with water was placed on the stove or water was sprinkled on the floor to increase humidity.

Light conditions: Natural lighting should be fully utilized to accelerate the growth and development of cubs. When the cubs slept, brightness was regulated to be soft in the chamber.

SUCKLING WITH A NURSING BITCH

Two-thirds of tiger cubs suckled by a nursing bitch survived. The maternal ability of the nursing bitch was dependent on her association with the cubs, established gradually through frequent contact with each other, including the information received from sights, sounds, and smells.

At first the nursing bitch was forced to nurse and the tiger cubs were unwilling to suckle, but they adapted to each other after one day.

Although the nutrient quality of a nursing bitch's milk was not better than that of the tigress, it was better than cow's milk. If the tiger cubs had not fully suckled the tigress' colostrum, they were supplemented with the colostrum of a nursing bitch. At 29 days of age the cubs were fed five grams of meat mash daily, and additional cow's milk day-by-day to prepare the cubs for weaning. In our experience, cubs ceased to suckle the bitch's milk at 45 days of age and switched to meat and cow's milk.

When the cubs changed from suckling bitch's milk to suckling tigress' milk, they became constipated. For example, the cub Xiaohua was constipated 14 days, but its spirit, appetite, growth, and development were still normal. Xiaozhu, Xiaodong, and Xiaoming were constipated for 15, 20, and 21 days, respectively.

FEEDING CUBS AFTER WEANING

Under wild conditions, the tigress nurses for about half a year. In our zoological garden, all cubs were weaned by about 45 days of age. At that time, the cubs were indeed small. The average body weight was about 3.5 kg (compared to an average birthweight of 1.0 kg). For underweight cubs, food dosage and composition was changed gradually as required. For example, the changes of daily intake within 28-90 days of age were tabulated for several cubs in Table 4.

Table 4. Weight increase of South China tiger cubs.

Cub	3	10	30	40	65	100
		Weight (kg) at Age (days):				
Xiao Hua	1	1.28	2.65	3.5	3.5	9.25
Xiao Zhu		2.05	3.28	4	5.6	9.5
Miao Ming		2	3.25	3.75	5.75	9.25
Xiao Dong		2	3.3	3.9	6	9.5

When carnivorous animals, especially young animals, are fed a single meat for a long time, they sometimes contract acute gastroenteritis which might be related to an "allergic reaction." Therefore, we used pork, beef, chicken, liver, and heart to feed the cubs alternately.

When the cubs were given supplemental feedings of eggs and cow's milk, they grew rapidly. If the nutrients provided were incomplete, they were unbalanced in growth and development. From 1.5 to 6 months of age, milk must be provided, because lactose is insufficient in meats and eggs, and because milk is one of the most important sources of calcium and phosphorus. Eggs were supplemented gradually after two months of age. If given too early, the cubs sometimes suffered indigestion. Eggs contain albumen, a little ovoglobulin, and vitellin. Vitellin is a complete protein and its iron content is higher than that of milk. Egg yolk is also a source of vitamin A, B_{12}, B_6, and D.

Because cubs are sensitive to low temperatures, meats in cold storage were preheated to 25°C before feeding. We used infrared lamps to preheat the meats. If the meat is too cold, it could cause the cubs to vomit.

Chicks and young birds supplemented the cubs' diets by 2.5 months of age. Bone growth was more rapid than muscle growth in the cubs for about the first year of life. The cubs received a single calcium supplement and still suffered from hypocalcemic tetany and calciprivic fractures easily. Bones of chicks and birds are chewed and digested easily. The chick's bones were pounded at the beginning of feedings, and the cubs were watched carefully to avoid accidents. Whole chicks and birds were gradually added to their diet at a later date.

In rearing the cubs, we found that when the cubs were separated from soil for a long time, and then returned to it, they licked it with all their might. We paid attention to this phenomena. It seemed to satisfy some requirement, but we were not clear on which microminerals were being ingested.

All meat, eggs, and milk quickly spoil in the cub's feed. When the cubs ingested <u>Bacillus</u> <u>botulinus</u>, <u>Bacillus</u> <u>dysenterae</u>, <u>Salmonella</u>, <u>Staphylococcus</u>, <u>Escherichia</u> <u>coli</u>, or other bacterial toxins, they took ill within several hours. Thus, the control of digestive tract diseases was critical for successfully rearing cubs.

Cubs could jump and were very active after 45 days of age. We made the cubs exercise outside daily for an hour in our zoological garden. Exercise accelerated ossification and growth of bone, improved blood circulation and muscular nutrient uptake, strengthened their viscera function, and accelerated gastrointestinal peristalsis.

29

South China Tiger Recovery Program

Xiang Peilon, Tan Bangjie, and Jia Xianggang

INTRODUCTION

In recent years, in view of the endangered situation of the South China tiger (Panthera tigris amoyensis), Chinese scholars and specialists have repeatedly appealed to authorities in China and abroad to pay more attention to the seriousness and urgency of the problem. For instance, Sheng Helin and Lu Houji (1979) reported that the "South China tiger is on the brink of extermination." Zhu Jing (1980) and Liu Zhenhe et al. (1983) emphasized that the "danger of extinction is imminent." They estimated the remnant population of wild South China tigers at 30-40 individuals. Xiang Peilon (1983) proposed that a South China tiger conservation meeting be called, and that a breeding center for the South China tiger be established. Tan Bangjie (1984), after stressing the close relationship between the South China tiger and mankind, especially the Chinese people, suggested that it should be rescued from the wild and the captive population should be better managed. He has appealed to the international circles to think highly of the South China tiger's dangerous situation through the WWF Monthly Report issued in March, 1983, and then emphasized once more the seriousness and urgency of this problem in his paper entitled "The Status of Felids in China", submitted to the symposium called by the IUCN/SSC Cat Specialist Group at Kanha, India in April, 1984.

While making these appeals, Chinese specialists also took steps to rescue the South China tiger. Under the guidance of Xiang Peilon, the Chongqing Zoo obtained valuable information on the breeding of South China tigers. A pair of tigers gave birth to nine litters - 21 cubs from 1979 to 1985; 16 of them survived and 11 were retained as breeders. At the same time, the Shanghai and Nanchang zoos also obtained certain achievement in breeding.

The Chinese Ministry of Urban and Rural Construction and Environmental Protection officially approved on June 22, 1984 the report on the establishment of a scientific research and breeding center for the South China tiger at the Chongqing Zoo. Under the joint auspices of the Zoological Society of Sichuan and the Wildlife Conservation Association of Sichuan, a symposium on the conservation of the South China tiger was called in Chongqing in

October, 1984. More than 20 Chinese scholars and specialists
participated in the symposium, during which they exchanged views
and information on the subject, studied the feasibility of a
conservation project, and passed the resolution of "an appeal to
the public on the conservation of the South China tiger." The
symposium suggested that for the conservation and recovery of the
South China tiger to be effective, international cooperation
should be adopted, and substantial international support should
be sought. In order to strengthen the resolution, Chinese
authorities decided in 1985 that the building and management of
the "South China tiger scientific research and breeding center"
be placed under the leadership of the Chinese National
Environmental Protection Agency.

PAST AND PRESENT STATUS OF THE SOUTH CHINA TIGER

Before the 1950's, the range of the South China tiger had
been widespread throughout the vast territory south of the
Changjiang (Yangtze) Valley; historically the provinces of Hunan
and Jiangxi had been the centers of its distribution, while the
neighboring provinces of Guangdong, Guangxi, Fujian and Guizhou
also had quite large populations and fair distributions. Outside
of that range, the provinces of Chejiang, Hubei and Sichuan had
also certain numbers, while the southwestern part of Henan and
the southern part of Shaanxi had still less. In addition, there
were occasional discoveries in southern Shanxi and eastern Gansu
(Fig. 1). Later on, owing to the higher density of human
population, more extensive economic and agricultural exploitation, continued reduction of forested area, and belated natural
conservation, there was a sharp decline of population and a great
reduction of range since the end of the 1950's. The present
population of wild South China tigers, according to our
preliminary estimate, is approximately 40. This figure means
that the basic viable population for the subspecies is on the
verge of extinction. It is already impossible to depend on the
wild population for a natural recovery.

SIGNIFICANCE OF SOUTH CHINA TIGER CONSERVATION

Precious wildlife resources are undoubtedly the most
valuable natural properties that belong to the whole of mankind.
Tigers, as one of such precious animals, had originally a great
population and a wide range reaching the Sakhalin Island to the
east, Turkey to the west, Java and Bali to the south and the
soviet tundra near 60° to the north. The first tiger subspecies
to disappear from earth was the Bali tiger (P. t. balica), which
became extinct late in the 1930's; then followed the Caspian
tiger (P. t. virgata), which was exterminated in the mid-1960's
or early 1970's; the few remnant Javan tigers (P. t. sondaicus),
are also doomed to finish before long (Seidensticker this
volume). For the remaining subspecies, P. t. amoyensis is on the
threshold of a life and death struggle. No matter what angle is
considered - historical relationship, wildlife conservation,
cultural, scientific and economic values - there is great
significance attached to the timely rescue of this tiger
subspecies from extinction.

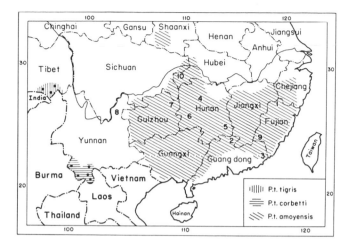

Fig. 1. Historical distribution of tigers in southern China.

RESCUE OF THE SOUTH CHINA TIGER

There are different ways to rescue the South China tiger, including the formulation of wildlife conservation law, so that there is a legal basis for the conservation of tigers and other wildlife. A second way is to investigate the present status of the South China tiger, especially their remnant groups and whereabouts in the wild, and to work out corresponding plans of protection based on the investigation. A third is to create new reserves or to strengthen the existing ones (such as Fanjingshan of Guizhou, Chebaling of Guangdong, Bamianshan of Hunan) where tigers are reported or known to exist. All of these are undoubtedly important and necessary steps in conserving the South China tiger. However, in our consideration, the most urgent, as well as effective, measure for the present time is to set up as quickly as possible a world central captive breeding group for the South China tiger. This idea originated from the experiences gained from works in saving a number of other very precious endangered animals in recent years. To establish a world's central breeding group, it is necessary to have international cooperation. We therefore propose to set up an international cooperative relationship following consultations between international bodies such as the IUCN or WWF and related Chinese governmental organizations, such as the Chinese National Environmental Protection Agency, and civilian bodies such as the Chinese Association of Zoological Gardens and the South China Tiger Coordinated Group for Conservation.

Feasibility of a Captive Breeding Group

The hope of rescuing the South China tiger from doom actually rests upon the 20-30 pure blood South China tigers now living at the zoos of Chongqing, Shanghai, Guiyang and Nanchang, especially the Chongqing Zoo, which owns three breeding pairs (one 13 and 12 years old, another seven and six years old, the

third four and three years old) and five younger ones.   These
pairs may form a nucleus of the proposed central breeding group.
Since South China tigers kept by the Shanghai Zoo have different
genetic make-ups because of a Fujian bloodline, it is appropriate
to set up different breeding groups at the Shanghai and Guiyang
Zoos besides the Chongqing Zoo, so as to maintain as much genetic
diversity as possible.

Conceptions of the Captive Breeding Group

Number of breeders to be kept: The central breeding group
at Chongqing, comprised of 20-22, tigers will be of the Guizhou
bloodline.   The second breeding group at Shanghai, comprised of
10-15, tigers will be of the Fujian-Guizhou bloodline.   The third
breeding group at Guiyang, comprised of 8-10 tigers, will be of
the Guizhou bloodline.

Basic facilities for housing the tigers: Secluded quarters
for breeding are necessary.   Each breeding unit should consist of
one male and two female tigers.   Each unit should have its own
indoor and outdoor cages and delivery room.   Enough installations
should be provided for at least 4-6 units.

Quiet nursery quarters should also be provided.   The key to
successful rearing of tiger cubs rests on careful and scientific
management, which demands separate cages to accommodate up to 20
cubs from 1-8 months old.   Besides indoor cages and outdoor
playgrounds, each must also have a quiet closet for rest.

Exhibition quarters for adult tigers open to the public
would have positive effects on the publicizing of wildlife
conservation as well as the spreading of scientific knowledge to
the common people.   It is therefore a good method to gain more
understanding and support from the public.   The exhibit quarters
should present the natural habitat of the South China tiger, so
they must be fairly spacious quarters and surrounded by a flooded
moat.   Some of the breeders may also be exhibited here.

The central breeding group should also be made available for
research.   Installations should be provided for scientific
workers who study genetics, physiology, ecology, nutrition,
pathology, etc. in these quarters.   Meeting rooms, bedrooms, and
other support facilities should also be provided.

Establish Studbooks for Chinese Tigers

The compilation of a studbook for all Chinese tigers in
captivity (P. t. amoyensis, P. t. altaica and P. t. corbetti) is
a prerequisite for good breeding.   Preliminary statistical
figures for captive Chinese tigers are shown elsewhere (Tan
Bangjie this volume).   However, to set up a complete formal
studbook requires much more investigation.

Supplementation and Redistribution of Breeders

In order to improve the bloodline of captive South Chinese
tigers whose lineages are rather inbred, additional wild tigers,
as supplement breeders, are needed.   However, this is not an easy
task and we need to be cautious not to deplete the remaining wild

populations. A more practical measure for the present is to redistribute and consolidate the existing breeding tigers kept by various Chinese zoos. As a final step, it is necessary to exchange breeders between the Shanghai lineage and the Chongqing-Guiyang lineage. The recovery of the Guiyang tiger breeding group would be beneficial in the long run.

RETURNING TIGERS TO THE WILD

To return the offspring of captive tigers to the wild and to reestablish an ecological system is another important step to be taken in the South China tiger recovery program.

Tigers stand on the summit of the food pyramid. It has been estimated that one South China tiger requires about 30 sq. km of forest, an area that may furnish approximately 75,000 kg of herbivore's meat. To conserve a tiger not only means the conservation of a great number of herbivores and other species of wildlife, but also a vast area of forest or jungle as well. Therefore, it is a significant undertaking to rebuild an ecological system and to return captive tigers to it. However, one must not under-estimate the difficulties in carrying out such a project in southern China where the human population is so dense, and forested regions so few. It is not a simple matter to smoothly coordinate the man-tiger relationship. There are many social and economic problems in addition to natural and technical ones. Nevertheless, we hope that through the combined efforts of all organizations concerned, it will be possible to solve this problem in 5-10 years.

DIFFICULTIES AND PROBLEMS

Financial Problems

Funds for construction: Zoological gardens have limited funds and material resources and they are insufficient for constructing a conservation center, especially for the conservation of such a large carnivore like the tiger. A large-scale construction requires a large fund that must be collected elsewhere.

Expenses for feeding and management: It would be too heavy a burden for any zoo to bear the cost of feeding a large group of tigers. Without financial support from natural conservation organizations, the continued maintenance of a large breeding group would not be feasible. Other expenditures for public relations, publications, research and other aspects of an overall conservation project of this scope would also have to be found elsewhere.

Technical Problems

International cooperation and technical interchange in the fields of reproductive physiology, breeding techniques, genetic studies, etc. would sometimes be necessary for the success of the project.

REFERENCES

Liu Znenhe and Yuan Xicai. 1983. Resources of South China
    tigers in China. <u>Wildlife</u> 4.
Sheng Helin and Lu Houji. 1979. South China tiger on the verge
    of extinction. <u>Wildl. Cons. Util</u>. Dec.
Tan Bangjie. 1984. South China tiger on the threshold of life
    and death. <u>Nature</u> 5.
Xiang Peilon. 1983. An appeal for the rescue of the South China
    tiger. <u>Wildlife</u> 4.
Zhu Jing. 1980. Worried at the mention of tigers. <u>Nature</u> 1.

30

The Genetic and Demographic Characteristics of the 1983 Captive Population of Sumatran Tigers (*Panthera tigris sumatrae*)

Jonathan D. Ballou and John Seidensticker

INTRODUCTION

In this paper, we review the demographic and genetic status of the 1983 captive population of Sumatran tigers (Panthera tigris sumatrae) and discuss the implications of these findings for the management of both captive and wild populations of Sumatran tigers.

Planned captive breeding efforts to aid in the conservation of the tiger have focused primarily on the Siberian (Amur) subspecies, Panthera tigris altaica. A coordinated international program has been established to maximize their long-term captive viability by maintaining genetic diversity in a demographically stable population. The program addresses the genetic and demographic consequences of inbreeding and the loss of genetic variation, the regulation of population growth rates through control of reproductive rates, and the distribution of individuals across age and sex classes (Foose 1986, Seal and Foose 1983, Seal and Foose 1984, Seifert and Muller this volume). The tiger breeding plan includes recommendations for tiger shipments, and for pairings and breeding schedules for each participating institution. Recommendations are based on an analysis and consideration of the captive population's present, past, and future genetic and demographic status. Similar captive breeding and management programs are recommended for other extant tiger subspecies in accordance with a Global Tiger Survival Plan (Seal et al. this volume).

A captive breeding program for the Sumatran tiger has not been formulated. As an essential first step, we examined the distribution and the demographic and genetic status of Sumatran tigers in captivity. The genetic and demographic characteristics of these individuals will determine the objectives and actions that need to be taken by a Sumatran tiger captive breeding program.

Census estimates of the Sumatran tiger subspecies indicate that fewer than 800 wild individuals remain, many living in small, isolated habitats (Santiapillai and Widodo, Seidensticker this volume). We believe that lessons from the analysis of this data set for captive Sumatran tigers have important bearings on the long-term conservation efforts of wild Sumatran tigers.

THE DATA SOURCE: THE INTERNATIONAL TIGER STUDBOOK

Data used in this analysis are from the 1984 edition of the *International Tiger Studbook* (Seifert and Muller 1984). The Studbook is a chronology of the captive population of Sumatran tigers. For each known captive Sumatran tiger that has been registered since the late 1930's, it contains information on each tiger's birth date, death date, parentage, current and past locations and both institutional and Studbook (standardized) identification numbers (see Seifert and Muller this volume). The 1984 *International Tiger Studbook* includes data through 31 December 1983. Information on current and historical genetic relationships and demographic parameters (survivorship and fecundity) were derived from these data.

DEMOGRAPHIC CHARACTERISTICS

Studbook Summary

The 157 tigers in the 1983 population (Table 1) are distributed in 51 institutions world-wide. Individual collections range in size from one (6 institutions) to 13 (Tierpark, Berlin). The 56 tigers listed as "status unknown" were at institutions, animal dealers or private facilities that failed to respond to the numerous requests by the Studbook Keeper, Dr. Seifert, for further information. Most are probably dead, even if alive, these tigers are not likely to be significant contributors to the future captive population. For our analysis, these individuals were removed from the population on the date they were last censused in the Studbook.

Table 1. A Summary of the 1984 Sumatran Tiger Studbook Data (from Seifert and Muller 1984).

| Number of... | Sex | | | |
	Males	Females	Unknown	Total
Registered:	256	197	40	493
Wild-Caught:	13	13	--	26
(Produced Offspring):	(8)	(10)	--	18
Origin Unknown:	2	2	--	4
Captive Born:	241	182	40	463
Dead:	144	96	40	280
Status Unknown:	36	20	--	56
Alive in 1983:	75	82	--	157

This population has grown from 26 known wild-caught tigers and four tigers, included as founders, of unknown origin. Only 18 (60%) of these 26 tigers ever reproduced. Fourteen of the original 30 are still alive or have living descendants in the 1983 population (Table 2). This percentage is probably an overestimate of the proportion of wild-caught tigers that bred. It is probable that many wild-caught tigers that failed to breed before dying were never registered in the Studbook (P. Muller, pers. comm).

Both wild-caught tigers and tigers of unknown origin have entered the population intermittently since 1937 (Fig. 1), and as consequence, the founders have had various degrees of influence, both genetic and demographic, on the population.   The number of

Table 2.   The reproductive history, sex, first captive site and current status and locations of both wild-caught Sumatran tigers and Sumatran tigers of unknown origin.

Stud-book No.	Sex	First Captive Location (Yr Acquired)	Total	No. Offspring That Have Reproduced	Location/ Status (Yr Removed)
1	M	ROTTERDAM (??)	0	–	Dead (??)
2	F	ROTTERDAM ('48)	1	1	Dead ('61)
3	F	ROTTERDAM ('48)	9	0	Dead ('66)
4	F	ROTTERDAM ('53)	8	1	Dead ('71)
5	F	ROTTERDAM ('48)	0	–	Dead ('57)
6	M	BERLIN (TP) ('56)	7	2	Dead ('62)
7	F	BERLIN (TP) ('56)	15	6	Dead ('71)
8	M	PRAGUE ('58)	5	4	Dead ('66)
9	F	PRAGUE ('58)	0	–	Dead ('60)
10	F	ROTTERDAM ('62)	13	5	Dead ('78)
11	M	TOKYO (UZ) ('64)	0	–	Dead ('81)
12	F	BERLIN (TP) ('67)	9	4	Dead ('80)
13	M	JAKARTA ('66)	1	0	Dead ('80)
14	M	JAKARTA ('68)	1	0	JAKARTA
15	M	JAKARTA ('75)	0	–	Unknown('80)
16	M	JAKARTA ('75)	0	–	Unknown('80)
17	F	JAKARTA ('69)	2	0	Unknown('80)
48	M	ROTTERDAM ('53)	1	1	Dead ('53)
49	F	ROTTERDAM ('70)	30	4	ROTTERDAM
147*	M	WROCLOW ('53)	7	2	Dead ('68)
148*	F	WROCLOW ('53)	7	2	Dead ('62)
168*	M	COLORADO SPRINGS ('37)	22	4	Dead ('52)
169*	F	COLORADO SPRINGS ('39)	22	4	Dead ('52)
259	M	FRANKFURT ('72)	4	0	LEIPZIG
369	M	JAKARTA ('70)	0	–	Unknown('70)
370	F	JAKARTA ('77)	0	–	Unknown('77)
371	M	JAKARTA ('80)	0	–	JAKARTA
372	F	JAKARTA ('80)	0	–	Dead ('80)
386	M	JAKARTA ('81)	0	–	JAKARTA
387	F	JAKARTA ('81)	0	–	JAKARTA

* Sumatran tigers of unknown origin.

offspring produced by the original tigers varies from 0 to 30 with a mean of 5.5. This variation in reproduction reflects differences in the reproductive potential of individual tigers as well as differences in institutional efforts with respect to both their intent to breed and their success in breeding tigers.

Population Growth

The captive Sumatran tiger population exhibited two distinct growth phases (Fig. 1). Neither phase shows the exponential growth seen in many captive populations (Ballou 1985, Foose et al. 1986). Between 1937 and 1966, the population experienced a slow linear growth of one tiger/year. Since 1966, the growth, still linear, increased to approximately seven tigers/year. These phases match those found in both the North American and Soviet populations of Siberian tigers with respect to timing and degree (Seifert and Muller this volume). Note that this constant growth indicates that the growth rate, i.e., the number of animals recruited as a proportion of the total number in the population, is decreasing over time.

Age Structure

Age structures are useful for illustrating the overall demographic stability of the population and for highlighting significant year-to-year differences in mortality and fecundity effects.   In 1970, the standing population consisted of 53

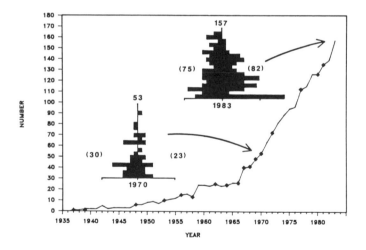

Fig. 1. The growth since 1937 and selected age structures for the captive population of Sumatran tigers. The points on the line indicate years in which wild-caught or assumed wild-caught tigers (founders) entered the population. The age structure for 1970 and 1983 shows the number of males (on the left of the vertical axis) and females (on the right) in different age classes (horizontal bars). The numbers in parentheses are the total number of males and females in the population. The number above the vertical axis is the total number in the population.

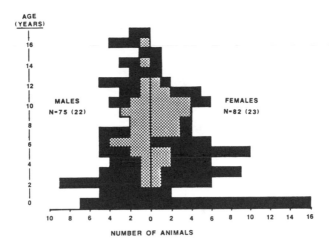

Fig. 2.   The 1983 age structure of the captive Sumatran tiger population.   The hatched area indicates tigers that have produced young.   The total number of males and females as well as the number that have produced young (in parentheses) is shown.

individuals: 30 males and 23 females (Fig. 1). Most tigers (22 males, 16 females) were of breeding age (over 2-3 years), but most of these were not proven breeders. The proportion of young tigers, the breeders of the future, was low. This age distribution reflects both the slow recruitment during the proceeding years and the low potential for future recruitment.

The 1983 standing age distribution is a more saturated age structure, although there is still much variation in the number of animals across age classes. This is not surprising given such a small population (Fig. 2).

Eighty percent of the 1983 population is of breeding age, only 29% (22 males and 23 females) are proven breeders. Since most adult tigers are paired, this low proportion of breeding animals is a reflection of non-breeding pairs rather than tigers maintained alone. It is not certain how many of the non-breeders are not reproducing because of institutional versus biological considerations. The status of this large number of non-reproducing animals must be evaluated as the captive breeding program is developed.

The overall age distribution does not show the wide-based pyramid characteristics of a stable growing population. The number of individuals in the pre-reproductive age classes is highly variable and relatively small in proportion to the breeding age classes. The future breeding population will show similar characteristics as these young tigers are recruited into the older age classes.

Sex Ratio

The sex ratio in the overall population and in the number of breeders is not skewed (Fig. 2). However, the 0-1 age class contains substantially more females than males because more females than males were born in 1983 (59% of the 42 cubs born were female), and mortality of the male cubs was greater than that of the females (44% vs. 30%).

There is a deficiency of both males and females in the 1-2 age class as a result of low birth rates and high infant mortality: the number of births in 1981 1982 and 1983 were 22, 29, and 42 respectively, and mortality rates of those born were 42%, 83% and 40%, respectively. The year 1982 had fewer births and significantly higher mortality rates than 1981 or 1983 (P < 0.001, Chi-squared test).

Age-specific Survivorship and Fecundity

The growth patterns in the population are clearly a function of both survival and fecundity rates. Thus, an understanding of both age specific survival and fecundity is necessary to project future population growth and to develop management plans.

Age-specific survivorship in the captive population of Sumatran tigers (Fig. 3) was calculated separately for male and female tigers and for the overall population using the methods described by Cutler and Ederer (1958). Overall survivorship to the first year of life was 67%, most of the mortality during the first year occurred during the first month. Once a tiger survived the first six months, it was likely to live several years and survivorship did not decrease sharply again until ages 10 and 11. At this age, mortality increases sharply; few tigers live over 15 years. The maximum longevity recorded was 18 years.

An accurate estimate of cub mortality rates is needed to develop recommendations regarding reproductive rates and popula-

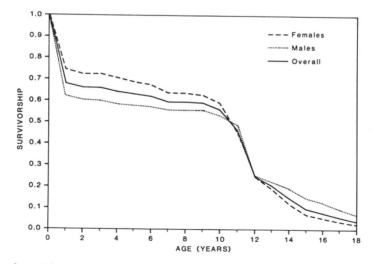

Fig. 3. The proportion of Sumatran tigers surviving from birth to different age classes for males, females, and the sexes combined.

tion growth control. If cub mortality varies highly from year to year, precision management is difficult. The percentage of cubs born that survive to one month of age for the years 1968 through 1983 is shown in Fig. 4. Before the mid-1970s, survivorship was quite variable. Overall survivorship has been decreasing steadily (at approximately 1% per year) and currently is approximately 60%. The reason for this downward trend is unknown. It may be due to the effects of increased levels of inbreeding (see below) or to a bias towards over-estimating survivorship in the early history of the population because institutions may not have reported stillbirths and early cub deaths. This trend needs further investigation.

Fecundity is the second component of a population's growth dynamics. Females (Fig. 5, A) begin to breed at ages 2-3 and show an increase in fecundity through age 5, a peak at ages 12 and 13, and a sharp decrease to age 17. There was no reproduction in females at age 17 or older. Males (Fig. 5, B) showed a similar pattern with a peak at ages 11, 12 and 13. However, males as old as 18 bred successfully.

GENETIC CHARACTERISTICS

Founders and Lineages

The entire captive population is descended from 18 individuals, 14 known to be wild-born and four of unknown origin (Tables 1 and 2). For the purpose of the genetic analysis of the population, the four tigers of unknown origin were assumed to be unrelated to each other and to the known wild-caught animals in the Studbook. All wild-caught animals were assumed to be

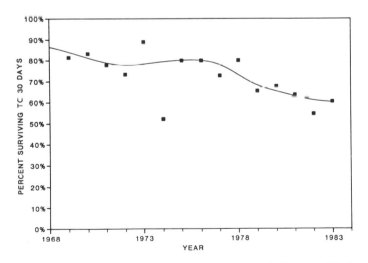

Fig. 4. The cub mortality (% cubs surviving to 30 days) for Sumatran tigers during the years 1968 through 1983. The points are the yearly cub mortality rates; the line reflects trends in the smoothed data.

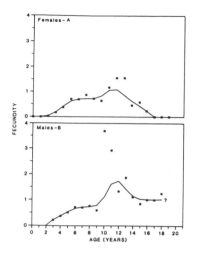

Fig. 5. The average number of offspring born to females and
sired by males of different age classes. The points are the data
values; the line reflects trends in the smoothed data.

unrelated to each other. Inbreeding coefficients and the genetic
contributions of founder animals to the gene pool were calculated
as described by Ballou (1983).

The genetic characteristics of the 1983 population were
determined by how these original founders bred among themselves
to form lineages and how these lineages crossed and combined. In
Fig. 6, we illustrate the lineages of the 18 founders who have
produced offspring. The captive population consists of three
separate lineages that we term the "Indonesian Line", the "North
American Line" and the "European Line".

The North American line was founded in the late 1930s at the
Cheyenne Mountain Zoo in Colorado by #168 and #169 - animals of
unknown origin. For forty years, this lineage consisted solely
of descendants of these two founders breeding among themselves.
The lineage remained at the Cheyenne Mountain Zoo until the early
1960s when animals were dispersed to other institutions. This
lineage has remained fairly small, growing to no more than 14
animals during the mid-1970s. There are now eight descendants
from this line-bred lineage. The European lineage is the largest
and was founded by 11 wild-caught animals and two animals of
unknown origin at the zoos in Rotterdam, Frankfurt, and Prague.
These founders entered the lineage at differing times and have
had various influences on the captive population. All founders
except #3 are represented in the 1983 population. Animals
descended from this lineage numbered 145 in 1983.

The North American and European lineages remained both
geographically and genetically distinct until the early 1970s.
At that time, animals from the European lineage were sent from
Europe to both North and South American zoos. Although the
lineages were no longer geographically distinct (some individuals
from each lineage were distributed in the same countries and, in
one case, even in the same zoo), they remained genetically

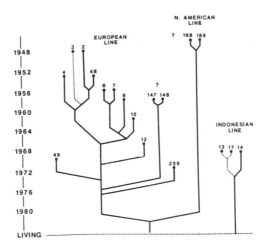

Fig. 6. Lineages of the Sumatran tiger founders who have produced young. Wild-caught and assumed wild-caught founders, represented by their Studbook numbers, are listed at the year in which they entered the population. Intersecting lines show the years when lineages were crossbred. The "?" indicate founders of unknown origin; the dotted lines represent lineages from founders no longer present in the 1983 population.

The Indonesian line, the smallest, was founded in Jakarta by three animals: #13, #14 and #17. These three animals bred among themselves, but no offspring survived and the only living member of the lineage is tiger #14 in Jakarta. The contributions of #13 and #17 have been lost from the population.

separate. It was not until 1980 that cubs were produced at the Cheyenne Mountain Zoo from a female of the North American lineage and a male from the European lineage. As a consequence, these lineages are no longer genetically distinct.

In Fig. 7, we illustrate the histories of the remaining 12 known wild-caught Sumatran tigers that failed to produce offspring. Three of these, #371, #386 and #387, are still alive in Jakarta. It should be noted that all of these wild-caught tigers (except #11) were, at one time or another, in institutions with other Sumatran tigers, but were either not paired for breeding or failed to breed before dying or being sent elsewhere.

Founder Contribution

The contribution of the founders to the gene pool has varied considerably from 1953 to 1983 (Fig. 8). It is not necessarily true that early founders have the highest contribution. For example, the founders of the North American lineage accounted for a relatively large proportion of the total captive gene pool until the late 1960s, but their 1983 contribution is very low (2.9% each).

The largest growth in contribution is from founders #6 and #7 (Fig. 9). Both entered the population just prior to 1958;

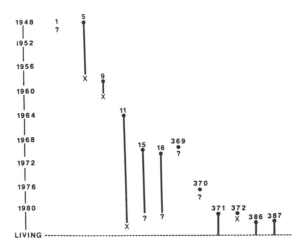

Fig. 7. The captive tenure of founders who have failed to produce offspring to the 1983 Sumatran tiger population. The vertical lines are identified by the Studbook numbers of the founders. The years of entry into and exit from the population are shown by the length of the vertical lines. An "X" indicates an exit by death, a "?" indicates that the status of the animal is unknown.

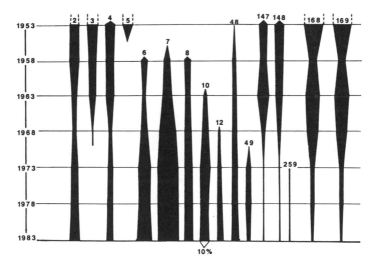

Fig. 8. The genetic contribution of the European and North American founders (Studbook #168 and #169) to the population between 1953 and 1983. Each bar is labeled by the Studbook number of the founders. Dotted lines at the top of the bars indicate that the founder was present in the population prior to 1953. The width of each bar corresponds to the proportion of that year's gene pool descended from each founder. A width corresponding to 10% is shown under the 1983 contribution of founder #10 for a point of reference.

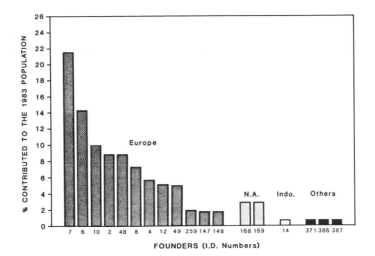

FOUNDERS (I.D. Numbers)

Fig. 9. The proportion of the 1983 captive Sumatran tiger gene
pool contributed by each of the extant founders. The founder
contributions are grouped by lineages: EUROPEAN = European
lineage; N.A. = North American lineage; INDO = Indonesian
lineage; OTHER = wild-caught tigers who have not yet produced
offspring.

their contribution increased steadily and stabilized in 1973.
They have remained the largest contributors to the gene pool.
Together, they have contributed over 35% of the 1983 gene pool.
The contributions from founders #14, #371, #386 and #387 simply
represent the living founders themselves.

Inbreeding

The small number of founders and their unequal contribution
to the gene pool are indicative of high levels of inbreeding in
the captive population. The increase in the proportion of the
population which is inbred and the mean level of inbreeding are
shown in Figs. 10 and 11. The number of inbred animals and the
mean level of inbreeding are greatest in tigers in the small
North American lineage. By 1965, all North American lineage
tigers were inbred and in 1983 the mean inbreeding coefficient
was 0.41. The European lineage, larger and descended from more
founders than the North American lineage, began inbreeding in
1966. In 1983, the mean level of inbreeding in the European
lineage was 14%.

Overall, the mean level of inbreeding in the Sumatran tiger
population has increased approximately 0.75% per year between
1950 and 1975 and approximately 1% per year since 1975. This
rapid increase in the levels of inbreeding is primarily the
result of the same individuals and their descendants breeding
from one year to the next.

The distribution of inbreeding coefficients in the 1983
population is shown in Fig. 12. Although 30% of the population
(49 animals) are still non-inbred, the inbred animals range in

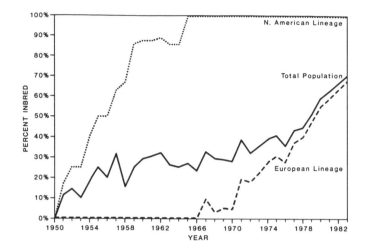

Fig. 10. The yearly increase in the percent of inbred animals in
the captive Sumatran tiger population. The values for the two
primary lineages and the total population are shown.

levels of inbreeding from 0.05 to 0.44. The largest class of
inbreeding is the 0.25 category and is equivalent to father/
daughter, mother/son, or full-sib matings.

    The overall increase in levels of inbreeding has had a
detrimental effect on the population through increased cub
mortality (Fig. 13). Inbred animals were defined as individuals
with inbreeding coefficients greater than zero, and non-inbred
animals as individuals with inbreeding coefficients equal to
zero. Cub mortality was defined as the proportion of cubs born
that died before reaching 30 days of age. Cub mortality was
significantly higher in the inbred cubs than in the non-inbred
cubs in both the European lineage (34% vs. 20%; p=0.003;
Chi-square test) and the overall population (31% vs. 22%;
p=0.036; Chi-square test). However, mortality was significantly
higher in the non-inbred cubs in the North American lineage (60%
vs. 9%; p=0.009; Chi-square test).

    These results suggest that the 13 European founders
contained enough genetic load to affect mortality when inbred
while the two North American founders either contained little
genetic load or the load was removed from the gene pool through
intense selection on the inbred animals. The high level of
inbreeding reached early in the lineage combined with the
observed level of inbred cub mortality being lower than both the
non-inbred and inbred mortality in the European lineage suggest
that they may have adapted to inbreeding. Further analyses will
be necessary to examine this relationship. But mortality in the
overall population indicates a deleterious inbreeding effect on
cub survival in the captive population.

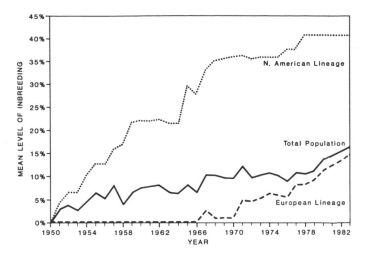

Fig. 11. The yearly increase in the mean level of inbreeding in the captive Sumatran tiger population. The values for the two primary lineages and the total population are shown.

DISCUSSION

The results of this demographic and genetic analysis of the captive population of Sumatran tigers are important for the management of Sumatran tigers, and all tigers, whether in captivity or in the wild.

The maintenance of genetic diversity is the primary goal of conservation-related captive propagation programs. Maintaining genetic diversity maximizes the probability of long-term conservation success (Hedrick et al. 1986). The implications of this analysis concern the effects a small, demographically unstable population, founded by only a few tigers, have on the population's capacity to maintain genetic diversity.

The strategy for maintaining genetic diversity in captive populations is outlined by Foose (this volume) and Foose et al. (1986). To summarize, the strategy entails:
  1) founding populations with enough wild-caught animals to adequately sample the genetic diversity (both hetero-zygosity and allelic diversity) of the wild population;
  2) minimizing genetic drift and loss of variation by maximizing the effective size of the population;
  3) increasing the population size to the captive carrying capacity as rapidly as possible and maintaining a demographically stable population at the carrying capacity;
  4) minimizing the level of inbreeding in the population;
  5) attempting to perpetuate equal genetic representa-tion of the founders in the population through time;
  6) avoiding selection; and
  7) if the overall population size permits, subdividing the population to maximize the total amount of diversity preserved.

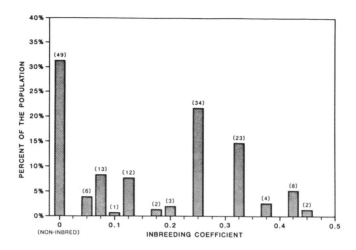

Fig. 12. The distribution of inbreeding coefficients in the 1983 population. The shaded bar indicates individuals descended from the North American lineage.

The captive population of Sumatran tigers fails to meet the specifications for several of these recommendations. Although 30 potential founders have been listed in the Studbook, only 18 have produced offspring and only 13 can properly be defined as "effective" in that they and their offspring have contributed portions of their genome to the 1983 population (Fig. 9). Five other potential founders (#14, #259, #371, #386, and #387) are still alive but have yet to be proven "effective."

Thirteen founders can retain as much as 96% of the heterozygosity found in the wild population, providing that they constitute an effective population size of 13 animals. However, an unequal sex ratio (here there were five males and eight females) and extreme variation in the number of offspring produced will reduce the effective size of the founders.

The variation in the number of offspring produced should be calculated using the life-time total number of "effective" offspring from each founder, i.e., the number of offspring that in turn lived to reproduce. Using the data from Table 2 and excluding founders that did not or have not (yet) produced breeding offspring, the variation in the founder offspring number was 2.7 and the mean number of offspring produced per founder was 3.1. The formula to calculate the effective population size (N_e) from variation in offspring number (Kimura and Crow 1963) is:

$$N_e = \frac{(N)(K) - 2}{K - 1 + (V/K)}$$

where N = the number of animals (here the number of founders; N = 13);

K   = the mean number of offspring per founder (K = 3.1);

V   = the variance in the number of "effective" offspring (V = 2.7).

We estimate the effective size of the founders to be 13. This agrees well with the actual number of founders because the sex ratio and the variation in family size were not highly skewed.

It is difficult to evaluate the success these founders have had in transferring heterozygosity from the wild population to the captive population. They arrived in the population at different times and it is difficult to determine the long-term genetic impact of recently acquired founders. Their levels of contribution to the 1983 population are highly skewed. Nevertheless, it seems likely that the founders have transferred a large proportion of the heterozygosity from the wild population to the captive population.

Genetic diversity can be measured in terms of heterozygosity and allelic diversity. Using heterozygosity as the sole criteria for evaluating retention of genetic diversity may be overly optimistic (Allendorf 1986, Fuerst and Maruyama 1986). Founding events and bottlenecks have a much larger impact on loss of allelic diversity than on heterozygosity (Nei et al. 1975, Denniston 1978). While we may be retaining much of the heterozygosity of the wild population with these 13 effective founders, we may in fact be losing substantial amounts of allelic diversity. Rare alleles (frequencies of 0.01 or less) will most certainly be lost as will many of the alleles of intermediate frequency (0.05 - 0.10). Effective founder sizes approaching 20 - 25 or even greater will be needed to secure the allelic diversity of the wild population (Allendorf 1986, Gregorius 1980).

The uncertain long-term contribution of recently acquired founders, the highly skewed genetic contribution of the effective founders to the current population, and questions relating to the inadequate sampling of the allelic diversity of the wild population by the current founders suggest the need for acquiring additional wild-caught founders for the captive population. The total number of effective founders should at least approach 20 for adequate genetic representation (both heterozygosity and allelic diversity) of the wild population.

The rate of loss of heterozygosity in this captive population is indicative of the population's overall effective population size. As evidenced by the steady rise in the mean level of inbreeding (Fig. 11), the captive population has, over the last 30 years, lost approximately 16% of its heterozygosity. This corresponds to an approximate 3.4% loss of heterozygosity per tiger generation (ca. 7 years) or an estimated mean historical effective population size of only 15 tigers. The rate of loss of heterozygosity has not been constant and has, at times, been substantially lower than 15. During the last tiger generation (the last seven years), the mean level of inbreeding has increased from 10% to 16%, indicating an effective population size of only 7.5 animals. During this period, the effective population size was approximately 5 to 8% of the actual number of tigers in the population.

This rapid rate of loss of heterozygosity is not surprising given the overall slow growth of the population, the unstable age structures observed in past years (Fig. 1), the relatively small proportion of breeding animals (Fig. 2), and the lack of efforts

to minimize inbreeding. A management program to retain genetic
diversity in this population will have to substantially increase
the population's effective size through both genetic and
demographic management.

Inbreeding in small populations has a dual effect on the
loss of genetic diversity. The first effect is a direct genetic
effect. Breeding related individuals causes a direct decrease in
the level of heterozygosity in the population, regardless of
whether or not inbreeding depression is present. For this reason
alone, inbreeding should be minimized in the captive population
if retention of genetic diversity is a primary goal (Foose this
volume).

The second effect is an indirect, demographic effect.
Inbreeding depression, the reduction in reproductive rates,
and/or the increase in mortality due to the unmasking of
deleterious recessive alleles in the homozygous state, affects
the demographic characteristics of the population. The observed
effect of inbreeding on cub mortality in the Sumatran tiger (Fig.
13) dampens the population's growth rate and should be considered
basically a demographic effect. Slower growing populations lose
genetic diversity at a faster rate than rapidly growing
populations, thereby increasing the levels of inbreeding in the
population. The deleterious effects of inbreeding become more
severe and further reduce the growth potential of the population.
The depression indirectly results in increased loss of diversity
by suppressing the growth of the population. In small
populations, this synergistic, circular effect could drive the
population to extinction or restrain its growth enough to make it
more susceptible to other unrelated demographic and stochastic
factors.

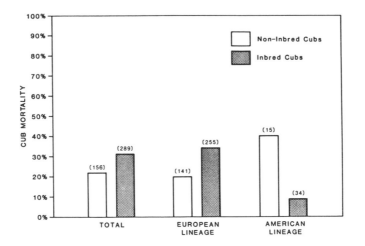

Fig. 13. The effect of inbreeding on cub mortality (% cubs dying
before 30 days) in the European, North American and combined
populations. The number of cubs born to each category are shown
in parentheses.

Although inbreeding depression is not universal, it does appear to be a very general phenomenon in captive mammalian populations (Ralls and Ballou 1983) and should be considered as a likely effect of inbreeding in both natural and captive populations of all tiger subspecies. The ramifications of inbreeding depressions in both small captive and wild populations are clear. Inbreeding should be minimized because of its direct effect on reduction in heterozygosity as well as its potential demographic effect on reduction in survivorship and reproduction.

CONCLUSIONS

The analysis of the 1983 captive population of Sumatran tigers revealed that:

1) The entire population has descended from 18 wild-caught or assumed wild-caught tigers that have entered the population intermittently since 1937. These 18 tigers formed two primary lineages, one in North American and one in Europe. The North American lineage, founded by only two tigers, remained very small (less than 14 animals) and is highly inbred. The European lineage, founded with 13 tigers, has accrued inbreeding at a relatively slower rate than the North American lineage but shows evidence of high mortality rates in inbred cubs;

2) The population has grown very slowly; recruitment was approximately one animal per year between 1940 and 1966 and seven animals per year since 1966;

3) The 1983 population of 157 tigers is distributed in 51 institutions world-wide and exhibits an unstable age distribution with low potential for population growth;

4) The slow population growth, the unstable age structures, the low number of founders, the historically small effective population sizes, and the increasing level of inbreeding indicate that the population is losing genetic diversity at a rapid rate.

These results suggest that the potential for the 1983 captive population to contribute significantly to the conservation of the Sumatran tiger subspecies is low. The small number of founders and their highly skewed genetic contribution to the living population suggests that the population may not have a representative sample of natural genetic diversity. The high levels of inbreeding and the low captive effective population size suggests that much of the diversity that the founders did transfer may have been lost.

If the captive population is to serve as an aid to the conservation of the Sumatran tiger subspecies, immediate action must be taken. Additional founder animals must be acquired for the captive population and a captive breeding program to maximize the retention of genetic diversity must be developed.

The goal of retaining genetic diversity should not be a banner carried only by captive breeding interests. The retention of genetic diversity is equally important for the long-term conservation of wild tigers and the management of wild tiger reserves must be cognizant of this goal.

Evidence of inbreeding depression in the captive population suggests that small natural populations of tigers may be susceptible to similar effects. The observation that inbreeding is related to offspring mortality in many mammalian populations suggests that the management of all tiger subspecies in small reserves consider, among other demographic factors, the demographic implications of higher, inbreeding-related cub death rates and lower reproductive rates on the population's long term survival.

ACKNOWLEDGEMENTS

We thank Ron Tilson for inviting us to participate in the tiger conference and Susan Lumpkin for reading and editing drafts of the manuscript. Our thanks also to Laurie Bingaman for editing and typing changes in the manuscript. Special thanks also go to Dr. Siegfried Seifert and Peter Muller, the International Tiger Studbook Keepers, for their work in compiling and publishing the International Tiger Studbook.

REFERENCES

Allendorf, F.W. 1986. Genetic drift and the loss of alleles versus heterozygosity. Zoo Biol. 5(2):181-90.

Ballou, J.D. 1983. In Genetics and Conservation: A Reference for Managing Wild Animal and Plant Populations, ed. C.M. Schonewald-Cox, S.M. Chambers, B. MacBryde and L. Thomas. Menlo Park, CA: Benjamin/Cummings Pub. Co.

Ballou, J.D. 1985. 1983 International Studbook for Golden Lion Tamarins. Washington, DC: National Zoological Park.

Cutler, S.J. and J. Ederer. 1958. Maximum utilization of the life table method in analyzing survival. J. Chronic Dis. 8:699-712.

Denniston, C. 1978. In Endangered Birds: Management Techniques for Preserving Threatened Species, ed. S.A. Temple. Madison Wisconsin: Univ. Wisconsin Pr.

Foose, T.J., R. Lande, N.R. Flesness, G. Rabb and B. Read. 1986. Propagation plans. Zoo Biol. 5(2):139-46.

Fuerst, P.A. and T. Maruyama. 1986. Considerations on the conservation of alleles and of genic heterozygosity in small populations. Zoo Biol. 5(2):171-79.

Gregorius, H.R. 1980. The probability of losing an allele when diploid genotypes are sampled. Biometrics. 36:643-52.

Hedrick, P.W., P.F. Brussard, F.W. Allendorf, J.A. Beardmore and S. Orzack. 1986. Protein variation, fitness, and captive propagation. Zoo Biol. 5(2):91-99.

Kimura, M. and J.F. Crow. 1963. The measurement of effective population size. Evolution 17:279-88.

Nei, M., T. Maruyama and R. Chakraborty. 1975. The bottleneck effect and genetic variability in populations. Evolution 29:1-10.

Ralls, K. and J.D. Ballou. 1983. In Genetics and Conservation: A Reference for Managing Wild Animal and Plant Populations, ed. C.M. Schonewald-Cox, S.M. Chambers, B. MacBryde and L. Thomas. Menlo Park CA: Benjamin/Cummings Pub. Co.

Seal, U.S. and T.J. Foose. 1983. Development of a masterplan for captive propagation of Siberian tigers in North American zoos. Zoo Biol. 2:241-4.

Seal, U.S. and T.J. Foose. 1984. Siberian tiger Species
    Survival Plan: A strategy for survival. <u>J. Minn. Acad. Sci.</u>
    49:3-9.
Seifert, S. and P. Muller. 1984. <u>1984 International Tiger
    Studbook</u>. Leipzig: Leipzig Zool. Garten.

31

Development of an International Captive Management Plan for Tigers

Siegfried Seifert and Peter Muller

INTRODUCTION

In 1973 we assumed responsibility for the development of an international tiger studbook begun at the Prague Zoological Garden. The first register, published in 1976, contained information on 1,197 Amur tigers (<u>Panthera tigris altaica</u>) and 227 Sumatran tigers (<u>P. t. sumatrae</u>) current up to December 31, 1975. Since that time we have been sending updated issues of the <u>International Tiger Studbook</u> to about 700 tiger keepers, scientific institutions and libraries every year. After every four annual reports containing acquisitions, births, movements and deaths of the last year, a revised <u>General Register</u> is published in the fifth year in order to provide a better survey and make the studbook easier to use. Furthermore, the annual copies of the studbook contain analyses of the data gathered, and other contributions on tiger management.

Since, to our surprise, at first no provable pure-bred Bengal tigers (<u>P. t. tigris</u>) were announced, we have been registering all white and all normal-colored Bengal tigers originating from whites since 1980 because their lineage may completely be shown.

In the following we present a survey of the registered tiger stock development by means of tables (1-9) and diagrams (Fig. 1-10) in which we have restricted ourselves to the Amur tiger.

The zoos of the Soviet Union, the main heart of the Amur tiger distribution, naturally have a large stock with a great number of wild-caught animals. That is why we chose a separate list for them. The fourth diagram contains the tiger-keeping institutions outside of the mentioned regions.

The first table of every region shows the annual stock changes, on the left side the growth of stock due to acquisition of wild caught animals and zoo-born animals from other regions, as well as offspring within the respective region itself, whereas on the right side the sale of zoo-born tigers into other regions and deaths divided into wild-caught and zoo-born animals is shown. From this results the growth and the entire stock at the

348

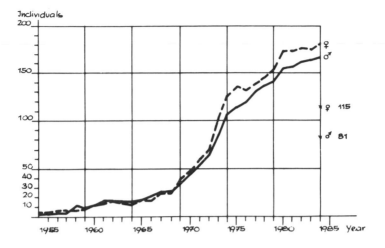

Fig. 1. Stock development of the Amur tiger in North America.

end of the year 1985, as shown in the last two columns. This figure is also shown in the last issue of the International Tiger Studbook as the official stock data. In the tables at hand and also in previous publications of the Tiger Studbook up to 1984, all those tigers are considered in the stock number which had to be taken as living according to the register but for which we had not received any report about their whereabouts or possible death. This part of the stock is without any meaning for the breeding management of the population because surely not all of the animals are still living, so we have assumed a reduced number of animals. Hence the lowest line of the table shows only the stock per January 1, 1985 confirmed by announcements of tiger

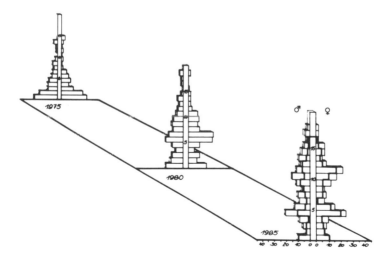

Fig. 2. Age structure of the Amur tiger stock in North America.

Table 1. Stock development of the Amur tiger (<u>Panthera</u> <u>tigris</u> <u>altaica</u>) in the North American region (Males/Females). Confirmed stock as of January 1, 1985 = 81/115.

Year	Increases			Decreases			Growth	Stock
	Born in Region	Acquired Wild-caught	Acquired Zoo-born	Moved from Region	Died Wild-caught	Died Zoo-born		
Pre-1955	5/12	1/0				4/8		2/4
1955		1/1				0/1	1/0	3/4
1956	0/1	1/1					1/2	4/6
1957	2/1 +2					2/1 +2		4/6
1958	8/3 +5	1/1	0/1			2/4 +5	7/1	11/7
1959	1/3			1/1	1/1		-1/+1	10/8
1960	2/3		0/1			1/1	1/3	11/11
1961	6/3 +1	1/0	2/2			3/3 +1	6/2	17/13
1962	0/4		1/2			0/2	1/4	18/17
1963	1/1			0/2	1/1		0/-2	18/15
1964		1/0			1/1	1/0	-1/-1	17/14
1965	1/3		2/1			1/0	2/4	19/18
1966	5/0 +2		1/0	2/0		0/0 +2	2/0	21/18
1967	8/9		1/2	0/1		3/2	6/8	27/26
1968	7/7 +4		1/0	1/4		5/2 +4	2/1	29/27
1969	8/11 +3		2/5	1/1		4/2 +3	5/13	34/40
1970	13/13		2/4	2/2		1/6	12/9	46/49
1971	13/12		1/1	1/0	1/0	3/2	9/11	55/60
1972	25/18 +1		1/2	1/2		16/7 +1	9/11	64/71
1973	33/37		7/8	4/2		14/12	22/31	86/102
1974	31/36 +1		3/4	1/3		10/12 +1	23/25	109/127
1975	17/17		1/0	0/2		11/6	7/9	116/136
1976	22/12			3/5		15/12	4/-5	120/131
1977	18/19 +2			0/1		7/9 +2	11/9	131/140
1978	27/17		1/1	4/1		16/12	8/5	139/145
1979	19/23			7/8		10/8	2/7	141/152
1980	24/39 +2			2/2		7/15 +2	15/22	156/174
1981	8/13 +1		1/0			7/13 +1	2/0	158/174
1982	11/12 +4		2/1			10/9 +4	3/4	161/178
1983	7/10		2/3	3/8		4/7 +1	2/-2	163/176
1984	13/12		0/1	2/3		8/5	3/5	166/181
	335/351 +29	4/3	32/39	34/46	4/3	167/163 +29		

breeders during the past two years. In future studbooks we will consider only these tigers.

Table 1 shows that in North America there have been only seven wild-caught animals; the last one died in 1971. However, in contrast to the following regions, there has been an exchange with other regions which was began early and intensified (columns 4 and 5).

A slow rise prior to 1969, a heavy growth between 1969 and 1980, and thereafter a very slight increase in the number of Amur tigers in North America can be noticed.

Table 2. Age structure of the Amur tiger (Panthera tigris altaica) stock in the North American region (Males/Females).

Age Yrs.	1960	1965	1970	Year[a] 1975	1980	1985	1985[b]
1	1/2		34/40	27/32	16/17	10/11	10/11
2	4/0	2/1	6/7	21/26	12/8	4/7	4/7
3	1/0	1/2	3/5	12/15	10/11	6/8	4/5
4	1/2	4/2	5/3	12/9	7/6	9/10	1/4
5		4/6	2/2	10/13	10/12	21/24	12/20
6	1/2	2/1		6/9	17/23	13/15	4/12
7	1/1	1/0	1/2	5/7	14/24	9/6	3/1
8		1/0	1/2	3/2	9/11	10/8	4/5
9		1/1	5/2	5/1	10/9	7/6	6/5
10			2/5	2/2	9/10	12/12	8/5
11	1/0	1/1	2/1	0/1	5/9	15/17	5/9
12			1/0	1/2	4/3	11/21	7/16
13				1/1	3/2	8/8	7/5
14				4/2	4/2	9/5	3/3
15				0/4	1/2	7/7	1/4
16			1/1		0/1	4/7	1/2
17					1/0	2/2	1/1
18					1/0	3/2	
19					2/1	2/2	
20							
	10/7	17/14	34/40	109/126	135/151	162/178	81/115

[a] Stock as of January 1.
[b] Confirmed.

Table 2 shows the age structure of the stock. For this purpose a five-year period was chosen. For 1985 both the theoretically existing and confirmed stocks are given.

For the graph, only the last columns of the table have been chosen since the stocks of the years from 1960 to 1970 were low, and thus rather uninformative. In 1975 there were a great number of one and two-year-old animals, which in 1980 were six and seven years old and thus in the best breeding age. This resulted in a great number of one-year-old cubs in 1980. From the information mentioned above, it can be seen that in 1985 two large age groups of five and six year olds and 10 and 12 year olds are striking. This coincides with the optimal ages for the procreation of the desired litters as stated in the World Breeding Plan. It is important, though, that in 1985-87 a greater number of cubs will have to be born in order to ensure propagation and a good age structure for the years to come. The stock confirmed for 1985 is especially emphasized in the graph.

In this compilation of tiger stocks some differences between European (excluding Soviet) and North American conditions are striking: in Europe the entire stock is essentially larger, more wild-caught animals have been included into breeding (altogether 21 animals, five of them still alive, kept in Leipzig), and the continuous delivery into other regions has begun. The acquisi-

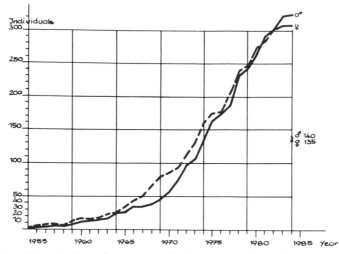

Fig. 3. Stock development of the Amur tiger in Europe (except the USSR).

tion of zoo-born animals from other regions started in 1965, due to a greater base of wild-caught animals in Europe. This particularly refers to the Hamburg, Rotterdam and Leipzig, lineages still over-represented in the present world stock.

The 1975 age structure of the European Amur tiger stock shows a broad base of young animals, and in 1980 a different but totally sufficient stock in all age groups. The relatively high number of five, seven, ten and 11-year-old tigers in 1985 is striking and comprise at present a broad breeding base. In

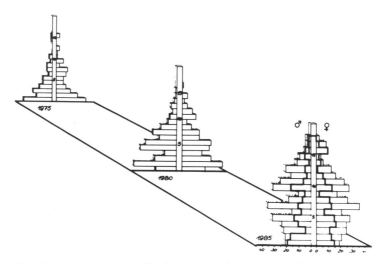

Fig. 4. Age structure of the Amur tiger stock in Europe (except the USSR).

Table 3. Stock development of the Amur tiger (Panthera tigris altaica) in the European region, except USSR (Males/Females). Confirmed stock as of January 1, 1985 = 140/135.

Year	Born in Region	Acquired Wild-caught	Zoo-born	Moved from Region	Died Wild-caught	Zoo-born	Growth	Stock
Pre-1955	1/0	1/2		1/0				1/2
1955		3/3					3/3	4/5
1956		2/3		1/0			1/3	5/8
1957	3/2				1/1		2/1	7/9
1958	2/0 +2			0/1		3/1 +2	-1/2	6/7
1959	4/6				0/1	2/0	2/5	8/12
1960	4/7			0/1		0/1	4/5	12/17
1961	3/2			2/2		0/1	1/-1	13/16
1962	6/8			1/2	0/1	2/2	3/3	16/19
1963	9/5 +3				2/1	4/0 +3	3/4	19/23
1964	9/?			2/0		1/1	6/1	25/24
1965	7/12	1/0	0/1	2/1	1/0	2/2	3/10	28/34
1966	9/8 +1		1/0			3/1 +1	7/7	35/41
1967	4/16 +1			1/2		3/5 +1	0/9	35/50
1968	14/18 +2		1/4	1/0		10/6 +2	4/16	39/66
1969	18/31 +2		1/1	2/7	0/1	10/10 +2	7/14	46/80
1970	27/24 +1	1/0	2/2	2/4	2/2	13/13 +1	13/7	59/87
1971	27/30 +1		1/0	1/1		11/22 +1	16/7	75/94
1972	47/40 +9		1/2	1/2	0/1	24/20 +9	23/19	98/113
1973	39/47 +3	0/1	5/3	9/10		26/24 +3	9/17	107/130
1974	58/49		2/3	3/4		27/19	30/29	137/159
1975	57/46 +11	1/0		2/2		31/28+11	25/16	162/175
1976	27/42 +1		2/2	2/4		16/37+1	11/3	173/178
1977	32/50 +3		2/2	3/4		17/18+3	14/30	187/208
1978	69/69 +4		1/0	4/7		22/30+4	44/32	231/240
1979	44/45 +6			1/5		32/31+6	11/9	242/249
1980	59/61			6/3		29/32	24/26	266/275
1981	49/48 +4		1/0	1/1		23/37+4	26/10	292/285
1982	54/61 +6	1/2	0/1	4/2		40/45+6	11/17	303/302
1983	41/46 +3			2/2	0/1	20/37+3	19/6	322/308
1984	44/37		2/1	1/2	1/0	42/35	2/1	324/309
	767/812 +63	10/11	22/22	55/69	7/9	413/458+63		

contrast to the current North American population, a good prerequisite for future breeding is provided by numerous one and two-year-old animals.

The Amur tiger stock of the Soviet part of the European region is characterized by a great number of wild-caught tigers. Forty-eight tigers are registered altogether, 26 of which were still alive in 1985. The strikingly late beginning of breeding in the Soviet Union may likely be explained by insufficient data from the period before 1968. Because of an extraordinarily good breeding base, few zoo-born tigers have been imported from other regions up to now. On the other hand, an increasing number of Soviet zoo-born tigers have been exported in recent years. The high percentage of non-confirmed stock in 1985 results from the

Table 4. Age structure of the Amur tiger (<u>Panthera tigris altaica</u>) stock in the European region, except USSR (Males/Females).

Age Yrs.	Year[a]						
	1960	1965	1970	1975	1980	1985	1985[b]
1	4/5	7/3	14/21	34/35	19/23	21/18	19/14
2		3/4	5/13	19/25	41/39	21/20	17/11
3	1/0	4/5	6/14	23/17	21/28	22/24	5/10
4		2/0	6/4	16/12	12/20	24/24	11/15
5	1/3	2/0	4/11	11/8	35/27	32/27	14/14
6	1/3	4/5	4/3	11/18	30/28	15/20	8/6
7	1/0		2/2	4/11	17/17	36/35	11/11
8		1/0	0/4	6/13	23/17	19/25	7/10
9	0/1			6/3	15/9	10/16	5/2
10		1/2	1/0	2/8	8/5	32/26	16/13
11		1/1	3/4	3/1	7/13	25/22	6/7
12				0/1	4/10	16/15	4/5
13				0/3	5/5	20/11	5/1
14		0/1			2/2	12/5	1/3
15			1/2	1/0	1/5	9/3	5/2
16				1/3	1/0	4/9	2/9
17					0/1	3/7	2/2
18						3/2	2/0
19			0/1				
20							
	8/12	25/21	46/79	137/158	241/249	324/309	140/135

[a] Stock as of January 1.
[b] Confirmed.

great number of registered tigers living in circuses which do not report to the <u>Tiger Studbook</u>.

The stock summarized in Tables 7 and 8 does not refer to a geographically connected area but includes Latin America, Africa, South and East Asia and Australia. Thus, it may not be taken as an uniform population. This area includes two native countries of the tiger, the People's Republic of China and the People's Republic of Korea. That is why there is a relatively high percentage of wild-caught animals (12 specimens), but, due to incomplete reports, we do not know how many animals are still living.

The stock increase evident from this diagram may be attributed to both the increased acquisition of zoo-born animals from North America and Europe beginning around 1975, and the development of their own tiger breeding at about this time.

In Table 9 the inbreeding coefficients of the different areas are compiled. Nearly half of all tigers living in North America have the inbreeding coefficient 0, whereas in Europe, with the exception of the Soviet stocks, it is only one-third of the animals. In Soviet zoos all Amur tigers, except for two animals, have an inbreeding coefficient of 0, and in areas outside the North American and European regions about one-fourth

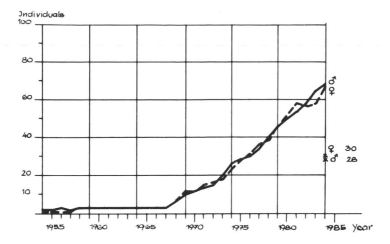

Fig. 5.   Stock development of the Amur tiger in the USSR.

of the animals have an inbreeding coefficient of 0, a preponderance of them females.

A look at the development of the tiger stock in the zoos of the world results suggests that it was necessary and worthwhile to establish an international studbook for this species, and to continue to keep it current.

It has, however, turned out that it is not sufficient to provide just information, even though it is comprehensive and revealing.   Impulse and instruction are needed in order to use the achieved preconditions for implementing desirable goals.

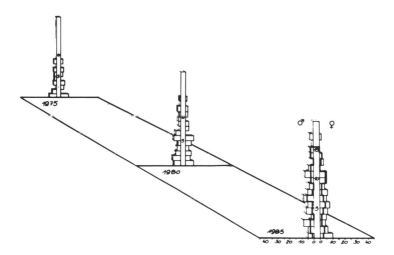

Fig. 6.   Age structure of the Amur tiger stock in the USSR.

Table 5.   Stock development of the Amur tiger (<u>Panthera</u> <u>tigris</u> <u>altaica</u>) in the European region, USSR only (Males/Females).   Confirmed stock as of January 1, 1985 = 28/30.

Year	Born in Region	Acquired Wild-caught	Zoo-born	Moved from Region	Died Wild-caught	Zoo-born	Growth	Stock
Pre-1955		2/1						2/1
1955								2/1
1956			1/0				1/0	3/1
1957					1/0		-1/0	2/1
1958		1/2					1/2	3/3
1959								3/3
1960		0/1			0/1			3/3
1961								3/3
1962	0/1				0/1			3/3
1963								3/3
1964								3/3
1965								3/3
1966								3/3
1967		1/0			1/0			3/3
1968	0/2	4/2					4/4	7/7
1969	1/0	2/2	0/2				3/4	10/11
1970	1/0	1/2				0/2	2/0	12/11
1971	2/6	1/0				1/2	2/4	14/15
1972	1/3	2/0		2/0	0/2		1/1	15/16
1973	4/2	2/4		1/1	0/1	0/2	5/2	20/18
1974	10/7	0/1		1/0	1/1	1/1	7/6	27/24
1975	6/5	1/2		1/0	1/1	3/3	2/3	29/27
1976	4/5	0/2	0/1	2/2	1/0	0/2	1/4	30/31
1977	13/8 +3			2/1		7/2 +3	4/5	34/36
1978	11/10 +3			1/0	0/1	4/6 +3	6/3	40/39
1979	10/10 +1	2/0			1/0	5/3 +1	6/7	46/46
1980	6/5	1/1			1/0	2/1	4/5	50/51
1981	11/13	1/1		1/0	1/1	7/6	3/7	53/58
1982	11/7 +2			0/1	1/1	5/7 +2	5/-2	58/56
1983	8/13	3/2		1/2	1/2	2/9	7/2	65/58
1984	5/13	0/1		2/2		0/3	3/9	68/67
	104/110 +9	24/24	1/3	12/9	12/10	37/51+9		

It is our feeling that not all tiger breeders have made adequate use of the International Tiger Studbook in the attempt to establish genetically optimal stocks.   That is why it is significant that U.S. Seal, based on the data available from our Studbook, drafted a World Breeding Plan for the tiger, and submitted it to the International Union of Directors of Zoological Gardens.   This document has been approved after extensive discussions.   The implementation depends on how many tiger breeders are ready to participate in the program.   In order to find this out we sent an inquiry sheet to about 600 institutions which keep or intend to keep tigers.   According to the responses we have received so far zoos agreed to include Amur, Sumatran and tigers of other subspecies.   Thus the

Table 6. Age structure of the Amur tiger (_Panthera tigris altaica_) stock in the European region, USSR only (Males/Females).

Age Yrs.	1960	1965	1970	Year[a] 1975	1980	1985	1985[b]
1			2/2	9/9	7/8	5/10	1/2
2			3/3	5/0	8/5	5/6	3/3
3	0/2	0/1		0/3	5/6	9/3	3/1
4			3/2	3/4	3/3	2/4	0/1
5	0/1		0/1	2/0	2/5	6/6	1/2
6				2/2	7/8	5/7	1/4
7				3/3	5/0	9/5	6/3
8		0/1	0/1		0/3	5/6	1/1
9	1/0			3/2	2/3	3/3	0/1
10	2/0	0/1			2/0	2/5	2/4
11					1/2	7/5	2/4
12					3/2	4/0	2/0
13			0/1			0/3	0/1
14		1/0			1/1	2/2	2/2
15		2/0	0/1			2/0	2/0
16							
17						2/1	2/1
18				0/1			
19			1/0				
20			1/0				
	3/3	3/3	10/11	27/24	46/46	68/66	28/30

[a] Stock as of January 1.
[b] Confirmed.

preconditions for the implementation of a World Breeding Plan for all tiger subspecies kept in zoos are met.

The object of this plan is to retain as much of the genetic diversity available in the founder stock as is possible with current space and resources. In order to achieve this, a program of bringing together the appropriate breeding animals will be developed and pursued on an international level, the development of an effective population size of the total tiger stock kept in zoos will be pursued.

These procedures will be directed by one international and several regional coordinators. For the time being, Europe and North America have been designated as regions in this sense. Until the establishment of further regions, breeding facilities from outside these two mentioned regions can participate in either the North American or European region.

The coordinators do not put forth their recommendations alone, but decide on these cooperatively with a committee consisting of nine elected representatives of zoological gardens participating in the Breeding Plan. The keeper of the International Tiger Studbook acts as International Coordinator, the Regional Coordinator for North America is Ulysses Seal, Minnesota, and Georgina Mace, London, assumes this task for

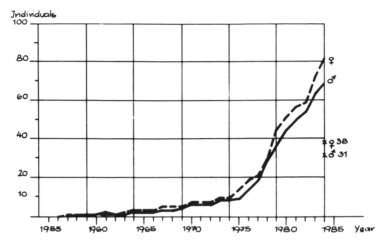

Fig. 7. Stock development of the Amur tiger outside North
America and Europe.

Europe. The election of the committee occurs after receipt of
the requested agreement to participate in the Breeding Program
for three years respectively. By circulars and publications in
the International Tiger Studbook of 1984, the World Breeding Plan
has become known to all interested parties. Once more, those
items concerning the breeders in practice are:

    1) A total population size of 300 individuals of Amur tigers
is planned for Europe and North America;
    2) New founder stock should be systematically incorporated

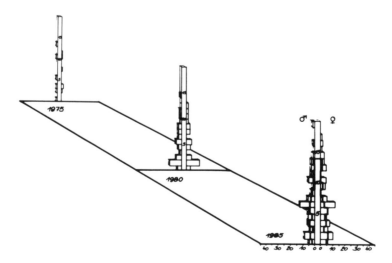

Fig. 8. Age structure of the Amur tiger stock outside North
America and Europe.

Table 7. Stock development of the Amur tiger (Panthera tigris altaica) outside the North American and European regions, (Males/Females). Confirmed stock as of January 1, 1985 = 31/38.

Year	Increases — Born in Region	Increases — Acquired Wild-caught	Increases — Acquired Zoo-born	Moved from Region	Decreases — Died Wild-caught	Decreases — Died Zoo-born	Growth	Stock
Pre-1955								
1955								
1956								
1957		0/1					0/1	0/1
1958		1/0					1/0	1/1
1959			1/1		1/1			1/1
1960								1/1
1961		1/0					1/0	2/1
1962	0/1			1/1			-1/0	1/1
1963		0/1					0/1	1/2
1964		1/1	1/0		1/0		1/1	2/3
1965		0/1		0/1				2/3
1966								2/3
1967	0/1	1/0	0/1				1/2	3/5
1968								3/5
1969	1/0						1/0	4/5
1970		2/2					2/2	6/7
1971	0/1				0/1			6/7
1972								6/7
1973			2/2				2/2	8/9
1974								8/9
1975	4/3		2/4	2/2		3/1	1/4	9/13
1976			5/8			0/2	5/6	14/19
1977	3/2		3/5	0/1		1/4	5/2	19/21
1978	2/2		7/7			0/1	9/8	28/29
1979	4/7 +1		8/13			3/4 +1	9/16	37/45
1980	0/2		8/5			0/1	8/6	45/51
1981	6/7 +1		0/1			1/2 +1	5/6	50/57
1982	13/5		4/2	2/1		11/4	4/2	54/59
1983	9/7		4/9			4/2	9/14	63/73
1984	3/5		3/5			0/1	6/9	69/82
Total	45/43 +2	6/6	48/63	2/3	3/4	25/23 +2		

into the international population, and the offspring of wild-caught animals should be reasonably distributed in order to broaden the founder base;

3) Each founder (wild-caught animal) should be bred with five different mates to produce two surviving litters from every single mate;

4) The contribution of over-represented founder lineages should be reduced, and at the same time the contribution of under-represented lineages should be expanded. Mating between closely related or even sibling individuals should be avoided;

5) In order to achieve and maintain the intended size and composition of the population, each zoo-born male and each zoo-born female should equally contribute to the next generation, at best in a sex ratio of 1:1. This means that each female has

Table 8.   Age structure of the Amur tiger (Panthera tigris altaica) stock outside the North American and European regions (Males/Females).

Age Yrs.	1960	1965	Year[a] 1970	1975	1980	1985	1985[b]
1		1/1	1/0	1/0	3/4	4/6	2/5
2		0/1			12/13	8/10	4/6
3	1/0	0/1	1/1	1/1	5/5	3/6	2/3
4		1/0		1/2	1/0	5/7	5/5
5	0/1		0/1		2/4	5/4	2/4
6			1/1	1/0	5/5	5/6	4/6
7			0/1		2/0	13/12	4/3
8				0/1	2/3	5/5	3/2
9			1/0		2/3	2/0	2/0
10					0/1	1/3	0/2
11				1/1	1/1	5/4	1/0
12				0/1	1/1	2/0	1/0
13					0/1	2/4	0/1
14				1/0		2/3	1/1
15					1/0	1/3	
16						0/1	
17						1/2	
18						0/1	
19							
20						1/0	
	1/1	2/3	4/4	6/6	37/41	65/77	31/38

[a] Stock as of January 1.
[b] Confirmed.

to produce approximately two litters with a total of four surviving offspring, and each male must therefore also sire two litters;

6) The first litter should be produced when the parents are four to six years of age, and the second at the age of nine to 11. For retention of genetic diversity, do not use phenotypic selection of animals for the breeding program;

7) Animals 12 to 13 years of age should be removed from the breeding program;

8) If a participant wishes to have more litters than determined in the plan, that will be left up to him. The additional offspring would then not be included in the breeding program but will continue to be registered in the International Tiger Studbook;

9) In regional plans there are established recommendations as to which animals are to be brought together for breeding in which years in order to achieve the objectives of equalizing founder representation, avoiding or reducing inbreeding, and stabilizing age structure and population size;

10) All tigers included in the breeding program will be given some designation.

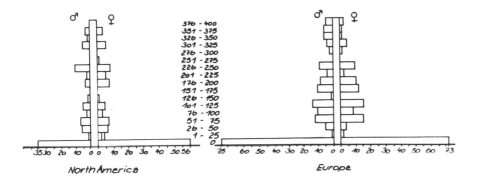

Fig. 9. Inbreeding coefficient of the Amur tiger stock in North America and Europe.

It is also important that attention be given to the other subspecies. This refers to the Sumatran (P. t. sumatrae), Bengal (P. t. tigris), Indochinese (P. t. corbetti) and South Chinese (P. t. amoyensis) tigers, for which populations capable of development will need to be built up. Close cooperation between the countries of origin of these subspecies will be particularly important.

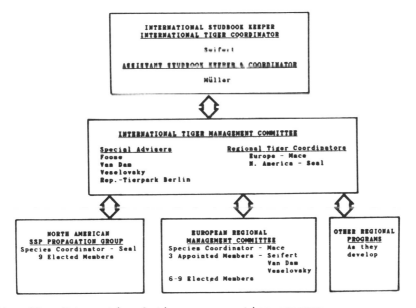

Fig. 10. International tiger propagation program.

Table 9.   Inbreeding coefficients of the Amur tiger (*Panthera tigris altaica*) stock as of January 1, 1985 (Males/Females).

Inbreed. Coeffic. (x 1000)	N. America		Europe (except USSR)		Europe (USSR only)		Outside N. America & Europe	
	No.	%	No.	%	No.	%	No.	%
0	35/56	43/49	48/44	34/33	27/29	96/97	6/11	19/29
8	1/1	1/1						
16	1/1	1/1	1/2	1/2				
20	0/2	0/2						
31	6/6	7/5	3/3	2/2				
35							0/1	0/3
47			2/0	1/0			1/1	3/3
55	1/0	1/0					1/1	3/3
63	3/6	4/5	12/16	9/12				
70	3/1	4/1						
78	0/2	0/2	4/3	3/2				
86							0/2	0/5
94	2/1	3/1	6/6	4/4			1/0	3/0
102			3/4	2/3				
109			5/8	4/3			5/4	16/11
125	5/5	6/4	6/4	4/3			2/0	7/0
133	1/1	1/1						
141			1/0	1/0				
152			2/5	1/4				
156			4/4	3/3				
172			4/4	3/3			1/1	3/2
188	3/7	4/6	7/10	5/7			3/4	10/10
200			1/1	1/1				
219	2/5	3/4	3/2	2/2			1/0	3/0
234	0/1	0/1					0/1	0/3
250	11/9	14/8	8/9	6/7	1/1	4/3	0/1	0/3
266	0/1	0/1						
281			2/0	1/0			8/6	26/16
297			0/1	0/1				
309			2/1	1/1				
313	4/6	5/3	2/4	1/3			1/4	3/11
336	1/0	1/0						
340			2/0	1/0				
344			3/1	2/1			0/1	0/3
375	2/4	3/4	5/1	4/1			1/0	3/0
379			4/2	3/2				
	81/115		140/135		28/30		31/38	

32

Possible Selection in Captive
Panthera tigris altaica

Nathan R. Flesness and Karen G. Cronquist-Jones

Because half or better of all extant <u>Panthera tigris altaica</u> are in captivity, and wild habitat limitations restrict the wild population(s) to some few hundred, the captive population needs to be managed with long-term conservation intent. It is an insurance policy against catastrophe in the wild, and also has potential for slowing the total loss of genetic diversity.

The value of captive populations for conservation depends, in considerable part, on how suitable they are as a source for stock for re-introduction of genes or individuals. An important unanswered question is how rapidly such populations are becoming genetically adapted to captivity. Such adaptation would make them easier to sustain and manage under captive conditions, but would presumably eliminate some genetic traits and variability useful in the wild.

An earlier worker made quite a start in the qualitative study of the selection that typically goes on in populations managed intimately by man. Charles Darwin, in 1868, published <u>The Variation of Animals and Plants under Domestication</u>. Darwin spent years talking with the animal breeders of his day, and found a couple of broad generalizations that seemed valid.

Darwin noted that species where breeding was controlled by man tended to show reduction in traits that caused managers difficulty - i.e. lowered aggression, shortened muzzles, etc. Darwin also noted that rare or newly mutant coat-color genes were positively selected by most human breeders, developing different colored and patterned individuals at much greater frequency than ever found in the wild.

Phenotypic studies of possible morphological changes in captive tiger populations are just beginning. Genotypic studies have only recently begun as well. It is too early to conclude a great deal from these. In the absence of detailed quantitative data on individual traits, it would be a useful start to set bounds on the amount of selection which <u>may</u> have occurred in the captive population.

Almost 30 years ago James Crow (Crow 1958) published an article entitled "Some Possibilities for Measuring Selection

Intensities in Man". He developed a "Selection Intensity"
parameter, denoted "I", which is a measure of the total possible
selection, i.e. an upper bound. To quote Crow,

> "There can be selection only if, through
> differential survival and fertility, individuals of one
> generation are differentially represented by progeny in
> succeeding generations. The extent to which this
> occurs is a measure of <u>total</u> selection intensity. It
> sets an upper limit on the amount of genetically
> effective selection."

I is the variance in progeny number divided by the square of
the mean progeny number:

$$I = V \, / \, W^2$$

It measures the upper bound on selection. Crow called it the
"Index of Total Selection". We will follow Cavalli-Sforza and
Bodmer (1971) and refer to it as the Index of the <u>Opportunity</u>
for Selection.

Another way to state the meaning of this quantity is to
imagine that all differences in fitness between individuals are
completely genetically determined. If this were so, the Index of
Opportunity for Selection becomes the Index of Selection, and
measures the relative change in fitness in one generation. It
would therefore measure directly the rate of evolution. Since
all fitness differences are certainly <u>not</u> genetically determined,
the actual amount of genetically effective selection is something
less, as is the actual rate of evolution.

Following Crow (1958), the total Index can be partitioned
into two components. One is pre-reproductive mortality (I_m), the
other the fecundity of survivors (I_f):

$$I = I_m + I_f/p_s$$

where p_s is the probability of survival to reproductive age.

It is important to note that this partition is an
approximate one when applied to a real population with mortality
occurring during reproductive ages. Nonetheless its values should
be useful as a rough guide. Real populations with overlapping
generations and finite mortality during reproductive years
violate to greater or lesser degree the assumptions used in its
derivation.

METHODS

Information on the captive population of <u>Panthera tigris
altaica</u> was obtained from the extremely useful International
Studbook (Seifert 1984). Information through the end of 1983 was
entered into a microcomputer database by Dr. Ulysses Seal.

Some International Species Inventory System (ISIS) data were
used to update these files, where reporting was known to be
incomplete. Nine animals were deleted from the file before

analysis because date information was too incomplete (Studbook #'s 2448-2455 + 2463).

Information on parentage and age at death were extracted from these files and analyzed with a Pascal program, which calculated the Index of Possible Selection Intensity and its components.

Potential Biases

At least three potential sources of bias may have effects on the measured indexes. One is that wild-caught tigers may be more likely recorded in the studbook if they left offspring. Of 84 Siberian tigers with both parents listed as wild, 65 are recorded as having left offspring.

Another biasing factor is pre-reproductive mortality for wild-caught animals. Most wild-caught individuals are of reproductive age (two years or greater), so the pre-reproductive mortality component hardly applies to this group.

A third biasing factor is that specimens born less than one lifetime before the data set cutoff at the end of 1984, have not all completed their reproductive lives. Their progeny number could yet rise, and those born in 1983 and 1984 could yet die before they reach reproductive age (two years). For example, mean progeny number for 1981-1983 born animals is only 0.17 as of the end of 1984; squaring this makes the denominator very small, and the index quite large (i.e. in the 70's). It is very likely that reproductive output from this cohort will rise later on.

Animals born after 1980 were therefore generally excluded in these analyses. The group born between 1976 and 1980 is included because of its management importance. However, results for this group are to be interpreted with caution as their reproductive output will likely rise further.

RESULTS

It's worthwhile to look at simple survivorship and fecundity patterns first. In Fig. 1, survivorship to age two is shown (i.e. the probability that an animal born in the year groups shown lived to age two). This has been fairly constant at about 0.6, but reaches its lowest figure (0.5) for the latest births—when the captive population is at captive carrying capacity.

In Fig. 2, the average number of births per animal surviving to age two is presented. Animals born between 1950 and 1970, which survived to age two, averaged about nine offspring each. This corresponds to the tremendous growth phase of the captive population. The next group, those born between 1971 and 1975, have almost finished their reproductive lives but averaged only 3.5 births each. The last group shown, those born 1976-1980, have thus far produced just over one birth each.

This is a result of a captive population which expanded rapidly, saturated available captive facilities, and which has since been roughly constant in size.

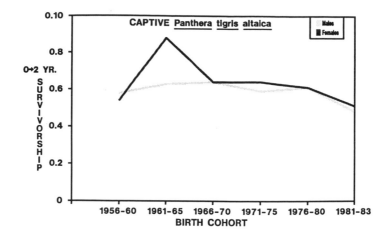

Fig. 1. Survivorships of captive born Siberian tigers from birth to 2 years by birth cohort.

Progeny Number Variance

Fig. 3 shows the progeny number variance for animals which survived to age two (onset of reproduction). This is the numerator term of the Index of the Opportunity for Selection. The variance is high overall, but exceptionally so for the male cohort born between 1956-60. Much of this single peak is due to the 87 progeny produced by a single male. This peak should not obscure the high overall variance, much of it due to individuals with zero progeny number.

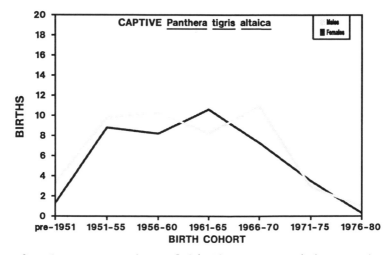

Fig. 2. Average number of births per surviving captive born Siberian tiger by birth cohort.

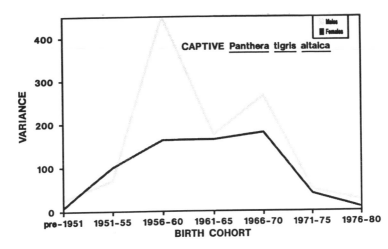

Fig. 3. Variance in progeny number of surviving captive born Siberian tigers by birth cohort.

Opportunity For Selection: Pre-reproductive Mortality

In Fig. 4, the calculated indices for possible selection intensity due to male and female mortality up to the age of two years are presented. The pattern is rather uniform, with the Index due to mortality about 0.6 (0.63 for males, 0.59 for females).

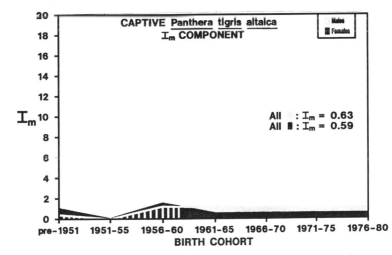

Fig. 4. Possible selection intensity, I_m component, for captive born Siberian tiger by birth cohort.

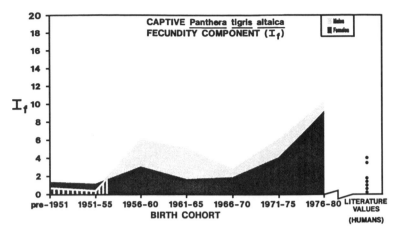

Fig. 5. Possible selection intensity, I_f component, for captive born Siberian tiger by birth cohort.

Opportunity For Selection: Fertility

In Fig. 5, the calculated indices for possible selection intensity due to male and female fertility are presented. The pattern is decidedly not uniform. The opportunity for selection acting through differential fertility began low, rose dramatically for specimens born 1956-1960, declined for those born in the next 10 years, and then rose sharply again.

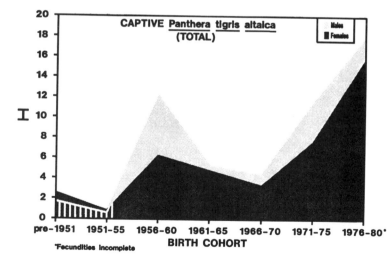

Fig. 6. Total possible selection intensity for captive born Siberian tiger by birth cohort.

The most recently born group included, those born between 1976 and 1980, may yet have more offspring which would lower this highest value.

Total Opportunity For Selection

In Fig. 6, the total Index of the Opportunity for Selection is presented, for males and females. The total figures are dominated by the fertility component. The cohort born between 1956 and 1960, and those born since 1971, show very large values. Of the seven pairs of figures, males have the highest value six times.

DISCUSSION

The upper bound on selection intensity in captive Panthera tigris altaica measured here is difficult to interpret, because there is little to compare the numbers against. The only other calculations of this quantity we are aware of are for humans (Crow 1958, 1962, Cavalli-Sforza and Bodmer 1971). We are making the calculations for other captive species.

Our results are typically an order of magnitude higher than the human values, implying that the rate of evolution is potentially an order of magnitude greater.

As another comparison, consider a imaginary, stable population with "random" (Poisson) distribution of progeny number. With a mean of two offspring, and a variance of two, the index value would 0.5. Such a population might approximate what eventual "Species Survival Plan" captive populations will be like (Flesness 1977, Foose, 1983), although the aim will be to reduce family size variance as much as possible.

Clearly, evolution may be proceeding in the captive Panthera tigris altaica population at perhaps 10 - 20 times the rate of such an idealized population. Does this matter?

We believe the answer to be a strong "yes". If captive populations are intended to contribute directly to the conservation of global biotic diversity (as they should), there is real reason to be concerned about genetic adaptation to captive conditions. Individuals with high captive fitness may well have low fitness in the wild, as conditions and stresses are so different. Slowing the rate of evolution in captivity will likely maximize the captive population's conservation value.

There is therefore reason to be concerned about the high opportunity for selection in this captive population. Coordinated management programs now underway (Seal and Foose 1984, and Seifert 1984) will attempt to reduce this opportunity. Ironically, this will not lower the next generation's index value much, because the management programs in effect wisely practice selection for those whose fitness has previously been low in captivity.

In the long run, such coordinated conservation-minded management should result in a lowering of the opportunity for selection in captivity, hopefully by a substantial amount.

ACKNOWLEDGEMENTS

     This work would not have been possible without the dedicated data-gathering of Studbook Keeper Seigfried Seifert, and the dedicated microcomputer work of Ulysses Seal and several volunteers.

REFERENCES

Cavalli-Sforza, L.L. and W.F. Bodmer. 1971. The Genetics of Human Populations. San Francisco: Freeman and Co.
Crow, J.F. 1958. Some opportunities for measuring selection intensity in man. Human Biol. 30:1-13.
Crow, J.F. 1962. Population genetics: selection. In Methodology in Human Genetics, ed. W.J. Burdette, San Francisco: Holden Day.
Flesness, N.R. 1977. Gene pool conservation and computer analysis. Int. Zoo Yearb. 17:77-81.
Foose, T. 1983. The relevance of captive populations to the conservation of biotic diversity. In Genetics and Conservation, ed. Schoenewald-Cox. Menlo Park, CA: Benjamin/Cummings.
Seal, U.S. and T. Foose. 1984. Siberian tiger species survival plan: a strategy for survival. J. Minn. Acad. Sci. 49:3-9.
Seifert, S. 1984. International Tiger Studbook. Leipzig, DDR: Zoological Garden of Leipzig.

PART VI

WHITE TIGER POLITICS

33

White Tiger: Phantom or Freak?

Edward J. Maruska

Mankind has always been fascinated by strikingly colored black and white animals. Perhaps it is the distinct contrast between the lightest and darkest of colors. I need only remind you of the exhibit value of the following mammals: giant panda, killer whale, zebra, Malayan tapir, colobus monkey or black and ruffed lemur, which is often referred to as the panda of the primate world. Among birds, we have penguins, ostrich, the larger black and white cranes and storks and many species of waterfowl and alcids. Albeit, to some degree, some species of birds may owe their popularity to their shape, as well as to their black and white color.

Another very popular black and white animal that has proven to have universal appeal among zoo visitors is the white tiger. The literature on this animal, both popular and scientific, is extensive (Gee 1959 and 1964, Reed, T.H. 1961, Thorton et al. 1966, Reed, E. 1970, Leyhausen and Reed 1971, Berrier et al. 1975, Sankhala 1977, Roychoudhury 1978, Thorton 1978, Simmons 1981, Iverson 1982, Anonymous 1983, Fay 1983, Roychoudhury and Acharjyo 1983, Isaac 1984). This mutation or color phase of the tiger _Panthera_ _tigris_ does not have the same appeal among all zoo directors. In addressing the general membership of the American Association of Zoological Parks and Aquariums, William Conway, General Director of the New York Zoological Society, stating his concerns for the captive space being utilized by white tigers that might otherwise be used for other tiger subspecies, made reference to white tigers as freaks. Conway can be quoted making the same statement from an article in Geo Magazine. He said, "White tigers are freaks. It's not the role of a zoo to show two-headed calves and white tigers." (Isaac 1984) Although the comment makes for interesting prose, his approach is not sound biology.

I would like to quote from Webster's Dictionary on the distinction between a freak and a mutation: "Freak - one that is markedly unusual or abnormal; a person or animal with a physical oddity who appears in a circus sideshow." I believe Conway's two-headed calf belongs in this category. Likewise, Webster's makes the following distinction: "Mutation - a significant and basic alteration; a relatively permanent change in hereditary material involving either a physical change in chromosome relations or a biochemical change in the codons that make up genes." As I will attempt to illustrate in my paper, the white tiger belongs in this category.

Sporadic reports of white tigers in India and Nepal have been recorded for more than 160 years. The Bombay Natural History Society states that between 1907 and 1933, 17 were shot in India. Nine white tigers have been taken in the Rewa district since the turn of the century. In an article in the Bombay Natural History Society "Albinism and Partial Albinism in Tigers," E. P. Gee (1959) records no less than 35, with comments of unrecorded numbers occurring in Assam. Rowland Ward's Records of Big Game also records a number of white tigers. Many of these animals were collected as adults, which indicates survivability. I know of no instance of a two-headed calf surviving in the wild. It is a popular belief that Mohan, captured by the Maharaja of Rewa on 27 May 1951 and kept at his palace courtyard at Govindgarh, was the first white tiger to be taken alive. However, Richard Lydekker (1894) recorded a white tiger that was exhibited at the old menagerie at Exeter Change about the year 1820. I have not been able to ascertain whether this was a live or mounted specimen. However, a fine specimen was exhibited alive at the Calcutta Zoo in 1920 (Iverson 1982).

White tigers are not albinos since albinism refers to a total lack of pigment and the pinkish eyes of true albinos are a result of the blood coursing through the pigmentless retina. White tigers have black or ash grey stripes on a varying cream to white background. They have a pink nose and pads and pale blue eyes and are, on average, slightly larger than the orange phase.

Why should these off-tone genetic specimens continue to sporadically appear and persist in the wild in an otherwise seemingly uniform population when they presumably would be at a distinct disadvantage in competing with their orange conspecifics?

One explanation might lie in the fact that presumably the tiger's chief prey are unable to differentiate colors. So, if a white tiger were to alter its hunting habits to strictly nocturnal hunting to avoid detection by the warning calls of birds and primates that perceive color, the mutant tiger would be at no disadvantage since it has the same stripes to break up its bodily contours as its orange counterpart (Cott 1957).

Leyhausen, in an article on white tigers, states that the more varied the gene pool, the better chance a species has to adapt to changing conditions and thus survive (Leyhausen and Reed 1971). He also states that it appears that a gene may influence more than one characteristic in an individual and refers to K.S. Sankhala, a former director of the Delhi Zoological Park who was a successful supervisor of white tiger breeding. Sankhala noted that, for example, the "whities" grow faster and attain larger size than the average orange tiger. So one function of the white coat gene may be to keep another gene—a body size gene—within the gene pool.

Enough discourse has occurred regarding the undesirability of white tigers, citing such reasons as abnormal coloration, premature death, low fertility, inbred, high occurrence of stillbirths, higher susceptibility to disease than their orange counterparts, reduced litter sizes and body deformities, eg.

shortened legs, crooked necks, arching of the backbone and eye
weakness.

Since 1970 the Cincinnati Zoo has bred 52 white tigers, more
than any other zoo in the world (Table 1). Here I will discuss
the above undesirable traits individually and relate them to our
experience in our collection as well as to white tigers in other
zoos. I will also attempt to demonstrate where the real weakness
in this mutant lies.

White color: The first trait is in reference to the
"abnormal" color--the white striped coat. White tigers have
survived in a natural state and still occur. Thus, it is
ironical to me that anyone would object to their coat color.
Zoos commonly exhibit melanistic phases of cats, such as black
jaguars, and I wonder how many have been recorded throughout
their range? As stated earlier, any powerful contrast of color
in any animal attracts human attention. These contrasting colors
have been selected for in domesticated species. There are black
and white dogs, cats, cows, horses, hogs, goats, sheep, and
poultry. Further, zoos would readily accept wild, white species
or white color phases into their collections, even though these
first appeared through a process of mutation that evolved further
when conditions were favorable. Some examples are white wolves,
caribou, Kermode bears (the white phase of the black bear) and
Dall sheep.

Premature death: We have not experienced premature death
among our white tigers. Forty-two animals born in our collection
are still alive. Mohan, a large white tiger, died just short of
his 20th birthday, an enviable age for a male of any tiger
subspecies since most males have shorter captive lives.
Premature deaths in other collections may be artifacts of captive
environmental conditions.

Low fertility: This may occur in other collections, but on
this topic, I stand on our breeding record to date (Table 1).

Inbred: In order to preserve any characteristic in any
animal or plant, a certain amount of inbreeding and/or line
breeding is necessary. I believe that most white tigers in the
U.S. and abroad are severely inbred. This is usually the result
of most zoos maintaining white tigers not having enough room to
house outcrosses which provide the necessary infusion of new
blood into their lines. Inbreeding is a problem that can be
controlled utilizing proper management.

Occurrence of stillbirths or a high percentage of neonatal
deaths: In 52 births, we have had four stillbirths, one of which
was an unexplained loss. We lost two additional cubs from viral
pneumonia, which is not excessive. Without data from non-inbred
tiger lines, it is difficult to determine whether this number is
high or low with any degree of accuracy.

Sumita, one of our prime breeding females, had successfully
delivered and reared 19 cubs up to 1984. In 1984, she
experienced a breech birth 24 hours after delivering three
healthy cubs. This last cub born was removed dead, and the
remaining three cubs were hand raised. In a subsequent
parturition, she gave birth to two healthy cubs, and another was
found in the birth canal and one more in the uterus, both of
which were dead.

Table 1. White tiger births recorded at the Cincinnati Zoo.

Birthdate	Sex	Tiger (Name/No.)	Dam/Sire	Death Date/Cause
20/06/74	M	Ranjit/M9569	Kesari/Ramana	
	F	Bharat/M9570		
	F	Priya/M9571		
27/06/76	M	Arjun/M9602	Kesari/Cuneo Tony	
	M	Bhim/M9603		
	M	Vir/M9604		
	F	Sumita/M9605		
09/07/79	M	Cheytan/M9664	Sumita/Bhim	
	F	White/M9665		24/07/79-Pyloric Sphincter Constriction
21/04/80	F	Shubhra/M9695	Sumita/Bhim	
	F	Sundari/M9696		
	F	Tapi/M9697		
16/05/80	M	Mota/M9701	Kamala/Bhim	
06/07/81	M	White/M9747	Kamala/Bhim	
05/09/81	M	White/M9752	Sumita/Bhim	07/09/83-Trauma from Tail Wound & Anesthetic. On Loan to Metro-Miami Zoo
	M	White/M9753		
	F	White/M9754		
14/11/81	M	White/M9757	Kamala/Bhim	
	M	White/M9758		
19/08/82	F	Kavali/M9782	Sumita/Bhim	
	F	Ambala/M9783		
	F	Sun/M9784		
04/04/83	M	White/M9799	Sumita/Bhim	
	F	White/M9800		
	F	White/M9801		

Table   1. Continued

Birthdate	Sex	Tiger (Name/No.)	Dam/Sire	Death Date/Cause
23/08/83	M	Lucknow/M9813	Sumita/Bhim	
	M	Jaipur/M9814		
	M	Ahmadabad/M9815		
	F	Kanpur/M9816		
	F	Chanda/M9817		
15/07/84	M	Indiana Jones/M9835	Sumita/Bhim	
	F	Sankara/M9836		
	F	Sheva/M9837		
	F	White/M9838		15/07/84-Stillborn Breech
23/02/85	F	White/M9848	Tapi/Mota	23/02/85-Maternal Neglect
	F	Rani/M9849		
04/07/85	M	Saafaid/M9879	Tapi	25/07/85-Viral & Bacterial Pneumonia
	F	Laal/M9880		
	F	Neela/M9881		
07/08/85	F	White/M9890	Sumita/Bhim	29/08/85-Viral Pneumonia
22/10/85	M	Bhut/M9894	Indira/Bhim	
	F	Jadugari/M9895		
14/12/85	M	Chiquillo/M9900	Sumita/Bhim	
	F	Sahiba/M9901		
		White/M9902		14/12/85-Stillborn
		White/M9903		14/12/85-Stillborn
		White/M9904		14/12/85-Stillborn
29/03/86	M	White/M9911	Tapi/Mota	02/04/86-Inanition
	M	White/M9912		16/04/86-Pneumonia
	M	White/M9913		
	M	White/M9914		
	F	White/M9915		

Higher susceptibility to disease than their orange counterparts: White tigers of the Cincinnati Zoo have proven to be relatively hardy and free of disease. Our cats are exposed to inclement weather and are given access to the outdoors every day of the year. They are routinely immunized once a year against several feline respiratory infections, and medicated for control of parasitic infections as well. Problems of susceptibility to disease that other zoos have encountered may lie in their management. For example, I visited the Bristol Zoo's white tigers in 1968, and at that time they did not have a vaccination program. The National Zoo reported (Berrier et al. 1975) a disease bearing some characteristics of Chediak-Higashi Syndrome (CHS) present in their white tigers. None of the signs or symptoms they reported have been manifested in our collection.

Reduced litter size: Our tigers have had litters from one to five, which falls within the normal range for the species.

Body deformities: arching of backbone, crooked necks, shortened tendons, legs. Other than a case of hip dysplasia that occurred in a male white tiger, we have not encountered any other body deformities or any physiological or neurological disorders. Some of these reported maladies in mutant tigers in other collections may be a direct result of severe inbreeding or improper rearing management of tigers generally.

In 1970, we housed a pair of heterozygous tigers "Ramana" and "Kesari" for the National Zoo, during which time they bred and Kesari gave birth to three white and one heterozygous cub. A period of days went by and Kesari lost interest in her litter which necessitated removing them for hand rearing. All are still alive at this time.

In a paper on inbreeding in white tigers, A.K. Roychoudhury and K.S. Sankhala (1979) commented on average litter size and mortality rates, "There is a tendency for the average litter size to decrease and the mortality rate to increase in the value of inbreeding coefficient as shown by the signs of the regression coefficient." All the offspring of Kesari and Ramana (a direct line from Mohan) in Washington survived and had great impact on the regression analysis. They state that if these offspring were not considered in the analysis the regression of mortality rate on inbreeding coefficient would have been higher. The authors' only error was that Ramana and Kesari were not bred in Washington, DC.

This example illustrates that captive breeding and rearing techniques may vary from one institution to another and that mixed results can have an important impact on statistics used to determine mortality in white tiger offspring.

Eye weakness: If there is one glaring undesirable trait in white mutant tigers it is ophthalmic disorders. Some specimens are cross-eyed, a malady in captive, normal colored lions and tigers, and there is also evidence of feline central retinal degeneration. This is possibly related to reduced pigment formation. In 52 white tiger births, there were four cases of strabismus, all from the four white offspring of Kesari and Tony. Bhim and Sumita (siblings) were retained and all of their offspring had normal set eyes except one male of their first litter.

Because strabismus is of rare occurrence and probably linked to the white coat gene, it is possible that it might be further reduced or even eliminated by selective breeding.

There is further evidence of eye problems in white mutants that may be genetically linked to the white coat. In 1985 three sibling white tiger cubs were born with cataracts. However, these cataracts occurred in cubs that were pulled from their mother dehydrated and hand raised. We are now evaluating whether the cataracts were nutritional or genetical in origin.

In summary, it is easy to rationalize the positive and negative aspects of maintaining white tigers as there are strong arguments on both sides.

On the positive side, white tigers have proven to be immensely popular with our visitors and have increased attendance in those institutions housing them. They have had an additional economic value to zoo budgets in that their offspring presently command high prices. In the view of education, they can be utilized in dramatic fashion and have a favorable impact on the teaching of basic components of Mendelian genetics.

On the negative side, there are some inheritable weaknesses, primarily in the eyes, that seem to be linked to the white coat. Another negative aspect is that early owners of white tigers chose to cross subspecies in order to achieve white offspring.

A real issue against keeping and uncontrolled breeding of white tigers in captivity, regardless of their public appeal and economic value to zoo budgets, is that they are mutants occupying space that should be made available to orange colored animals.

To debate whether or not white tigers should or should not persist in zoos is pointless, because I believe it is already a foregone conclusion. The facts as they appear are that these mutants have at least some universal appeal among zoogoers. Zoo managers are less than united as to whether they have a place in our collections or not. Because this enigmatic variation is already well established in several zoos, and because of their appeal and uniqueness, I advocate we include a limited number of white tigers in our overall tiger breeding strategy.

REFERENCES

Anonymous. 1983. The rare propagating white tigers of the Cincinnati Zoo. Marathon World No. 2:18-21.
Berrier, H.H., F.R. Robinson, T.H. Reed and C.W. Gray. 1975. The white tiger enigma. Vet. Med./Small Anim. Clin. 69(April):467-72.
Cott, H.B. 1957. Adaptive Coloration in Animals. London: Methuen & Co., Ltd.
Fay, J. 1983. White tigers, A rare cat makes zoo news. 3-2-1-Contact, Feb.:4-8.
Fox, H.M. and G. Vevers. 1960. The Nature of Animal Colours. New York: Macmillan and Co.
Gee, E.P. 1959. Albinism and partial albinism in tigers," J. Bombay Nat. Hist. Soc. Dec.:1-7.
Gee, E.P. 1964. The white tiger. Animals. 3(11):282-86.

Heran, I. 1976. Animal Coloration: the Nature and Purpose of Colours in Vertebrates. Feltham: Hamlyn Publ. Group Ltd..

Isaac, J. 1984. Tiger Tale. Geo 6(August):82-86.

Iverson, S. 1982. Breeding white tigers. Zoogoer (publication of Friends of the National Zoo) 11(1):4-9.

Leyhausen, P. and T.H. Reed. 1971. The white tiger: care and breeding of a genetic freak. Smithson. 2(1):24-31.

Lydekker, R. 1894. The Royal Natural History. London: Fredrick Warne & Co.

Reed, E. 1970. White tiger in my house. Nat. Geo. 137(4):482-91.

Reed, T.H. 1961. Enchantress! Queen of an Indian palace, a rare white tigress comes to Washington. Nat. Geo. 119(5):628-41.

Roychoudhury, A.K. 1978. A study of inbreeding in white tigers. Sci. Cult. 44:371-72.

Roychoudhury, A.K. and K.S. Sankhala. 1979. Inbreeding in white tigers. Proc. Indian Acad. Sci. pp. 311-23.

Roychoudhury, A.K. and L.N. Acharjyo. 1983. Origin of white tigers at Nandankanan Biological Park, Orissa. Indian J. Exp. Biol. pp. 350-52.

Sankhala, K. 1977. Tiger! The Story of the Indian Tiger. New York: Simon & Schuster.

Simmons, J. 1981. White tiger enchantment. American Way, October:82-84.

Thorton, I.W.B., K.S. Sankhala and K.K. Yeung. 1966. The genetics of the white tigers of Rewa. Univ. of Hong Kong, Depart. of Zool. pp. 127-35.

Thorton, I.W.B. 1978. White tiger genetics, further evidence. J. Zool. (London) 185:389-94.

34

White Tigers and Their Conservation

A.K. Roychoudhury

INTRODUCTION

White tigers are not true albinos in the strict sense, for their chalky-white coats are marked with black or chocolate stripes and their eyes are icy blue, not pinkish. The white tiger originated in India, presumably the result of a gene mutation from normal orange color to white coat color. If two tigers mate and each has a mutant gene, one in four cubs is likely to be white. There are at present 103 white tigers in the world: India has 25, the United States has 74 and England has four. Recently the Cincinnati Zoo sold a pair of white tigers to the Tokyo Zoo and one white female to the Toronto Zoo.

In 1951, a white male cub, nine months old, was caught in the forests of Rewa, Madhya Pradesh (central India) by the Maharajah of Rewa. Before the cub was caught, white tigers had been irregularly observed in the area for 50 years, as well as in the forests of Bihar, Orissa, and Assam (Gee 1964a, b; Sankhala 1978).

The white male cub, named Mohan, was taken to the Maharajah's palace zoo at Govindgarh, near the town of Rewa. When mature, Mohan was mated to a orange female, Begum, also captured in the forests of Rewa. She produced ten cubs in three litters during 1953-56, all orange. The Maharajah made another attempt to produce white tigers by pairing Mohan with one of his orange daughters, Radha. In 1958, she gave birth to one white male, Raja, and three white females, Rani, Mohini, and Sukeshi (Fig.1).

These breedings, and subsequent breedings by others, were helpful in determining the principle of inheritance of coat color in white tigers. The gene for white coat in tigers is inherited as an autosomal recessive to normal orange color (Thornton et al. 1967, Thornton 1978, Roychoudhury 1978, Roychoudhury and Sankhala 1979).

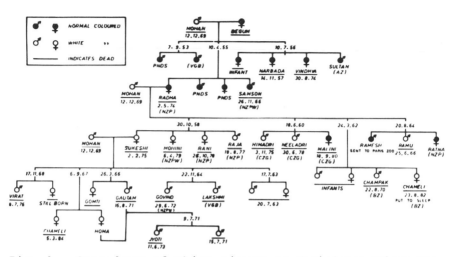

Fig. 1. Genealogy of white tigers at Govindgarh Palace Zoo,
Rewa. AZ - Ahmedabad Zoo, Ahmedabad; BZ - Bristol Zoo, Bristol,
England; CZG - Calcutta Zoological Garden, Calcutta; NZP -
National Zoological Park, Delhi; NZPW - National Zoological Park,
Washington, D.C., USA; PNDS - Sold to P.N. Das & Sons, Calcutta;
VGB - Victoria Garden, Bombay. The line under a symbol or name
of a tiger indicates dead.

WHITE TIGERS IN INDIA

    National Zoological Park, New Delhi: Raja and Rani, a pair
of white tigers of Radha's first litter produced twenty white
cubs in seven litters at the Delhi Zoo. Intensive inbreeding
resulted in reduced fertility and increased infant mortality.
Since 1963, 38 white tigers have been bred, and ten (26%)
survived. One male was sent to the Bristol Zoo, U.K., and one
female each, to Kanpur, Hyderabad, Bhubaneswar, and Mysore Zoos,
India. There are two white females and three white males
remaining at the Dehili Zoo (Fig. 2).

    Calcutta Zoological Garden, Calcutta: Two white males,
Neeladri and Himadri, and an orange female, Malini, of Radha's
second litter, were brought to the Calcutta Zoo in 1963. The
white tigers of Calcutta, like those of Delhi, are highly inbred
and suffered high mortality. Of 30 white tigers born, only four
(13%) survived. One female is in the Gauhati Zoo, Assam, and two
females, Tara and Himadri Jr., and one male, Barun, are in the
Calcutta Zoo (Fig. 3).

    Nandankanan Biological Park, Bhubaneswar: Three white cubs
with black stripes were born to a pair of orange tigers in 1980
in the Nandankanan. Apparently, the parents had no biological
relationship with Mohan or with his descendants. They produced
seven more cubs in three litters, two of which were white. A
complete history of the ancestors of the orange parents is not
available. Presumably, the male grandparent of the white tigers
carried the recessive gene for white coat color and transmitted
it to a male offspring, Deepak. When Deepak was bred to his

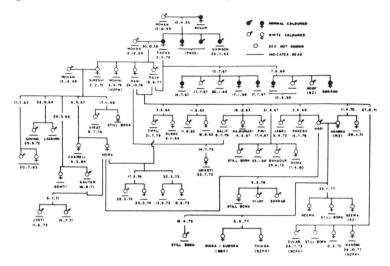

Fig. 2.   Genealogy of the white tigers at National Zoological Park, Delhi.   The offspring of Sukeshi and Mohan and those of Homa and Gautam born at Govindgarh are shown here for the sake of reference.   KZ – Kanpur Zoo, Kanpur; MZ – Mysore Zoo, Mysore; NBP – Nandankanan Biological Park, Bhubaneswar; NZPH – Nehru Zoological Park, Hyderabad.   Other abbreviations are the same as those of Fig. 1.

daughter, Ganga, the white cubs were born (Roychoudhury and Acharjyo 1983).   The Nandankanan Biological Park has the largest collection of white tigers in India, consisting of eight females and five males (Fig. 4).

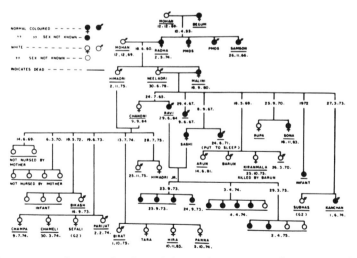

Fig. 3.   Genealogy of the white tigers at Calcutta Zoological Garden, Calcutta.   GZ – Gauhati Zoo, Gauhati.   Other abbreviations are same as Fig. 1.

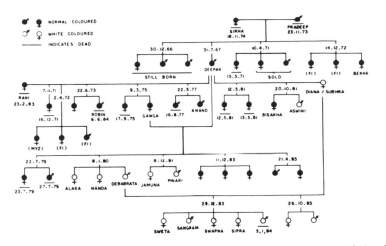

Fig. 4. Genealogy of white tigers at Nandankanan Biological
Park, Bhubaneswar. FI - Sold to Fauna of India, Mysore; MYZ -
Mysore Zoo, Mysore.

Sri Chamarajendra Zoological Garden, Mysore: In 1984 two
female cubs, one white and one orange, were born to a pair of
orange tigers at the Mysore Zoo in south India.   The grandmother,
Thara, was acquired from Nandankanan Biological Park in 1972 and
is the sister of Deepak, the progenitor of all white tigers at
Nandankanan.   The Mysore Zoo has two white females, Priyadarshini
and Ashima, the latter was brought from the Delhi Zoo.

WHITE TIGERS IN THE USA

National Zoological Park, Washington, D.C.: Mohini, a white
tigress born to Radha and Mohan in their first litter, was
purchased from the Maharajah of Rewa, and brought to the National
Zoological Park in December, 1960.   When mature, she was paired
with Radha's orange brother, Samson, also brought from India,
producing five cubs in two litters, one of which was white (Figs.
5 and 6).

Cincinnati Zoo, Ohio: The Cincinnati Zoo first acquired
white tigers when Kesari, a carrier female of the National
Zoological Park, and one white male, Tony, of the Hawthorn Circus
in Illinois arrived.   They belonged to different lineages and
produced a litter of five cubs: four white and one orange in 1976
(Fig. 5).

Two of these offspring, Sumita and Bhim, set a world's
record by producing 25 white cubs in eight litters within a
period of six and one half years.   Among 25 cubs of Sumita, five
are nearly stripeless; faint stripes are on their hind legs
suggesting that the striped character is dominant over
stripeless.

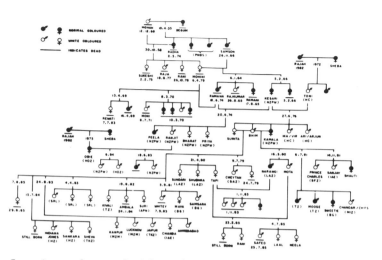

Fig. 5. Genealogy of white tigers at National Zoological Park, Washington, D.C. and Cincinnati Zoo, Ohio. APN - Audubon Park Zoological Garden, New Orleans; BG - Busch Gardens, Tampa, Florida; HC - Hawthorn Circus, Illinois; HDZ - Henry Doorly Zoo, Omaha; HZ - Houston Zoo, Houston; IAE - International Animal Exchange, Michigan; LAZ - Los Angeles Zoo, Los Angeles; MIZ - Milwaukee Zoo, Wisconsin; MZM - Metro Zoo, Miami; SAZ - San Antonio Zoo, Texas; SFZ - San Francisco Zoo, California; SRL - Siegfried and Roy, Las Vegas; TZ - Toronto Zoo, Canada; TKZ - Tokyo Zoo, Japan. Other abbreviations are same as Fig. 1.

The Cincinnati Zoo has raised 47 white tigers, of which 36 are still living as of March 1986, and a number of orange tigers carrying white genes (Fig. 5). A number of their white tigers have been either sold or loaned to zoological parks in the USA, Canada, and Japan, to the International Animal Exchange, Michigan, and to Siegfried and Roy's Magic and Illusion Show, Las Vegas, Nevada.

Hawthorn Circus, Illinois: The Hawthorn Circus now has 28 white tigers; the largest collection in the world. An orange tigress named Sheba, from the International Animal Exchange, was in the company of five orange tigers and produced a number of cubs, including two white, Bagheera and Frosty, in two different litters, indicating that she was carrying the white gene. The sire of the two white cubs is unknown (Thornton 1978). In 1974 the Hawthorn Circus purchased a two year old white tiger named Tony born to orange parents, Sheba and Raja, from the Shrine Circus in Sarasota, Florida. Sheba, the mother of Tony, is not the same Sheba that is the mother of Bagheera and Frosty.

Both Sheba and Rajah, who gave birth to Tony, were the offspring of a Bengal tigress, Susie, and a Siberian tiger, Kubla, at Sioux Falls Zoo, South Dakota. Susie was from an unidentified west coast zoo and Kubla came from the Como Zoo, St. Paul, Minnesota. Thus, all descendants of this line are Bengal/Siberian hybrids.

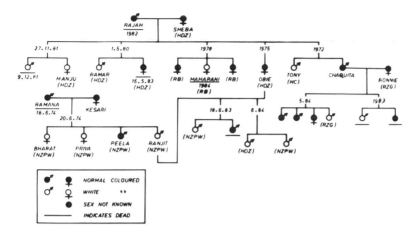

Fig. 6. Genealogy of white tigers at Henry Doorly Zoo (HDZ),
Omaha, Nebraska and Racine Zoological Garden (RZG), Racine,
Wisconsin. HC - Hawthorn Circus, Illinois; NZPW - National
Zoological Park, Washington, D.C.; RB - Ringling Bros.,
Washington, D.C.

Henry Doorly Zoo, Omaha, Nebraska: In 1978 the Henry Doorly
Zoo purchased Sheba and Rajah and their three year old orange
daughter, Obie, from the Shrine Circus. The genealogy of the
white tigers at this zoo is shown in Fig. 6.

Racine Zoological Garden, Racine, Wisconsin: In 1984 a
white female was born to a pair of orange tigers, Chaquita and
Bonnie, at the Racine Zoological Garden, suggesting that both
parents carried genes for whiteness. The male parent is the
brother of white Tony at the Hawthorn Circus. The female parent
proved to be heterozygous for the white gene, since she was once
accidentally bred with her father and produced a white cub.
Unfortunately, nothing is known about the ancestry of her father
(Fig. 6).

WHITE TIGERS IN ENGLAND

Bristol Zoo, Bristol: A pair of white tigers, Champak and
Chameli, born to Radha and Mohan in their third litter, was
purchased from the Maharajah of Rewa and brought to the Bristol
Zoo in 1963 (Fig. 7). Later, when there were no white males at
Bristol, one of the white females was exchanged for a white male,
Roop, born to Radha and Raja at the Delhi Zoo. The Bristol Zoo
has not raised any more white tigers since those born in 1977.
There are at present two white females, Sumati and Nanda and two
white males, Roop and Akbar II at the Bristol Zoo (Fig. 7).

Fig., 7.Genealogy of the white tigers at Bristol Zoo, Bristol,
England.   The abbreviations are the same as those of Fig. 1.

DISCUSSION

All the white tigers found at the zoos of Calcutta, and
Delhi in India, Washington, DC in the USA and Bristol in England
originated from two founder animals, namely Mohan, a white male
and Begum, a orange female.   Since these two animals were caught
in the forests of Rewa, Madhya Pradesh, all their descendants
belong to the Rewa lineage.   During the last two decades the
white tigers of this lineage suffered significant losses,
probably due to their high coefficients of inbreeding which lead
to reduced fertility and to increased neonatal mortality
(Roychoudhury and Sankhala 1979).

The white tiger should be viewed as a gift of Nature.   Its
conservation is as important as that of the normal tiger.
Initially, matings between close relatives had been practiced in
order to increase the number of white tigers in captivity.   One
consequence was reduced fertility and early mortality became a
common feature of white tigers.   Also, average litter size
decreased (Roychoudhury and Sankhala 1979).   White tigers that
die prematurely are found to be more susceptible to diseases than
normal tigers.   Some of these diseases are pneumonia, trauma of
the abdomen, congestion of the lungs, feline distemper and
negligence by the mother.   Weakening of the eyes and shortness of
the legs of one tigress in Washington, D.C., neck twisting of
some white cubs in Dehli, lack of development of a kidney in one
male and recurrent miscarriages of female white tigers in
Bristol, arching of the backbone of two male tigers in Calcutta
and frequent occurrence of stillbirths in Dehli and
Washington,D.C. have also been observed.   White tigers at
Calcutta suffer from sexual malaise.   All these defects and
diseases may not be due to the effects of inbreeding, but at
least a part of them must reflect increased homozygosis.

According to E.J. Maruska (this volume), it is incorrect to assume all early mortality of white tigers is due to inbreeding. Rather, poor animal husbandry may be partly responsible for some early mortality.

If the object is to conserve the gene for white coat, normal colored tigers should be mated with white tigers. Their offspring would be heterozygous for the white gene. If they bred with white tigers, white cubs would result. A certain degree of inbreeding is therefore necessary. By adopting such procedures, new bloodlines could be introduced into highly inbred strains and the level of inbreeding could be reduced. Alternate outbreeding and inbreeding could thus contribute to the conservation of white tigers.

A dearth of appropriate mates is a problem in the breeding of white tigers. As mentioned earlier, the Bristol Zoo did not have a white male. One of its white females was exchanged for a white male of the Delhi Zoo. Due to the lack of a young white or heterozygous male, the Calcutta Zoo could not undertake a breeding program for a long time. In other cases, behavioral incompatability of the mates precludes breeding this rare animal. Consequently, authorities are sometimes forced to adopt undesirable matings between homozygous and heterozygous colored tigers. The progeny are all orange and thus create a problem in differentiating heterozygous cubs from homozygous ones. A similar problem of differentiation of colored cubs arises when both parents are heterozygous for the white gene. Until new methodology (e.g. DNA technology) is explored, it will be difficult to identify heterozygous cubs from homozygous ones.

CONCLUSIONS

To reduce the level of inbreeding, white tigers should be allowed to outcross with normal orange tigers and the offspring thus produced are to be backcrossed with white tigers. If outcrossing and backcrossing are conducted alternately, white tigers can be produced with the lowest possible level of inbreeding.

ACKNOWLEDGEMENTS

I am grateful to the authorities of Zoological Parks in India, USA, and England for supplying relevant information of white tigers, and Mr. John F. Cuneo, Jr., Hawthorn Corporation, Illinois, and Ms. Jean Schmidt, Cincinnati Zoo, Ohio, for updating the information on white tigers in the USA. I am indebted to S. Chakraborty for drawing the genealogical charts and W.J. Schull for improving an earlier draft of the manuscript.

REFERENCES

Gee, E.P. 1964a. The wildlife of India. London: Collins.

Gee, E.P.   1964b.   The white tigers. _Animals_ 3:282-86.

Roychoudhury, A.K.   1978.   A study of inbreeding in white tigers. _Sci. Cul._ 44:371-72.

Roychoudhury, A.K. and K.S. Sankhala.   1979.   Inbreeding in white tigers. _Proc. Indian Acad. Sci._ 88:311-23.

Roychoudhury, A.K. and L.N. Acharjyo.   1983.   Origin of white tigers at Nandankanan Biological Park, Orissa. _Indian J. Exper. Biol._ 21:350-52.

Sankhala, K.S.   1978.   _Tiger! The Story of the Indian Tiger._ London: Collins.

Thornton, I.W.B.   1978.   White tiger genetics - further evidence. _J. Zool._ 185:389-94.

Thornton, I.W.B., K.K. Yeung and K.S. Sankhala.   1967.   The genetics of white tigers in Rewa. _J. Zool._ 152:127-35.

35

White Tigers: The Realities

Lee G. Simmons

Quite likely from the first time a white tiger was seen roaming the forests and certainly since the capture of Mohan in Rewa, India in 1951 these great white cats with chocolate brown or black stripes have enjoyed a unique fascination with hunters, biologists and the public.

Unfortunately, this fascination may well be responsible for the fact that much of the genetic management of these sports has been designed to produce the maximum number of white cats in the shortest possible time. The absence of background pigment is the result of a pair of simple recessive genes. This gene was possibly the result of a mutation which occurred at least a hundred years ago. This gene when paired in the heterozygous condition produces a normal-colored orange tiger that carries a single white gene. Breeding strategies that followed have often been to cross fathers with daughters or brothers with sisters (Roychoudhury and Acharjyo, this volume). This deliberate inbreeding to increase the pairing incidence of the white gene has also apparently increased the incidence of undesirable traits (Roychoudhury and Acharjyo, this volume).

In all likelihood, all North American white tigers are descendants of Mohan and a normal-colored wild-caught female (Roychouldhury and Acharjyo, this volume). This coupled with the introduction of Siberian tiger blood into a very prolific line carrying the white gene has resulted in the situation in North America today where there appears to be only two pure-blood white Bengals, both probably nonreproductive, and normal-colored orange carriers which appear to be Siberian X Bengal hybrids with high coefficients of inbreeding.

Omaha's Henry Doorly Zoo's entry into the white tiger business began in 1977 in a cooperative agreement with the National Zoological Park (NZP) (Washington D.C.). We were opening a new cat complex with room to hold and manage sizeable numbers of animals and we wanted a spectacular centerpiece. Both Omaha and NZP were interested in expanding the overall gene pool in which the recessive white traits were carried, while at the same time increasing the numbers of white tigers. Our program expanded rapidly from one white Bengal male and five orange Bengal females to include 11 normal-colored offspring carrying the white gene, five white Siberian X Bengal hybrids and orange hybrid carriers. At this time, NZP decided to close out their part of the program. Add to this eight Siberian tigers plus 29 other large cats representing seven other subspecies, and we are nearly up to our hip pockets in cats before the second phase of the "white Bengal" program has begun.

Despite the fact that this large collection has provided a resource for a great deal of research, it is fairly obvious that unless a number of other institutions adopt a vigorous program, such as the one the Omaha Zoo originally embarked on, of out-crossing white Bengals and Bengals carrying the white gene, followed by a program of crossing the resulting normal-colored carriers to ultimately produce white "Bengal tigers", white tigers can have no significant place in an Species Survival Program (SSP) of the American Association of Zoological Parks and Aquariums (AAZPA). Such a program would require a substantial commitment of space and resources to hold Bengal tigers carrying the white gene.

How then can the expenditure of space and other scarce resources on a race of animals the majority of whom are highly inbred hybrids be justified? The answer is simple. You justify white tigers in exactly the same way you justify traveling giant pandas, koalas or any other high-visibility animal which, through the ability to catch the public fancy, significantly enhances public support and therefore the financial well-being of your institution. Currently, a dozen zoos display white tigers. Another six having displayed white tigers for a short time. The results are readily measurable in increased revenues.

The bottom line realities of life are that long term conservation and propagation programs are accomplished only with stable financial and public support. Institutional survival and species survival may be as tightly linked as any two genetic traits.

36

White Tigers and
Species Survival Plans

Katherine Latinen

The Species Survival Plan (SSP) program document published by the American Association of Zoological Parks and Aquariums (AAZPA) describes the program as an effort to protect vanishing species. Zoos must agree on which species are most in need of protection and then scientifically manage the captive populations of those species in an effort to reduce genetic and demographic problems. Criteria for selecting an SSP species include establishing a breeding nucleus of the species or subspecies, documenting that their existence is in peril as defined and identified by the International Union for the Conservation of Nature (IUCN) and U.S. Fish and Wildlife Service (USFWS), and finally, organizing a committee of captive-propagation professionals who will work with the species.

The white tiger is not a true subspecies, but only a morphological variation of the Bengal subspecies (Panthera tigris tigris). In captivity the stock has been bred with another subspecies at some facilities, thus making it a "hybrid" (as defined by the studbook keeper and by presentations delivered at the tiger symposium) and therefore undesirable. There has been no documentation by the IUCN or USFWS that this color morph merits protection in the wild. Only a few North American zoos currently have white tigers and no committee exists for overseeing their long-term breeding strategy. Thus, all three basic criteria for inclusion in an SSP program cannot be met by the white tiger in the North American captive population.

Four points in the International Tiger Plan (1986) sponsored by the International Union of Directors of Zoological Gardens (IUDZG) seem to be in conflict with some of the current breeding programs for white tigers. These are: 1) breedings between closely related animals should be avoided; 2) unless a founder, each individual should only produce two litters per lifetime; 3) individuals should be removed from the breeding program at the age of 12-13 years unless they are founders or additional breeding is needed and; 4) phenotypic selection should not be used in the breeding program. Although the second and third points could potentially be met, the first and fourth points are impossible to comply with in any white tiger breeding program.

In the draft for the Global Tiger Conservation Plan (Seal et al. this volume), criteria are defined for evaluating management units for inclusion in captive-breeding programs. These are: 1) availability of founders; 2) systematic relationship to other populations and; 3) the need to establish populations outside the country of origin. The current white tiger population in captivity is based on very few founders with the white gene (two to four) and has been "hybridized" in some programs (Roychoudhury this volume). The scientific need to continue to breed the white tiger can conceivably include the need to represent the white pelage trait in a Bengal tiger breeding program, but probably not at the scale of current breeding programs. However, the reason the white tiger was originally brought into captivity was the novelty of the specimens. Today this has developed into an economic justification for North American facilities (Maruska this volume, Simmons this volume).

The problems with subspecies "hybrids' has surfaced in other SSP programs, notably the orangutan and Indian lion SSP's. The option taken in both situations was to discontinue breeding of the subspecies hybrids at least on a temporary basis. The need to work with tiger subspecies hybrids may exist in the future, but does not appear necessary at this time.

What benefits are there to be gained for the Tiger SSP by including white tigers? If the intent is to control the white tiger population size, based on the surplus problems which now exist for the Siberian tiger SSP, is it realistic to think that more spaces would be made available for other legitimate subspecies?

The disadvantages to the Tiger SSP seem to outweigh any potential benefit of their inclusion. By including the white tiger in the SSP, the basic premises of the program would be contradicted. The validity of the entire SSP program could then be challenged.

Although the issue of using the white tiger as a marketing strategy for increasing attendance is a separate issue, by inclusion in the SSP we would in essence be condoning the use of animals for such purposes whether that was our intent or not. Like many other management practices which are currently debated, there will probably never be a consensus of opinion within the AAZPA on these issues.

The SSP programs have been touted both within the AAZPA and to the public as our efforts to organize the management (in North America) of captive forms threatened with extinction. The message offered to the public by a white tiger exhibit is merely that an unusual color morph was maintained and bred in captivity based on the interest and whimsy of royalty of another country. There is no significant or meaningful conservation message rendered by an exhibit of white tigers. The association of the SSP with such a form would legitimize the breeding and exhibition of these unusual morphologic variations and this is clearly contradictory. It also implies that our efforts with other SSP species such as gorillas, rhinos, lions, etc., might also be based on whimsy or novelty and financial gain.

If the primary reason for including white tigers in the SSP is to control the population size so that other subspecies such as Sumatran and South China tigers can be better managed, then it seems that the problem is being approached through the back door. As pointed out at the recent SSP meeting, there are 25 (eight male, 17 female) surplus tigers born prior to 1974 that are no longer part of the SSP.   Until these animals and other surplus tigers are removed and additional spaces created for SSP animals, the breeding rate of the various subspecies will be less than optimal.   If options other than placement of surplus animals outside of the SSP population, such as euthanasia, were viewed as feasible by AAZPA institutions, the surplus problem could be solved and breeding could continue as needed to maintain the population.   Should not the Species Survival Plan Subcommittee, the Wildlife Conservation Management Committee and the AAZPA governing board be addressing this problem in a more realistic manner?

PART VII
CONSERVATION STRATEGIES

37

Tigers in the Wild:
The Biopolitical Challenges

Christen Wemmer, James L.D. Smith, and Hemanta R. Mishra

Although the establishment of an international reserve system for the tiger has been a remarkable and perhaps unprecedented achievement (Panwar, 1984), the long-term survival of the species remains debatable. If environmental and human conditions have achieved a steady state in tiger-harboring nations then the species' future is assured. Few would agree though that the developing nations of the world have stopped changing. If you share the view that Third World conditions will continue to change, then the future of the tiger remains far from certain.

Western biologists initially address conservation problems in the developing world with missionary zeal; but the experience is one of sobering enlightenment. The cherished methods of Karl Popper somehow do not always serve the best purpose in dealing with conservation in the broad sweep, and like Alice in Wonderland the conservation-committed biologist abroad may find himself swept away in a separate reality. Although the scientific method is a necessary component of nearly every conservation effort, practical solutions to the most pressing problems of wildlife and environmental preservation are, in fact, intangibly beyond the realm of science. We are a "mixed bag" of administrators and biologists whose paths have crossed during the past decade in pursuing research and conservation of the tiger in the terai ecosystem. We recognize that like most of the participants here, our experience is limited to a particular time and place. To us the future of the tiger in South Asia is not assured. In the following paragraphs we examine observations of parks and reserves, research, people and aid organizations in light of their effects on the continued survival of the tiger.

SACRED GROVES AND THE WOODCHOPPERS BALL

The preservation of species diversity and wilderness areas has been an inspired topic of discussion, particularly in academic circles of western developed nations. From theoretical considerations of population genetics, population theory and island biogeography, reserve design has emerged during the past decade as a sub-discipline of conservation biology (Soule and Simberloff 1986, and Shaffer 1985). The value of discussions on the factors which must be evaluated and incorporated into reserve design cannot be questioned; no doubt they will receive more detailed consideration as new reserves are developed. But as important as these biological considerations may be, designing reserves without regard to the surrounding environment is like commemorating the sacred grove at the woodchoppers ball. The basic assumption that ideal reserve design can secure the continued existence of ecological diversity is an unrealistic simplification of political and economic reality. The greatest challenges in wildlife and land management deal with the conflict between human interests and wildlife (Milton and Binney 1980); and it is unfortunate that in the real world much of the park-design debate is academic (Mishra 1982, this volume).

The degree to which tiger reserves are subject to "human-mediated erosion" varies greatly between areas. The pace of habitat destruction is so swift in some cases that the demise of certain parks can probably be forecast within a decade. On the other hand, devastation seems remote, an outrageously pessimistic notion, in view of the situation in certain Indian Parks. But as Mark Twain noted, "The pleasant labor of populating the world" ...goes on..."with prime efficiency". Even in the most pristine sanctuary the tiger's fate will hinge on the success of managing the constricting cocoon of humanity. Because the solution to this basic problem of reserves can not be reduced to a unitary set of biological factors, it has attracted little academic interest.

OASES OF DOOM

Parks and reserves often become attractive nuisances, containing the seeds of their own destruction. Consider a worse case example of environmental succession. The ecological wealth of a forest reserve confers economic benefits to its human neighbors, and in due course it spawns a growing community of concessions which cater to foreign tourists, and a hodgepodge of ever-expanding villages. The consequence for the park is obvious. Trail systems honeycomb the boundary lands, and illegal grazing and wood gathering reduces the habitat to weeds, spurring deeper human encroachment. The land's ability to sustain wildlife diminishes, and the wildlife increasingly raids the encroaching crop fields. It is in these marginal habitats frequently used by man that conflict between man and tiger often takes place (Khan this volume, McDougal this volume, Sanyal this volume).

As soon as the park-people conflict has escalated to this stage, the problem is nearly insurmountable. Zoned development in the Third World is nearly non-existent or unenforceable, and once the village becomes entrenched, the government must offer comparable or better land as a substitute. Even if a substitute is offered, the proposal may not be accepted. A bleak landscape can be envisioned when the process is projected three or four decades into the future. The problem is a complex consequence of social, political, and cultural factors which usually differ greatly between countries and often vary regionally within countries.

Clearly, the human dimension must be more effectively factored into the design of reserves in the future than it has been in the past. "Social forestry" and "buffer zones" near parks often fall short of ameliorating human erosion, because the village and the adjacent reserve are managed separately. Integration of the better funded social forestry programs with park and reserve objectives through an independent coordinator might be a suitable solution to the problem. Incorporating programmed community development as an essential phase of park development is another possible approach. Human population and subsistence activity would have to be highly regulated through contractual agreement in an area designated as a multi-use zone within the park boundary. One thing is clear; the process and mechanism for such a system can not be generalized, but must be developed on a regional basis. This is not a new idea. The activities of villagers living within National Parks in the developing world are already regulated closely in many instances. But many social problems have also developed with time. In this regard, an experimental approach might prove useful in determining those factors which will create a steady state favoring long-term ecological stability.

WELL-INTENTIONED OGRES

The role of international agencies in addressing human and environmental welfare is complex, and embarrassingly inefficient when scrutinized. The topic merits much more detailed analysis than we offer here, but several points deserve mention. There are a large number of agencies from western and developing nations committed to improving the standard of living in certain Third World countries, and naturally there are political consequences of the programs if not explicit motives. The independent agendas of aid programs within and between agencies is often counterproductive.

Primary agenda items are linked in a chain of cause and effect, but are bureaucratically independent. Improved health care stimulates fecundity, which increases the need for enhanced food production; education and family planning march to the beat of yet a different drum. The amount of aid for wildlife and environmental conservation is but a fraction of that channeled into programs of economic development.

Second, most developing countries suffer from the "capital city syndrome". This means that international agencies must have their offices situated where the bureaucratic action takes place. While field excursions indeed occur, agency employees receive most of their experience and information from a rather non-representative segment of the indigenous population...capital city dwellers living remote from the rhythm of the village. Similarly, there is often a wide gap in knowledge and experience between those working in the field and the agency headquarters overseas. Timely coordination of activities and communication can be greatly impeded by time lags brought about by distance.

Lastly, international agency work is considered hardship, and employees are required to matriculate between countries. The consequence of shuttling new players into a program is obvious. Bureaucratic process perseveres much more easily than problem solving, especially when the latter requires the continuous education of new employees. The ultimate and costly result is often a shallow analysis of critical problems, and unintegrated and inefficient attempts at solution. Furthermore, by successively employing the same individuals in different national contexts, agencies do not discriminate between situations. What proved successful in Kenya in the 1960's may not reap similar results in Sumatra in the 1980's. The players in this game are probably no less dedicated than in other walks of life; it's just that the system doesn't work very well. How does one bring about change in a confederacy of ogres? The need to correct these problems are recognized by many beneficiaries of the system, but the solution seems beyond the reach of the visionaries.

THE QUESTIONED RELEVANCE OF RESEARCH

Research is like religion. One's commitment is usually conditioned by the amount of exposure and the frequency with which it is practiced. Recent converts embrace it as a panacea and a total way of life. "Old hands" may view its usefulness in a more moderate way. These facts have a lot to do with how wildlife and environmental research is conducted and perceived in the Third World. The relevance of research to tiger conservation in the wild has been contested. In a vast, pristine, and unmanaged ecosystem it is arguable that research findings have little application. After all, knowledge of the system's dynamics is meaningless if a hands-off policy prevents intervention. It is generally conceded however that knowledge of the systems' interrelationships may be necessary to maintain the ecological character of smaller reserves. Much of the controversy over the utility of research in park and wildlife management has been brought about by differences between developed and developing nations in motives, economics, and perceived benefits. The structure and process of international agencies has played a major role in setting the stage for several kinds of relationship to develop, as follows:

The Imperialist Pattern: In this situation the research is proposed or conducted by foreign citizens of a developed nation, often a former imperial power. Circumstances favoring this pattern are policies of international agencies which restrict funding to foreign experts, and special provisions for international cooperation in science and technology. The initiative for the study may arise in the host country, in which case it often aims to solve a problem. Alternatively and more often, the problem is identified by the foreigner. As a co-author of the published results the host country counterpart gains some measure of prestige, while the foreign expert may receive an advanced degree. The host country researcher learns that competition between westerners is keen for these positions, and the rewards are considerable.

The Nationalist Pattern: This pattern is perceived by westerners as a backlash to historical imperialism and international aid programs which advance the careers of foreign experts. The basic position of the host can be summed up as "We'll do the research ourselves, or it won't be done". In many cases the work is not done because the funding opportunities don't exist in the host country, political bases are transitory, or a bureaucratic game of "who's most qualified" slowly dampens the ardor of the candidates.

The Mercenary Pattern: When the host country values the financial benefit of foreign aid more than the results of research, the attitude can be characterized as "You can do anything you want as long as the price is right". This situation is most likely to develop when the host detects an almost desperate craving by foreign experts to conduct a particular project. Lucrative returns are possible to a clever host, but if the mercenary motive becomes too apparent, the experts may be scared away.

The Pattern of Negotiated Equality: When host and foreign interests deliberate at sufficient length to appreciate one another's objectives, methods and benefits, a balanced and mutually rewarding research program can be developed. This pattern is less common than the others, but is likely to develop when a host country becomes familiar with the international system and the motives of the foreigners. A dynamic tension develops as a result of the equal status of the counterparts. Contention over cleverly disguised personal issues can not threaten the program as long as overall goals are reinforced as the foremost concern.

Our own experience with tiger research in Royal Chitwan National Park can be classed in the last category. As in any field project disputes were common, but as the major objectives were shared by the participants and the professional benefits were equalized, the program was rarely threatened seriously (Wemmer et al. in press). The benefits of the program fell into three categories. The enhanced credibility of Nepal's conservation program was one of the most obvious. The project resulted in international recognition of Nepalese expertise. Awareness was indirectly enhanced by tourism, and was directly improved through professional activities in the international conservation community. Publicity through mass media further promoted the project and generated funding. Second, information generated on tiger populations was critical to justifying the establishment of a national park extension and dispersal corridor (Smith et al. this volume). In the political arena, scientific facts often make a much stronger case than theory or anecdote. Last, research activity significantly supplemented the regulatory activities of the National Park's staff by increasing awareness and interest in the eyes of villagers. In summary both the process and results of research reaped rewards. But the question remains, are there any challenges to the preservation of the tiger that further research in the wild can genuinely assist?

LEARNING TO MANAGE TIGERS IN SMALL RESERVES

We do not really know the size of most wild tiger populations. It is likely that the number of tigers in many reserves is less than the best estimates of minimum viable population (MVP), and it is certain that most reserves have minimal likelihood of immigration. Almost our entire knowledge of population dynamics is based on the Chitwan population, which may or may not be typical. As no conflicting data exists from other areas there is no other choice than to use this population

as the model. We have learned several important facts. First,
even in an area approaching 1,000 sq. km the number of tigers
ranged from approximately 25 to 60. Second, at times the
effective population was very small (Ne = 25) and the average
degree of relatedness was as high as 0.35 between females (Smith
et al. this volume). Third, social unrest and reproductive
cessation lasted nearly three years when one of the two breeding
males died (McDougal unpub. data). It is currently estimated
that as much as 60-75 % of the wild tiger population lives
outside reserves in the Indian subcontinent, where continued
survival is doubtful (Panwar this volume, Smith et al. this
volume). If reserves are indeed the last chance for wild
populations of tigers, and their continued protection can be
assured, the future still looks bleak. As little as 25 % of the
extant tiger population may survive following the demise of
non-reserve habitat at some point in the future.

A FAIRY TALE WHICH COULD COME TRUE

Let us imagine a world much like our own in which tiger
biologists work diligently and cooperatively toward a common
goal...ensuring the continued viability of tiger populations.
Research and monitoring of known individuals has revealed an
ominous pattern; the effective population size of nearly all
small reserves is below the minimal acceptable level established
by population biologists. It is found that for extended periods
a very small number of males in four reserves has monopolized the
breeding female population with the result that the degree of
female relatedness approaches 0.50. Theory dictates that the
population must be "genetically scrambled". Within two weeks the
tiger specialists convene and conclude that "scrambling" can be
most easily accomplished by increasing the rate of male turnover.
Ingenious plans are discussed, but the population biologists are
vexed by the field worker's disregard for theory, and the field
workers are appalled by the dogmatic inability of the former to
recognize practical problems. The population biologists maintain
that the males should be randomly assigned to new reserves, while
the field men insist that the resulting strife between unfamiliar
colonizers will wreak havoc. As a compromise it is decided to
rotate males between reserves but maintain neighbors together.
The New York Times commends Peter Jackson and U.S. Seal for
decisive action. Even more amazingly, the required funds are
immediately appropriated by several national conservation
foundations, and matching logistical support is promptly mustered
by the host country bureaucracies.

The plan is implemented and the released tigers are
monitored using radio-telemetry. Let us imagine some possible
results. The first is considered "A brilliant success". The
males immediately establish territories; within two years several
females give birth, and within five years the coefficient of
relationship between resident breeding females is less than 0.20.
A flurry of publications follows. A routine system of management
evolves based on male rotation, and the IUCN Cat Specialist Group
becomes an international exemplar of successful problem solving
in wildlife conservation. And the probability rating of this
scenario? Less than 0.05.

The second result is "Success debatable". Half of the released males disperse, but stable territories are established by unknown interlopers after three years of inter-male rivalry by equally unknown contenders. Successful reproduction resumes five years after the release of the males, which is a sore point in certain factions of the group. The males released in one reserve are killed by other tigers within six months. Only one introduced male becomes established in a small corner of the last reserve. Two brothers born to a prolific old female divide most of the reserve between themselves. The population biologists are on the whole dissatisfied with the outcome, because the field workers obviously lack the necessary grasp of the field situation. The field workers refer to the population biologists as "air-heads", and contend their involvement in the program is an unnecessary complication. The probability of this particular scenario is significantly greater than that above.

The third is considered "Sad results". In one reserve two of the males die during capture, while 18 months are required to capture the cats in the second reserve. One tiger released from the third reserve is found dead three months later from an infected plywood splinter in its mouth. In the fourth reserve three tigresses die from poisoning following the disappearance of two village children within a fortnight of the release of the males. In all reserves reproduction plummets, and male rivalry runs rampant. Social stability comes first to reserve number 4, fully three years after the release. A strange one-eyed tiger appears on the scene. He is distinguished by enormous pugs and an extra toe on the right forefoot. One year after his entree there is a tiger boom. The population biologists are extremely displeased with the results and openly question the competence of the field team. The latter begin to organize a regional task force to deal with tiger management directly and without the involvement of population biologists, or "theorists" as they are now referred to. In the midst of all this, a third interest group with strong bio-political ties emerges with a rally cry of "Hands off!". The probability of a scenario like this one is not as small as the first situation. In the final analysis, we still do not know enough about the social dynamics of tiger populations to predict the results of manipulating population structure. If you are an optimist it doesn't really matter. But if the time comes that we must genetically manipulate wild populations the exercise could prove to be a costly form of shock therapy.

A limited continuing effort could fill the gaps in our knowledge now, as a natural experiment is already going on with replicates in numerous small reserves. But the questions have yet to be asked. The costs of acquiring the information are not that great. The estimates provided in Table 1 are probably inflated, because some initial capital expenses would not necessarily be required. National Park and Tiger Reserve vehicles for example are not used full time, and maintenance costs and fuel could probably be absorbed within many park and reserve budgets. A microcomputer on the other hand would have great time-saving value for storing and analyzing data, but would also become indispensable for other park business. We have not included the costs of capturing tigers. Reliable methodology already exists (Smith et al., 1983), but considerable manpower and elephants are required. Where there are foreign-owned tourist concessions, funding for tiger monitoring and research could be required as a condition within the park-concession

Table 1. Estimated annual costs of maintaining a small tiger-monitoring team in South Asia, based on the Smithsonian-Nepal Tiger Ecology Project budget. Estimates are based on a full-time work schedule.

	Cost	Subtotal
Personnel		
Trackers @ $720 x 2	$1,400	
R~search Biologist @ $2,000	$2,000	
Mahouts (2/elephant) @ $600 x 2	$1,200	$4,600
Transportation		
Petrol/motor oil @ 40 km/day	$2,000	
Vehicle maintenance	$500	
Vehicle purchase	$8,000*	
Elephant fodder/tack @ $1,000 x 2	$2,000	$12,500
Equipment		
Radio-receivers @ $1,700 x 2	$3,400*	
Transmitters (24/2 yrs) @ $400 ea	$9,600	
Immobilization drugs @ $300	$300	
Immobilization gun/darts/charges	$2,500*	
Microcomputer	$4,000*	
Microcomputer supplies/software	$200	$20,000
Total		$37,100

N.B. Capital equipment purchases total $17,900; after the first year annual running costs would average $19,200. * = initial expenses only.

agreement. On its own initiative Tiger Tops Jungle Lodge invests $2,000 annually to cover the costs of two full time tiger trackers, petrol and vehicle maintenance, and related supplies (McDougal pers. comm.). Information generated by such activities can greatly enhance the tourist experience and promote the concession, as well as contribute to the data base for tiger management.

At this crossroads a coordinated research plan with specific objectives and a time plan could greatly benefit tiger management. There are several important needs. The first is that a prototype database should be developed for fundamental ecological information needed for tiger population management within parks and reserves . This should include: a) A refined and universally used system of population estimation that would permit reliable comparison of population trends between diverse areas; b) A simple but reliable system of monitoring principle prey species populations, and possibly availability of water and other limiting resources; and c) A geographic information system to evaluate land use and habitat changes adjacent to and within protected areas. (Smith et al. this volume). The second is that to determine variation in ecology, social organization and demography further long-term studies of tiger ecology based on known individuals are needed in ecosystems other than the terai.

The means to achieve such a program to an extent already exist within state and federal forestry and wildlife department budgets; so the economic requirements are not overwhelming. The reintroduction of captive bred tigers is particularly relevant to the topic of field research. The economic obstacles to successful reintroduction on a biologically meaningful scale however are almost overwhelming. The best solution to this challenge, if indeed it is worth considering, probably remains to be invented. An enormous gulf separates captive and wild tiger populations, and ultimately we must decide whether a causeway is needed. The demographic management of captive populations has been thoroughly analyzed and planned; if cooperation between institutions can be fostered to achieve the objectives, the tiger will at least be preserved in its alien captive dimension. But let us not be pessimists: the tiger's image burns bright in the minds of men, and probably more so than any other endangered land animal. Its ability to stir human emotion can also motivate reaction to the most steadfast of obstacles. With resolve and unified direction we can also succeed in conserving tigers in the wild.

CONCLUSIONS

The challenges to conserving the tiger can be viewed in two ways. Identifying the problems and defining the methods are relatively simple processes which can be handled at an organizational level which is reasonably effective. Executing the resultant plan on the other hand is embedded in bureaucratic process at so many more levels that organizational inertia often prevents timely results. Means of short-circuiting mega-bureaucratic inertia will be necessary to bring about efficient solutions on time. Innovation in managing human population as a critical element of the ecosystem is the second and most important challenge to preserving environment in general and the tiger in particular. Given the complexity of the human condition and the rapid rate of environmental change it is an open question whether the challenges will be met in time.

ACKNOWLEDGEMENTS

The views shared herein have been shaped over the years by discussions with several individuals deserving credit, namely Jim and Lou Ann Dietz, Eric Dinerstein, Alfred Gardner, Ullas Karanth, Chuck McDougal, Sunil K. Roy, Rudy Rudran, John Seidensticker, Ross Simons, and Mel and Fi Sunquist. Scott Derrickson, Ullas Karanth, and Ross Simons reviewed the manuscript and provided critical commentary.

REFERENCES

Milton, J.P. and G.A. Binney. 1980. Ecological planning in the Nepalese terai. A report on resolving resource conflicts between wildlife conservation and agricultural land use in Padampur Panchyat. Threshold, International Center for

Environmental Renewal.

Mishra, H.R. 1982. Balancing human needs and conservation in Nepal's Royal Chitwan National Park. <u>Ambio</u> 11:246-51.

Panwar, H.S. 1984. What to do when you've succeeded: Project Tiger, ten years later. In <u>National Parks, Conservation, and Development</u>, ed. J.A. McNeely and K.R. Miller. Washington, DC: Smithsonian Inst. Pr.

Shaffer, M.L. 1985. Assessment of application of species viability theories. A report prepared for the Office of Technology Assessment, Congress of the United States. Washington, DC.

Smith, J.L.D., M.E. Sunquist, K.M. Tamang and P.B. Rai. 1983. A technique for immobilizing and capturing tigers. <u>J. Wildl. Man.</u> 47:255-59.

Soule, M.E. and D. Simberloff. 1986. What do genetics and ecology tell us about the design of nature reserves. <u>Biol. Cons.</u> 35:19-40.

Terborgh, J. 1975. Faunal equilibria and the design of wildlife reserves. In <u>Tropical ecological systems: trends in terrestrial and aquatic research</u>, ed. F. Golley and E. Medina. New York: Springer Verlag.

Wemmer, C., R. Simons, and H.R. Mishra. In press. Case history of a cooperative international conservation program: the Smithsonian-Nepal Tiger Ecology Project. Proceedings of the Centenary Conference of the Bombay Natural History Society.

38

Developing International Tiger Conservation Programs: U.S. Fish and Wildlife Service Cooperation with India and the U.S.S.R.

David A. Ferguson and Steven G. Kohl

Operating under a variety of statutory mandates, including international treaties and conventions as well as bilateral and multilateral agreements, the U.S. Fish and Wildlife Service has entered into a number of cooperative affiliations with counterpart organizations abroad to carry out work in conservation of endemic fauna and flora. The underlying premise is the development of activities beneficial to wildlife and their habitats.

Through the Endangered Species Act, passed by Congress in 1973, the Fish and Wildlife Service was given not only the responsibility to determine the status of threatened or endangered animal and plant species worldwide, but also to enter into agreements with foreign countries to foster the conservation of these species. Such agreements can take several forms. For example, in countries such as India where the balance of payments for foreign debts favors the United States, excess or special foreign currencies have been made available to the Service by the Endangered Species Act to fund joint projects through Public Law 83-480 (Agriculture Trade Development and Assistance Act of 1954). In the case of the Soviet Union, the Service participates in exchanges under the intergovernmental U.S. - U.S.S.R. Environmental Agreement, signed during the Nixon-Brezhnev summit meeting of 1972 and since renewed every 5 years. Funding for these projects is on a reciprocal, "receiving-side-pays" basis, whereby the host country covers all expenses except international round trip airfare.

All cooperative efforts between the United States and India are overseen by a formal body known as the Indo-U.S. Joint Commission. Specific areas of interest are approved at Joint Subcommission meetings, which in theory convene annually in alternate countries but actually do so every 2 to 3 years. Because the administration of wildlife matters in India was the domain not of one, but rather of several government agencies, the overall subject of species conservation was formally addressed first by the Subcommission on Science and Technology in 1978 and subsequently by the Subcommission on Agriculture in 1979. Extensive lobbying by conservation interests resulted in formal pronouncements recognizing wildlife as a priority area of cooperation and agreeing to collaborate in the identification of projects of mutual interest.

The Service compiled a list of threatened and endangered species and approached India wildlife authorities about their interest in joint activities relating to the listed species and their habitats. India responded by proposing a number of multi-year projects for U.S. consideration. Those which appeared to further the goals of the Endangered Species Act were endorsed by the Fish and Wildlife Service and forwarded through Indian bureaucratic channels for review and evaluation by appropriate segments of the scientific community and eventual approval by the Ministry of Finance.

As strategies for project selection were worked out, it became apparent that the review process for major projects might take up to two years. The Service therefore undertook a dialogue with a number of wildlife-related organizations both within and outside the Indian government to pinpoint "targets of interest" representing low-budget, short-term activities with a visible payoff. This process was aimed at building ties with the overall Indian conservation community, establishing confidence that the Service's goals would be beneficial to India's wildlife, and, most importantly, identifying those indispensable elements, the doers and decision makers.

The two countries eventually agreed upon a number of these activities and slowly built a rapport. The Service sponsored or supported a series of international symposia held in India on endangered species and their habitats; the Service participated in local, regional and national efforts to produce environmental education materials for use in secondary schools; research was encouraged and a grants program introduced to facilitate cooperation between American and Indian scientists; short-term training and orientation activities were arranged for selected individuals; and a large quantity of U.S. books and other publications was donated to a number of Indian institutions to help build their reference base.

Once it appeared that the Service's credibility and overall objectives had been generally accepted, the focus of joint work gradually shifted from short-term to long-range programs. One of these, the conservation project for Indian tigers (Panthera tigris), also known as Project Tiger, is a success story of the greatest significance and a major credit to the Government of India. The role of the World Wildlife Fund has also been monumental, and since the establishment of Project Tiger in 1973, the tiger population in protected areas has doubled. The Fish and Wildlife Service's contribution includes support of two immobilization and telemetry workshops, as well as assistance in the distribution of two training films that were produced as a result.

The Service first became affiliated with Project Tiger ten years ago, when L. David Mech traveled to India together with Ulysses S. Seal (University of Minnesota) to teach a course on capture, immobilization, and radio-tracking for tiger reserve personnel. Over the next four years, new strategies and techniques were developed and refined, so that by 1980 a second, month-long course was ready. Titled "Wildlife Capture, Immobilization and Radio-Tracking," the course was taught at field locations in several national parks and forest reserves.

Students became familiar with darting techniques, including drug handling, loading and use of dart guns, and safety procedures. Methods of attaching radio collars and tracking tiger movements by use of telemetry units were also taught and demonstrated.

Another aspect of the 1980 course was the production of a 16 mm instructional film on animal immobilization techniques for specialists working in the country's tiger preserves. A second film on methods of drug delivery was completed in 1984. Both films were made available to Project Tiger personnel and to research and training institutions throughout India.

More recently, the Service has been working with the Columbus Zoo to obtain young Indian tigers from the wild to act as new founder animals, and to secure either the breeding loan of a white tiger from India or semen from an adult tiger for artificial insemination of white tiger females in the United States. The Service has also supported attendance of American biologists at the April 1984 International Union for the Conservation of Nature (IUCN) Cat specialist Working Group meeting in India, and contributed funds for Indian participation in this Tiger Conservation Symposium in Minnesota.

One of the more exotic activities in which the Service has been involved was the 1985 National Geographic Society film special on tigers. The Service had engaged the husband and wife team of Stan and Belinda Breeden to produce an ecological film on Keoladeo National Park in India in 1979. That film, "Birds of the Indian Monsoon", was premiered at the February 1981 C.I.T.E.S. (Convention of International Trade in Endangered Species of Wild Fauna and Flora) conference in New Delhi and has been shown all over the world. The Breedens were then commissioned to do the National Geographic special in 1982. The Service acted as liaison between National Geographic and the Breedens and facilitated the shipment of film and equipment between India and the U.S. The result in early 1985 was the award-winning documentary, "Land of the Tiger." Probably viewed by more people than any other film of this type at its initial showing, it is being used continually as an educational tool throughout India by the Forest Department, schools and Ministry of Tourism. The Service continues to facilitate this kind of educational use.

Besides tigers, the Fish and Wildlife Service is active in projects involving two other endangered cats in India: the snow leopard (Panthera uncia) and Asiatic lion (Panthera leo). In 1983 Dr. Helen Freeman, Curator of Interpretive Services at the Woodland Park Zoo in Seattle and President of the International Snow Leopard Trust, contacted the Service to solicit support for a proposed project to reintroduce snow leopards from American zoos back into their native habitat.

Dr. Freeman suggested conducting a survey of snow leopards and their habitat. This would entail selection and monitoring of a suitable study site for long-term observations of predator-prey interaction, with a reintroduction program to follow only if justified by the survey results. A joint field research project on snow leopards is currently underway in the Himalayas with U.S. and Indian personnel supported by the Government of India, the Fish and Wildlife Service, and the International Snow Leopard Trust, with the contribution and participation of many other organizations. A snow leopard symposium is scheduled for October 1986 in Srinagar, Kashmir.

The entire wild Asiatic (Indian) lion population is presently restricted to about a 500-square-mile tract of the Gir Forest in the west Indian State of Gujarat. Recent suspicions have arisen as to the genetic purity of the captive population of this sub-species in zoos, which nearly outnumbers that in the wild. In March 1985 the Fish and Wildlife Service sponsored a team from the National Institute of Health, Chicago Zoological Society, Brookfield Zoo and Lincoln Park Zoological Society to India to obtain samples of blood and semen from known pure Asiatic lions in the Sakkarbaug Zoo. The blood samples were subsequently analyzed at the National Institute of Health laboratory along with blood taken from pure east African and South African lions and representative samples from animals in question maintained in North American zoos. Using electrophoretic techniques to compare differences, a test is being developed to screen for evidence of hybridization in captive Asiatic lions.

To the north of India lies the U.S.S.R., where the Service has played a key role in arranging for American zoos to receive Siberian tigers and snow leopards from Soviet zoos for the purpose of enhancing the genetic diversity of animals held in North American zoos.

U.S.-Soviet cooperation in the field of nature conservation and wildlife protection is carried out as a subsection of the overall Environmental Agreement between the two countries. The Fish and Wildlife Service administers exchanges pertaining to threatened and endangered species of fauna and flora, arid and arctic ecosystems, biosphere reserves, marine mammals, and fish husbandry, with the participation of other U.S. agencies, private organizations and academic institutions as well.

The idea for exchanges of captive-bred animals dates back to 1980, when, in response to inquiries from the San Diego and Bronx Zoos, the Service approached its Soviet counterpart agency, the U.S.S.R. Ministry of Agriculture, to discuss the possibility of establishing a joint project to exchange captive-bred animals of species where the frequency of inbreeding was undesirably high. Because the lead agencies were already affiliated under the Environmental Agreement, the proposal was adopted with a minimum of red tape, and was set up as a non-commercial venture, with all animals to be exchanged for the cost of transportation only.

In collaboration with the Bronx, Omaha and Indianapolis Zoos, the Service in 1982 sought the help of Moscow Zoo Director Vladimir Spitsin in obtaining three Siberian Tigers (Panthera tigris altaica) for introduction of new genetic material into American zoos. When the Moscow Zoo replied affirmatively, the U.S. side negotiated to receive three, two-year-old Siberian tigers (one male and two females) born in captivity to wild-captured parents. The two females would be bred with males in American zoos, while the male would be shared by several facilities in rotation. Once the individual tigers had been selected and the August 1983 transfer date agreed upon, both countries set about securing the necessary export-import permits, blood tests, shipping containers, and immobilization equipment. The tigers were immobilized at the Moscow Zoo by Ulysses Seal and then examined, vaccinated and transferred to crates specially constructed for the 8,000 km journey to the United States. The entire transfer process was also filmed by a camera crew representing U.S. public broadcasting, and eventually presented as a segment on the "Smithsonian World" television series.

After the arrival of the tigers in New York, the male was sent to the Henry Doorly Zoo in Omaha, with the two females going to the Bronx and Indianapolis. One of the females, Alica and Astra, has reproduced, and the male, Tulip, sired four offspring in 1985.

In May 1983 the Fish and Wildlife Service arranged for Helen Freeman, Dan Wharton of the Bronx Zoo, snow leopard studbook keeper Leif Blomqvist of the Helsinki Zoo, and researcher Kathleen Braden of Seattle to travel to the Soviet Union to visit major snow leopard holdings at the Moscow and Alma Ata Zoos as well as to discuss an exchange of these animals. After the delegation's return to the U.S. there began lengthy negotiations which resulted in the May 1985 exchange of Bisser, a 5-year-old male from the Novosibirsk Zoo, for Oona, a 3-year-old female from the Bronx.

Looking forward to the future, plans are currently underway for a 1986 U.S.-U.S.S.R. exchange of Pallas cats (Felis manul). In coming years, the Fish and Wildlife Service will expand its ties with zoos in the People's Republic of China to include the Chinese tiger (Panthera tigris amoyensis), which is endangered and found only in the Yangtze River valley.

The commitment of the Fish and Wildlife Service to working with endangered species of large cats is now firmly established as a priority for its Office of International Affairs, and the Service will continue to promote their effective management through exchanges of research personnel and breeding stock, training of host country personnel, and promotion of educational programs.

39

An Outline Strategy for the Conservation of the Tiger (*Panthera tigris*) in Indonesia

Kenneth R. Ashby and Charles Santiapillai

INTRODUCTION

At the turn of the century there were three subspecies of the tiger (<u>Panthera tigris</u>) in Indonesia; <u>Panthera tigris</u> <u>sumatrae</u> in Sumatra, <u>P. t. sondaica</u> in Java and <u>P. t. balica</u> in Bali. Today only the Sumatran tiger survives. According to the consensus of scientific opinion, both the Javan and the Bali tigers are extinct. The last stronghold of the Bali tiger (the smallest of the three subspecies) was the 20,000 ha Bali Barat Game Reserve (Fig. 1) from which it is presumed to have been extirpated by 1950 (Seidensticker this volume).

Nevertheless, it is not at all uncommon to hear reports of the animal being sighted by local inhabitants. According to Jackson (1963), the last tiger in Bali was shot in 1937. The population of Javan tigers declined rapidly since the turn of the century. Hoogerwerf (1970) reported that by 1930 the animal was already very rare or extinct outside Udjung Kulon, Baluran and Meru Betiri Game Reserves, while Seidensticker and Suyono (1980) have suggested it was still present in a number of localities in Java outside these reserves in 1940 (Fig. 1). Regardless of the precise course of decline, by the mid 1970s, the tiger was still present only in the Meru Betiri Reserve in eastern Java. Seidensticker and Suyono (1980) believed there were five tigers present in 1976. Angst (1980) estimated that three were present in 1979. According to Blouch (1982), even if tigers were still present in Meru Betiri, it was unlikely that the subspecies could be saved from extinction.

Hoogerwerf (1970) devoted five chapters in his monograph on Udjung Kulon to the Javan subspecies and the reasons for its decline. Seidensticker (this volume) reviews the current status of the tiger in Bangladesh and Indonesia and its prospects for the future in these countries. Santiapillai and Widodo (this volume) assess its current status in Sumatra, concluding that while it is still widely distributed in the island (Fig. 3), fragmentation of its habitat threatens its continued survival there in the long term.

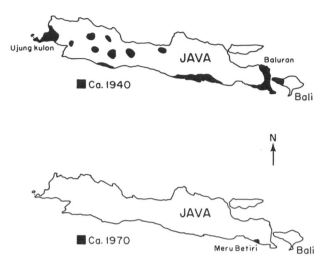

Fig. 1. The decline of tiger in Java and Bali (from Seidensticker and Suyono 1980).

BASIC CONSIDERATIONS

Tigers range over large areas and need medium-sized or large herbivores available as prey. They cannot, therefore, be maintained at high density even in the most favorable habitat. So tiger reserves must be substantial in area; there is a well defined upper limit to the number of tigers which can live in a given area (Santiapillai and Widodo 1985). We may assume 60 sq. km as the minimum area needed per male plus female tiger, as this is considered the standard territory size of the male (Sunquist 1981). Where prey density is low or variable, the area must be larger. We suggest perhaps 30 adults (15 pairs) as the minimum population size in the core area (the part actively protected) of any reserve for tigers which will be viable in the long term as a breeding unit, giving a minimum size of the core area required of about 900 sq. km, or 35 km in diameter if roughly circular. Seidensticker (1986) argues that a minimum population size of some hundreds and a minimum reserve area of 25,000 sq. km are needed. However, a number of species of mammals have escaped extinction and ultimately flourished after their population sizes were for many years in single figures.

Examples include the European bison, the Texas population of American bison, Chillingham cattle, Przewalski's horse, Pere David's deer and the Manipur brown-antlered deer. Likewise, since their invasion of Isle Royale in Lake Superior about 40 years ago, wolves have maintained a stable, healthy, totally isolated population of approximately 20 individuals occupying an area of 360 sq. km (Mech 1970). This example is particularly pertinent, as the species involved is another top predator.

With Seidensticker's figures as minima, there would be no future for tigers in such reserves as Wang Kambas and Berbak in Sumatra. Their respective areas are 1,300 sq. km and 1,900 sq. km and both currently contain healthy populations of tigers. In fact, prospects would be poor for maintaining more than at most a single tiger reserve in Sumatra, given that the island's total population of tigers at present is only in the hundreds (and falling) and that no large national park exists. The largest conservation area in Sumatra, the Kerinci-Seblat National Park, is 14,850 sq. km in area, which is little more than half the minimum suggested by Seidensticker for the conservation of tigers. The success of Project Tiger in India provides further reason for believing that tiger reserves only 1,000 sq. km in area may be potentially viable. Reserves there vary from a maximum of 3,500 sq. km to as little as 350 sq. km.

However, apart from any section which is flanked by a secure boundary such as the sea, a tiger reserve will need a substantial buffer zone that is good habitat, but without specific national park designation and protection, to protect the core reserve from incompatible land uses such as dense human agricultural settlement. The success of the tiger reserves in India owes much to the existence of forests around these reserves. If a radius of 20 km is assumed for a buffer zone, the total area needed for a reserve becomes, if it is circular, 4,400 sq. km in area and 75 km in diameter, and would hold perhaps twice the number of tigers suggested above for the core area.

Both primary forest and that which has been logged form good habitat for tiger. In fact, prey densities may be markedly higher in regenerating forest than in forest which has not been disturbed. Sustained yield exploitation of the forest is therefore compatible with tiger conservation. What is not compatible is dense human settlement, or human intrusion associated with road construction on an extensive scale. The abundance of natural prey will be depleted and attacks on domestic animals, and perhaps also on man, are bound to become frequent. Apart from the damage and loss of life resulting, such conflict is sure to be damaging to the image of conservation. As with elephant, planning for tiger conservation requires as basic premise that the animals and new settlements need to be kept apart. This fact requires recognition by agriculturalists and settlers alike, that settlements should occur only in clearly defined zones.

STRATEGY FOR SUMATRA

In spite of losses of habitat to date, there is still time for substantial tiger habitat to be maintained, provided that it is not fragmented into inviable units. In Sumatra, a large proportion of the tiger's geographic distribution coincides with the chain of volcanic mountains along the western coast (Fig. 3). In such forest formations, tigers are thinly distributed (Seidensticker 1986). We suggest the following strategy:
    1. Identify and designate key areas for tiger conservation based on existing national parks and wildlife reserves, and their available buffer zones. These should include one or two large areas approaching the size recommended by Seidensticker (1986);
    2. Adopt the policy to keep as much unfragmented habitat elsewhere, as is practicable on a long-term basis, as a multi-purpose forest (presumably logged on a sustainable yield basis), water and wildlife resource. The policy should be put forth as being in the long-term interest of the nation from all viewpoints

Fig. 2.   Major conservation areas in Sumatra.

Providing that good feeding conditions and cover are maintained over substantial areas, the tiger should be able to survive outside specially protected areas.   Hoogerwerf (1970) considered the critical element in the decline of the tiger in Java to be poisoning, almost certainly the work of nearby agricultural settlers for whom tigers were unwelcome neighbors.   It is vital for wildlife conservation that areas chosen for new settlements be identified in terms of compatibility with nature conservation as well as agricultural suitability, and that such uses of habitat as logging without thought for regeneration be brought under control.

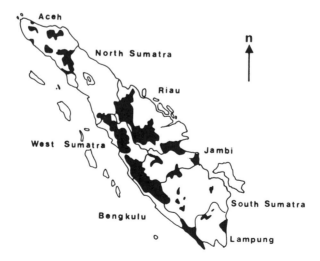

Fig. 3.   Estimated current distribution of Sumatran tiger (from Seidensticker and Suyono 1980).

JAVA AND BALI

We suggest that for the foreseeable future both these areas should be studied in terms of lessons to be learned from recent history and not as potential sites for tiger reintroduction, which has been suggested and undoubtedly has the attraction of being a proposal that could obtain world-wide publicity. We suggest the main factors are:
1. There is no possibility of recreating the indigenous subspecies, which were markedly smaller than the Sumatran subspecies;
2. The final disappearance of the tiger in Java and Bali was rapid and deliberate even though it occurred at a time when conservation was already the accepted national policy. The policy's nominal protection had little influence on settlers living near areas where tigers remained. There is no reason to believe that local antipathy has declined since then. One may draw a parallel with the wolf in Sweden, where the antipathy of reindeer herdsmen has prevented this species from surviving even though good habitat still abounds in the country and the human population as a whole is sympathetic to the objectives of conservation;
3. While there are still extensive areas of secondary forest on the mountains, there is no continuous tract of good, unsettled habitat within which there are, or could be, well-protected national parks of adequate size. Of the two such existing parks with adequate prey, Baluran in East Java has settlements right on its boundaries and Ujung Kulon is located on a peninsula with no possibility of having an adequate buffer zone;
4. The potential benefits of reintroduction may not compare with the probable strong opposition from local inhabitants and the substantial financial resources which would be required.

REFERENCES

Angst, W. 1980. Annual Report for 1979. WWF/IUCN Project No: 1024.
Blouch, R.A. 1982. Proposed Meru Betiri National Park Management Plan 1983-1988. A World Wildlife Fund Report, Bogor, Indonesia.
Hoogerwerf, A. 1970. Udjung Kulon: The Land of the Last Javan Rhinoceros. Leiden: E.J. Brill.
Jackson, P. 1963. The Tiger Lives On. Ambio. 12(5):278-79.
Mech, L.D. 1970. The Wolf. Garden City, NY: Am. Mus. Nat. Hist. Press.
Santiapillai, C. and S.R. Widodo. 1985. On the status of the tiger (Panthera tigris sumatrae Pocock, 1829) in Sumatra. Tigerpaper 12(4):23-29.
Seidensticker, J. and I. Suyono. 1980. The Javan Tiger and the Meru Betiri Reserve, a plan for management. Gland, Switzerland: IUCN. 167 pp.
Seidensticker, J. 1986. Large Carnivores and the Consequences of Habitat Insularization: Ecology and Conservation of Tigers in Indonesia and Bangladesh. In Proc. Int. Cat Symp., Texas A & I Univ., Kingsville, TX.
Sunquist, M.E. 1981. The social organization of tigers (Panthera tigris) in Royal Chitwan National Park, Nepal. Smithson. Contrib. Zool. 336:1-98.

40

Managing Tigers in the Sundarbans: Experience and Opportunity

John Seidensticker

THE SUNDARBANS INTERNATIONAL WILDLIFE REFUGE: A PROPOSAL

When the water drains off the higher ground with the ebb and steel-gray mud banks are exposed and shimmering under a scorching mid-day sun, the place to look for a tiger is in the shade at the mouth of small, side khals. And if you are truly fortunate you see it: head raised, lying half-submerged, intently watching as you slip by in the launch - a classic Sundarbans tiger.

The tiger is in its element in the Sundarbans and looks it. The conditions and opportunities for long-term tiger conservation in the Sundarbans are unparalleled. The Sundarbans forest is big (>10,000 sq. km) so a large population (effective population size, Ne, >100) of tigers has been and can continue to be maintained for the next 50, 100 even 200 years from now. We are not dealing here with a few cats living in a small fragment of habitat, a small island surrounded by agriculture and who knows what other threats, or some less than ideal remnant habitat that nobody could find much use for and didn't want. Both are all too frequently the norm where conservation planning for endangered species is concerned. However, there are no corridors that allow immigration from or to the Sundarbans from adjacent areas; there are no tigers in adjacent areas, only people. Any tiger that ventures out is dead. The tiger's survival hinges on those there now and how we manage for their needs under ever-increasing pressure.

It is not in the economic cards to set the Sundarbans or a significant portion of it aside where tigers can live apart from man. I propose that the entire Sundarbans, India and Bangladesh, be designated the Sundarbans International Wildlife Refuge. So there is no confusion, I am not proposing that India or Bangladesh give up any management control of their forests or that an International body manage the Sundarbans. What I am proposing is a way to emphasize and include the ecologic, demographic, and genetic needs of a tiger population and those of other wildlife species into the forest management systems of both countries.

This proposal seeks to provide a mechanism to refine management capabilities to include all components of the Sundarbans ecosystem that contribute to the protection of <u>vital ecosystem function</u>. Wildlife management and forest management activities should have a synergistic effect towards achieving a management goal of <u>best achievable ecosystem function</u>. In this context, an increased investment in wildlife management will be cost effective by contributing to additional management capability for the entire Sundarbans land management system.

An examination of the ecologic character of the Sundarbans and its history of outstanding professional forest management provides the necessary background for the refuge proposal. I will focus my discussion on the Bangladesh side of the delta because that is where I have been involved in conservation planning (Seidensticker and Hai 1983).

ECOLOGICAL CHANGES AND FOREST MANAGEMENT

Bangladesh is a land born of great rivers. The Ganges, the Brahmaputra, the Meghna join here and flow into the Bay of Bengal, forming the largest delta in the world. Monsoon rains,

Fig. 1. The Sundarbans in India and Bangladesh, largest mangrove forest in the world, is at the interface of two huge ecological systems - the Bay of Bengal and the watersheds of the Ganges and Brahmaputra Rivers. (Photo by J. Seidensticker 1980).

the rise and fall of river levels, floods, alluvium, and changes
in river courses form the substrate of the cultural and physical
personality of this land.

More than 87 million people live in Bangladesh (Library of
Congress 1980). Less than 9% of the nation remains under forest
cover (Library of Congress 1980). But in Bangladesh, at the edge
of the land that is not quite sea, there remains one of the
world's great wild places - the labyrinth of channels and
beautiful forests of the Sundarbans. This 6000 sq. km storm-
lashed, coastal zone is the only remaining habitat of the tiger
(Panthera tigris tigris) in Bangladesh's Bengal Basin.

Coastal zones are among the world's most ill-treated
ecosystems; usually they are a no man's land where every man
takes what he can. Properly maintained, many coastal zone areas
are the most productive on earth but they are particularly
vulnerable to man-induced stresses (Clark 1974). Bangladesh's
Sundarbans Forest Division has been managed with care; the long-
term sustained outturn of forest products speaks well for the
professionals responsible. And the tiger has survived.

Let's look briefly at this fortunate situation. The
Sundarbans Forest Division is the only large block of forest left
in Bangladesh. It is the world's largest mangrove forest. It is
the focus of happenings in two huge ecosystems - the Ganges/
Brahmaputra watersheds and the Bay of Bengal (Fig. 1). The
tides, floods, and intense and frequent cyclonic storms make this
an open, dynamic resilient, heterogenous ecological system. It
is a pulse-stable system (Odum 1971) that is resilient to
disturbances from within the forest and waterways but sensitive
to those from the outside. Variations in the flow of fresh water
and the tides are compartmentalized by the tributaries of the
Ganges and tidal rivers, and productivity of the system is thus
compartmentalized (Hendrichs 1975).

Two hundred years ago, the region's chief consumable
resource was its undeveloped land and the forest was about double
its present size. The British government assumed property rights
in 1828. Regulated clearing and cultivation went on until about
1875 (Curtis 1933). It was during this development phase that a
decline in the diversity of large mammals occurred as their
habitat went under the ax and plow (Table 1). The switch from
land transformation to managed forest utilization occurred in the
1870's when the population of Bangladesh was under 22 million
people (Rashid 1977). When the first efforts to increase the
number of mangrove species in production occurred in 1931, the
population was 36 million (Curtis 1933). In 1960, the starting
date of the last working plan (Chaudhury 1968), the population
was 55 million. At the present rate it will be 100 million by
the year 2000. In the next few years as many people will be
added to the population of Bangladesh as were present when the
curves of supply and demand for timber and other products of the
Sundarbans crossed. It was then deemed necessary, in support of
economic development, to switch from land transformation and the
pushing back of the forest to the sustained management of forest
utilization.

If there is one suture in this hundred-year history of the
management of the Sundarbans that has resulted in success, it has
been a continuing effort to improve the management system in the
face of ever increasing demands. Had this not been the case

Table 1.   Changes in the distribution of felids, canids, and
ungulates on the Ganges Delta (from Hendrichs 1975, Mukherjee
1975).

Extant in the Sundarbans Forest:
<table>
<tr><td>Panthera tigris</td><td>Sus scrofa</td></tr>
<tr><td>Felis bengalensis</td><td>Muntiacus muntjak</td></tr>
<tr><td>Felis viverrina</td><td>Axis axis</td></tr>
</table>

Recent invaders into the mangrove forest from newly converted
adjacent lands:
Felis chaus
Canis aureus
Vulpes bengalensis

Recently extinct from the lower Ganges Delta:
<table>
<tr><td>Panthera pardus</td><td>Cervus duvauceli</td></tr>
<tr><td>Rhinoceros sondaicus</td><td>Bubalus bubalis</td></tr>
<tr><td>Axis procinus</td><td>Bos gaurus</td></tr>
</table>

---

there would be no forest. The process of forest management has
been a series of incremental acts to bring more food, fiber, and
material into production. This has been done with care and
control. And the tiger has survived.

However, there has been no concurrent development of
management mechanisms to maintain optimal ecosystem function
(Clark 1974) for the Sundarbans. The major strength of the
management system has been that an entire coastal region has been
placed under the control of one government unit, the Forest
Department. The future development of the management system will
necessarily have to reflect the natural rhythms and dynamics of
the system if the full component of values is to be maintained.

The value of the forest as a major producer of timber,
firewood, and golpatta (Nipa fruticans) is well known. Certainly
the people living in adjacent areas appreciate the protection
value of the forest against the fury of cyclones. No less
important but much less appreciated is the work this forest
performs for man as a nutrient producer for the fisheries, as a
water purifier, as a sediment trap, as a shore stabilizer, and as
an energy storage unit. It is also an area of great aesthetic
attraction and an essential wildlife habitat.

The continued survival of the tiger pivots on the well-being
of its prey and of the forest. Accordingly, the tiger will
survive only in the sphere of wildlife management which is a part
of forest management that ensures sustained production from this
ecosystems of goods and services to provide for the needs of the
people of Bangladesh. A simple linkage in theory; a complex task
in practice.

TIGER NEEDS AND TIGER MANAGEMENT

The performance standard for management of tigers in the

Sundarbans conservation system could be stated: ensure the tiger
is established in such numbers and is so distributed so as to
provide a reasonable likelihood of its surviving in its native
habitat with its genetic spectrum intact. The management
decisions required to sustain tiger populations will necessarily
be derived from knowledge about its life history processes. There
are three areas of primary management concern: 1) numbers and
age-specific reproduction and survival rates, 2) habitat size and
effective population size, and 3) habitat quality.

Numbers and Reproductive and Survival Rates

The Forest Department estimates there are several hundred
tigers living in the Sundarbans. If we use crude density
estimates of tigers in other areas of good tiger habitat such as
one adult/40 sq. km in the Royal Chitwan National Park, Nepal
(Sunquist 1981) indeed there is room for about 250 adults to live
there. And this is what is so important about the Sundarbans:
space.

I know of no census with reasonable confidence limits to
tell us how many tigers actually live in the Sundarbans. It is
important that we develop a suitable scheme for monitoring this
population of tigers. As in the management of other large
solitary carnivores (Knight and Eberhardt 1958), the techniques
used must be sufficiently sensitive to anticipate and guide
necessary management actions.

The key parameters are population size and age-specific
reproductive and survival rates. Ecologists studying long-lived
animals with low reproductive rates have learned that perturbations of the age and sex structure of such populations can
influence population dynamics for decades (Knight and Eberhardt
1958). I would expect serious perturbations to this tiger
populations from two sources: 1) political events, such as the
war for the independence of Bangladesh and subsequent political
turmoil, that give rise to conditions where tigers are killed,
and 2) the removal of problem tigers (man-killing animals).

Problem tigers are one of the most significant issues facing
the conservation of tigers today, and the impact of their removal
should be known. Between 1971 and 1985, 260 people were reported
killed in the Bangladesh Sundarbans (Khan this volume).
Protecting people working in the forest from tiger predation must
be a priority research issue.

Do tigers follow the Eberhardt (1977) model of population
regulation? This model postulates a sequence with changes in
survival of young animals, followed by changes in age at first
reproduction, followed by changes in reproductive rates, and
finally by changes in adult survival? We do not know how tiger
populations are regulated, and thus, can not predict the impact
of a particular schedule of morality or perturbation on the
population. I do not know a non-invasive way to obtain age-
specific reproductive and survival rates on tigers other than
through longitudinal studies of known individuals. These
longitudinal studies should be initiated as a management
priority.

Habitat Size and Effective Population Size

The reason that the space provided by the Sundarbans is so important to the conservation of wild tigers is because there is room to maintain a population with a panmictic effective population size of >100 [Ne = (4 x Nr females x Nr males) / (Nf + Nm), Allendorf 1986]. In small populations, genetic diversity is lost each generation at a rate inversely proportional to the effective size of the populations (Ralls and Ballou 1986). We also know that inbreeding depression, common in small captive populations (Ralls and Ballou 1983), does affect tiger population demography (Ballou and Seidensticker 1983). Soule and colleagues (1986) report extensive evidence suggesting that even a modest decrease in heterozygosity can reduce fitness, as estimated by physiological efficiency, growth rates, and developmental (morphological) stability. Where else do we have the space and the possibility of maintaining this large an Ne for a tiger population other than in the Sundarbans? In Sumatra perhaps, for Panthera tigris sumatrae (Seidensticker 1986), but I know of no other area of suitable habitat large enough within the Bengal tiger's range.

Habitat Quality

As reliable density estimates of tiger populations become available, it is clear that density varies between areas and even within relatively small areas (Sunquist 1981). We have determined that tiger density and home range size are strongly and positively correlated with the biomass of larger cervid prey (Sunquist 1981). Tigers have co-evolved in mammal assemblages as the largest predator of cervids and Sus (Seidensticker 1986). The biomass of this essential prey is strongly and positively correlated with the extent of forest/grassland edge available (Eisenberg and Seidensticker 1976). The tiger needs cover to hunt and radiotelemetry studies have documented how sensitive tigers are to this habitat dimension (Sunquist 1981).

Energy yield relative to mobility is one way to measure habitat suitability. A tigress with young cubs is not as mobile as a resident, adult male, and thus, the energy yield pattern from an area must be different for each to survive. Another essential dimension to the structure of tiger habitat is that patches which support cervid prey at densities sufficient for a female tiger to rear cubs must be generously dispersed throughout the area.

Within an area, man can diminish habitat suitability by decreasing and/or altering the spatial and temporal patterning of resource availability. The three principle modes of doing so are to alter the structure of the vegetation, or to divert productivity away from the tiger's food chain, or man can simply degrade the ecosystem's ability to produce to the point where it simply can not support a population of big predators. Establishing an extensive monoculture plantation of exotic trees can destroy sites of locally high prey productivity and availability that are critical for the survival of the population seasonally or in difficult times. The problem is amplified in small areas.

Forestry and Tiger Habitat in the Sundarbans

A primary reason for managing the entire Sundarbans as a wildlife refuge is because in the Sundarbans, habitat suitability is in a state of constant flux as deltic processes and cyclones irregularly present new combinations of environmental variables. Added to this environmental template are forest harvest schedules and practices. In the very productive, pulse-stable environment of the Sundarbans, forestry operations have been widely dispersed over the forest and are based on about a 20-year cutting cycle. So far these conditions have resulted in suitable tiger habitat.

REFUGE VS RESERVE MANAGEMENT FOR SUNDARBAN TIGERS

The ecological and socio-economic character of the Sundarbans is such that habitat needs for wildlife can best be met not through benign neglect of a small reserve within the forest but through good protection and by identifying and protecting vital areas of habitat critical to wildlife. The management response will require the capacity to ensure that ecological processes which are essential for the wildlife community are not disrupted in the process of resource exploitation. We do not have the knowledge needed to identify all the vital areas or ecological processes which will require protection. The accumulation of this knowledge and its infusion into the management system must be the essential strategy of any wildlife conservation program for the Sundarbans.

The conventional approach to wildlife conservation in South and Southeast Asia has been to provide legal protection and to conserve selected population of species and their habitats in strictly protected reserves, a mechanism known as the core-buffer concept (Leopold 1933). This approach rests on the assumption that inviolate core areas surrounded by restricted-use buffer zones can ensure the survival of species and communities; it assumes that within these reserves the life-cycle needs of species and communities will go on in a "natural" way if they are safe from man-induced environmental perturbations. This undoubtedly is the best strategy for many ecological and socio-economic situations. However, the application of this land management regime for tiger conservation in the Sundarbans could be self-defeating in the long-term.

There are three reasons why this is so. I have made the case above for the space needed to maintain a large effective population size.

At the interface of land and sea, the Sundarbans coastal zone ecosystem is strongly influenced by external factors. Furthermore, this zone is a recent formation and continually changing in its configuration. A wildlife conservation strategy must recognize the dynamic nature of the system, the overwhelming impact of storms, the nature of the processes of change, and the factors limiting the movements, distribution and numbers of animals. Providing for the ecological needs and populations processes of the wildlife community over the long-term would require a huge portion if not the entire Sundarbans coastal zone. If the total wildlife management effort was directed only towards isolated, small inviolate areas, a single storm or change in water regime might well render those areas inappropriate as habitat. Such factors are the environmental norm for the Sundarbans.

Finally, it simply is not economically possible to set aside all or a major portion of the Bangladesh Sundarbans as an inviolate reserve. By one estimate, one-third of the population of Bangladesh, directly or indirectly, depends upon the mangrove forests for their livelihood (Library of Congress 1980).

Long-term tiger conservation in the Sundarbans can be achieved through strict protection and by adopting the concept of a wildlife refuge in which critical habitats and vital areas, where essential species functions occur or where essential ecological processes occur or originate, are managed. Vital areas would include "hot spots" of high local resource production. The sea face meadows are resource hot spots for deer and their predator, the tiger (Fig. 2). However, in the refuge concept, it is recognized that these hot spots constitute only a small fraction of the area needed to ensure that vital population processes for the deer and tiger are not confounded, a recognition that will necessarily have to be integrated into the overall system of forest management as a management standard (Kusler 1980).

I make the case for space for tigers above. Another example is water birds. With their dependency on patchy food supplies, this group must integrate resources over large areas: their nesting and roosting sites are vital areas that require protection (Kushlan 1979).

Rather than attempting a strategy of habitat management that involves setting aside areas where the entire life-cycle needs of a wildlife community can be met, as in the core-buffer regime, the genetic, demographic, and ecologic needs of wildlife are linked to the overall resource management system of the Sundarbans. In so doing, the genetic, demographic, and ecologic processes upon which wildlife depend would become integral values in the management matrix.

MANAGEMENT AXES

The concept of the Sundarbans as a wildlife refuge must develop along a number of axes. The wildlife program must be integrated at the policy, implementation, and control levels of the forest management system. In addition to control and protection, this will require the capacity to analyze, monitor and infuse findings. There is sufficient information to determine initial wildlife management priorities and to set the refuge program in motion. Further refinement will require more detailed knowledge to identify vital habitats and critical ecological processes, to identify opportunities for optimizing the resource mix at minimal ecological cost, to avoid disruptions, and to provide a more complete framework from which real costs and benefits can be estimated.

Communication is a second axis. To infuse information into policy and decision-making will require developing communication formats and processes which force analyses to be responsible,

Fig. 2:   The Sundarbans East Sanctuary, Bangladesn: a resource
hot spot for axis deer and tigers. (Photo by J. Seidensticker
1980).

usable, and transferable (Holling and Clark 1975). The wildlife
refuge program is a focal point where conservation principles
that establish a healthy relationship between man and his
environment can be demonstrated.

     Tiger management must be a part of a continuing effort to
refine the management matrix to include all components of the
Sundarbans ecosystem. It should be based on the principle that
best achievable ecosystem function will provide the greatest
benefit in terms of natural goods and services to society in the
long-run. The tiger will survive only if ecosystem function is
maintained. And only a management expertise that can preserve
sensitive wildlife species has the problem in hand. The power to
do so must develop over time.

ACKNOWLEDGEMENTS

     I made three trips to the "beautiful forests" of the
Sundarbans, as a guest of the government of West Bengal, India,
in 1974, and of Bangladesh in 1978 and 1980. It is with great
pleasure that I received these invitations through the World
Wildlife Fund to come and look into the issues concerning
wildlife conservation in this coastal zone. The main body of our
findings from the Indian Sundarbans were reported in
Seidensticker et al. (1976). Results of work in Bangladesh were
published in The Sundarbans Wildlife Management Plan:
Conservation in the Bangladesh Coastal Zone (Seidensticker and
Hai 1976). I have included sections of this plan in this report.
The views expressed are my own.

REFERENCES

Allendorf, F.W. 1986. Genetic drift and the loss of alleles versus heterozygosity. Zoo Biol. 5(2):181-90.

Ballou, J.D. and J. Seidensticker. 1983. In International Tiger Studbook, ed. S. Seifert and P. Muller. Zool. Garten Leipzig: Leipzig.

Chaudhury, A.M. 1968. Working Plan of the Sundarbans Forest Division for the Period 1960-61 - 1979-80, Vol I. Dacca: Govt. of Bangladesh, Forest Depart.

Clark, J. 1974. Coastal Ecosystems. Washington, DC: Cons. Found.

Curtis, S.J. 1933. Working Plan for the Forests of the Sundarbans Division. Calcutta: Bengal Govt. Press.

Eberhardt, L.L. 1977. Optimal policies for conservation of large mammals, with special reference to marine ecosystems. Environ. Cons. 4:205-12.

Eisenberg, J.F. and J. Seidensticker. 1976. Ungulates in southern Asia: a consideration of biomass estimates for selected habitats. Biol. Cons. 10:293-308.

Hendrichs, J. 1975. The status of the tiger Panthera tigris (Linne 1785) in the Sundarbans Mangrove Forest (Bay of Bengal). Saugetierd. Mitt. 23:161-99.

Holling, C.S. and W.C. Clark. 1975. In Unifying Concepts in Ecology, ed. W.H. van Dobben and R.H. Low-McConnell, pp. 247-51. The Hague: W. Junk.

Knight, R.R. and J.L. Eberhardt. 1958. Population dynamics of Yellowstone grizzly bears. Ecology 66:323-34.

Kusler, J.A. 1980. Regulating Sensitive Lands. Cambridge: Ballinger Publ. Co.

Kushlan, J.A. 1979. Design and management of continental wildlife reserves: lessons from the Everglades. Biol. Cons. 15:281-90.

Leopold, A. 1933. Game Management. New York: Charles Scribner's Sons.

Library of Congress, Science and Technology Division. 1980. Draft Environmental Report on Bangladesh Washington, DC: U.S. Man and Biosphere Secret., Dept. State.

Mukherjee, A.K. 1975. The Sundarbans of India and its biota. J. Bombay Nat. Hist. Soc. 72:1-20.

Odum, E.P. 1971. Fundamentals of Ecology, 3rd ed. Philadelphia: W.B. Saunders Co.

Ralls, K. and J.D. Ballou. 1983. In Genetics and Conservation: A Reference for Managing Wild Animals Plant Populations, ed. C.M. Schonewald-Cox, S.M. Chambers, B. MacBryde and L. Thomas. pp. 164-84. Menlo Park, CA: Benjamin/Cummings Pub. Co.

Ralls, K. and J.D. Ballou. 1986. Preface to the proceedings of the "Workshop on Genetic Management of Captive Populations". Zoo Biol. 5(2):81-86.

Rashid, H. 1977. Geography of Bangladesh. Bangladesh: Univ. Pr. Ltd.

Seidensticker, J. 1986. In Proceedings of the International Cat Symposium, ed. S.D. Miller and D.D. Everett. Washington, DC: Nat. Wldlf. Fed.

Seidensticker, J., R.K. Lahiri, K.C. Das and A. Wright. 1976. Problem tiger in the Sundarbans. Oryx 13:267-73.

Seidensticker, J. and M.A. Hai. 1983. The Sundarbans Wildlife Management Plan: Conservation in the Bangladesh Coastal Zone. Gland, Switzerland: IUCN.

Soule M., M. Gilpin, W. Conway, and T. Foose. 1986. The millenium ark; how long a voyage, how many staterooms, how many passengers? <u>Zoo</u> <u>Biol</u>. 5(2):101-14.

Sunquist, M.E. 1981. The social organization of Tigers (<u>Panthera</u> <u>tigris</u>) in Royal Chitawan National Park, Nepal. <u>Smithson.</u> <u>Contrib.</u> <u>Zool.</u> 336:1-98.

41

Managing the Man-Eaters in the Sundarbans Tiger Reserve of India— A Case Study

P. Sanyal

In the Sundarbans Tiger Reserve of India, wildlife management also includes management of man-eaters, and measures have been taken to end this aberrant behavior of tigers.

The mangrove forests of the Sundarbans, at the Ganges-Bramhaputra river delta, form an excellent dynamic habitat for the Bengal tiger (Panthera tigris tigris). The nutrient recycling system of the estuary, which is fed both by the river system and the sea, renders the ecosystem highly productive. Both marine and riverine bacterio-fungal decomposers thrive by converting the mangrove litter into protein (Jadav and Chowdhury 1985), which is subsequently consumed by invertebrates. These ultimately become food for the fish population, and the food chains end with the tiger and the Estuarine crocodile. Thus, the key factor of this highly productive ecosystem is the contribution of plant litter which forms the primary energy source. In spite of being such a highly productive system, the mangrove is a unique, specialized and fragile ecological association which is very delicately balanced in nature, and is sensitive to even minor disturbances (Sanyal 1983). For example, a change of river salinity from 24.3 ppt to 28 ppt in the Saptamukhi estuary of the Sundarbans initiated a fall of gross primary production from 36.8 mgC/M hr to 18.8 mgC/M hr (Chowdhury 1983).

In terrestrial measurements, the biomass productivity has been measured to be 212 metric tons/ha in the case of Avicennia-Sonneratia association in the Sundarbans (Chakrabarty 1985). Most of the mangrove leaves, like those of Sonneratia excoecaria or Avicennia species, are good fodder for deer and monkey; roots are available in plenty for wild boar. This has resulted in an ungulate biomass of 1250 kg/sq. km (Sunquist 1981), which is comparable to the best peninsular Indian habitat, such as the Kanha Tiger Reserve. Another advantage the Bengal tiger has in the Sundarbans is a greater aquatic preybase, including species like fish, crab, water monitor, marine turtle, etc, which extends over 35% of the Indian Sundarbans. Tis rich preybase, and recent conservation efforts are reflected in the area's increased tiger population. The population of the Sundarbans Tiger Reserve, established in December 1973, increased from 135 tigers to 181 in 1976, 205 in 1979, and 264 in 1983, showing an average increase of 10.5% per year.

Special Factors to be Considered in the Sundarbans Tiger Reserve Mangrove Habitat

The mangrove is distinct from other types of tiger habitats in the following considerations:

Disturbances of the habitat (and total ecosystem) by man include: 1) Over-felling of trees; 2) conversion to agriculture; 3) diversion of fresh water; 4) conversion to aquaculture; 5) conversion to salt ponds; 6) conversion to urban developments; 7) construction of harbors and channels; 8) mining extraction; 9) liquid waste disposal; 10) solid waste disposal; 11) spillage of oil and other hazardous chemicals, and; 12) exploitative traditional uses such as honey collection, fishing, etc.

Reclaimed agricultural land produces only a single crop. Sweet-water irrigation is extremely difficult and the local people have to eke out their living during the idle period either by fishing, collecting honey or cutting wood. In India, out of 9630 sq. km of intertidal forests, 5366 sq. km have been converted to the above uses.

The everyday tidal changes give rise to: 1) diurnal shrinkage of habitat, and; 2) obliteration of the territorial urine markings of tigers.

These two factors offset the 'Home Range' concept of tigers in the Sundarbans (Sanyal 1981), and may explain why they have been reported freely swimming across wide rivers, from block to block.

Most of the available drinking water for the animals is saline. This may irritate the carnivores by causing damage to their kidneys and liver (Hendrichs 1971).

Tigers mate in the Sundarbans late in the monsoon season, which is spring for peninsular Bengal tigers. Of course, tigers may mate at any time of the year in captive situations.

Honey collection season is during the months of April and May when the nectar-bearing Mangrove plants bloom and Apis dorsata, the rock bee, form large beehives yielding an average of about 10 kg of honey per hive (Chowdhury and Chakrabarty 1972). So far, all attempts to attract the bees to artificial beehives have failed.

When a honey collector obtains a permit and enters the forest, a tigress with young cubs may attack the human being. Although the number of fishermen entering the Sundarbans is small, there may be frequent attacks on them. These 'circumstantial man-eaters' evolve by repetition into 'designed man-eaters', and presumably stalk humans when in search of prey. The cubs of such designed man-eaters probably also stalk human prey. Possibly, when human prey resists, but is over-powered by a designed man-eater, the tiger evolves into an 'aggressive man-eater'. In the Sundarbans Tiger Reserve of India, no aggressive man-eaters are known.

A study of tiger-human interaction in the Indian Sundarbans revealed the following: 1) tigers did not engage in man-eating sprees and attack men outside the forests; 2) fatal bites by tigers in all cases were on the right nape; 3) fishermen

Table 1.   Record of entry by permit holders into the Sundarbans
Tiger Reserve (monthly averages from 1977-1981).

Month	Number
January	512
February	637
March	380
April	215
May	204
June	358
July	385
August	321
September	459
October	444
November	762
December	803
Total	5480

accounted for 70% of human entry and 82% of total casualties
(Chowdhury and Sanyal 1985); 4) from 1975-1982, an average of 45
people were killed by tigers per year, and; 5) although fewer
people entered the Sundarbans during April and May than any other
time of the year (Table 1), both man-eating frequency (Table 2)
and salinity of local water sources (Table 3) peaked at this
time.   This indicates that man-eating by tigers is positively
correlated with the salinity of creek water, but is not
correlated with the number of human entries.

Management of the Mangrove Habitat

The wildlife management of the mangrove habitat primarily
envisages a total conservation approach.   A tiger reserve
encompassing 2585 sq. km has been created in the Indian
Sundarbans.   A core area of 1330 sq. km is legally protected as a
National Park, and no exploitation is allowed.   The remaining
area acts as a buffer zone, and caters to the needs of the local
people by allowing systematic, limited exploitation.

Primary production is restored by: 1) curbing the export of
forest timber and firewood to metropolitan areas; 2) creating a
contended buffer population along the forest fringe by catering
to the needs of the local people for fuel and small wood, thereby
lessening the pressure on the forest; 3) guarding against wood
and animal poaching, and; 4) artificially assisting the breeding
of endangered animals such as the Estuarine crocodile and the
Olive Ridley marine turtle in order to keep the food web
equilibrium intact (Sanyal 1984).

Management of Man-eating Tigers

A very important step in the management of the mangrove
habitat is minimization of man-animal confrontation.   Endeavors
in the Sundarbans Tiger Reserve of India have included the
following:

Table 2. Monthly record of man-eating by tigers in the Sundarbans Tiger Reserve (1975-81).

Month	1975	1976	1977	Year 1978	1979	1980	1981	Total	Average
January	10	9	0	7	3	4	3	36	5.0
February	7	2	2	6	4	3	2	26	3.7
March	5	4	4	4	3	6	4	30	4.3
April	6	4	6	5	7	4	5	37	5.3
May	11	4	6	6	11	10	3	51	7.3
June	2	0	3	6	4	3	1	19	2.7
July	7	2	0	2	2	1	1	15	2.1
August	2	1	1	0	4	2	1	11	1.6
September	1	1	1	4	2	3	2	14	2.0
October	2	1	3	1	3	3	2	15	2.1
November	4	4	6	3	6	6	2	31	4.4
December	6	8	5	4	3	5	2	33	4.7
Total	63	40	37	48	52	50	28	318	45.4

1) Because an analysis of time data for man-killing by tigers showed that most accidents took place during the early morning, late afternoon and at midnight, an attempt was made to change the time of day people entered the forest, but it was not practicable (Chakrabarty 1979);
2) Permits to collect Phoenix paludosa, which forms the preferred den for the tigers, have not been granted since 1979, and an arrangement has been made to provide alternate fuel;
3) As revealed by studies of the pugmarks of tigers conducted over the last four years, local people have been informed of the fact that only 5% of all the tigers were found to be man-eaters;
4) Tranquilization and removal of stray tigers has had a salutary effect on the villagers;
5) Because all attacks on humans have been on the right nape, permit holders were instructed to carry strong Ceriops sticks on their right shoulders;
6) Three-piece tiger-guard head gear made of fiberglass covering the neck, nape and chest has been devised, and iron spikes have also been fitted on the back piece. So far, no one has been attacked while wearing this gear;
7) Farm-bred wild pigs have been released in the microlocalities of the buffer zone at the rate of one per 10 sq. km/year;
8) Because both Hendrichs (1971) and evidence presented in Tables 1-3 indicate a positive correlation between the salinity of local water sources and tigers' man-eating proclivity, 11 sweet-water ponds have been dug in the Sundarbans Tiger Reserve.

Animals which failed to adapt to the changed saline conditions surrounding the Indian Sundarbans as a result of a recent geological change (Morgan and McIntire 1959), including the Javan rhino, water buffalo and swamp deer, have become extinct. Presumably tigers, which drink saline water in the Sundarbans, have developed a salt excretory or ultra-filtration mechanism (like most mangrove plants) in order to survive. However, the response of the animals to the ponds has been striking;

Table 3.   Temporal salinity variation in the mangrove biotope.
Record of seasonal variation of peak salinity in the creek waters
of the Sundarbans Tiger Reserve.

Month	Highest Salinity of Water (% salt)
January	1.90
February	2.35
March	2.35
April	2.55
May	2.83
June	1.90
July	1.80
August	1.80
September	1.25
October	0.96
November	1.05
December	1.50

9) The latest innovative measures taken in the Sundarbans
Tiger Reserve have been aimed at conditioning tigers to associate
pain with human beings.   In this conditioning, clay models of
actual professional permit holders, dressed in used garments to
impart a strong human smell, are placed inside buffer zone
forests.   Wire is wrapped around the neck of a model honey
collector or woodcutter, and along the gunwale of the boat of a
model fisherman.   This is charged to 230 volts by an energizer
and a 12-volt battery source.   Only man-eaters attack such
dummies and receive the shock.   A one ampere safety fuse and a
limited 20-25 milliamps of current provide adequate safety for
the animal.   This experiment began in 1983.

RESULTS

The effect of the measures taken to manage the man-eating
tigers in the Sundarbans Tiger Reserve started becoming palpable
in 1981.   Although sweet-water ponds were dug long before (during
1974-1975), the decrease in tiger victims did not become
conspicuous until 1981, when only 29 people were killed.   This
was the lowest recorded since the inception of the Tiger Reserve,
and may be an effect of the discontinuance of permits to collect
P. paludosa in 1979.

Control of man-eating has been most effective since 1983.
Human casualties in the Sundarbans Tiger Reserve are 21 in 1983,
16 in 1984 and 28 in 1985.   Thus, the average number of victims,
which used to be 45 per year (1975-1982), has dropped to 21 per
year since the introduction of electrified human clay models in
1983.

So far, 11 dummies have been attacked.   The first attack
took place in June 1983 on a dummy fisherman placed in a country
boat.   The boat capsized while the shocked tiger was trying to

Table 4. Measures taken in the Sundarbans Tiger Reserve to prevent man-eating, and their effects on the number of human casualties per year.

Year	Measures	Human Casualities	Average
1975	Sweet-water pond	63	
1976		40	
1977		37	
1978		48	47
1979	Stoppage of _Phoenix_ Permit	52	
1980		50	
1981		29	
1982		41	43
1983	Electrified dummy	21	
1984		16	
1985		28	22

escape, and was salvaged the following year. A dummy woodcutter was placed at Netidhopani, the ill-famed man-eating block, during November 1983, and was attacked by a tiger in February 1984. The tiger's canines pierced the back of the right arm base. In May 1984, a dummy honey collector, which was confirmed to have a profuse human smell, was attacked by a sub-adult tiger. Only claw marks were noticed on the back of the overturned dummy. The fourth attack was on a dummy fisherman placed at Kalichar. The dummy had been set in December 1984, and the position and garments were periodically changed. Between 6 a.m. and 4 p.m., the energizer for the model was routinely switched off. In February 1985, the tiger pounced on the model at 8 a.m., tore the dummy into pieces, roared and kept the torso by its side for three hours in view of the assigned group of staff. In March 1985, a similar incident took place involving the Jharkahli fisherman's model. The same reaction occured in May 1985 at the site of the Sudhanyakhali honey collector's model. The last attack occured in January 1986, when a tigress near the

Table 5. Number of honey collectors entering the Sundarbans, and amount of honey harvested, from 1980-85.

Year	Honey Procured (quintal)	No. Honey Collectors
1980	408	---
1981	473	---
1982	405	---
1983	487	484
1984	615	651
1985	777	971

Sajnakhali Tourist Lodge came close to a more animated dummy of a woodcutter and received a shock. Six days later, this same model was again attacked by another tiger, but during the daytime when the model was not energized.

All attacks on the dummies took place in the buffer zone. The main constraint has been to energize the dummy continuously in remote forests, where batteries must be replaced every three or four days if it is to be kept charged during the daytime. From the reaction of the man-eating tigers, the need for daytime charging is evident. At present, attempts are being made to use a solar panel kit for continuous charging, and to make the dummy more animated in order to deceive the man-eater.

DISCUSSION

Among the measures taken in the Sundarbans of India to control man-eating by tigers, the most effective has been the use of electric shocks to deter aberrant tigers from preying on man, without removing the tigers from their habitat.

Discontinuing the issuance of P. paludosa collection permits from the preferred den areas of the tigers has been another effective step. People have also become more cautious when in the forest, and are carrying strong sticks on their right shoulders as a protective measure.

It is, perhaps, too early to draw the conclusion that some man-eaters have been electrically shocked out of their habit, but a considerable drop in the incidence of man-eating has been observed since model use began. A summary of the measures taken to prevent man-eating appears in Table 4.

The greatest benefit, however, has been the salutary effect on local inhabitants, on whom the relevance of wildlife management technique has become more significant. A better sense of security for honey collectors may have encouraged more people to pursue this work, and may have resulted in the improved harvest since 1983 shown below (Table 5).

It is apparent from this study of the Sundarbans Tiger Reserve that tigers should survive with their carnivorous instincts in the wild, and man can manage them (even for their man-eating aberration) to reach a peaceful co-existence without disturbing the ecological balance.

REFERENCES

Chakrabarty, K. 1985. Sundarbans mangroves biomass productivity and resource utilization on Mangrove - an in-depth study. Proc. Nat Symp. on Biol.: Utilization and Conservation of Mangrove 17.
Chowdbury and Chakrabarty. 1972. Wildlife biology of Sundarbans forests honey production and behavior pattern of the honey bees. Sci. Cult. 38.

Chowdbury and Chakrabarty. 1979. A decade study on the behavior pattern of tiger. Intern. Symp. on Tiger, New Dehli. Unpublished report.

Chowdbury, M.K. and P. Sanyal. 1985. Use of electroconvulsive shocks to control tiger predation on human beings in Sundarbans Tiger Reserve. Tigerpaper 12(2):1-5.

Chowdhury, A.M. 1983. Project report on mangrove ecosystem of Sundarbans. Unpublished report.

Hendrichs, H. 1971. Report on Project 669. Tiger: Study of the man-eater problem in the Sundarbans. Unpublished report.

Jadav and Chowdhury. 1985. Litter production in mangrove forests, Lothian Island, Sundarbans West Bengal India. Proc. Nat. Symp. on Biology - Utilization and Conservation of Mangrove. Dacca.

Morgan and McIntire. 1959. Geology of the Bengal Basin. Bull. Geol. Soc. Am. 70:31-342.

Sanyal, P. 1981. Sundarbans Tiger Reserve - an overview. Cheetal 23(1):5-8.

Sanyal, P. 1983. Mangrove Tiger Land the Sundarbans of India. Tigerpaper 10:1-4.

Sanyal, P. 1984. Saving the Ridley turtle in Sundarbans of India. Hemadryad 9(3):21-23.

Sunquist, M.D. 1981. Social organization of Panthera tigris in Royal Chitawan National Park, Nepal. Smithson. Contrib. Zool. 336:1-98.

42

The Man-Eating Tiger in Geographical and Historical Perspective

Charles McDougal

INTRODUCTION

Although the evolution of the tiger in Asia during the Pleistocene was contemporaneous with that of man in the same region, human beings do not constitute part of the tiger's natural prey. The normal tiger exhibits a deep-rooted aversion to man, with whom he avoids contact. We can only speculate on the reasons for the development of this behavior. At some stage during the tiger's prehistorical interaction with humans, avoidance of bipedal man became an adaptive behavioral strategy. Although the advent of modern firearms may have reinforced this behavior, it was not the cause, which stems from deeper in the past.

There have always been exceptions to the rule. Other dictates may override the antipathy to man. Throughout recorded history some tigers have killed and eaten human beings and they have done this under a variety of circumstances. This paper will examine the phenomenon of man-eating tigers from the different regions where this behavior has been recorded. Essentially we are dealing with situations in which man and tiger come in conflict, during the course of which the natural predator-prey balance is disturbed, almost always by human interference. Man-eating is the ultimate expression of this conflict. The available statistics in most cases refer only to the number of persons recorded or estimated to have been killed as the result of attacks by tigers, and not to the number actually devoured by these cats. Nevertheless, the figures serve as an index of the degree of man and tiger conflict.

REGIONS WITH A LOW INCIDENCE OF MAN-EATING

There are certain regions where conflict between the two species has been historically minimal, and cases of man-eating comparatively rare.

435

Caspian Region

Man-eaters appear to have been almost non-existent among the Caspian, or Turanian, race of the tiger, at least in Iran. Kennion (1911), who hunted tigers on the southern edge of the Caspian Sea around the turn of the century, and who interviewed local hunters, stated that man-eaters were unknown in Mazanderan, the stronghold of the tiger in that region. In those days the coastal plains consisted of marshy reed-beds where wild prey, especially pigs, (Sus scrofa) were abundant. Not many tigers were killed there, although they were more actively hunted in Russian territory. By the middle of the present century almost all of the tiger's preferred habitat had been reclaimed for cultivation, with the result that the survivors retreated to the mountain forests, where the last recorded Caspian tiger was shot in 1959. Although tigers preyed on domestic livestock during the final phase of their existence, man-eating tigers never became a problem.

Southeast Asia

Burma: Instances of man-eating were sporadic and in some areas almost absent (Burton 1933). "One does not hear of many man-eaters," wrote Evans (1911) of upper Burma in 1911. In the Tenasserim region, tigers were numerous but cases of attacks on humans almost unknown. Local people had no qualms about wandering around in the heart of the tiger habitat, even when alone (Burton 1933). In 1877, when human fatalities attributed to tigers in British India, excluding the Native States, numbered 798, in Burma only 16 persons were killed by these cats (Burton 1933). There was a local outbreak in the Arakan region during the Second World War, when tigers that fed on the corpses of soldiers who perished during the 1942 retreat, subsequently killed and ate humans (Perry 1964).

Thailand: With local exceptions, man-eating tigers have never presented a problem, and of those that did occur, most were promptly destroyed before claiming many victims (Perry 1964).

Vietnam: Baze (1957) refers to occasional outbreaks, and both he and Sutton (1927) note instances in which villages were abandoned for this reason. Nevertheless, man and tiger conflict never became general. The tribal people of the interior had little fear of the tiger. "The women and children going to and from the river or stream often meet a tiger on the way, but he never does them any harm, in fact he disappears at once," comments Baze (1957). During the Vietnam War there were instances not only of tigers scavenging corpses on the battlefield, but also of attacking live soldiers (Jackson 1985).

Malaya: In 1930, the beginning of a decade when up to a thousand or more persons were killed annually in British India, there were only 15 fatalities in the entire Malay Federated States (Burton 1933). Locke (1954) who hunted down two man-eaters in Trengganu State, remarked on their comparative scarcity in peninsular Malaya. He emphasized that tigers that begin to kill people rarely last long, owing to the resolute action on the part of the local people as well as the ready assistance rendered by experienced hunters. Only two cases were recorded in 1985, none in 1984.

Sumatra: During the 1880's tigers were described as being very abundant in certain areas, judging by pugmarks and other signs, but few were killed or even sighted. At that time, man-eating was a rare event (Burton 1933). During the 1920's the loss of human life from tiger attacks averaged only a dozen per year for the entire island (Perry 1964). Borner (1975), who made a three-year survey of Sumatra on behalf of World Wildlife Fund (WWF) in the early 1970's, could only confirm three cases of man-eating.

Java: The situation in Java is less clear. No data is available prior to the second half of the 19th century, by which time tiger habitat had disappeared from most of the island. Man-eaters were present, but they were usually abruptly terminated by the prompt and concerted action of the local people, hundreds of whom turned out to hunt down the culprit, to judge by two cases cited by Burton (1933) and Hamilton (1892).

With the above exception, tiger habitat was extensive in all the regions described, and natural prey was plentiful. The density of the human population adjacent to tiger habitat was low, and the reclamation of the latter for cultivation and other purposes was gradual, minimizing conflict. Despite the fact they were only occasionally hunted in some of these localities, tigers nevertheless avoided contact with man. Where man-eating did occur, it was a localized and occasional problem.

FORMER PROBLEM AREAS

Historically, there are three places where man-eating was a serious and persistent problem which was eventually resolved, either through the elimination of the tigers or their being confined to remote areas with low human population density.

South China

As early as the 13th century, Marco Polo (1958) relates that in the northern part of what is now Fukien Province, tigers (he calls them lions) "of great strength and ferocity, are a danger to wayfarers." The neighborhood of the city of Fu-chau was "infested" with them. By the first quarter of the present century tiger habitat in Fukien had been reduced to a few wooded ravines in otherwise barren hills bordering cultivated land. Attacks on livestock and humans, especially children, were common. Sixty persons were killed in one village during the course of a few weeks in 1922. Very little natural prey remained, and tigers had become so bold they sometimes entered houses to secure prey (Caldwell 1925). Clearly time was running out for this remnant population, forced by the almost complete destruction of its habitat to depend largely on domestic livestock and human prey. Against all odds a few representatives of Panthera tigris amoyensis still exist (Huji this volume).

Singapore

When Raffles first visited Singapore in 1819 he found it entirely covered with swamps and jungles, aside from a tiny Malay settlement of 100 houses (Hoefer 1983). At some time tigers had succeeded in crossing over the strait from the mainland and established themselves on the island. As the settlement of Singapore rapidly expanded, and more and more land was reclaimed for plantations, conflict between man and tiger became inevitable. By the early 1840's an average of 200-300 persons, mainly Chinese coolies, were killed by tigers each year, this figure being considered an underestimation by informed residents of the period (Burton 1933). On the small islands lying between Singapore and the Malay mainland man-eaters claimed 600-800 victims per year. The high incidence of man-eating was ascribed to the scarcity both of natural prey and domestic livestock (Burton 1933). According to Perry (1964), 15 persons were killed by tigers in Singapore as late as 1929. The problem persisted for 20 years until the tigers were brought under control by the payment of bounties, the use of pit-traps, and the sharpshooters of the "Tiger Club."

Manchuria

There were frequent confrontations between man and the tiger in Manchuria and neighboring Korea during the late 19th century and first part of the present one, and man-eating was prevalent. Marauding tigers caused weekly markets to shut down over a distance of 150 kilometers, and interfered with the collection of fuel wood for the winter (Perry 1964). Across the border in Korea 18 people were killed by tigers from one village alone (Cavendish 1891). A regiment of Cossacks were employed as guards to protect laborers during the construction of the western portions of the Chinese Eastern Railway during the 1890's (Sowerby 1923). It is reported that during this period man-eaters entered huts to carry off Chinese and Russian settlers. This was a time of rapid development, and there was extensive clearing of the forests for settlement, especially with the opening up of the country by the railroads (Prynn 1978). Although man-eating tigers appeared with frequency, individuals were usually eliminated before accounting for many victims, owing to the active part taken by local hunters in promptly dealing with the problem (Perry 1964). The conflict was resolved in favor of man. The remaining tigers survived only in the remotest areas, only 20-30 being left in Russian territory by the end of the 1930's (Zhivotchenko and Trofimenko 1985). It was only during the 1950's that protection was afforded the tiger and reserves set up in the USSR. During the latter half of the same decade an average of 50 per year were still being killed by hunters in Northern China (Mazak 1967).

No recorded cases of man-eating occurred in the USSR between 1917 and 1962 (Abramov 1962). Two isolated instances occurred in 1976 and 1981. The tiger population now is estimated at about 290. Although Matjushkin et al. (1981) believed that in cases of man and tiger confrontation, "The probability of a tragic issue is equal most likely to the hundredth of thousandth parts of the percent." Zhivotchenko and Trofimenko (1985) suggest that conflict between the two species may be on the increase. In March 1986 a tiger was shot on the streets of Vladivostok.

THE INDIAN SUB-CONTINENT

This region has served as a stage for intensive man and tiger confrontations. This is hardly surprising in the view of the fact that the region affording habitat where tigers have achieved their highest densities is also one which has housed one of the most concentrated and rapidly expanding human populations.

Background

"Tigers infested certain parts of the country up to the beginning of the nineteenth century to such an extent as to make them almost uninhabitable," remarks Hewett (1938). This statement is supported by Brander (1923), commenting that, "At one time in parts of India at the beginning of the last century, tigers (sic) were so numerous it seemed to be a question as to whether man or the tiger would survive." Conflict was inevitable, and Forsyth (1872) emphasized, "The obstacle presented by the number of these animals to the advance of population and tillage."

It became the official policy to encourage the killing of tigers as rapidly as possible, rewards being paid for their destruction in many localities. Agricultural expansion proceeded with the runaway growth of the human population under the pax britannica. In some areas the results were obvious even at an early date. Mundy (1833), who visited India in 1827-28 and who hunted tigers, described how, "In these modern times..., the spread of cultivation and the zeal of English sportsmen have almost exterminated the breed of these animals."

The man and tiger conflict was thus resolved in many parts of India by killing large numbers of tigers and at the same time bringing their former habitat under human use. Thus, in Khandesh District of the Bombay Presidency, where in the year 1822 no fewer than 500 persons and 20,000 head of cattle had been killed by wild beasts, chiefly tigers, by the 1880's man eaters were considered rare (Gouldsbury 1915). Sanderson (1878) reported a similar situation for Mysore, as did Burton (1931) for Hyderabad. Sportsmen began to complain that the supply of tigers for future shooting was becoming depleted. Eventually most of the forest areas of British India were brought under management, and a system of reserved forests established; shooting blocks and closed seasons gave tigers a large measure of protection in these localities so set aside. In others, rewards continued to be paid for their destruction. Tiger habitat had also been preserved in the States, so that the rulers and their guests might enjoy the sport of killing tigers on a sustained yield basis.

Nevertheless, in two regions in particular, Bengal and the Central Provinces, conflict between the two species continued unabated throughout the 19th century and well into the present one. In another, the United Provinces, which supported large numbers of tigers in the submontane terai portion of the state where man-eating had been uncommon, marauding tigers began to take a toll of human life in the hill tracts beginning only in the latter part of the 19th century. The depredation of these animals, made world famous by Corbett's book The Man-Eaters of Kumaon (Corbett 1944), continued until the Second World War.

Bengal

        In Bengal (including present West Bengal and Bangladesh) the
situation was indeed serious, and over large parts of the region
villagers protected themselves by building high stockades.
Casualties were heavy on both sides. In the year 1877, of the
798 persons killed by tigers throughout British India (excluding
the States), no fewer than 347, or 43 percent, lost their lives
in Bengal, where an additional 617 persons were reported killed
by unspecified animals, some of which may have been tigers. On
the other side of the coin, almost twice as many tigers were
killed as were humans, namely 1,579; of these 426 tigers, or 27
percent were destroyed in Bengal (Burton 1933). In 1908 the
reported human fatalities by tigers in British India totaled 909,
of which 455 were killed in Bengal (Burton 1933, Berg 1934). The
most notorious part of Bengal for man-eating tigers had always
been the Sundarbans delta where the Ganges meets the Bay of
Bengal (Khan this volume). Nevertheless, judging by the
statement of Eardley-Wilmot (1910) that 60-70 persons were killed
by tigers each year in the Sundarbans, deaths from tiger attacks
in this tract constituted only 15 percent of Bengal's total.

        In Bengal there was not only a dense, expanding, human
population, but likewise a high, decreasing density tiger
population. The two species were on a head-on collision course.
Baker (1887) pointed out that local shikaris or hunters killed
off the tiger's natural prey with the result that predators
resorted to humans instead, and argues that, "To the axe and the
plough, and not to the gun in the hands of villagers, will be due
ultimately the protection of the country from the ravages of
these terrible pests." The result was that tigers survived
ultimately only in the Sundarbans tidal mangrove forests and in
the remote tracts at the base of the Himalayan foothills.

The Central Provinces

        Second only to Bengal as a theatre for man and tiger
conflict was the Central Provinces (including present Madhya
Pradesh and part of Maharashtra), which accounted for 21 percent
of the human fatalities from tiger attacks in 1877 (Burton 1931).
Especially hard hit were certain districts, such as Mandla, where
an estimated 200-300 persons were killed by tigers annually in
the middle of the last century (Hewett 1938), where as late as
1906 a special forest officer was appointed for the sole purpose
of destroying man-eating tigers (Hicks 1910). Chanda and Seoni
Districts were also notorious for man-eating tigers.

        Scarcity of natural prey was also a factor correlated with
man-eating. Hicks (1910), who served as a forest officer in the
Central Provinces from 1866 till the end of the century
considered that, "Man-eaters are almost invariably females." In
his opinion these animals took to human prey because of the
difficulty of feeding their cubs during times when natural prey
was scarce, having to temporarily reside in marginal habitat or
because of periodic drought conditions. Incidentally, Hicks felt
that such man-eaters might stop killing humans when circumstances
returned to normal, although they would harbor a proclivity to
resume the practice if things again changed for the worst.
Hewett (1938), who acted as Chief Commissioner for the Central
Provinces just after the turn of the century shared Hick's views,
remarking that, "There are a number of places where at certain

times of the year there is hardly any game, and no cattle are about, and a tigress with cubs finds herself driven to preying on human beings." At the time there were three notorious man-eaters operating in the province, all of them tigresses. Gouldsbury (1915) records that a tigress operating along the Jabalpur to Nagpur road where it crossed the Satpura Range was reported to have killed 700 humans before she was destroyed. During the late 1880's a tigress with cubs killed 4 persons in three years in one locality along the Bengal Nagpur Railway (Burton 1931).

Man-eaters appear to have been most common in those parts of the province being rapidly opened by railroads, roads, and the development which took place in their wake. Most of the localities, once notorious for man-eaters, have since been brought under cultivation and most of the tigers have disappeared. Madhya Pradesh still has the largest number of tigers in India, but their distribution now is coterminous with the lowest density of human settlement, with the result that man and tiger conflict has become minimized, man-eating cases being only occasional and sporadic. Neighboring Orissa, especially the districts of Kalahandi and Koraput, had a bad reputation for man-eating tigers until the 1960's but tigers have now almost completely disappeared form those localities.

The United Provinces

At the same time that contemporary writers cited the scarcity of natural prey as a factor underlying the man-eating problem in Bengal and the Central Provinces, its abundance was related to the rarity of man-eating tigers in the lush, lowland habitat of the terai extending in a belt along the base of the Himalayan foothills in the United Provinces (Hewett 1938). Excellent tiger habitat also existed in the interior Doon Valley above the outer foothills. "It is no doubt....owing to the abundant supply of animals upon which the tiger preys that a man-eater is so rarely found in the Doon," wrote Fife-Cookson (1887).

Nevertheless, the hill tracts of the province began to be subjected to the depredations of man-eaters during the latter part of the last century. One of the first two cases on record is that of the Mandala Tigress, an animal that in 1876 suddenly appeared in the hills of the Dehra Dun District in a locality where tigers had hitherto been unknown. She frequented a ridge nearly 3,000 meters in elevation, where natural prey was not only in scarce supply, but difficult to hunt. She began by attacking flocks of sheep and goats, and then, in 1880, when she had three almost full-grown cubs, began to prey on humans, eventually becoming so bold as to enter the shelters of herdsmen in search of victims. When finally shot in 1889, all save one of her canines were worn down to stumps, and she was generally in miserable condition (Hewett 1938). Another man-eating tigress appeared in the Garhwal District at about the same time, and was destroyed in 1881 (Eardley-Wilmot 1910). The first of the Kumaon man-eaters, the Champawat Tigress, is said to have crossed over from Nepal, where she had already claimed 200 victims, in 1905; before her career was ended she added another 236 in India (Sanderson 1878) (Corbett 1954). Man-eaters continued to appear, from time to time, in these hills up until the Second World War. They were in fact most numerous during the 20s and 30s, and during this period there were two, the Talla Des Tigress and the

Lohaghat, each of which accounted for approximately 150 victims (Hewett 1938) (Corbett 1954).

Why the sudden advent of man-eaters in the hill tracts, and why the increase in frequency of their appearance over time? According to the explanation offered by Turner (1959), and one which seems to be in accord with the facts, these were animals that were pushed into marginal habitat where tigers had formerly not been known or where they existed only at very low density, by an expanding population of more vigorous animals occupying the prime habitat in the lowlands, where there was high prey density and good reproduction. The dispersers, who had no where else to go, since the prime habitat was bordered on the south by cultivation, are thought to have followed back the herds of domestic livestock that wintered in the plains when they returned to the hills in the spring, and then been left without prey when the herds dispersed back to their respective villages. At the time there was continuous forest cover, allowing these animals to migrate upward. These tigers were the old, the young and the disabled. Of 16 notorious man-eaters of the period, 10 were females. All suffered from some disability, mainly caused either by gunshot wounds or porcupine quills. I agree with Singh (1984) that these tigers probably only attacked porcupines out of desperation when other prey was scarce or difficult to hunt.

The reasons that these migrants to the hills became more numerous is that the well-managed habitat in the lowlands facilitated good reproduction. Hewett (1938) notes that: "In the report on forest administration in the United Provinces for 1935-36 it is stated that tigers were tending to increase in numbers and that it is almost inevitable that..., there will be a migration of surplus tigers into the surrounding forests."

The Current Situation

Sankhala (1975) suggests there were approximately 30,000 tigers in India at the beginning of the Second World War. Until that time the human death toll from tiger attacks remained high. During the nine-year period 1902-1910 fatalities per annum averaged 851, the range being 698 to 1,046 (Berg 1934). The figure for 1922 was 1,603, that for 1927, 1,033 (Burton 1933). Hewett (1938) records that during "a recent period of five years, 7,000 deaths were reported to have been caused by tigers" in British India, excluding the Native States, but including a few cases in Burma.

The drastic reduction both of tigers and their habitat immediately following the war effectively solved the general problem of man-eating, except in certain relatively unaffected localities such as the Sundarbans. Sankhala (1978) remarks that in the states of Kerala, Tamil Nadu, Rajasthan, Assam and Bihar, no man-eating cases had been reported for the last 30 years; there had only been one from Andhra Pradesh; and none in Karnataka since 1959.

A new situation is now developing. For well over a decade tigers have received total environmental protection in a series of special reserves and national parks, and in many cases these have become reproductive centers, from which there is dispersal to surrounding forests. The trouble is that in some cases the latter have become degraded and disturbed, and there is little

natural prey available for the dispersers. The result is attacks
on domestic livestock, and eventually man. In Kheri District of
Uttar Pradesh, 128 persons were killed by tigers during the seven
year period 1978-1984, during which time 10 tigers were destroyed
and two others captured (Project Tiger Directorate 1985). This,
together with the Sundarbans, are the only places in India and
Bangladesh where man-eating remains a persistent problem.

A common factor in all the cases so far discussed has been
the scarcity or difficulty in obtaining natural prey, either due
to the disturbance of tiger habitat by humans or the dispersal by
tigers into areas of peripheral habitat. A common sequence
commented on by some authors (Sanderson 1878, Gouldsbury 1915) is
that from natural prey to domestic livestock to man. The partial
dependence on cattle not only increases the tiger's familiarity
with man, but also makes humans, in the form of herdsmen,
immediately available as alternative prey. Disability and
consequent handicap in capturing natural prey may also be a
contributory factor, although many disabled tigers never become
man-eaters and many man-eaters have no obvious disabilities
(Eardley-Wilmot 1930). Hunger is most likely the factor which
overrides the tiger's aversion to man.

Although the majority of cases can be explained in this
manner, the explanation does not cover the totality of cases. On
the basis of the available evidence most of the tigers that kill
and even devour humans in the Sundarbans do not do so out of
hunger, i.e., because there is a shortage or difficulty in
securing natural prey. Cases from Royal Chitwan National Park in
Nepal are not correlated with a scarcity of natural prey,
although in at least one case temporary incapacity to capture
wild animals may have led to man-eating.

The Sundarbans: The one region in India where man-eating
has been endemic since the earliest times is the Sundarbans
delta, a 10,000 square kilometers tidal mangrove forest covering
a multitude of islands formed by numerous channels emptying into
the Bay of Bengal. This region, 60 percent of which now belongs
to Bangladesh and the other 40 percent to India, presently
harbors the largest single, contiguously distributed tiger
population in the sub-continent, possibly as many as 300 adults
(Seidensticker this volume). The ferocity and man-eating habits
of the Sundarbans tigers, that even board small boats in search
of human prey, have been described for three centuries (Khan this
volume). The point to make is that this behavior dates from a
time when the Sundarbans was "little frequented by man" (Burton
1931). Baker (1887) stated that, "Tigers, deer, wild hog,
rhinoceros, and buffalo are abundant and probably are increasing
since comparatively few are ever shot, and the decrease in the
forest area is so slow as to be unlikely to affect the number of
its wild denizens." At the same time he notes that its tigers,
"have the reputation of being the most fearless and confirmed
man-hunters and man-eaters of any in India." The rhinos and wild
buffaloes have disappeared, but Axis deer (Axis axis) and wild
pigs (Sus scrofa), as well as tigers, are still plentiful. Human
pressure has increased, and the Sundarbans are exploited for a
variety of forest products as well as fish (Seidensticker and Hai
1983).

The man-eating problem in the Sundarbans (what is now the
Bangladesh portion) was investigated by Hendrichs (1975) in 1971.

Although his research was cut short by the Indo-Pakistan war, he did establish some important ecological correlations.

Essentially Hendrichs made a distinction between two kinds of man-eaters: (1) the confirmed, or dedicated ones, that go out hunting especially for human prey, and (2) the opportunistic ones that do not go out of their way to search for humans, but that will, if they encounter a man, attack, kill and devour him. As Hendrichs puts it, the latter are simply "ferocious tigers" as opposed to the former, "the real man-eaters." The dedicated man-eaters, constituting only one percent of the total population, are distributed uniformly throughout the Sundarbans. The opportunistic man-eaters, on the other hand, constituting 30 percent of the total tigers, increase in frequency from northeast to southwest. Their distribution does not vary with the density of main prey species, Axis deer and wild pigs, but with the height of the water level and the degree of salinity in the water. In those areas where opportunistic man-eaters are found, the killing of humans is correlated with their availability, most victims beings claimed during the honey gathering season.

Hendrichs (1975) suggests that the salinity of the water may result in physiological changes, causing tigers resident in such localities to become more ferocious and prone to attack humans. This has not been proven, and given the same set of correlations, alternative explanations are possible. One hypothesis, highly speculative in view of our limited knowledge, is that because the high tide level renders ineffective intra-specific communication by means of chemical and visual marking, more overt patterns of spacing behavior may be used, rendering tigers more than ordinarily aggressive. Most of the man-eaters recorded by Hendrichs (1975) were males (10 out of 13), and they accounted for 86 percent of the victims.

In the Indian Sundarbans the man-eating problem has now been tackled in a resolute and imaginative fashion by the Project Tiger authorities, using electrified dummies and fences, as well as providing fresh water supplies in saline areas. This has resulted in a sharp decline in human fatalities, from an average of 45 per annum between 1975 and 1982, to an average of only 22 during the last three years, 1983-85 (Khan this volume).

Nepal: Historically, the incidence of man-eating by tigers in Nepal has been only sporadic. Man-eaters were usually quickly dealt with by personnel deputed from the Narayani Shikar Phant attached to the Royal Palace.

No cases of man-eating were recorded in Royal Chitwan National Park prior to 1980. Since that time 13 persons have been killed and eaten in the park and its immediate environs. Man-eating appears to have been opportunistic and, in the majority of cases, to have been related to an escalation of intra-specific competition among male tigers. The identity of the tiger was established in all but two cases.

One male, Number 127, was involved in five cases of man-eating, two of them in association with a tigress, Number 118, that was destroyed after the third human was killed. At the time, October 1980 to April 1981, male 127's home range was not stable, but was in the process of being displaced westward in response to pressure from a large male to the east. The latter appropriated part of the range which 127 formerly occupied, containing an adult female with which he had regularly associated. Each successive human kill took place some distance to the west of the last, reflecting 127's shift in that direction. All these took place in prime riverine habitat, containing a variety of prey animals at high density. Subsequently, 127 established a stable home range, including two reproductive tigresses, which he maintained for a period of 10 months, at the end of which he suddenly disappeared and was replaced by another male. During that entire time he killed no other people, despite the fact that the park was opened to thousands of grass-cutters in early 1982.

A second male, Bange Bhale, maintained essentially the same home range from mid-1981 until the first part of 1984, when it was gradually usurped by his newly established western neighbor, again a larger male, who gained control of the two tigresses resident there. On two occasions Bange Bhale was driven off a kill he had made by the other male, who appropriated it. After one of these encounters Bange Bhale was observed noticeably limping. In May, 1984, he killed a grass-cutter inside the park, and later two additional victims on the park's periphery, in a locality well outside his former range. Finally, in December he made an unsuccessful attack on a man cutting fodder for an elephant, was immobilized, and removed to the Kathmandu Zoo.

A third male, Kanchha Bhale, the tiger that replaced 127, killed a man who had gone to the jungle to cut wood at the end of January 1986. This tiger had maintained the same home range from early 1982 until late 1985, when three fights took place between Kanchha Bhale and his newly established western neighbor. The latter began to associate with the three tigresses formerly monopolized by Kanchha Bhale. After being injured in the last encounter, Kanchha Bhale left his home range and remained near some villages at the edge of the park, where in three weeks he killed nine cattle and buffaloes, before claiming his first human victim. Two other persons were subsequently killed; the man-eater was shot in March.

When competition among males escalates, it would seem reasonable to suppose they become more than usually aggressive, even towards humans, the more so if wounded. In all three cases the males had been worsted by rivals. The human victims were probably just at the wrong place at the wrong time. Almost all were grass-cutters, the most available potential victims, penetrating as they do into dense cover harboring tigers.

DISCUSSION

Historically, a low incidence of man-eating has been correlated with localities where there was an adequate supply of natural prey and extensive habitat, into which human encroachment was only gradual. In some areas tiger habitat has been completely destroyed without man-eating becoming a problem, for instance in Iran or in parts of the lowlands of Nepal following resettlement by hill people. In such cases the tigers were killed off at approximately the same rate that their habitat was removed, and there was no breeding reservoir of tigers to restock the peripheral habitat being created by human activities.

A high incidence of man-eating has been associated with situations in which tigers from areas of good reproduction, where numbers are increasing, were forced to occupy marginal habitat. In Bengal, for example, not only was the habitat being rapidly reduced but large numbers of tigers were being killed at the same time; nevertheless, so long as there was sufficient good habitat for tigers to reproduce and multiply, peripheral areas were restocked and man-eating persisted. In the United Provinces, tiger numbers were increasing in the lowland habitat which was bounded on the south by cultivation. Dispersers had nowhere else to go except into the hills.

In most of the contexts examined, the availability of natural prey was the proximal limiting factor. Tigers forced to occupy areas where their normal food was in short supply supplemented the latter with domestic livestock and also humans. In such cases a tigress with dependent young to feed or a disabled tiger suffered an additional handicap. The ultimate limiting factor in these cases, of course, is space, the lack of which caused tigers to move into sub-optimal habitat.

Opportunistic man-eating that takes place in localities where natural prey is plentiful cannot be attributed to the same factors as the classic cases of man-eating which arise through necessity or, as Corbett (1944) put it, "the stress of circumstances." The operative factors may vary in different ecological contexts, but in all cases they will be ones that cause the tiger to subordinate its normal inclination to avoid man and make it more prone to attack him. There is no hard and fast line between the opportunistic man-eater and the dedicated one, since the former can turn into the latter.

"Of the many man-eaters I have known none have been aged or decrepit animals driven to feed on human beings because they could not obtain other food. They lived in a country full of game and where cattle were plentiful; but had lost their fear of man." These are the words of Eardley-Wilmot (1930), whose long experience in the Indian Forest Service spanned the last quarter of the 19th century.

Greater familiarity with man is probably a factor increasing the likelihood of a tiger opting for human prey. Man has been penetrating tiger habitat since time immemorial, and there are no places, in the Indian subcontinent at any rate, where tigers are unfamiliar with man. It is probably a matter of degree. Increased frequency of contact must not only make man more available as alternative prey but at the same time lessen the tiger's reluctance to attack him.

As the existing habitat outside parks and reserves, where tigers are multiplying, becomes progressively fragmented and degraded owing to ever greater human encroachment, we may expect an increased incidence of problem tigers.

REFERENCES

Abramov, V. 1962. A Contribution to the biology of the Amur tiger, Panthera tigris longipilis (Fitzinger, 1868). Vest. Cesk. Spol. Zool. 26:189-203.

Baker, E.B. 1887. Sport in Bengal. London: Ledger, Smith and Co.

Baze, W. 1957. Tiger! Tiger! London: Elek Books.

Berg, B. 1934. Tiger und Mensch. Berlin: Halltorp.

Borner, M. 1975. Status and conservation of the Sumatran tiger (Panthera tigris sumatrae): A survey of WWF International. Unpublished report.

Brander, A.A. Dunbar. 1923. Wild Animals in Central India. London: Edward Arnold and Co.

Burton, R.G. 1931. A Book of Man-Eaters. London: Hutchinson and Co.

Burton, R.G. 1933. The Book of the Tiger. Plymouth: Mayflower Press.

Caldwell, H.R. 1925. Blue Tiger. London: Duckworth.

Cavendish, A.E.T. 1891. Korea and the Sacred White Mountain. London.

Corbett, J. 1944. The Man-Eaters of Kumaon. London: Oxford Univ. Pr.

Corbett, J. 1954. The Temple Tiger and More Man-Eaters of Kumaon. London: Oxford Univ. Pr.

Eardley-Wilmot, S. 1910. Forest Life and Sport in India. London: Edward Arnold.

Eardley-Wilmot, S. 1930. Leaves from Indian Forests. London: Edward Arnold.

Evans, G.P. 1911. Big Game Shooting in Upper Burma. London: Longmans, Green, and Co.

Fife-Cookson, J.C. 1887. Tiger-Shooting in the Doon and Ulwar. London: Chapman and Hall.

Forsyth, J. 1872. The Highlands of Central India. London: Chapman and Hall.

Gouldsbury, D.E. 1915. Tiger Slayer by Order. London: G. Bell and Sons.

Hamilton, D. 1892. Records of Sport in Southern India. London: R.H. Potter.

Hendrichs, H. 1975. The status of the tiger Panthera tigris (Linne, 1758) in the Sundarbans mangrove forest (Bay of Bengal). Saugetierkd. Mitt. 23.161 99.

Hewett, J. 1938. Jungle Trails in Northern India. London: Methuen and Co.

Hicks, F.C. 1910. Forty Years among the Wild Animals of India. Allahabad: Pioneer Pr.

Hoefer, H.J. and S. Black, eds. 1983. Insight Guides: Singapore. Singapore: Apa Productions.

Jackson, P. 1985. Man-Eaters. Int. Wildl. 15(6):4-11.

Kennion, R.L. 1911. By Mountain, Lake and Plain: Being Sketches of Sport in Eastern Persia. Edinburgh: Blackwood and Sons.

Locke, A. 1954. The Tigers of Trengganu. London: Museum Pr.Ltd.

Matjushkin, E. N., V.I. Zhivotchenko and E.N. Smirnov. 1980. The Amur Tiger in the U.S.S.R. Morges, Switzerland: IUCN. Unpublished report.

Mazak, V. 1967. Notes on the Siberian Long-haired Tiger, Panthera tigris altaica (Temminck 1844). Mammalia 31(4):537-73.

Mundy, A. 1833. Pen and Pencil Sketches being the Journal of a Tour in India. London: John Murray.

Perry, R. 1964. The World of the Tiger. London: Cassell.

Polo, M. 1958. The Travels. London: Penguin Books.

Project Tiger Directorate. Bikaner House, New Delhi: Dept. of Environment.

Prynn, D. 1978. Siberian tiger. Wildlife 20(9):398-402.

Sanderson, G.P. 1878. Thirteen Years among the Wild Beasts of India. London: W.H. Allen and Co.

Sankhala, K. 1978. <u>Tiger! The Story of the Indian Tiger</u>.
　　London: Collins.
Sankhala, K. 1975. <u>Tigerland</u>. London: Collins.
Seidensticker, J. and M.A. Hai. 1983. <u>The Sundarbans Wildlife
　　Management Plan</u>. Gland, Switzerland: IUCN.
Singh, A. 1984. <u>Tiger! Tiger!</u> London: Jonathan Cape.
Sowerby, A. de C. 1923. <u>The Naturalist in Manchuria</u>. Vol. 2,
　　Mammals. Tientsin.
Sutton, R.L. 1927. <u>Tiger Trails in Southern Asia</u>. London: Henry
　　Kimpton.
Turner, J.E. Carrington. 1959. <u>Man-Eaters and Memories</u>. London:
　　Robert Hale.
Zhivotchenko, V. and S. Trofimenko. 1985. The Tiger Irony. <u>BBC
　　Wildlife</u> May:229-30.

43

Tigers in Nepal: Management Conflicts with Human Interests

Hemanta R. Mishra, Chris Wemmer, and J.L.D. Smith

INTRODUCTION

The "Save the Tiger" campaign has been one of the most successful conservation efforts on both national and international venues. This effort provided impetus for governments to set aside areas as national parks or reserves, and to obtain funds for the areas' protection. A survey of current literature and governments leaves no reason to doubt that populations of large mammals, including those of the tiger, have more than doubled within the last decade. However, these increases have been associated with heavy damage to agriculture crops, loss of livestock, and even of human life, bringing conservation in direct conflict with the local people (Mishra 1981, 1982 and 1984). Most of the countries that subscribed to and participated in the campaign of "Operation Tiger" have achieved remarkable success in saving the tiger from extinction. Yet much of two decades of effort is threatened with disintegration if biologists and conservationists fail to grasp the crucial fact that without meeting the basic needs of food, fuel, fodder and shelter for impoverished farmers outside boundaries of tiger sanctuaries, there may be no tigers or wilderness areas left to protect (Sherpa 1979, Mishra 1982, 1984, Hinrichsen et al. 1983). Thus, the question is - Is the future of tigers secure beyond the turn of this century in developing countries, like Nepal, where the plethora of economic and social problems is staggering?

This is a case study. Its objective is to outline the sources of conflicts between tiger conservation and human interests. This study is from the perspective of the poor rural family that has to dwell and struggle for existence in the vicinity of large tracts of land set aside for saving the tiger and other species such as the rhino. The aim of this paper is to illustrate the harsh socio-economic realities in Nepal in contrast to the abundance of wildlife. It also describes some of the approaches undertaken in Nepal to minimize the antagonistic attitude of the people towards the tiger and its habitat. Finally, the paper highlights the creation of a new and unique institution which aims at striking a balance between nature conservation and human needs by illustrating that the relationship between the two is symbiotic and not antagonistic (Shah 1984a, b and c).

"OPERATION TIGER" : A HISTORICAL PERSPECTIVE

Mountfort (1981) summarized the history of the campaign to bring the tiger back from the brink of extinction. In September 1972, "Operation Tiger" was launched at an international press conference by His Royal Hi⁻hness Prince Bernhard, President of the World Wildlife Fund. The media gave wide coverage on the needs for saving the tiger, an animal which has remained unchallenged as one of the most popular and readily recognized in the world.

Bureaucrats in the small Himalayan Kingdom of Nepal could not remain idle and let this momentum wither, particularly since some tracts of tiger habitat still existed in Nepal. Thanks to the vision of the late King Mahendra, most of the tiger country in the southern Terai of Nepal has been afforded some protective status as Shikar (Hunting) Reserves since the early 1960's. This action provided insurance against the upsurge of clear felling, cultivation and agriculture development programs that were initiated by various international aid agencies. The current Sovereign of Nepal, His Majesty King Birendra Bir Bikram Shah Deva, proclaimed the country's first National Parks and Wildlife Conservation Act in 1973 to ensure that nature conservation formed a vital part of the government's socio-economic development plans. Consequently, Nepal's first national park was established in Chitwan, an area already well known from bygone days for its great tiger hunts (Smythies 1942). Legal, administrative and financial provisions by His Majesty's Government of Nepal have made this former hunting reserve into one of the most outstanding tiger habitats in southern Asia.

Parallel to the declaration of the Royal Chitwan National Park, the Nepalese authorities realized that effective management plans could not be formulated without sound scientific data. In 1973, His Majesty's Government of Nepal asked the Smithsonian Institution of the United States to provide technical and financial assistance to undertake long-term ecological research on the tiger, its prey and habitat requirements. This request materialized as the Nepal Tiger Ecology Project. Funds for the project were generously provided by the Smithsonian Institution and the World Wildlife Fund. The project continues with the financial assistance of the Smithsonian Institution, and has thus far provided much information on the biology and behavior of the tiger (Seidensticker 1976, Tamang 1979, Sunquist 1981, Mishra 1982, Smith 1984). These findings were instrumental in enlarging the park from 544 sq. km in 1973 to its present 1,040 sq. km. Knowledge of dispersal and movement patterns of tigers, and management recommendations made by field biologists (Smith and Mishra 1981, Smith 1984) resulted in the creation of the new Parsa Wildlife Reserve in 1985, which adjoins the Royal Chitwan National Park.

Besides Chitwan, two other reserves were established in Nepal by 1975 under the "Operation Tiger" scheme; the Royal Sukla Phanta Wildlife Reserve in far western Nepal and the Royal Bardia Wildlife Reserve in the mid-western region. Both were formerly a part of the Royal Shikar (Hunting) reserve, and were known for high densities of tigers. In addition, the small Koshi Tappu Wildlife Reserve was also established in eastern Nepal in 1975. This reserve supports the last remnant population of wild Asiatic buffalo (<u>Bubalus bubalis</u>). It has no resident population of tigers, though one or two occasionally enter the reserve from adjoining forests.

REVIEW OF THE CURRENT STATUS OF TIGERS IN NEPAL

The exact number of tigers in Nepal is not known. With the exception of Chitwan, there are no programs to monitor their densities and movement patterns on a systematic basis. Nevertheless, thanks to the personal involvement of influential and dedicated persons like His Royal Highness Prince Gyanendra Bir Bikram Shah, Chairman of the King Mahendra Trust for Nature Conservation, the tiger sanctuaries of Nepal are matched by none (Fig. 1). Areas such as Royal Chitwan National Park have proven to be classic examples of success stories where government determination, backed by local and international expertise, changed depleted habitats into some of the most unique wilderness areas in the world. A decade of effective conservation measures has triggered increases in the populations of not only the tiger, but other large mammals in the National Parks and Wildlife Reserves of Nepal as well.

In 1974, the Chitwan tiger population was estimated to be less than 25 (Poppleton and Mishra 1974), but was over 60 in 1980 (Smith and Mishra 1981). The former number appears to be an underestimation. Nevertheless, including the extended areas and forests adjoining the reserves the number of tigers (including the young cubs) could easily exceed over 100. Here, the ungulate crude biomass is estimated to have increased from about 2,000 kg/sq. km in 1974 (Seidensticker 1976) to about 5,000 (Mishra 1982). From 1968, the population of rhinos increased from 100 and could well exceed 350 by 1988.

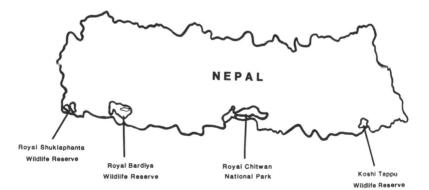

Fig. 1. Map of Nepal showing the location of tiger reserves.

In the Suklaphanta Wildlife Reserve (155 sq. km), the number of tigers was once estimated to be less than 15 (Poppleton and Mishra 1973). Now there are at least 50 - 75 tigers in the district of Kanchanpur, with the highest density concentrated in the core of the reserve (currently an area of 355 sq. km). Here, the Swamp deer (Cervus duvaucelli) population, believed to have dwindled to less than 500 in the 1960's, is now reported to be more than 2,000 (Warden pers. comm.).

At the onset of "Operation Tiger," the number of tigers in the Royal Bardia Reserve (368 sq. km, excluding extensions), was estimated to be less than 20 (Poppleton and Mishra 1973). Now, including the extension of 968 sq. km in 1978, there may be more than 100 tigers. This takes into account the adjoining forests of Banke and Dhangarhi districts (Warden pers. comm.). In Bardia the density of prey species for the tiger, particularly of Chital (Axis axis), is one of the highest in the predator's range (Dinerstein 1979).

THE NEPALESE SCENE

The Setting

The Kingdom of Nepal is a land of unique ecological contrasts and unexcelled beauty. Within a short span of about 200 km, the altitude changes from less than 100 m above sea level to the highest point on the Earth's surface (8,848 m) and contains some of the most outstanding wilderness areas in the world.

The country's 147,000 sq. km include four distinct ecological zones. Approximately 23% of the area is comprised of the hot and humid lowland Terai, an extension of the fertile Indo-Gangetic plains. The Terai harbors nearly all of the tiger population. The midlands, a central region of rugged mountains and terraced farmlands, cover nearly 44% of the land. The rest of the country is dominated by the Himalayas - a largely uninhabitable area with permanent snow - and the Trans-Himalayan region, characterized by the treeless steppes of the Tibetan plateau.

Land Use Pattern

It is an unfortunate paradox that over 90% of the people live off the land in a country where only 20% of the area is ecologically sound for farming (Bhattarai 1983). In comparison to other mountainous countries, the patterns and trends are more of 'abuse' rather than 'use' of the land. Definitive data on the tenure system do not exist and land capability surveys have not been undertaken. Nevertheless, figures obtained from the National Planning Commission (1981) indicate that 29% of the land is forested and 22% is under agriculture. Natural pasture occupies 13%, while 18% is classified as barren. Water bodies, including the large snow-fed rivers, form 3% of the surface area. Urban areas encroach a mere 0.2%, while the gigantic peaks with permanent snow command 15% of the total land mass of Nepal. The most alarming news is that forests are being destroyed at a rate of nearly 3%/year (Mishra 1984).

The Socio-economic Dilemma

Population pressures: Population growth in Nepal is among the highest in the world.   Two-thirds of the people dwell in climatically hostile and rugged mountainous terrain that produces only one-third of the food required.   The estimated population is 16.6 million human beings, and is increasing at the alarming rate of 2.6%/year.   Nearly 40% of the population is less than 15 years of age.   The density is 472 persons per sq. km of cultivable land.   More than 90% of the people are subsistence farmers who depend upon depleted forests for fuel and fodder (Asian Development Bank 1982).

The fertility rate is one of the highest in the world; it is common for a woman to have 5-7 children.   Family planning programs have been quite active, yet few couples practice birth control.   Thus, the population, which took 60 years to double in 1971, may now take less than 27 years to double again.

Food: Between 1975 and 1980 a total of 15,000 sq. km of natural habitat, including 7,000 sq. km of highly prized virgin forests deemed to be tiger habitat, were converted into agriculture (Asian Development Bank 1982).   During this same period, the per capita food production decreased as the population exploded.   In 1980 grain production increased by 3%, while the population increased by 14%.   It has been estimated that by the year 2020 the population will have expanded by over 50%, while the grain output will have increased by only 5% (Nepal National Conservation Strategy 1983).

The livestock situation at present is not encouraging.   There are an estimated 14 million head of hoofed stock in Nepal.   While they contribute 15% of the Gross Domestic Product (GDP), they consume 50 million tons of green plants, mostly new growth (Bhattarai 1983).   Over 50% of the cattle may be suffering from disease or malnutrition and many become feral.

Other socio-economic indicators: The United Nations has classified Nepal as one of the least developed of the developing countries (LDC).   Infant mortality is 133/1,000 live births and life expectancy is only 44 years.   Adequate health care for most Nepalis is lacking.   There is one doctor for every 32,000 persons, and one hospital bed for every 5,000 patients (Asian Development Bank 1982, Bhattarai 1983).

Despite government efforts to provide free primary education to all children, the literacy rate is a mere 23% (Manandhar 1982); only half of the eligible primary school-aged children enroll in schools.

Aside from tourism, industries are very much underdeveloped; they employ about 60,000 people and provide only 4% of the GDP.   Although the government has heavily emphasized cottage industries, their average turnover is a mere $150/year.   The per capita income of $120/year is one of the lowest in the world (Manandhar 1982, Asian Development Bank 1982).   Nearly 65% of the 1.3 million rural labour force is unemployed or underemployed (Asian Development Bank 1982).

In spite of these hardships, outsiders regard the people of Nepal as hardworking, friendly and tolerant.   The Kingdom's

Tibeto-Burman and Indo-Aryan ethnic groups form a mosaic of rich and diverse cultures (Bista 1976).

The Ecological Situation

It took only five years to lose 15% of Nepal's forests (Asian Development Bank 1982), most of it prime tiger habitat. This situation, combined with cultivation of slopes over 30°, or land with shallow soil fueled by heavy monsoon rains, has radically increased the processes of erosion. Soil loss per annum is between 20 and 50 tons/hectare. This is 20 times more in weight than the amount of rice produced from the same-sized land (Nepal National Conservation Strategy 1983). Some of the erosion can be attributed to the geologically young mountains, but over half is caused by human activities (Joshi 1981). Soil loss in overgrazed pasture is estimated to be nearly 40 metric tons/year; and the top soil in crop lands is being reduced at a rate of 25 tons/year (Bhattarai 1983). Each year the four major rivers, with over 6,000 tributaries, export 240 million cubic meters of Nepal's precious soil into the Bay of Bennal (Joshi 1981).

The demand for commercial timber is expected to increase from 292,000 cu. m. in 1980 to 736,000 cu. m. in the next two decades. The loss of these renewable resources, which during 1980-81 contributed $16.3 million in export earnings (Asian Development Bank 1982), shall have grave economic consequences. Consequently, it may not be long before Nepal becomes even more heavily deforested.

The consequences of environmental degradation do not only impede human prosperity, but are also directly linked with the problems of conservation and management of those large mammals, like tigers, that need a very large home range.

TIGERS AND MAN: AREAS OF CONFLICTS

Land Use and Agriculture

In comparison to the countries of the Third World, conflicts of land use between wildlife and humans have been largely diminished in North America. There, efficiency in agriculture and industrial development have left large tracts of land uneconomical or unsuitable for farming. In the United States less than 10% of the population produces food. Yet this output is more than enough to feed the whole country, as well as producing vast quantities of surplus for export. In contrast, Nepal is a net importer of agricultural products despite the fact that over 90% of the people are farmers.

Land use planners often suggest parks and reserves for wildlife should be established only in "marginal land" where they will be secure from human use. Adherence to this idea would wipe out most of the tiger sanctuaries, not only in Nepal, but elsewhere in Asia. The fact is that most parts of our tiger reserves, at present, could also be utilized to some degree for farming, cattle grazing and other immediate human needs. Yet they are also the only areas left where endangered species, such as the tiger, have a chance for survival.

Biologists have often elaborated upon the relationship between the shape, size and other parameters of reserves and the maximization of species diversity and genetic variability (Diamond and May 1981, Smith and Mishra 1981). But the hard fact is that existing tiger sanctuaries were simply established where land was available. They were created by compromising between political and human interest groups in Kathmandu, hardly at the rural level! Thus, the concept of "National Parks and Wildlife Reserves" as a sanctuary for rare species such as the tiger inherently induces conflicts in land use. The local people, who traditionally used the land for grazing, firewood and even shifting cultivation, have now been prohibited from their major resource and way of life.

Loss of Livestock

Traditionally, it has been the practice in Nepal to graze cattle in public forest lands, as private pastures are non-existent in southern Nepal. Villagers often let the cattle out in the morning to return home in the evening. The cattle often wander inside reserve boundaries, for the cultivated land offers little graze. Besides being illegal, these stray animals often fall prey to large carnivores. This immediately induces conflicts with human interests. An example is an incident which occurred in Royal Chitwan National Park. One villager bought a buffalo with his savings, the reward of nearly four years of hard labor. Unfortunately, the animal could not resist the grasses across the river and inside the park. While wandering inside the park boundary, it was devoured by a tiger. Adding "insult to injury" - the villager could have been fined for letting cattle into the park. He has no legal recourse, and worse, he will hold a grudge against the tiger as long as he lives (Mishra 1982). Unlike leopards, tigers rarely venture into farmlands, but they do take a heavy toll of domestic cattle that enter their parks and reserves. In one area in the Royal Chitwan National Park, it has been estimated that domestic cattle constituted nearly 30% of the tiger kills (Tamang 1979, Sunquist 1981). Tigers generally do not prey on cattle because of low density of prey population. Yet, due to the vulnerability of domestic animals which find their way into the park, the tigers do not let a good meal go by (Mishra 1982)! This problem is becoming more severe as the tiger population increases and dispersing tigers are forced into marginal habitats along the fringes of farm lands.

Loss of Human Life

McDougal (this volume) has reviewed the problems of man-eating tigers in geographical and historic perspective. Humans are not tigers' natural prey. Thus, man-killing is rare, and is considered to be an exceptional aspect of the behavior of the species. But it becomes an ultimate expression of the conflict between man and the tiger (McDougal 1986). The result of this conflict is an outburst of emotional and psychological reactions against the tiger.

We have seen examples in Chitwan, where the actual incidence of man-killing has been increasing over the last seven years.

One morning in December 1978, a school teacher from a village on
the border of the park encountered a young but lame male tiger
and was killed instantaneously. This rare incident caused the
local villagers to riot (Mishra 1981). Fortunately, the tiger,
one of the study specimens of the Tiger Ecology Project, carried
a radio-collar, and was captured alive and translocated within 24
hours to the Kathmandu Zoo. This operation calmed the local
population. In Chitwan, to date, another 13 persons have been
killed and/or eaten by tigers. This includes three kills made
this year by a tiger named Kancha Bhale. During the last eight
years, a total of seven tigers, four males and three females,
have been involved in man-eating or life-threatening
confrontations. One of the females was an old tigress, Madi
Baghini, with no previous record of man-killing. However, she
eventually mauled one villager and terrorized local farmers by
her nightly prowling within the village area. Another female,
Bangi Pothi, was found near a tourist lodge where she was
reported to have attacked one of the staff. This particular
tigress appeared weak and was not deemed able to harm a human.

The number of deaths from tigers is significantly less than
from other natural and accidental causes. Nonetheless, such
statistics are meaningless to a bereaved family, particularly if
the deceased happened to be the head of the household. For
example, the school teacher that was killed by the tiger was
caring for two orphaned children - the mother had died six months
earlier! It may be naive to assume that these children, and
relatives of others that have been devoured, would be active
supporters of any campaign to save the tiger!

Conflicts Arising from Legal Provisions

Conflicts arising from legal measures to protect the tiger
and its habitat are best described by another example from
Chitwan (Mishra 1982, 1984): For years, villagers used to graze
their cattle inside the forests. While there, some villagers
collected firewood, wild fruits and other forest products, fished
or collected honey. Since enforcement of National Park
regulations, they have had to graze their cattle on the open
banks of the Rapti River. It is illegal for the animals to stray
across. Little do they realize that the National Park cannot
sustain grazing pressures, nor would there by any forest left if
they continued to collect fuel wood and timber. They regularly
see strange-looking foreigners going through the jungle on
elephant back, and are confused as to why an area is set aside
for the recreation of these outsiders, seemingly at their
expense. They resent the fact that their movements are
restricted from dawn to dusk even along the traditional public
pathways. Most villagers fail to understand why the law protects
animals such as rhinos and tigers, which are pests to their crops
and livestock, particularly since no one uses these animals for
food or as beasts of burden. The villagers simply want to
survive. They believe their lives and the welfare of their
families are more important than "saving wildlife."

Thus, legislative measures based upon intellectual or
aesthetic values for conservation of endangered species such as
the tiger are meaningless to villagers struggling day-to-day for
their existence. They accept the conservation legislation as one
of the many hardships they must face as a part of their survival.
Avoiding being caught by park guards, scheming to get away with a

minimum penalty when they are caught, or harassment, seems to be
their only reasonable recourse.

TIGERS AND MAN: AN APPRAISAL OF MUTUAL BENEFITS

Tourism

Among the three main tiger sanctuaries of Nepal, Chitwan is
the most popular as a tourist destination. The facilities and
infrastructures are therefore relatively well developed.   In the
last ten years the number of visitors has increased from less
than 1,000 in 1975 to over 14,000 in 1986.

In Bardia, attempts to develop tourism have been made
recently, but the number of visitors, including Nepali citizens,
has not exceeded 500 in the last three years.

In Suklaphanta, the number of visitors has not exceeded a
hundred in any given year, as the reserve is quite isolated from
the primary tourist circuit within Nepal.

Benefits from tourism in Royal Chitwan National Park to the
indigenous peoples are minimal (Mishra 1982).   Park revenue
constitutes less than 25% of the government's financial inputs.
Increased tourist facilities may generate more employment
opportunities, but will be detrimental to the fragile ecosystem
of the national park.   Except for visitors that go to Tiger Tops,
a luxurious jungle lodge, it is quite difficult to argue for any
positive relationship between tourism and tiger conservation.   At
Tiger Tops, visitors have the opportunity to observe tigers from
a viewing platform, which are baited with domestic buffalo.
Visitors to other parks are not shown tigers in a similar
fashion.

This method of showing tigers to tourists has recently been
criticized due to sensational publicity on man-eating.   The fact
that at least two man-eaters were frequent visitors to Tiger Tops
bait sites has generated conflicting views.   The ethical question
of showing tigers to tourists by baiting is being reviewed by
government authorities despite the fact that there has been no
specific evidence to relate man-eating to baiting practices.

Employment Generation

National parks, reserves and their concession-aires are the
largest employers of local people. Research projects such as the
Smithsonian Tiger Ecology Project provide local employment as
well.   However, as with tourism, most local farmers do not
believe there is any relationship between the tiger and direct
monetary benefits from increased job opportunities or other
sources.

Soil and Water Conservation

In the fragile ecosystem of southern Nepal, the most
important long-term benefit is the role protected areas play in
the conservation of soil and water.   Since tigers have extensive

home ranges (Sunquist 1981, Smith 1984), large tracts of land
must be set aside. Now, with controlled grazing and consequently
decreased habitat destruction, the banks of major rivers and
streams have stabilized. This is particularly important during
the torrential monsoon rains, but these benefits are neither
realized nor appreciated by local communities. Floods and
landslides are believed to be acts of God, not the results of
land mismanagement. The lack of an effective conservation
education and publicity program is responsible for the local
people's failure to realize the importance of the parks in
conserving soil and water.

EFFORTS TO MINIMIZE CONFLICTS

Removal or Destruction of Problem Tigers

Among the problem tigers described above, two males and one
female (tiger 119, Bange Bhale, Bangi Pothi) have been captured
alive and translocated for confinement in Kathmandu, the
whereabouts of one (tiger 127) is not known, and three were
destroyed on site (females 118 & Madi Pothi, and male Kancha
Bhale). These actions were necessary to pacify the villagers.
As is often the case, rumors were spread that park authorities
deliberately maintained man-eaters to keep people out of the
park. The translocation and destruction of problem tigers were
always witnessed by several villagers - an act vital to
demonstrate that, despite the full legal protection afforded
these animals, the park management cares about human lives and
the problems of the village people.

Community Forum

In most of the countries of Asia, managers of protected
areas are foresters or biologists who are more at home in the
forests than they are at dealing with the community's related
problems. (Mishra 1982). Training and experience in public
relations programs at a community level is lacking because such
programs are difficult to devise. Furthermore, there are
important social and perceptual differences from area to area,
even within the same country. The result is a general lack of
communication with the local people. Most local people view the
staff of protected areas as watch dogs. Conversely, the
experience has taught park or reserve staff to regard most
villagers with suspicion.

Since 1977, protected areas' authorities have assembled
groups of villagers, school teachers and leaders of local
communities once a year to discuss the relationship of the
problems of the community and the needs for nature conservation.
They meet at park or reserve headquarters, where they are housed
and fed by the government. They are also given a tour of the
area. During a series of meetings, park officials listen to the
complaints and grievances. Attempts are made to solve or provide
answers to every issue raised. The staff also explains why
requests for grazing or timber exploitation in the national park
cannot be granted.

This program is more of a public relations campaign than anything else. The biggest impact of these meetings has been psychological. The local people are beginning to feel that they are being involved in management procedures. These gatherings are envisaged to allow staff to learn the real problems faced by local people; they have also been useful to illuminate the complexities of conflicts between nature conservation and human needs. These gatherings have also given the local people chances to "blow off steam" against the government conservation programs, particularly on the aspects that affect them directly.

The Annual Grass Harvest

Since 1978, local villagers in Chitwan, Bardia, Suklaphnta and Koshi Tappu have been permitted to enter the park to harvest grasses. Most houses in the Nepalese Terai are roofed with thatch grass. Canes of tall elephant grass have been used traditionally to construct walls and partitions both outside and inside the house. These materials are now available only in the parks and reserves. A nominal fee (less than US $0.05) is charged for each permit. The purpose of the fee is to keep a record of people entering the park rather than to generate revenue. One permit allows a villager to harvest and remove as much grass as he or she can. The collection period lasts for 15 days in January. The villagers spend the first week harvesting thatch grass, after which they set fires. Elephant grass canes are removed after the fire burns most of the leaves and other dry parts. Motorized vehicles and bullock carts are not permitted, so both rich and poor have equal opportunity to take as much as they require. Many of the latter also make a living by selling surplus cane, or making crafted products from it.

The government authorities view this program in three ways. First, any system in the midst of a densely populated area must consider the local community as a part of management. Second, by allowing the local people limited access to a resource central to their livelihood, the program illustrates one principle of conservation in terms that villagers can easily understand. Most villagers who farm a few hectares do not understand the aesthetic or intellectual arguments for nature conservation. However, they do recognize that most tall grasses outside protected areas have disappeared, and that the reserve protects those that remain. Furthermore, they are aware that the tall grasses never matured to their full height before the implementation of effective conservation measures, for the grasses were constantly cropped by domestic cattle. Thus, this grass cutting program has been one of the most powerful educational and public relation tools for saving the tiger's habitat. The villagers have access to a resource, essential to their livelihood, that had virtually disappeared. Third, the grassland is of a character of successional stages that has been maintained for hundreds of years by human activities - cutting and burning for example (Wharton 1968). The annual grass cutting program may be the most effective and economical way of maintaining this ecosystem - an ecosystem that appears to have 'naturally' disappeared elsewhere. In the four reserves of the Nepal Terai, over 20,000 people are mobilized without any financial burden to the government treasury.

Obviously, the situation in the Terai is somewhat unusual since we are dealing with a rapidly renewable natural resource.

The application of the same principles for fuel wood or timber collection would be self-defeating. The short and long-term effects of this massive periodic influx of people on the fauna is not known. However, it is an effective means of compensation to the local population for the restrictions imposed by legislation and the consequent damage to their crops or property. In short, it is a practical "trade-off" (Mishra 1982).

DISCUSSION

Aldo Leopold was an outstanding naturalist and an American of great vision. Even during the earlier half of this century, he understood the relationship and interdependence between human needs and wildlife conservation (Mishra 1982). Although Leopold's (1933) ideas were concerned with the issues and problems of game management in the United States, his ideas have universal application in any country, irrespective of socio-economic status, for formulating conservation strategies for any mammal, including large carnivores such as the tiger. More recently, Kellert (1979) and others (Mishra 1982, Hatley and Thomson 1985) stressed the need to understand the human dimensions of conservation programs. This requires recognition and knowledge of the interaction between natural resources and those who use (or abuse) them. In Nepal, seasoned leaders of the conservation movement have often voiced the opinion that no conservation program can succeed without support and acceptance by the society that is affected by its aims and objectives (Shah 1984c). Furthermore, experience has proven that programs to preserve large mammals such as the tiger are more complex than previously understood.

Talbot (1979) pointed out that as conservation efforts become successful, they automatically conflict with other human interests. Problems once perceived as being of a biological nature become political, economic and social. The socio-economic scene in Nepal and the conflicts between humans and tigers verify these predictions. Strategies to save the tiger cannot be viewed in isolation. They must incorporate the harsh realities of poverty, hunger, illiteracy and other human miseries. If not, the tiger will not survive in perpetuity, especially in poor countries like Nepal.

The clarion call of today is to strike a balance between the needs of the growing population and the need for wildlife conservation (World Conservation Strategy 1980, Nepal National Conservation Strategy 1983, Shah 1984a, b and c, 1985). It was precisely for this reason that the elected legislators of Nepal created a new institution, the King Mahendra Trust for Nature Conservation. The Trust is an autonomous, nongovernmental and nonprofit organization. By its very nature, it is a people-oriented institution that aims at mobilizing popular support through programs that involve active participation and benefits to the local communities (Shah 1984a, b and c, 1985). It endorses the principles of the World Conservation Strategy of the IUCN (World Conservation Strategy 1980) and believes that nature conservation is not an obstacle, but a vital compnonent of economic development.

Harrowing accounts of happenings in Africa have tragically proven the warnings of many field experts. Despite $ millions spent on research, the plight of large mammals such as elephants and rhinoceros is serious, despite the fact that countries like Uganda once had the best systems of national parks and wildlife conservation programs in the world. Why? - Because neither years of research nor management made any attempts to investigate, involve or motivate the indigenous people; nor were programs geared toward fulfilling the aspirations and needs of the poor rural communities. It is easy to blame revolutions or bad government, or even nature, for the tragic state of large animals in countries like Uganda or Ethiopia, but if research or management plans were well conceived and programmed, they would survive not only one, but a few revolutions or bad governments. One could also ask - Why was the wolf eradicated from the United Kingdom during the latter part of the last century? This happened despite the fact that Britain was at its peak of social and economic development. The answer is quite simple; the farmers in the highlands and islands of Great Britain did not have a vested interest in saving this carnivore. Planners and managers of tiger sanctuaries in Asia must learn from the mistakes made in Africa and elsewhere. Otherwise, it may be that the future of the tiger in Asia beyond the turn of the century is, at best, uncertain. Creation of mechanisms that make saving the tiger "profitable" for the people surrounding its habitat are vital to reverse the trend. This will require new approaches, research priorities and management options.

Experienced decision makers in Asia, such as His Royal Highness Prince Gyanendra have often asked, What is conservation, if not for the people? It must be viewed only as a means, the end being the improvement of quality of human life (Shah 1985). Consequently, any strategy to save the tiger and its habitat must parallel programs aimed at providing better alternatives and economic gains to the poor rural communities. Man must obtain benefits in the form of monetary gains, jobs, markets, prestige, sentiments, or even thatch grasses, or the tug of war between the tiger and man will continue. Historical trends elsewhere indicate that the former may end as the loser. After all, it is people, not the deer and the tiger, that elect decision makers.

ACKNOWLEDGEMENTS

It would not have been possible for me to attend the Tiger Symposium without the approval and encouragement from my august Chairman, His Royal Highness Prince Gyanendra Bir Bikram Shah. I wish to express my sincere gratitude and respect to His Royal Highness for his guidance and support. The World Wildlife Fund - USA, generously provided funds for my travels and I am grateful for their support. Most of the information in this paper is based on my field research in the Royal Chitwan National Park that has been supported by the Smithsonian Institution. Dr. William Thompsom, Bijaya Katel, Charles McDougal and Udaya Sharma reviewed the draft. Last but not least, I am grateful to Madhav Parajuli and Tika Ram Bhattarai who spent hours at night in editing mistakes and getting the word processor fixed.

REFERENCES

Asian Development Bank. 1982. Nepal Agriculture Sector Strategy
     Report, Part 2. Manila: ADB.
Bhattarai, S. 1983. State of Environment in Nepal, Report No.2.
     Environmental Impact Study Project. Kathmandu: HMG/N.
Bista, D.B. 1976. People of Nepal. Kathmandu: Dept. Publicity.
Diamond, J.M. and R.M. May. 1981. Island biogeography and the
     design of natural reserves. In Theoretical Ecology:
     Principles and Applications, ed. R.M. May. Oxford:
     Blackwell.
Dinerstein, E. 1979. An ecological survey of Royal Karnali
     Bardia Wildlife Reserve, Nepal:
     --Part 1. Vegetation, modifying factors and successful
     relationship. Biol. Conserv. 15:127-50.
     --Part 2. Habitat and animal interactions. Biol. Conserv.
     16:265-300.
Hatley, T. and M. Thompson. 1985. Rare animals, poor people and
     big agencies: A perspective on biological conservation and
     rural development in the Himalaya. Mountain Res. Dev.
     5(4):365-77.
Hinrichsen, D., P.H.C. Lucas, B. Coburn and B.N. Upreti. 1983.
     Saving Sagarmatha. Ambio 12(34):203-05.
Joshi, M.D. 1981. Environment of Nepal. Kathmandu: Dept. Solid
     & Watershed Man., HMG/N.
Kellert, S.R. 1979. Some objectives for human dimension in
     wildlife management. Paper presented to the Symposium, Human
     Dimension in Wildlife Management. Yale University, School of
     Forestry.
Leopold, A. 1933. The Conservation Ethic. In Readings in
     Conservation Ecology, ed. G.W. Cox. New York: Appleton
     Century Crofts.
Manandhar, P.K. 1982. Introduction to policy, legislation and
     programmes of community forestry development project.
     Kathmandu: HMG/UNDP/FAO/CFDP.
Mishra, H.R. 1981. Wir fingen einen, Morder. Gnade fur den
     Tiger. Das Tier 6(21):47-51.
Mishra, H.R. 1982. Balancing human needs and conservation in
     Nepal's Royal Chitwan National Park. Ambio 11(5):246-51.
Mishra, H.R. 1984. Ecologie. Dans le parc des tuers d'homs. GEO
     Nouveau monde la Terre. 65(7):92-93.
Mishra, H.R. and D. Mierrow. 1974. Wild Animals of Nepal.
     Kathmandu: Ratna Pustak Bhandar.
Mountfort, G. 1981. Saving the Tiger. New York: Viking Pr.
Nepal National Conservation Strategy. 1983. Nepal National
     Conservation Strategy: A Prospectus. Gland, Switzerland:
     HMG/IUCN.
National Planning Commission. 1981. 1980-1985: The Sixth Plan.
     Kathmandu: HMG/N NPC.
Poppleton, F. and H.R. Mishra. 1973. "Operation Tiger." A
     preliminary project for conservation. Kathmandu: Wildlf.
     Conserv. Project.
Seidensticker, J. 1976. Ungulate population in Chitwan Valley,
     Nepal. Biol. Conserv. 10:183-210.
Shah, His Royal Highness Prince Gyanendra Bir Bikram. 1984a. A
     word from the Chairman in Conservation for Development. In
     An Introduction to the King Mahendra Trust for Nature
     Conservation. Sano Thimi, Kathmandu: JEMC Pr.
Shah, His Royal Highness Prince Gyanendra Bir Bikram. 1984b.
     Statement at Smithsonian Institution, National Zoological
     Park. Washington, DC.

Shah, His Royal Highness Prince Gyanendra Bir Bikram. 1984c. Statement at the Royal Nepalese Embassy. London.

Shah, His Royal Highness Prince Gyanendra Bir Bikram. 1985. Keynote Address to the International Workshop on the Management of National Parks and Protected Areas in the Hindukush Himalaya (May 6). Kathmandu: KmTNC/ICIMOD.

Smith, J.L.D. 1984. Dispersal, communication, and conservation strategies for the tiger (Panthera tigris) in Royal Chitwan National Park, Nepal. Ph.D. thesis. Univ. Minnesota, Minneapolis.

Smith, J.L.D. and H.R. Mishra. 1981. Management recommendations for the Chitwan tiger population. The Parsa - extension and the Bara Hunting Reserve. Smithson. Inst./WWF Project 1051.

Sherpa, N.W. 1979. A report on fire wood use in Sagarmatha National Park; Khumbu Region. Kathmandu: Dept. Nat. Parks Wildlf. Conserv.

Smythies, E. 1974. Big Game Shooting in Nepal. Calcutta: Thacker, Spink and Co.

Sunquist, M.E. 1981. The social organization of tigers (Panthera tigris) in Royal Chitwan National Park. Smithson. Contrib. Zool. No. 336.

Talbot, Lee. 1979. The tip of the iceberg: A biopolitical perspective on the Endangered Species Act. School of Forestry/Environmental Studies, Yale Univ.

Tamang, K.M. 1979. Population characteristics of the tiger and its prey. Ph.D. thesis. Michigan State Univ., East Lansing.

World Conservation Strategy. 1980. Living resources conservation for sustainable development. Gland, Switzerland: IUCN/UNEP/WWF.

Wharton, C. 1968. Man, fire and wild cattle in south east Asia. Proc. Tall Timber Fire Ecology Conf. 107-67.

William Rice. "Indian Game," (from quail to tiger). Published by W.H. Allen and Co., Waterloo Place, Pall Mall, W. London. 1884 (line drawing from page 213).

44

A Tiger Geographic Information System: The First Step in a Global Conservation Strategy

J.L. David Smith, Chris Wemmer, and Hemanta R. Mishra

INTRODUCTION

The forest and grassland habitats required by the tiger (Panthera tigris) were once distributed widely across Asia. However, with the rapid increase in human populations during the past several decades there has been a corresponding expansion of agriculture into forested areas, so that the present range of tigers is now confined to a few, small, remnant forest sanctuaries. Since 1900, the number of tigers in India and Nepal has dropped from 40,000 to approximately 3,000 (Holloway et al. 1976). There have been similar decreases elsewhere in the tiger's range; local extinctions have occurred in Bali and Java (Seidensticker this volume) and is imminent in China (Bangjie this volume).

Today, many tiger populations are near minimum viable size; below this, inbreeding, loss of genetic variability, and random ecological and demographic events may pose serious threats to survival even where the remnant numbers and habitats are securely protected. Because we are faced with a future in which extinction of small local populations is likely to be a very real and not uncommon event, it is important to recognize that a strategy to prevent the extinction of the tiger must be, in essence, a plan to conserve a series of local populations. The purpose of this paper is to describe an inventory and monitor system for tigers that uses a geographic information system (GIS) to store and manipulate data. This approach will document the number, size and location of existing tiger populations, which we believe should be one of the first steps in a global tiger conservation strategy. Furthermore, it will provide a comprehensive assessment of land resources needed for tiger survival. To demonstrate the value of this approach we will: 1) review conservation theory, especially as it relates to the management of small tiger populations; 2) examine the case history of a tiger population, and; 3) describe the elements of a tiger GIS which will provide the information needed to develop and monitor a global conservation strategy.

CONSERVATION AND SMALL POPULATIONS

The causes of extinction fall into two categories, systemic and stochastic factors (Shaffer 1986). Systemic factors exert steady pressure on a population. Loss of habitat, for example, causes continual fragmentation of large tiger populations into smaller and smaller population units. Any attempt to cope with the problems facing small populations must first assess the degree to which this systemic pressure can or is being alleviated. In India and Nepal, efforts over the past 12 years of Project Tiger have succeeded, at least temporarily, in halting the decline in tiger numbers. However, in many areas the decline continues as a result of unchecked forest degradation. We believe there is danger that wildlife biologists are relying on the unfounded assumption that they can save the tiger when the only tiger habitat left is that which occurs within the presently existing parks and reserves. The problem with this perspective is that by the time tigers have disappeared from areas outside parks and reserves, 60 to 75% of the tigers that are estimated to still exist in Asia will also have disappeared, and the remaining populations may be too small to ensure long-term survival. Furthermore, social, economic, political, and bureacratic inertia may be too strong to suddenly reverse patterns of over-utilization of forests.

Once the continued loss of habitat has been successfully addressed, the remaining habitat must be managed to encourage recovery of critically small populations. To do this we will review how popoulation structure and stochastic events affect critically small populations and examine some of the important ideas of conservation theory as they relate to the management of small tiger populations.

Genetic structure: Population genetic structure refers to the spatial relationship and size of populations, and their degree of isolation. The probability of local extinction is a function of population size; the smaller the population, the more likely it will go extinct. The probability of extinction is influenced also by the degree to which populations are connected and the rate of exchange of individuals between them (i.e. gene flow). When a group of local populations is related, the entire unit is often referred to as the metapopulation; management efforts must begin to focus at this higher level.

Variations in population genetic structure range from discrete populations, with little or no gene flow, to clustered populations where boundaries are more difficult to define. Finally, animals may be continuously distributed in large population units. It is important to be able to characterize local tiger distribution in relation to these patterns. Considerable heterogeneity exists even in large continuous blocks of habitat, which implies discontinuous distribution. Data on the movements of subadults in Royal Chitwan National Park, Nepal, demonstrate that dispersal is often short-ranged (Smith 1984). Genetic viscosity, therefore, creates small neighborhoods of closely related individuals (Wright 1978).

Within individual populations there are a variety of factors that infuence the probability of survival.

Genetic stochasticity: Chance can lead to an increase in deleterious genes in small populations. Although there is debate about the extent to which selection will counter these sampling problems in wild populations (Hemmer, O'Brien this volume), it would be unwise to ignore the potential for inbreeding depression. Population geneticists have attempted to estimate a minimum population size to ensure genetic viability and ecological resilience. Franklin (1980), Soule (1980), and Frankel and Soule (1981) recommend that there should be no more than 1% inbreeding per generation to avoid reduced fertility, decreased resistance to disease, and high juvenile mortality. Based on this estimate, populations probably should not drop below an effective population size of 50. This number is not derived from theory or conclusive data, but is the best available estimate based on recent analyses of breeding histories from zoos (Seal 1977, Ralls et al. 1979, Ballou and Ralls 1982, Ralls and Ballou 1982), experimental population genetics (reviewed in Wright 1978), and domestic animal breeding programs (Falconer 1981). Fifty breeders is not offered as the critical threshold for all species; it is simply a reasonable guideline. Some species persist successfully at less than 50, while others show deleterious effects well above 50 (Soule 1980). In the case of tigers, as with most wild animals, information to calculate the specific inbreeding threshold does not exist.

A less immediate problem is loss of genetic variability within a species and thus of the species' ability to adapt to changing environmental conditions. Soule (1980) and Franklin (1980) recommend effective population sizes of not less than 500 to avoid a long-term decrease in genetic diversity within a species. Lande and Barrowclough (1985) have emphasized that the loss of genetic diversity is a complex issue. For example, there are three types of genetic variability; single gene or Mendelian, quantitative or multigenic, and karyotypic. The relationship of each type of variability to effective population size is somewhat different. Lande and Barrowclough (1985) estimate that the minimum population size needed to prevent long-term loss of genetic variability is several hundred individuals, but at present a more precise estimate is impossible to calculate.

Demographic stochasticity: Sampling can lead to skewed demographic structure. In a small population normal variation in social structure, reproduction, sex ratio, and survival can lead to a situation in which the population size declines due to sampling chance alone. Just as a gambler can have a run at the roulette table, so can a population of 20 breeders have a losing streak. For example, the birth of a series of litters composed predominately of males. In Royal Chitwan National Park the chance death of a large male who was still in his prime led to a drop in mean reproduction in the Park for a period of three years. Following his death, competition by several adult males for females in his former territory, and subsequent infanticide of existing litters, resulted in a decline from a mean of 2.6 to 1.1 subadults that survived to dispersal age (Smith and McDougal pers. observ.).

Environmental stochasticity: Fluctuations in habitat quality, weather, food and disease can affect the size of a local population. When a population is confined within a small area, chance environmental events can have catastrophic consequences. Floods, fires, and hurricanes can cause sudden drastic declines and extinction in local populations. Human caused catastrophies can also be added to this category; oil spills are one of the more publicized examples of this type. To design reserves of adequate size to preserve the species that live within them, it is important to be able to estimate both the environmental variability and the probability of catastrophic events which may lead to local extinction (Soule and Simberloff 1986).

Minimum viable population size (MVP): A minimum viable population is one that has a given probability of persisting for some stated period of time (Shaffer 1981). The theory of MVP is still developing. A number of investigators have sought estimates of MVP with respect to genetics (Soule 1980, Franklin 1980, Wilcox 1980, Frankel and Soule 1981), Lande and Barrowclough 1985). Shaffer (1981, 1983), and Shaffer and Samson (1985) have focused on how demographic and environmental factors affect MVP, and Roff (1974a, 1974b), and Vance (1984) have analyzed the effect of spatial structure (i.e. the metapopulation dynamics) on MVP. What are we left with after marshalling all this theoretical fire power? A beginning. As yet, no one has integrated these components. Furthermore, there has been little in the way of field validation of these problems. The message is clear: if we are to conserve wild tigers for the future we need information on the sizes and dynamics of existing populations.

A CASE STUDY: ROYAL CHITWAN NATIONAL PARK, NEPAL

In Royal Chitwan National Park, Nepal, where one of the largest and most productive tiger populations exists today, long-term ecological studies of tigers have provided information on their distribution, density, social organization and genetic structure, as well as on some prey species (Seidensticker 1976a, 1976b, McDougal 1977, Sunquist 1981, Mishra 1982, Tamang 1982, Smith 1984). Smith (1984) developed methods for censusing tigers, determining tiger distribution and evaluating habitat quality. Now that these techniques have been refined, they can be used to collect comparative data from other populations. To assess the Chitwan tiger population structure we searched along roads, trails, dry stream beds, and ridge tops for tiger sign such as tracks, scent marks, and kills; in addition, we made over 6,000 radio locations of tigers (see Smith 1984, Smith et al. this volume, for details on study area and methods). To determine the distribution of the Chitwan population we mapped location data onto satellite imagery. Using this information we not only discovered where tigers were, but we determined also what habitat they did not use. Tigers did not go into the densely settled and denuded middle hills of Nepal, nor did they enter cultivation (Smith and Mishra 1981, Smith 1984). The border between forest cover and human habitation was usually distinct, making it easy to map areas of potential tiger distribution. The zone of continuous forest cover extended from Butwal in the west to the Bhagmati river in the east (Fig. 1). At these places cultivation extended from the Gangetic Plain north into the middle hills creating a barrier to tiger dispersal. The Chitwan population was therefore isolated from its nearest neighbors 250 km west and 150 km east. The area of continuous forest cover was 2,700 sq. km but resident breeding tigers occupied only 2,200 sq. km of these forests (Fig. 1).

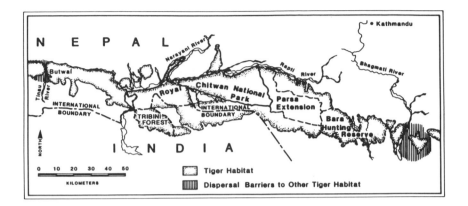

Fig. 1. Map shows the distribution of the Chitwan tiger
population. Dispersal barriers at Butwal and the Bhagmati River
isolate the Chitwan tigers from populations 150 km east and 250
km west.

Within the area occupied by the Chitwan tiger population we
censused tigers using radio telemetry data and pug mark tracking.
Our estimate ranged from 45-60 resident breeding adults. We
examined tiger habitat quality in the forest zone by measuring
three variables: an index of prey abundance, horizontal cover,
and human use. Discriminant function analysis of these variables
allowed us to classify habitat into breeding and non-breeding
habitat (Smith 1984). Classification of habitat quality is
essential for anticipating changes in tiger populations and
assessing management programs aimed at forest conservation and
sustained multiple use management.

Local tiger distribution in Chitwan can be summarized as
follows: the habitat is discontinuous, forming a habitat island;
to the north are the densely settled middle hills of Nepal and to
the south is the virtually treeless Gangetic plain. There are
dispersal barriers to other tiger habitat to the east and west
(Smith and Mishra 1981, Smith 1984). The Chitwan tigers,
therefore, must be managed as an isolated population. To
accomplish this, His Majesty's Government of Nepal approved a
plan to manage the entire area as a Tiger Reserve System in which
the Park (1,040 sq. km) is the primary component or core area
(Smith and Mishra 1981). The other components of the system
include the recently established Parsa Wildlife Reserve (545 sq.
km), which is adjacent to the Park in the east, and the Tribini
Forest (265 sq. km) in India, which was recently approved for
conversion from commercial forest management to a wildlife
reserve. In addition, the Bara Forest (540 sq. km) is under
consideration for multiple use management as a fuel wood forest
and/or a hunting reserve.

Essential for the long-term success of the Chitwan Tiger
Reserve System is the preservation of all remaining habitat and
management that contributes in some way to the goal of tiger

conservation. Some of these land uses may directly affect the
survival of tigers while others may only indirectly contribute to
tiger conservation, but may be essential to meeting local human
needs and thus reduce pressure on prime habitat. For example,
experimental multiple use programs are being planned in buffer
zones adjacent to Parsa and the Bara Forest to find alternative
patterns of forest use that will not lead to forest degradation.
If these forests can not be managed on a sustained yield basis it
is likely that as they decline the amount of illegal human
activity in the Park will increase.

A GEOGRAPHIC INFORMATION SYSTEM FOR TIGERS

     What is a geographic information system (GIS)? First, a GIS
is commonly computer based and is used for handling spatial data.
Second, it can process geographic data on land use patterns to
facilitate decision making. Third, a GIS can produce a set of
machine readable maps and computer algorithms to analyze these
maps to aid in developing solutions to specific land use
problems. And fourth, the system can store and manipulate
data. At the simplest level of analysis a GIS provides a means
to overlay and manipulate data of different types (Fig. 2).

     The primary goal of tiger conservation is to maintain
natural populations of tigers throughout their range. The
objective of a tiger GIS is to provide the information needed to
accomplish this goal. The individual objectives are: 1) to
determine the number of the population; size and distribution; 2)
to assess the quality of habitat so that we can anticipate where
tiger population declines may occur as well as measure the
results of habitat restoration efforts, and; 3) to record human

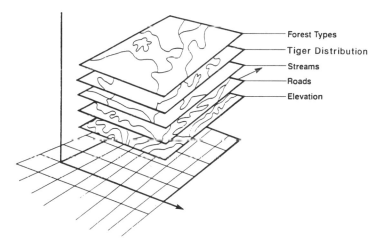

Fig. 2. A geographic information system provides a means of
manipulating data from a series of overlay maps to produce
syntheses of the input data as a series of new maps and spatial
analyses.

use patterns in, and adjacent to, tiger habitat so that we can
make informed decisions (for example, by using analysis theory
described by McGuire this volume). It is vital that multiple use
lands be used in a way to maximize the benefits for both wildlife
and the local populations which depend on them (Mishra 1982b).

Karanth (this volume) has pointed out the difficulty in
using census data to monitor tiger population changes. A GIS,
based on tiger distribution and habitat assessment, provides an
alternative means of estimating tiger numbers, the size of
individual populations and their degree of isolation. When the
habitable area is reduced, so too are the numbers of breeding
animals. By monitoring the quantity and quality of tiger
habitat, changes in tiger numbers can be tracked.

The advantage of this method is that it requires fewer field
skills than a complete tiger census. For example, the presence
or absence of tiger tracks can be sampled at a number of
localities, and thereby provide a more consistent and objective
data set than can be obtained using current methods. This
approach also requires less manpower and is more cost effective.

The first step is mapping tiger distribution as was done at
Chitwan (Smith and Mishra 1981). This method uses the
distribution of forest cover, as determined from satellite
imagery, to make a base map; then the presence or absence of
tiger tracks at a point location is mapped and tiger presence is
extrapolated to the surrounding locality based on known home
range sizes. Therefore, by judiciously surveying for tiger
presence at important locations, distribution over wide areas can
be mapped economically (Fig. 3).

Measuring habitat quality is a second, critical part of a
GIS system for tigers. Knowledge of this variable provides a
method to predict population trends before changes in population
numbers actually are detected. Furthermore, it allows for the

Fig. 3. Distribution of forest cover (potential tiger habitat)
in the lowlands of Nepal.

assessment of management practices intended to rehabilitate low quality forest habitat. One difficult requirement is that habitat information is needed over extensive areas. Analysis of wildlife habitat is usually restricted to a limited study area and often involves gathering much more information than is actually needed in the final analysis. GIS techniques provide an approach to habitat assessment whereby limited ground sampling can be extrapolated to wide areas using satellite and aerial photography to assess uniform habitat units.

A third component of a tiger GIS is information on how human populations use forest lands. This is important because 60 to 75% of the tiger's distribution is in forest lands used intensively by local people. A key to the conservation of tigers is to manage these lands to provide for the needs of humans as well. By coupling human use data with information on habitat quality, multiple use strategies compatible with tiger management can be devised.

For the most part, foresters are caretakers of wildlife in much of the developing world. Indeed, if they are not directly responsible for wildlife bureaucratically, their policy and decisions have major impact on it. However, forestry in much of the developing world is a tradition-bound practice, albeit highly successful in certain areas such as India in carrying out its objectives of sustained yield management. The challenge before us is to present a compelling arguement for adopting a methodology that is quite alien to the domain of forestry or wildlife management as presently practiced.

WHY IS THE GIS APPROACH IMPORTANT?

The reasons for adopting the GIS for tiger management are clear: it is objective, comprehensive, and relatively economical. Most important, it allows the conservationist to assess the critical issue of the survival of man and tiger, i.e. the interface of their habitats. There is also a political reason for adopting this method; the sophistication and scope of the data would be far more convincing in the political arena where decisions inimical to conservation are often made. The details of implementation of the method are beyond the scope of this paper. The technology exists in nearly all the tiger-harboring nations, but the will to use the technology needs to be developed.

Clearly defined land management plans are needed to prevent conservation objectives from competing with other land use objectives. Land management must be carefully regulated in and around each and every tiger population. These plans should specify how every block of land contributes to the conservation of that population. To do this efficiently over wide areas, spatial information on the distibution of tiger populations needs to be overlayed onto maps of forest cover. The next step in the evaluating process is to overlay human use of all the land components (e.g. reserves, reserve forests, multiple use forests, pasture areas). A GIS provides a method for compiling and analyzing this information so that the importance of land units can be prioritized both for providing human needs and for contributing to the requirements for preserving a local tiger population.

As decisions are made on which lands are to be developed and which are to be preserved, it is essential that wildlife biologists and managers contribute to the process of land use planning. When plans for new roads, dams, fuel-wood plantations, and paper mills are developed regionally, tiger conservationists must have an articulated strategy based on empirical evidence for preserving ecosystems and populations of minimum size to ensure their survival. However, there is also an important need to be able to integrate habitat preservation with competing land uses. This will require prioritization of land needs for tiger conservation on a regional and broad geographic basis, and the ability to integrate multiple use patterns that meet both development and conservation goals.

CONCLUSIONS

An essential and basic requirement for a global tiger strategy is unequivocal information on the number of populations, their sizes, and their spatial relationships. Our view is that all tigers left in the world are important to the long-term goal of saving the species. Even small remnant populations which scientists "know" are doomed can generate valuable information to guide future decisions. Why should we be interested in preserving all populations? The concept of MVP, for example, is not yet based on sufficient data. Although some of the components have been investigated, theoretically they never have been integrated into a comprehensive model. The reality is that we know very little about what a minimum viable population is for any species. To learn, we need in addition to theory, the validation of that theory in the real world. Let us not throw away the chance to learn from these "doomed" populations. Knowledge of the size and degree of isolation of the existing tiger populations and continued monitoring of these populations, especially the smallest and most "hopeless", is critical to developing an understanding of the extinction process, the degree to which inbreeding depression, stochastic events, and the loss of heterozygosity affects tiger populations.

ACKNOWLEDGEMENTS

Mr. B.B. Shah, Mr. J.J. Dutta, H.S. Panwar helped develop ideas in this paper, and D. Meisner, M. Bauer, and Kip Willet provided advice on mapping and remote sensing. We thank D. Miquelle, P.B Rai and the Tiger Ecology Project staff for their help in the field. F. Cuthbert, P.A. Jordan provided critical comments. The Financial support for the project was provided by the Smithsonian Institution and World Wildlife Fund - U.S. Appeal.

REFERENCES

Ballou, J. and K. Ralls. 1982. Inbreeding and juvenile mortality in small populations of ungulates. Biol. Conserv. 24:239-72.

Falconer, D.S. 1981. An Introduction to Quantitative Genetics, 2d ed. London: Longman Publ.

Frankel, O.H. and M.E. Soule. 1981. Conservation and Evolution. Cambridge: Cambridge Univ. Pr.

Franklin, I.A. 1980. Evolutionary change in small populations. In Conservation Biology: an Evolutionary-ecological Perspective, ed. M.E. Soule and B.A. Wilcox. Sunderland, Mass.: Sinauer Assoc.

Holloway, C.W., P. Leyhausen and M.K. Ranjitsinh. 1976. Conservation of the tiger (Panthera tigris tigris L.) in India. World Wildlife Fund Report.

Lande, R and G.F. Barrowclough. In press. Effective population size, genetic variation, and their use in population management.

McDougal, C.W. 1977. The Face of the Tiger. London: Rivington and Andre Deutsch.

Mishra, H.R. 1982a. The ecology and behavior of chital (Axis axis) in the Royal Chitwan National Park, Nepal. Ph.D. thesis, Univ. Edinburgh.

Mishra, H.R. 1982b. Balancing human needs and conservation in Nepal's Royal Chitwan Park. Ambio 11:246-52.

Ralls, K., K. Brugger and J. Ballou. 1979. Inbreeding and juvenile mortality in small populations of ungulates. Science 206:1101-03.

Ralls, K. and J. Ballou. 1982. Effects of inbreeding on juvenile mortality in some small mammal species. Lab. Anim. 16:159-66.

Roff, D.A. 1974a. Spatial heterogeneity and the persistence of populations. Oecologia 15:245-58.

Roff, D.A. 1974b. The analysis of a population model demonstrating the importance of dispersal in a heterogenous environment. Oecologia 15:259-75.

Seal, U.S. 1977. The Noah's Ark problem: multigeneration management of wild species in captivity. In Endangered Birds, Management Techniques for Preserving Endangered Species, ed. S.A. Temple. Madison, WI: Univ. Wisconsin Pr.

Seidensticker, J.C. 1976. On the ecological separation between tigers and leopards. Biotropica 8:225-34.

Shaffer, M.L. 1981. Minimum population sizes for species conservation. Bioscience 31:131-23.

Shaffer, M.L. 1983. Determining minimum viable population sizes for the grizzly Bear. Int. Conf. Bear Res. Man. 5:133-39.

Shaffer, M.L. and F.B. Samson. 1985. Population size and extinction: a note on determining critical population sizes. Am. Nat. 125:144-53.

Smith, J.L.D. 1984. Dispersal, communication, and conservation strategies for the tiger (Panthera tigris) in Royal Chitwan National Park, Nepal. Ph.D. thesis, Univ. Minnesota, St. Paul.

Smith, J.L.D., M.E. Sunquist, K.M. Tamang and P.B. Rai. 1983. A technique for capturing and immobilizing tigers. J. Wildl. Man. 47:255-59.

Soule, M.E. 1980. Thresholds for survival: maintaining fitness and evolutionary potential. In Conservation Biology, ed. M.E. Soule and B.A. Wilcox. Sunderland, Mass.: Sinauer Assoc.

Soule, M.E. and D.S. Simberloff. 1986. What do genetics and ecology tell us about the design of nature reserves? <u>Biol. Conserv.</u> 35:19-40.

Sunquist, M.E. 1981. The social organization of tigers (<u>Panthera tigris</u>) in Royal Chitwan National Park. <u>Smithson. Contrib. Zool.</u> No.336.

Tamang, K.M. 1982. Dynamics of tiger prey population in Royal Chitwan National Park. Ph.D. thesis, Michigan State Univ., East Lansing.

Vance, R.R. 1984. The effect of dispersal on population stability in one-species, discrete-space population growth models. <u>Am. Nat.</u> 123:230-54.

Wright, S.J. and S.P. Hubbel. 1983. Stochastic extinction and reserve size: a focal species approach. <u>Oikos</u> 41:466-76.

Wright, S. 1978. <u>Evolution and the Genetics of Populations, Vol. 4. Variability Within and Among Natural Populations.</u> Chicago: Univ. Chicago Pr.

45

Decision Analysis: A Tool for Tiger Conservation and Management

Lynn A. Maguire

INTRODUCTION

A Difficult Problem

Decisions about how to conserve and manage tiger (Panthera tigris) populations are difficult for a variety of reasons. First, there are multiple, and sometimes conflicting, objectives for tiger conservation. Some primary objectives guiding tiger management include: 1) maximizing probability of survival for tiger populations, and 2) maximizing retention of genetic diversity in tigers. For both, questions arise about whether these objectives should be pursued for the species as a whole, for selected subspecies, for wild populations, or for captive populations. A third primary objective may be to maintain, or even expand, the range of tigers in the wild.

Tiger conservation decisions are complicated by a context of secondary objectives that include: 1) promoting conservation education among rural and urban populations of both developing and industrial countries; 2) promoting international cooperation in resource conservation; 3) minimizing disruption of rural development activities; 4) minimizing financial costs of tiger management; and 5) protecting species and ecosystems associated with tigers. As examples of potential conflicts among objectives, strategies that are best for conserving tigers as a species may not be best for individual subspecies; exhibiting tigers in zoos may promote conservation education among urban populations, but not among rural people.

A second source of difficulty is failure to specify measurable criteria for evaluating alternative management strategies. Examples of measurable criteria that might be applied to maintenance of genetic diversity include: 1) empirical measures of heterozygosity, obtained by analysis of enzymatic diversity in tissue samples; 2) calculation of inbreeding coefficients from pedigree records; and 3) inference about inbreeding levels from population records indicating sources of origin and population bottlenecks.

Third, there are many choices to be made about tiger management in captivity and in the wild. How should the available tiger spaces in North American and European zoos be allocated among the five extant subspecies? Should additional tiger reserves be designated; should existing reserves be expanded in size? Should captive populations be subdivided; how large should the subpopulations be; what rate of migration should be maintained among subpopulations; should the subpopulations ever be merged? How many new founders, from what wild or captive populations, should be added to captive breeding programs, and at what times? How closely should wild tiger populations in reserves be managed; should there be intentional culling, fertility control, or translocation of tigers among reserves? Is it better to concentrate on protection of a particular subspecies in the wild or to capture animals for captive breeding? Where, when and how might tigers be reintroduced to the wild, or might sparse populations be augmented?

Fourth, and most difficult, these decisions must be made in the face of uncertainty about environmental and sociopolitical conditions that determine success or failure of tiger management actions. Extreme weather, diseases of tigers or their prey, and stochastic fluctuations in age or sex structure of populations are biological events that influence tiger population dynamics under different management strategies. Less widely recognized are the sociopolitical factors affecting tiger conservation; such as rural resettlement schemes, fluctuating government support for conservation efforts, campaign promises restricting export of endangered species, and civil unrest. Both environmental and sociopolitical factors may be largely beyond the control of those responsible for tiger management.

A Useful Tool

The purpose of this chapter is to introduce formal methods for analyzing decisions under uncertainty that have been developed in business management (Raiffa 1968) and to demonstrate how these could help design strategies for tiger conservation. Maguire (1986) and Maguire et al. (1986) describe previous applications of these methods to management of endangered wildlife. I will illustrate these methods here by analyzing the question: Should wild tiger populations in reserves be managed more closely than they are now, and should existing reserves be expanded in size?

In preparing these examples, I have attempted to provide plausible estimates for the parameters used in the analyses, but I have not asked experts in tiger biology and management to comment on these estimates. Therefore, the emphasis of the chapter is on how the methods of analysis enable conclusions to be drawn, rather than on the content of particular conclusions about tiger management. In a realistic application, parameter estimates would be obtained by structured questioning of those responsible for the management decision, as described below.

MANAGEMENT OF RESERVES

The Problem

The designation of tiger reserves, as in India for Panthera
tigris tigris and in eastern Russia and northeastern China for
Panthera tigris altaica, has led to apparent increases in tiger
populations in those areas (Srivastava 1979, Bangjie 1983, Panwar
1982). However, recent success of these efforts does not
guarantee that the reserves will be large enough to sustain
viable populations of tigers over the long term with current
management strategies. Increasing conflicts between tigers and
human activities bordering reserves (Srivastava 1979, Hendrichs
1975, Kahn, McDougal and Simgh this volume) suggest that some
reserves may be "full" already. It may be necessary to expand
the size of existing reserves, or to manage the tiger populations
in them more closely, to maintain viable populations without
unacceptable conflicts with humans.

A Decision Tree

Management Actions: A decision tree (Raiffa 1968, Behn and
Vaupel 1982) provides a framework for laying out the elements of
the reserve management question: possible courses of action,
uncertainties affecting outcomes, and possible results. Possible
courses of action that might be selected by a resource manager
are represented as branches of the tree emerging from a square,
denoting a "decision node" (Fig. 1). These choices are under the
manager's control. The alternative actions examined in this
example are: 1) "status quo" management (i.e. protect the
reserve from development and tigers from illegal killing); 2)
active management of tiger populations within reserves (e.g.,
culling, fertility control, translocation of tigers among
reserves or between reserves and captive populations, and
stringent removal of tigers outside reserve culling, fertility
boundaries); and, 3) expansion of existing reserve boundaries.
Either of the latter two options would entail costs in excess of
what is currently being spent for reserve management.

Uncertainties and Outcomes: The success of each action may
be affected by one or more sources of uncertainty, whether
biological or sociopolitical. The sources of uncertainty, or
random events, are shown on the decision tree (Fig. 1) as
branches emerging from a circle, denoting a "random node." The
resource manager has no control over these random events. In
this example, the major source of uncertainty affecting the
success of "status quo" management is whether the reserve is too
small to sustain a viable population of tigers over the long
term. If it is too small, conflicts between tigers and humans
bordering the reserve will increase and, ultimately, the tiger
population will decline because of habitat limitations and
control killing. If the reserve size is adequate ("OK"), tigers
will be maintained at a small, but stable, population level.

The major source of uncertainty affecting the active
management option concerns the potential of active management to
disrupt normal population processes. For example, immobilizing
and operating on tigers for fertility control may cause excess
mortality; translocating tigers among reserves may disturb the
social structure of home ranges, causing both excess mortality

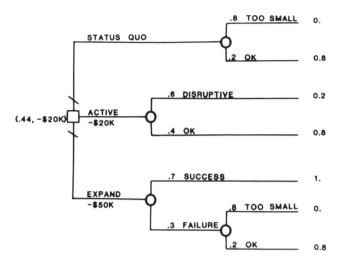

Fig. 1.   Decision tree for analyzing alternatives for managing tiger populations in reserves.

and decreased reproduction (Seidensticker 1976, Sunquist 1981). If active management does prove disruptive, the population will decline.   If active management can be carried out successfully, however, the population can be maintained at a small, but stable level, even if the reserve is too small to sustain an independent tiger population without such intervention.

The main uncertainties influencing reserve expansion are sociopolitical.   Existing human activities bordering the reserve; rural development and resettlement policies, including plans for alternative land uses such as forestry or agriculture; and government commitments to conservation will determine whether a planned expansion can be achieved in practice.   If it is successful, the larger reserve presumably will support a larger, stable population than the existing reserve.   If expansion fails, the status quo uncertainties and outcomes will apply.

Parameter Estimates

A formal analysis of the decision tree in Fig. 1 requires estimates of the values associated with different population outcomes, the probabilities of the uncertain events affecting those outcomes, and the costs of each management action.

Objectives and Criteria: For this example, I will assume that the objectives are to obtain the most favorable status for the reserve tiger population and to minimize financial costs of management actions.   To translate the verbal descriptions of possible population outcomes into a measurable criterion, I employed the methods outlined by Behn and Vaupel (1982) and Maguire (1986) for assigning relative values to outcomes.   The most desirable outcome, in this case a larger, stable population, receives a value of 1.   The least desirable outcome, in this case a declining population in conflict with human activities, has a value of 0.   The remaining outcomes, a declining population without conflicts and a small, but stable, population, receive values between 0. and 1, reflecting their merit relative to the best and worst outcomes. The resource manager responds to a structured series of hypothetical decisions to articulate his relative preferences for these intermediate outcomes. In this example, a declining population without conflict has a value of 0.2; and a small, but stable, population has a value of 0.8. These numerical assessments of value are shown at the ends of the branches of the decision tree (Fig. 1).

The costs associated with each management alternative may be entered either at the ends of the branches of the tree, or on the branches associated with each alternative, as has been done for this example (Fig. 1). Status quo management has no cost beyond what is already being spent. The entries on the branches for active management and expansion of the reserve represent the increases in cost above the status quo. Expanding a reserve is likely to be much more expensive than active management, since it entails not only the costs of acquiring and managing additional property as a reserve, but also the opportunity costs of foregone development alternatives, such as logging. The costs of $20 thousand for active management and $50 thousand for reserve expansion may be low, even for annual costs, but they will serve to illustrate the procedure.

Probability Estimates: Making a sensible decision under uncertainty requires estimates of the probabilities of random events that influence the outcomes of management actions. In some cases, objective information may be available for estimating these probabilities. For example, long term weather records can be used to estimate probabilities of severe storms. Theoretical models for population genetics can be used to estimate probabilities of losing certain percentages of genetic diversity.

In other cases, objective knowledge should be supplemented with careful use of intuitive information about random events to obtain subjective probability estimates (Raiffa 1960, Behn and Vaupel 1982). Again, these subjective estimates are developed through a series of structured questions to experts on tiger biology and management. In the reserve management example, objective information about home range size of male and female tigers, breeding and dispersal, could be combined with more intuitive assessments of habitat quality in a particular reserve to estimate the probability that the reserve is "too small." The probability estimates for each random event are entered on corresponding branches of the decision tree (Fig. 1). The particular values were chosen to illustrate the method of analysis and may not apply to any particular tiger reserve.

Analysis of the Decision

Calculating Expected Values: With all the parameter estimates in place on the decision tree (Fig. 1), an analysis of the best action can be made. The principles of decision making under uncertainty (Raiffa 1968, Behn and Vaupel 1982) suggest that the best action to take when outcomes are uncertain is the one with the best expectation of success. To employ this criterion, I calculated expected values for the population outcome at each random node, beginning at the right side of the tree. For example, the expected value at the upper right hand random node is $(.8)(0.) + (.2)(.8) = 0.16$. Carrying out this process for each of the three management alternatives gives expected values for population outcome of 0.16 for status quo management, 0.44 for active management, and 0.75 for expanding the reserve.

Choosing the Optimal Action: At each decision node, the action with the highest expected value should be chosen. If population outcome were the only decision criterion, the best action would be expanding the reserve, with an expected value of 0.75. However, minimizing financial cost is an objective, too. The action with the best expected value for population outcome is the most costly, and vice versa. To choose the best action, considering both criteria, requires an analysis of the tradeoff between improving population outcome and increasing cost.

Analysis of Tradeoffs

Is it worth $20 thousand to the resource manager to raise the expected value of population outcome to 0.44 from 0.16 (active management versus status quo)? Is it worth an additional $30 thousand to raise the expected value to 0.75 from 0.44 (expansion of reserve versus active management)? To answer these questions, the resource manager responds to a series of hypothetical decisions (Behn and Vaupel 1982, Maguire 1986) that identify pairs of expected population outcomes and financial costs to which he is indifferent. For example, one series of questions may reveal that he has no preference between an action that has an expected population value of 0.16 and no additional financial cost (status quo), and an action (hypothetical) with an expected population value of 0.9 and $50 thousand extra cost. Therefore, he would not pay $50 thousand to raise the expected population outcome to 0.75 by expanding the reserve. Similarly, if he is indifferent between an action with an expected population outcome of 0.16 at no extra cost, and an action raising the expected population value to 0.44 that costs $25 thousand, he must prefer active management, which costs only $20 thousand, to the status quo.

In this way, the tradeoff between population outcome and cost is resolved in favor of active management, with expected population outcome 0.44 and $20 thousand cost, as noted to the left of the decision node in the tree (Fig. 1). The alternatives that have been rejected in the course of the analysis, status quo and reserve expansion, are marked on the tree by slashes through the respective branches.

Sensitivity Analysis

Since many of the parameter estimates in a decision analysis may be in doubt, it is worth asking what changes in these parameters could shift the optimal choice from active management to another alternative. To do this, any of the probability estimates, values assigned to population outcomes, or costs can be changed, and the calculations and analyses described above repeated for the new values.

For example, what if the probability that the reserve is too small is 0.4, rather than 0.6? In this case, status quo management has an expected population outcome of 0.64, active management remains at 0.44, and expansion of the reserve has an

expected value of 0.89. The costs of each action remain as
before. In this case, active management can be rejected
immediately since it has both lower expected population outcome
and higher cost than status quo management. If the same
tradeoffs between population outcome and costs apply, the manager
would reject expansion of the reserve in favor of the status quo,
since he would not be willing to pay $50 thousand to raise the
expected value of population outcome from 0.64 to 0.89.

Analyzing the sensitivity of the decision to the parameters
of the tree reveals the relative importance of various features
of the decision. For example, changing the probability that the
reserve is too small may have far more impact on the decision
than changing the probability that active management is
disruptive to population processes. The resource manager may
want to calculate the probability that the reserve is too small
for which active management and the status quo are equally
desirable. Then, if he feels sure that the true probability that
the reserve is too small is below this "breakeven" probability,
he can reject active management with confidence, even if he does
not know the probability exactly.

Using sensitivity analysis, the manager can focus attention
on those feature of the decision that have the most impact on
choice of action. For example, if disruption of population
processes seems most likely to tip the choice of action, research
to develop new technologies for population management should be
emphasized.

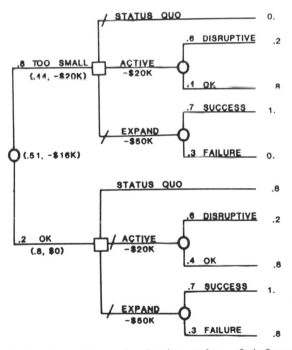

Fig. 2. Decision tree for calculating value of information about
whether reserve is too small.

It is equally important to notice which features of the problem have little impact on choice. For example, expanding the reserve always has a higher expected population outcome than the status quo, regardless of the probability of successful expansion or the probability that the existing reserve is too small. The tradeoff between population outcome and financial cost is what may make reserve expansion undesirable. A minimal amount of formal analysis makes such conclusions obvious, yet they are overlooked too often in unstructured decision making.

Value of Information

As a supplement to sensitivity analysis, calculating "value of information" for a particular aspect of the decision affords more specific guidance on where research efforts would be most helpful and what budget for research can be justified. For example, what would it be worth to know for sure whether a reserve is too small? To help answer this question, the decision tree in Fig. 1 is restructured with the random events "too small" or "OK" preceding the choice of management action, to show that the decision is to be made after finding out whether the reserve is too small (Fig. 2).

Table 1. Expected values of population outcome and costs for value of information calculations from the decision tree in Fig. 2. Starred actions are optimal.

Reserve Size	Action	Expected Outcome	Cost x 10^3
Too small	Status quo	0.0	$ 0
	* Active	0.44	20
	Expand	0.7	50
OK	* Status quo	0.8	0
	Active	0.44	20
	Expand	0.94	50

As above, expected values are calculated at each random node, starting on the right side of the tree, with the results noted in Table 1. At each decision node, the action with the best combination of population outcome and cost should be selected. If the tradeoff between population outcome and cost is as described above, the resource manager will choose active management if the reserve is too small (with expected population outcome 0.44 and cost $20 thousand, as noted on the appropriate branch in Fig. 2), and status quo management (with expected population outcome 0.8 and no extra cost) if its size is adequate.

To calculate the expected result of making the reserve management decision with full knowledge of whether the reserve is too small, the previously estimated probabilities that the reserve is too small or not are used to weight the results obtained under each random event: (.8)(.44, $20 thousand) + (.2)(.8, $0.) = (.51, $16 thousand). This means that, as an expectation or a long term average, reserve management decisions based on information about whether the reserve is too small would have an expected population outcome of 0.51 and a cost of $16 thousand. Comparing these results with the choice made under uncertainty about reserve size (active management with expected population outcome 0.44 and cost $20 thousand) reveals an average savings of $4 thousand and an improvement in population outcome from 0.44 to 0.51. This is the "value" of information about reserve size, or the degree to which decisions could be improved by obtaining information about reserve size. This calculation expresses the maximum it would be reasonable to pay for research to determine whether a reserve is too small, i.e., $4 thousand plus whatever financial value might be assigned to raising the expected population outcome from 0.44 to 0.51. Of course, if a single study of tiger habitat, home range size, etc., could be applied to decisions on management of several similar reserves, then the costs of the study can be distributed over anticipated improvements in management of all the reserves.

This example of value of information deals with the case where research determines with certainty whether the reserve is too small. The value of information in such cases is referred to as the "expected value of perfect information" (Raiffa 1968). In many cases, research will reduce uncertainty about whether a reserve is too small, but will not completely eliminate it. The value of this improved, but not perfect, information is termed the "expected value of sample information," and it can be calculated by methods similar to those used above (Raiffa 1968). The expected value of sample information will always be less than the expected value of perfect information about the same source of uncertainty.

SUMMARY AND CONCLUSIONS

Summary

Using the management of wild tigers in reserves as an example, this paper illustrates the main features of formal methods for decision making under uncertainty. Components of any decision analysis include: 1) the objectives for management actions (best tiger population outcome, minimum financial cost); 2) the alternative management actions to be considered (status quo, active management and reserve expansion); and 3) the biological (adequacy of reserve size, disruption of population processes by active management) and sociopolitical (success or failure of attempted reserve expansion) uncertainties affecting the outcomes of those actions.

To make a formal analysis, the possible outcomes of management actions and uncertain events must be expressed in terms of measurable criteria (a population rating scale obtained by questioning tiger experts, and dollars). The probabilities of the uncertain events must be estimated using both objective and subjective information. The best alternative is then identified by calculating expected values of the decision criteria (population outcome) at the random nodes on the decision tree, and taking the best expected value at each decision node, moving from right to left on the tree. In cases where the decision criteria give contradictory results (e.g., the dilemma between maximizing expected population outcome and minimizing cost), a structured series of questions may be used to assess the resource manager's willingness to trade off extra cost for improved population status. Using these tradeoff relationships, the best choice for the combined criteria can be determined for each decision node.

The potential for changes in any parameters of the analysis (cost, population outcome, probabilities) to change the optimal management action can be explored using sensitivity analysis. This process increases confidence in the decision and focuses research on important features of the decision. In conjunction with sensitivity analysis, calculating value of information about uncertainties affecting outcomes (adequacy of reserve size) helps guide research direction and expenditures.

Benefits of Decision Analysis

Clarification of Objectives: Why are formal, structured approaches to decision making under uncertainty likely to be beneficial to tiger management and conservation? First, a formal process encourages clarification of objectives and specification of measurable criteria. When representatives from institutions with widely differing resources and perspectives (e.g., North American zoos, governments of developing countries, universities) gather to synthesize a management plan, disagreements result from attempts to satisfy multiple, and conflicting, objectives that have not been articulated. Attempts to anticipate the likely results of future management, or to evaluate the results of past management, are frustrated by the absence of operationally defined criteria (e.g., percentage heterozygosity in captive populations, per capita reproductive rates in wild populations).

Recognition of Uncertainty: Uncertainty is a pervasive feature of nearly all conservation decisions. Psychological explorations of human decision making have shown that unaided intuition often draws incoherent and indefensible conclusions from probabilistic information (Kahneman et al. 1982).

Decision analysis provides a framework for incorporating uncertainties in tiger management decisions in a consistent way.

Use of Subjective Information: Although solid scientific information about tiger biology is accumulating, many phenomena vital to tiger conservation remain unstudied. In addition, many conservation decisions, such as specification of boundaries for a particular reserve, are unique. Where objective information is lacking and where unique circumstances prevail, exploitation of intuitive knowledge of local experts is essential to good decision making. A structured approach to developing subjective probabilities and relative ratings for outcomes (e.g., population outcome) ensures careful use of intuitive information and, by expressing it in a numerical form, makes it accessible to scrutiny by others.

Resolution of Tradeoffs: Conservation decisions typically have conflicting objectives. Nowhere is this more striking than in decisions to commit financial and land resources to tiger conservation in countries where needs of human populations may go unmet. Using a framework of hypothetical decisions, the tradeoffs among conflicting objectives can be analyzed from the

perspective of local decision makers. This process helps ensure
that the resolution of tradeoffs reflects the values of those
responsible for the decision, rather than the constraints of
strictly economic cost-benefit analyses.

Direction of Research: In most tiger conservation
decisions, additional information would be helpful, but resources
and time for obtaining it are limited. Both sensitivity analysis
and value of information focus attention on features of a
decision problem where additional information or effort would
have most benefit. In the reserve management example, such
analyses can reveal whether adequacy of reserve size, disruptive
potential of active management, or sociopolitical factors
affecting reserve expansion have the most critical impact on the
decision. In the first two cases, research on tiger habitat use
and on management techniques might be the best investment of
effort. In the third case, attempts to better understand, and
even to influence, the political processes affecting reserve
expansion would be most beneficial.

Communication and Negotiation: Tiger conservation decisions
are the joint product of researchers, resource managers, zoo
curators, and government officials, among others. Formal
decision analysis, and decision trees in particular, provide a
means for communication and negotiation among individuals with
divergent objectives and beliefs. Seeing a decision problem
expressed in the explicit format of a decision tree (Fig. 1) will
not resolve all sources of disagreement. However, it will let
different parties see more clearly the features of the decision
and the parameters on which the discussion is based. In
intuitive decision making, much effort is wasted disagreeing
about points that are misunderstood by both sides.

Finally, decision analysis should not be embraced as a tool
for automated decision making. Its uncritical use can lead to
decisions just as faulty as those made using intuitive reasoning.
Rather, it is a tool for decision making that augments the innate
capabilities of humans to synthesize information and make
decisions. It is the process of examining the features of a
decision problem and exploring the impacts of changing
parameters, more than the simple calculation of an optimal
action, that realizes the maximum benefit from formal analysis.

REFERENCES

Bangjie, T. 1983. Tigers in China. Tigerpaper 10(4):27-9.
Behn, R.D. and J.W. Vaupel. 1982. Quick Analysis for Busy
    Decision Makers. New York: Basic Books, Inc.
Hendrichs, H. 1975. The status of the tiger Panthera tigris
    (Linne, 1758) in the Sundarbans Mangrove Forest (Bay of
    Bengal). Saugetierkd. Mitt. 23(3):161-99.
Kahneman, D., P. Slovic and A. Tversky, ed. 1982. Judgment
Under Uncertainty: Heuristics and Biases. Cambridge: Cambridge
    Univ. Pr.
Maguire, L.A. 1986. Using decision analysis to manage
    endangered species populations. J. Environ. Manage. In
    press.
Maguire, L.A., U.S. Seal, and P.F. Brussard. 1986. Managing
    critically endangered species: the Sumatran rhinoceros as a

case study. In <u>Viable</u> <u>Populations</u>, ed. M.E. Soule. Cambridge: Cambridge Univ. Pr. In press.

Panwar, H.S. 1982. What to do when you've succeeded: Project Tiger ten years later. <u>Ambio</u> 11(6):330-37.

Raiffa, H. 1968. <u>Decision</u> <u>Analysis</u>. Reading: Addison-Wesley.

Seidensticker, J. 1986. On the ecological separation of tigers and leopards. <u>Biotropica</u> 8:225-34.

Srivastava, B.P. 1979. Status of tiger in India. <u>Tigerpaper</u>, 6(2-3):24-27.

Sunquist, M. 1981. The social organization of tigers (<u>Panthera</u> <u>tigris</u>) in Royal Chitwan National Park, Nepal. <u>Smithson.</u> <u>Contrib.</u> <u>Zool.</u> 336:1-98.

46

A Global Tiger Conservation Plan

Ulysses S. Seal, Peter Jackson, and Ronald L. Tilson

INTRODUCTION

This draft outline of a species management plan was prepared during discussions at the Tiger Symposium. It was the consensus of all present that the need for a species-oriented survival plan for tigers is necessary if we are not to loose the biological integrity of the species and its recognized subspecies in the near future. The erosion of numbers is occurring in most countries for all subspecies. The numbers of animals necessary for long-term survival and to retain the capability to evolve in the face of changing environments are larger than the current populations for at least three of the subspecies. Further, there is probably only one natural population of one subspecies that is large enough to meet this criterion for survival. Therefore it is necessary to develop a plan, a strategy, the technology, and the collaboration required to ensure that our diverse efforts and resources are put to best use to achieve our shared goal of survival of this magnificient species.

In addition to the outline we have included a Table of the national parks, wildlife reserves and animal sanctuaries, and the populations of tigers occupying them, for which some data and information may be found in this book. This should help in the task of filling the many gaps in our knowledge. We have also included a copy of the Draft Manifesto on Cat Conservation in preparation by the IUCN/Cat Specialist Group and a copy of the Captive Breeding Policy for Conservation prepared by the IUCN/Captive Breeding Specialist Group as supporting documentation and for your information.

Table 1.   Geographic distribution of tigers and known protected
areas by subspecies and country with detailed notes on wild
status.

Site	Area (km^2)	Number
P.t. amoyensis (approx. 40)		
Peoples Republic of China (endemic)		
(tigers by province: Chejiang ?; Fujian 4;		
Guangdong 4-6; Guangxi 10; Guizhou 4;		
Hubei 4; Hunan 8; Jiangxi 4; Shaanxi 4)		
Bamianshan Nature Reserve	200	"traces"
Chebaling Nature Reserve	?	?
Fanjingshan Mountain Reserve	?	"a few believed"
Jiangganshan Nature Reserve	159	?
Meihuadong Nature Reserve	?	?
Shixing Nature Reserve	?	4-6
P.t. altaica (approx. 200-300)		
People's Republic of China (no est.)		
(tigers by province: Heilongjiang ?;		
Jilin 16-24)		
Changbaishan Nature Reserve	?	?
Qixianlize Nature Reserve	330	?
People's Republic of Korea (no est.)		
Soviet Union (early 1980's est. 200)		
Sikhote-Alinsky Wildlife Reserve	?	?
P.t. balica (EXTINCT)		
Bali		
Bali Barat National Park	200	?
P.t. corbetti (approx. ?)		
Burma (est. 750)		
Thailand (no est.)		
Laos (no est.)		
Kampuchea (no est.)		
Vietnam (no est.)		
Peninsular Malaysia (est. 600-650)		
(tigers by state: Pahang 150; Terengganu 84;		
Kelantan 87; Perak 60; Johore 72; Kedah 21;		
Selangor 15; Negeri Sembilan 24;		
Taman Negara 72 = 505, + relict populations)		
(5,661 km^2 protected areas)		
Batu Feringgi Wildlife Reserve	6	?
Batu Frazer Wildlife Reserve	25	?
Batu Kutu Wildlife Reserve	?	?

Table 1. Continued.

Site	Area (km^2)	Number
Peninsular Malaysia (cont.)		
Belum Wildlife Reserve	?	?
Cameron Highland Wildlife Reserve	825	?
Endau-Rompin National Park	635	?
Gerik Wildlife Reserve	675	?
Gunong Belumut Wildlife Reserve	175	?
Gunong Tampurong Wildlife Reserve	25	?
Kerau Wildlife Reserve	562	?
Kuala Gula Bird Sanctuary	6	?
Kuala Selangor Wildlife Reserve	6	?
Pahan Tua Bird Sanctuary	12	?
Pulan Wildlife Reserve	150	?
Selama Wildlife Reserve	225	?
Sg. Dusun Wildlife Reserve	43	?
Sg. Nenggeri Wildlife Reserve	400	?
Surgkai Wildlife Reserve	37	?
Taman Negara Wildlife Reserve	4,737	?
Tasek Bera Wildlife Reserve	300	?
Tasek Chiai Wildlife Reserve	50	?
Templer Park	56	?
Ulu Lepar Wildlife Reserve	62	?
Ulu Muda Wildlife Reserve	1,675	?
Ulu Terengganu Wildlife Reserve	1,062	?
People's Republic of China (no est.) (tigers by province: Yunnan ?)		
P.t. _sondaica_ (EXTINCT)		
Java, Indonesia		
Meru Betiri National Park	500	0-3
P.t. _sumatrae_ (approx. 650)		
Sumatra, Indonesia (endemic) (92,637 km^2 protected areas, few large enough for tiger populations)		
Barisan Selatan National Park	3,568	?
Berbak Game Reserve	1,900	?
Gunung Leuser National Park	8,025	80-90
Kerinci-Seblat National Park	14,846	?
Kerumutan Wildlife Reserve	?	0
Padang-sugihan Wildlife Reserve	?	?
Way Kambas Game Reserve	1,300	30-40
P.t. _tigris_ (approx. ?)		
Bangladesh (est. 300-430)		
Burma (est. 750) (pers. comm. J. Blower 1983)		
Patolon-Taungdwin Forest Reserves	1,450	"tracks in practically every streambed"

Table 1. Continued.

Site	Area (km^2)	Number
Burma (cont.)		
Mulayit Wildlife Sanctuary	?	?
Piduang Wildlife Sanctuary	?	?
Shwe-U-Daung Wildlife Sanctuary	?	?

India (1984 est. 4,000)
(tigers by state (not all states covered):
West Bengal 352; Andhra Pradesh 164;
Meghalaya 125; Arunachal Pradesh 219;
Tamil Nadu 97; United Provinces 698;
Assam 376; Karnataka 202; Maharashtra 301;
M.P. 786; Bihar 138; Kerala 89; Orissa 202;
Rajasthan 96; Nagaland 104)
(100,000 km^2 protected areas as of 03/86,
>1,100 tigers in 15 tiger reserves)

Site	Area (km^2)	Number
Bandipur Tiger Reserve	690	53
Banhavgarh Wildlife Reserve	105	15
Buxa Tiger Reserve	745	15
Corbett National Park	521	90
Dudwa Wildlife Reserve	613	74
Indravathi Tiger Reserve	2,084	38
Kanha National Park	940	89-109
Keoladeo National Park	?	?
Manas Tiger Reserve	2,840	123
Melghat Tiger Reserve	1,572	80
Nagarahole Wildlife Reserve	640	27
Nagarjunasagar Tiger Reserve	3,000	65
Namdapha Tiger Reserve	1,808	43
Palamau Tiger Reserve	930	62
Periyar Tiger Reserve	777	41-44
Ranthambore Tiger Reserve	414	38
Sariska Tiger Reserve	800	26
Simlipal Tiger Reserve	2,750	71
Sundarbans Tiger Reserve	2,585	264
Tribini Wildlife Reserve	265	?

Site	Area (km^2)	Number
Nepal (est. 230)		
Parsa Wildlife Reserve	545	?
Royal Chitwan National Park	1,040	60-100
Royal Bardia Wildlife Reserve	1,336	100+
Royal Suklaphanta Wildlife Reserve	355	50-75
Koshi Tappu Wildlife Reserve	?	0-2
Peoples Republic of China (no est.)		
Nangun Nature Reserve	70	?

P.t. virgata (EXTINCT)

(no details given)

GOALS

The goal of tiger conservation is to maintain sufficiently large populations of each subspecies in natural habitat in each country of origin to provide a high probability of long-term survival. To this end it is essential that guidelines be established and action be taken to describe and classify the current status of each wild population and its prospects for both short-term (five years) and longer-term (50 years) survival.

Specific Objectives

1. Construct a detailed distribution map and listing of all known populations by country and indicate the known or estimated census for each population. This survey and mapping should indicate: a)the location and size of reserves and protected areas with their estimated tiger populations, b)the area and tiger populations present in reserve buffer zones where such zones exist, c)the distribution and number of tigers occurring outside of these areas, d)areas of critical habitat for tigers and, e)an assessment of dispersal possibilities between habitat areas.

2. Describe and assess threats to these populations and assess the viability of each population. This would include estimating the probability of extinction for the next five, 20 and 50 years of each mapped wild tiger population if no further actions are taken or changes made in current management. If the probability of extinction is 5% or greater for the 50 year projection, then the primary reasons for this possibility of extinction and some possible actions that might increase the likelihood of survival should be described.

3. Where multiple tiger populations and reserves exist, conduct a protected areas systems analysis with particular attention to: a)habitat types represented, b)numbers or density of available prey species, c)soil and vegetation mapping, d)relationship to local peoples and economies, e)land-use planning needed to provide sustained protection for the reserves, f)current, planned and needed management; g)planned local projects which will impinge upon the reserve or habitat (roads, dams, timbering, etc.) and, h)human-tiger conflicts.

4. Develop working agreements in each country on actions to be taken in the event of a decline in numbers below an established level for each age and sex. These agreements should be prepared for each population and made a part of the Global Tiger Conservation Plan prior to catastrophe or a decline below numbers such that removal of founders from the wild contributes to the crisis. These agreements need to be formalized to reduce the inevitable stress of decision-making at the time of crisis.

Clearly state criteria for when interventions are to be undertaken, including: a)current probability of extinction, b)planned destruction of habitat, c)numbers below a critical range for a viable population for 200 years, d)catastrophic event, e)decline in reproductive rate below mortality rate for three years and, f)numbers in the range making a probability of catastrophic loss greater than 5% during the next 20-50 years.

5. If the probability of extinction of a population is 5% or greater for the 50 year projection, then the the availability of replacement animals from other populations of the same subspecies needs to be established. The use of a captive breeding population to supply animals for recolonizations, when sufficient animals are not available from other wild populations, needs to be a part of the planning.

Evaluate the feasibility of establishing captive populations in terms of: a)availability of founders from the wild, b)systematic relationship to other related wild populations, c)presence of a population in captivity and its genetic and demographic status, d)availability of resources to establish a captive population locally (country of origin) and, e)need for captive populations outside the country of origin.

6. Develop a strategy and the technology for exchanging genetic material between wild populations to increase their effective population size. Where necessary for reasons of numbers or accessiblity, utilize carefully managed captive populations to increase the size of the gene pool and as a source of genetic material for exchange with small wild populations. If captive breeding is chosen as a technique to assist in the preservation of a particular tiger population or group of populations, develop a plan for the use of the captive population to support the continued existence of the wild population(s).

7. Develop a clear statement (scientific, managerial, and policy) for genetic and demographic integration of wild and captive populations at an early stage in the establishment of the captive populations.

8. Support the necessary research to achieve the objectives of the plan with special emphasis on the development of: a)a geographical information system; b)techniques for reducing tiger attacks on people and livestock; c)techniques for introducing genetic material into wild populations without the disruption of the local population structure and, d)techniques for releasing tigers into habitat presently unoccupied by tigers, including translocation of wild-caught and captive-bred animals.

9. Initiate detailed information gathering on individual tiger populations and establish a working group with members of the Cat Specialist Group, the Captive Breeding Specialist Group, and Wildlife officials from each of the countries of origin to prepare a detailed plan by countries and subspecies for submission to the SSC/IUCN for review and endorsement.

10. Develop public education programs at the local, national and international level and develop a strategy for making local tiger ecosystems a part of regional economic planning.

APPENDIX 1: DRAFT MANIFESTO ON CAT CONSERVATION

This manifesto has been prepared by the Cat Specialist Group of the International Union for Conservation of Nature and Natural Resources (IUCN) as a contribution to the World Conservation Strategy.

Preamble

Cats have been a part of the environment, culture and mythology of human beings for thousands of years. They were revered in ancient Egypt, and the lion, in particular, was widely used as a symbol of royalty and state to the present day. In pre-Columbian civilizations in Mexico and Central America, the jaguar had high ritual significance. The tiger has figured in the art and culture of the great civilizations of Asia, and the domestic cat has rivaled the dog as a beloved companion of man.

Nevertheless, some forms of wild cats are already extinct because of human impact, others are on the brink, and most are declining seriously in numbers.

The extinction of species of wild cats would be an inestimable loss to the world, not least because of their ecological role as predators, and it behooves us to make every effort to prevent it, not least because human activities are largely responsible for their deteriorating status.

Why Cats Should Be Conserved

Human beings have no right to eliminate other species. Indeed, in view of the extent of human domination of the natural environment, we have a responsibility and obligation to them and to our descendants to perpetuate their existence. Extinction is forever.

The decline of a carnivore generally alters the ecological balance of its biological community. Cats are linked through predation to herbivores, which are in turn, linked to each other through competition and to plant communities by their foraging. They are particularly sensitive to environmental disturbance, and the decline or disappearance of these vulnerable species serves as an indicator of changes in their ecosystem, which may be the result of natural phenomena or, as is increasingly the case in present times, of the impact of human activities. These changes frequently involve a deterioration in the human environment, such as the loss of forests and grasslands and their valuable animal and plant products, or impairment of water supplies essential to human life and agriculture. Furthermore, large cats, being at the pinnacle of the food chain, need considerable space, and are, therefore, key species in determining the area required to define an appropriate ecosystem.

In addition to the ecological consequences of the disappearance of these top carnivores, many people feel a sense of inner loss when such magnificent and mysterious animals are gone from the wild.

Problems Faced by the Cats

Accelerating loss of habitat has now reached a critical stage as the human population continues to soar. In many cat ranges, remaining habitat represents but a small percentage of what existed in the past, and what remains could be wiped out in the near future.

Cats have long been hunted. They are killed because they have been viewed as competitors for prey. They are killed because they have taken livestock. They are killed for sport, and even to use their body parts as medicine. Young cats are captured for pets. And some, especially spotted cats, are killed for their attractive pelts.

At the same time, the disappearance of natural prey has frequently resulted in depriving cats of their vital sustenance and contributed to conflict with humans and their livestock with the inevitable reprisals.

Where cat populations have been reduced to small numbers they are increasingly vulnerable to extinction due to fortuitous local events, such as epidemics, fires and floods. Some scientists also fear the possiblity too of deterioration through inbreeding depression and loss of genetic diversity in the long-term, which might reduce the ability of small populations to adapt to changes in the environment.

The Decline of the Cats

Cat populations have long been in decline and today every indicator suggests that declines are accelerating and have reached a critical stage in some cases.

The Asiatic lion is a classic example of decline because of human impact. Ranging 2,000 years ago from Asia Minor to central India, it was hunted and exterminated, so that by the beginning of this century only a few survived in India's Gir forest. Fortunately, conservation efforts have succeeded in maintaining the Gir lion population, which now numbers over 200.

In 1947 the last recorded Asiatic cheetahs in the Indian sub-continent were shot.

The Bali tiger is thought to have already become extinct before 1940, and during this present decade of the eighties, its neighbor, the Javan tiger, has passed into oblivion. No trace of the Caspian tiger has been found for several decades, and reports suggest that the South China tiger, which is endemic to China, is on the verge of extinction, and there may be no tigers in the wild there by the end of the century.

Among the small cats, the Iriomote cat, endemic to a small Japanese island east of Taiwan, is nearly extinct because of destruction of its habitat and human over-exploitation of its natural prey.

These few examples of the decline of the cats and of suitable habitat are representative of the general situation throughout their world range.

Problems of Cat Conservation

There is still only limited knowledge of the distribution, numbers, biology and behavior of almost all species of cat. Research to increase understanding of these factors is essential to the planning and implementation of effective conservation measures.

Economic planners and decision-makers rarely recognize the importance for human welfare of wild lands, including ecosystems of which cats are a part. Consequently, development programs are carried out with little or no consideration of the long-term impact, which may result in the decline and extinction of many species, including cats, as well as impoverishing the human environment.

As a result of increasing fragmentation of habitat and the pressure of human activities in their vicinity, large cats may become problem animals, particularly through livestock predation, and in rare cases taking human life. Demands may then arise for elimination, not only of the offending animals, but of all of the large cats in the area.

Insufficient resources are made available to pursue necessary research, and to implement protective measures and conservation management of natural habitats of cat, often because the necessary political will is lacking.

How Cats Can Be Conserved

Protected habitats of sufficient size and productivity to support viable populations of cats must be preserved.

The distribution of each species and the habitat available to it needs to be established in detail down to the level of discrete populations.

Legislation to ensure long-term conservation measures must be taken to protect cat species and their prey.

Conservation of cats has to be reconciled with the needs of humans. Some conflict may be inevitable in areas where agriculture or livestock farming impinges on cat habitats, but it should be minimized. For many cats, and particularly large cats, parks and reserves may not be adequate. Land-use patterns in adjacent areas need to be designed so that they are compatible with use by both humans and cats.

Local people must feel that efforts are being made to protect their interests. Information about the role of cats and ways to conserve them should be part of conservation education at all levels of the community, from children to the politicians, officials, industrialists and businessmen who are the decision-makers.

Establishing a captive propagation program may be an important precaution in some cases so that future reintroduction may be carried out if it is feasible.

All these measures should be included in an overall conservation strategy for each species to ensure its survival.

Conclusions

Species need not be lost provided action is taken to conserve them. Experience has shown that seemingly desperate situations can be reversed if protection is given to species and their ecosystems.

The Cat Specialist Group is pledged to do all in its power to achieve the conservation of all cats and appeals for the cooperation of all people to ensure that these magnificent animals continue to coexist with humans as they have through the ages.

APPENDIX 2:   DRAFT IUCN POLICY STATEMENT ON CAPTIVE BREEDING
(Revised 15 October 1986)

The following was prepared by the IUCN/SSC Captive Breeding
Specialist Group.

Problem Statement

IUCN data indicate that about 3% of terrestrial Earth is
gazetted for protection. Some of this, and much of the other
97%, is becoming untenable for many species, and remaining
populations are being greatly reduced and fragmented. From
modern population biology one can predict that many species will
be lost under these conditions. On average, more than one
mammal, bird, or reptile species has been lost in each year of
this century. Since extinctions of most taxa outside these
groups are not recorded, the loss rate for all species is much
higher.

Certain groups of species are at particularly high risk,
especially those forms with restricted distribution, those of
large body size, those of high economic value, those at the top
of food chains, and those which occur only in climax habitats.
Species in these categories are likely to be lost first, but a
wide range of other forms are also at risk. Conservation over
the long-term will require management to reduce risk, including
ex situ populations which could support and interact
demographically and genetically with wild populations.

Feasibility

Over 3,000 vertebrate species are being bred in zoos and
other captive animal facilities. When a serious attempt is made,
most species breed in captivity, and viable populations can be
maintained over the long-term. A wealth of experience is
available in these institutions, including husbandry, veterinary
medicine, reproductive biology, behavior, and genetics. They
offer space for supporting populations of many threatened taxa,
using resources not competitive with those for in situ
conservation. Such captive stocks have in the past provided
critical support for some wild populations (e.g. American bison,
Bison bison), and have been the sole escape from extinction for
others which have since been re-introduced to the wild (e.g.
Arabian oryx, Oryx leucoryx.)

Recommendation

IUCN urges that those national and international
organizations and those individual institutions concerned with
maintaining wild animals in captivity commit themselves to a
general policy of developing demographically self-sustaining
captive populations of endangered species wherever practicable.

Suggested Protocol

What: The specific problems of the species concerned need to be considered, and appropriate aims for a captive breeding program made explicit.

When: The vulnerability of small populations has been consistently underestimated. This has erroneously shifted the timing of establishment of captive populations to the last moment, when the crisis is enormous and when extinction is probable. Management to best reduce the risk of extinction requires the establishment of supporting captive populations much earlier, preferably when the wild population is still in the thousands. Vertebrate taxa with a current census below one thousand individuals in the wild require close and swift cooperation between field conservationists and captive breeding specialists, to make their efforts complementary and minimize the liklihood of the extinction of these taxa.

How: Captive populations need to be founded and managed according to sound scientific principles for the primary purpose of securing the survival of species through stable, self-sustaining captive populations. Stable captive populations preserve the options of reintroduction and/or supplementation of wild populations.

A framework of international cooperation and coordination between captive breeding institutions holding species at risk must be based upon agreement to cooperatively manage such species for demographic and genetic stability. The IUCN/SSC/Captive Breeding Specialist Group is an appropriate advisory body concerning captive breeding science and resources.

Captive programs involving species at risk should be conducted primarily for the benefit of the species and without commercial transactions. Acquisition of animals for such programs should not encourage commercial ventures or trade. Whenever possible, captive programs should be carried out in parallel with field studies and conservation efforts aimed at the species in its natural environment.

Summary

Habitat protection alone is not sufficient if the expressed goal of the World Conservation Strategy, the maintenance of biotic diversity, is to be achieved. Establishment of self-sustaining captive populations and other supportive intervention will be needed to avoid the loss of many species, especially those at high risk in greatly reduced, highly fragmented, and disturbed habitats. Captive breeding programs need to be established before species are reduced to critically low numbers, and thereafter need to be coordinated internationally according to sound biological principles, with a view to the maintaining or re-establishment of viable populations in the wild.

Species Index

Acinonyx jubatus - 187, 255, 266–
    267, 270, 275–276, 312,
    494
Axis axis – 424, 443–444, 452
Axis deer (see Axis axis)

Bali starling (see Leucopasar
    rothschildi)
Danteng (see Bos javanicus)
Bos javanicus - 5

Canis aureus - 121
Cave lion (see Panthera spelaea)
Cervus duvaucelli - 452
Cervus elaphus xanthopygus - 64
Cervus nippon hortulorum - 64
Cervus timorensis - 5
Cervus unicolor - 87
Cheetah (see Acinonyx jubatus)
Chinese muntjac (see Muntiacus
    reevesi)
Chinese paleotiger (see Felis
    palaeosinensis)
Chital (see Axis axis)
Clouded leopard (see Neofelis
    nebulosa)
Crocodilus palustris - 121
Cuon alpinus - 87, 121-122, 124–
    125

Dhole (see Cuon alpinus)
Dinofelis - 29

Elaphas maximus - 121
Elaphodus cephalophus - 74
Elephant (see Elaphas maximus)

Felis concolor - 187, 266–267,
    275, 312
Felis iriomotensis - 494
Felis palaeosinensis - 29–31
Fossil lion (see Panthera atrox)

Golden jackal (see Canis aureus)
Grus grus - 31

Hare (see Lepus nigricollis)
Hyaena hyaena - 121
Hyena (see Hyaena hyaena)
Hystrix indica - 121

Indian muntjac (see Muntiacus
    muntjak)
Indian rhinoceros (see Rhinoceros
    unicornis)
Iriomote cat (see Felis iriomotensis)

Jaguar (see Panthera onca)

Subject Index

Fanjingshan Nature Reserve - 135,
147, 325, 488
Feline central retinal degeneration -
151, 196, 208, 210, 377
Feline infectious peritonitis - 179,
195, 214
Feline leukemia virus - 173, 197,
199
Feline urinary syndrome - 149,
157
Feline viral rhinotracheitis - 172,
214
Fenbendazole - 175
Follicle stimulating hormone
(FSH-P) - 270-271, 273-
274, 276, 290, 292
Fossil record - 1, 14, 17, 29-31,
53-57
Furazolidone - 232

Gastroenteritis - 67, 191, 216,
218-221, 322
Gastrointestinal disorders - 192
G-banded karyology - 12, 15-17,
46
"General adaptation syndrome" -
192
General Register - 348
Genetic
bottleneck - 32, 44, 343, 475
degree of relatedness - 97,
104-107, 401
drift - 32, 46, 305, 307, 341
founder effect - 44, 343, 345
management - 305-311, 401
migration - 107, 114, 116, 305,
307-308, 465, 492
mutation - 10-11, 307, 363,
372-374, 377-378, 380,
389, 391-392
selection - 44-45, 143, 305,
307-308, 340-341, 360,
363-369, 374, 378, 391,
466
selection intensity - 364, 367-
369
variability - 304-305, 308, 313,
329, 341, 343-345, 363,

369, 421, 455, 466, 475,
492, 494
Gentamicin - 235-237, 241
Geographic information system
(GIS) - 403, 464, 469-472,
492
Geriatrics - 155, 158, 160, 189-190,
218
Gerik Wildlife Reserve - 83, 489
Global Tiger Conservation Plan -
329, 392, 464, 472, 487-498
Glomerulonephritis - 191
Gonadotropin releasing hormone
(GnRH) - 292
Greater Hinggan Mountains - 72
Griseofulvin - 195
Guidong Nature Reserve - 147
Gunong Belumut Wildlife Reserve -
83, 489
Gunong Tampurong Wildlife Re-
serve - 489
Gunung Betiri National Park - 3, 6
Gunung Leuser National Park - 88-
90, 414, 489

Habitat degradation - 5-7, 64, 71,
73, 75-76, 82, 85-90, 110-
111, 115, 118, 136, 140, 317,
324, 397-398, 411, 413-414,
418, 421, 428, 436-438, 442,
446, 450, 452-454, 458, 465,
469, 491, 493-495
Habitat reclamation - 112-117
Halothane - 188
Heartworm - 196
Hordeolum - 210
Human chorionic gonadotropin
(HCG) - 270, 274, 288, 290-
292
Hydrazones - 300
Hydrocephalus - 205, 208
Hyperthermia - 189
Hypothermia - 207-208, 219

Iman river - 66-67
Immunity - 184
Immunodeficiency - 196, 220, 223